Chromatography: Techniques and Instrumentation

Chromatography: Techniques and Instrumentation

Editor: Carol Evans

NY RESEARCH
P R E S S

New York

Published by NY Research Press
118-35 Queens Blvd., Suite 400,
Forest Hills, NY 11375, USA
www.nyresearchpress.com

Chromatography: Techniques and Instrumentation
Edited by Carol Evans

Cataloging-in-Publication Data

Chromatography : techniques and instrumentation / edited by Carol Evans.
   p. cm.
Includes bibliographical references and index.
ISBN 978-1-63238-627-4
1. Chromatographic analysis. I. Evans, Carol.
QD79.C4 C47 2019
543.8--dc23

Contents

Preface

Chromatography is a chemical process used to separate mixtures into their constituents. It is used as a technique across a number of industries such as ink factories, milk production, etc. As a method of separating mixtures, it can be performed in a variety of ways with different kinds of apparatus. Some popular techniques of chromatography include gas, column, liquid, ion exchange, thin layer, two-dimensional and reversed-phase chromatography among many others. Some of the topics included in this book elaborate the varied forms of chromatography. This book includes contributions of experts and scientists which will provide innovative insights into this field. The aim of this book is to present researches that have transformed this discipline and aided its advancement. Scientists and students actively engaged in this field will find this book full of crucial and unexplored concepts.

Significant researches are present in this book. Intensive efforts have been employed by authors to make this book an outstanding discourse. This book contains the enlightening chapters which have been written on the basis of significant researches done by the experts.

Finally, I would also like to thank all the members involved in this book for being a team and meeting all the deadlines for the submission of their respective works. I would also like to thank my friends and family for being supportive in my efforts.

Editor

Analysis of Citalopram in Plasma and Hair by a Validated LC – MS/MS Method

Pinto J[1], Mendes VM[1], Coelho M[1], Baltazar G[2], Pereira J LGC[3], Dunn MJ[4], Cotter DR[5] and Manadas B[1]*

[1]CNC - Center for Neuroscience and Cell Biology, University of Coimbra, Portugal
[2]CICS - UBI - Health Sciences Research Centre, University of Beira Interior, Portugal
[3] Departamento de Química, Universidade de Coimbra, Portugal
[4]Proteome Research Centre, UCD Conway Institute of Biomolecular and Biomedical Research, University College Dublin, Portugal
[5]Department of Psychiatry, Royal College of Surgeons in Ireland

Abstract

A simple method for quantification of citalopram in mice plasma and hair was developed and validated using liquid chromatography tandem mass spectrometry (LC–MS/MS). The procedure involves a protein precipitation extraction of citalopram and desipramine (internal standard) with methanol from mice plasma. On the other hand, hair samples were incubated overnight with methanol at 45°C followed by µ-SPE (OMIX Tip). The analysis was performed by resolving analytes in a Gemini® C18 column with a gradient of 0.1% formic acid in water and 0.1% formic acid in acetonitrile, at a flow rate of 250 µL/min and with a total run time of 9 min. The mass spectrometer was operated in multiple reaction monitoring (MRM) by monitoring citalopram transition 325.3→109.0, and internal standard transition 267.3→72.2 for quantification. The qualifier transitions 325.3→83.1 and 267.3→190.8, respectively, were also monitored. Linearity was observed from 32.4 to 973.2 ng/mL and the limit of quantitation achieved was 32.4 ng/mL. Also, the intermediate precision, repeatability and accuracy were below the acceptance limits of 15%. This method was applied to plasma and hair samples that were collected from mice submitted to a treatment with citalopram for different days. The plasma concentration–time profile of citalopram showed a tendency to stabilize, approaching zero as samples were collected 24 hours after the last drug administration. In contrast, the concentration-time profile in hair increased over the period of 30 days.

Keywords: Citalopram; Plasma; Hair; LC-MS/MS; Analytical method validation

Introduction

Citalopram (1-[3-(Dimethylamino)propyl]-1-(4-fluorophenyl)-1,3-dihydro-5-iso-benzofuran carbonitrile) is considered one of the most prescribed drugs being a selective and potent serotonin reuptake inhibitor [1]. This antidepressant is widely used because it is safer, less toxic and more tolerable than the tricyclic antidepressants and monoamine oxidase inhibitors. However it is worth noting that it may be involved in overdose deaths, particularly when combined with other drugs [2]. Liquid chromatography-mass spectrometry (LC–MS) has established itself as the clear leader in the quantification of antidepressant drugs in biological samples [3]. Some LC–MS [4,5] and LC–MS/MS [6,7] methods for the determination of citalopram have been developed in plasma samples. Nevertheless, in most cases they require plasma volumes by more than 200 µL. Hair testing is a valuable way to increase the window of drug detection, and due to simple sample collection and storage, it has become an alternative matrix for drug testing. Methods regarding identification and screening of citalopram in hair have been publised previously [8,9], with only two reports addressing quantification [10,11]. However these either entail amounts of hair greater than 10 mg or the procedure is demanding, requiring several extractions in opposition to incubation followed by a single solid phase extraction. Hence, we developed and validated a method for the determination and quantification of citalopram in mice plasma and hair with a small quantity of sample and with a simple extraction procedure, in contrast to the current literature which is mostly based on methods for humans. The small sample amount used for citalopram analysis in mice is significant for studies using this animal model. This method for both matrices was successfully used to characterize the time course of changes in plasma and hair concentrations of citalopram following intraperitoneal injection for different periods of time in mice.

Experimental

Reagents and standards

Citalopram hydrobromide (purity 99.8%) was purchased from BIOTREND® Chemicals AG. Desipramine hydrochloride (Internal Standard, IS) (purity ≥ 98%) was purchased from Sigma-Aldrich®. Acetonitrile (ACN) and methanol (MeOH) (LC-MS Grade) were obtained from Biosolve. Formic acid (FA) and dichloromethane (LC Grade) were obtained from Sigma Aldrich® and water (LC Grade) from VWR®. Standard stock solutions for citalopram and desipramine were prepared in acetonitrile and were both stored at -20°C. In each working day, freshly diluted calibration standard solutions were prepared by diluting the stock solutions in 2% ACN: 0.1% FA.

Animal study

Young black male C57BL/6J mice (n=30, weight range 20-25 g), were purchased from Charles River Laboratories International, Inc., Spain. The animals were sacrificed after 1, 2, 4, 8, 15 and 30 days of daily intraperitoneal injections of citalopram at doses of 10 mg/kg, after 1 week of habituation to needle stick. They were divided into

*Corresponding author: Manadas B, Center for Neuroscience and Cell Biology,University of Coimbra, Largo Marquês de Pombal 3004-517 Coimbra, Portugal, E-mail: bmanadas@gmail.com

six groups according to treatment time, then they were weighed and anesthetised with a mixture of ketamine and xylazine 24 hours after the final injection.

Blood samples were collected by cardiac puncture and placed in EDTA-coated tubes (BD Vacutainer® tubes, Becton Dickinson). Followed by centrifugation at 12,000×g for 2 minutes, the plasma was recovered to another tube (BD Microtainer', Becton Dickinson) with protease and phosphatase inhibitors (Roche). Regarding hair samples, they were pulled out with tweezers and placed into a microcentrifuge tube. Both plasma and hair samples were fast frozen in liquid nitrogen and stored at -80°C until further use. The thawed plasma and hair samples were spiked with IS solution and processed as mentioned in the sample preparation section.

Sample preparation

To each 70 μL of plasma, 10 μL of IS working solution (151.4 ng/mL) were added. After vortexing, 210 μL of MeOH (three times the plasma volume) were added and the samples were rehomogenized, followed by continuous agitation for 5 minutes at 1000 rpm's in a thermomixer (Comfort, Eppendorf). Samples were centrifuged at 14,000×g for 10 minutes, then the supernatant was collected, evaporated at 60°C during approximately 1 hour (Concentrator Plus, Eppendorf), reconstituted in 50 μL of 2% ACN: 0.1% FA in water.

To each sample of hair, containing between 0.7-9.7 mg (corrections were made to calculate the concentrations), 1 mL of dichloromethane was added in order to remove any blood residues or external contamination. Samples were mixed for 2 minutes at 750 rpm's in the thermomixer (Comfort, Eppendorf). Dichloromethane was discarded and this procedure was repeated. Then, the hair was incubated overnight (17 hours) in 1 ml of MeOH at 45°C. Methanol was recovered and evaporated to dryness at 60°C, for approximately 1 hour (Concentrator Plus, Eppendorf). Hair samples were cleaned up by using μ-Solid phase extraction (OMIX Tip C18) (Agilent Technologies). To the evaporated sample, 100 μL of 2% ACN: 1% FA were added and the samples sonicated (Sonicator VibraCell™ 75041, Sonics') for 2 min using a cuphorn (20% amplitude with 1 sec on 1 sec off cycle). After conditioning of the C18 SPE-tip with 200 μL of 50% ACN, the tip was equilibrated with 300 μL of 2% ACN: 1% FA, subsequently the sample was allowed to interact with the tip for 5 cycles. The tip was rinsed with 100 μL of 2% ACN: 1% FA and elluted with 400 μL of 70% ACN: 0.1% FA. These samples were evaporated to dryness at 60 °C, then resuspended in 50 μL of 2% ACN: 0.1% FA and sonicated.

HPLC operating conditions

The chromatographic analysis was performed using an Ultimate 3000 LC system (LC Packings, Dionex™). Separation was performed with a Gemini' C18 column (3 μm, 110 Å, 50 × 2 mm) coupled with a Security Guard™ cartridge Gemini' C18 (4 × 2 mm).

The mobile phase was composed of 0.1% formic acid in water (A) and 0.1% formic acid in ACN (B). Gradient conditions were 10-20% B (0-0.5 min), 20-30% B (0.5-0.6 min) and 30-99% B (6.0-9.0 min), with a flow rate of 250 μL/min. This step gradient was chosen in order to separate other interferents, causing them to elute at former retention times than citalopram and desipramine.

Cleaning and re-equilibration steps (between samples) are performed by a blank injection with a chromatographic gradient 0-90% B (0-1.9 min) and 90% B (1.9-8 min). Also between batches, three blank injections were introduced (solution of 0.1% FA in ACN), with the same program and the same volume of injection as the one

used for the plasma samples. The volume that was injected was adjusted for each condition, 1 μL for hair samples and 20 μL for plasma samples. The blank injection between samples was of 10 μL.

Mass spectrometry operating conditions

A 4000 QTRAP' mass spectrometer (ABSciex™), equipped with a turbo V™ source was operated in electrospray positive mode (ESI+) and under the following optimized settings: curtain gas (CUR), 30 psi; ion source gas 1 (GS1), 30 psi; ion spray voltage, 5500 V; source temperature, 450°C. Compound's parameters were optimized for multiple reactions monitoring (MRM) mode by direct infusion of a standard solution of each analyte, with a concentration of 42.2 ng/mL for citalopram and 342.2 ng/mL for desipramine at 9 μL/min. A dwell time of 30 ms, entrance potential (EP) of 10 eV and collision gas (CAD) of 8 psi were used. The values of declustering potential (DP), collision energy (CE), collision exit potential (CXP), and transitions are presented in the Table 1. Data acquisition was performed by Analyst' 1.5.1 (ABSciex™).

Method validation

The method was validated for plasma and hair in terms of selectivity, linearity, accuracy, intra-day precision, intermediate precision, extraction efficiency, matrix effect, and carry-over. The different validation parameters and the values for accepting the range of validation parameters were in accordance with international guidelines, as International Conference on Harmonisation (ICH) and Food and Drug Administration (FDA).

Selectivity

Plasma's selectivity was determined by analyzing six individual blank samples and each one was divided in two aliquots (70 μL each). One aliquot was spiked with 20 μL of citalopram (8.1 ng/mL) and desipramine (7.6 ng/mL) and the other was not fortified with any compound. Similarly, to evaluate the selectivity in hair samples, six individual blank samples were divided in two aliquots (approximately with 3-4 mg each). One aliquot was spiked with 20 μL of citalopram (162.2 ng/mL) and desipramine (151.4 ng/mL) and the other was not fortified with any compound. Then, all samples were subjected to the analytical procedure developed for the extraction of drugs from plasma and hair, respectively.

Linearity and analytical limits

To study the linearity, the calibration curve was established between 32.4 and 1621.5 ng/mL (corresponding to 0.10 and 5 pmol/μL) and were prepared on five different occasions. Apart from other tests, the criteria for acceptance included a r^2 value of at least 0.99, and the calibrators accuracy within 80-120%.

According to the FDA, the limit of quantification (LOQ) is the lowest amount of an analyte in a sample that can be quantitatively

Compound	Transitions (m/z)	CE (eV)	CXP (eV)	DP (eV)
Citalopram	325.3→109.0	39	0	66
	325.3→261.9	27	24	
	325.3→83.1	91	4	
Desipramine	267.3→72.2	27	4	56
	267.3→208.0	33	16	
	267.3→190.8	83	14	

Collision energy (CE), collision exit potential (CXP) and declustering potential (DP). Transitions used for quantification are underlined.

Table 1: MRM parameters for each transition of citalopram and its internal standard.

determined with suitable precision and accuracy. Moreover, the signal-to-noise ratio should be investigated in order to determine whether it reaches a ratio of 10:1 [12].

Precision and accuracy

The precision of the method was evaluated in two levels: repeatability (or intra-day precision) and intermediate precision (or inter-day precision). To evaluate the repeatability, triplicates of quality controls (QCs) were prepared at three concentration levels (40.5, 243.2 and 891.8 ng/mL corresponding to 0.125, 0.75 and 2.75 pmol/μL) and were characterized in terms of percentage of coefficient of variation (%CV) after performing an one-way analysis of variance (ANOVA). The intermediate precision was also expressed in terms of %CV and it was determined by analyzing the QCs over five different occasions Accuracy was evaluated in terms of percentage of mean relative error (%MRE) between the theoretical and the measured spiked concentrations for the calibrators [13,14]. An acceptable variability was set at 15% for all the concentrations, except at the LOQ, for which 20% was accepted [15,16].

Extraction efficiency

To evaluate the extraction efficiency of citalopram and IS in plasma, two levels of concentration were selected (162.2 and 973.2 ng/mL corresponding to 0.5 and 3 pmol/μL) and each concentration prepared in triplicate. For each level, two aliquots were prepared with 70 μL: one was spiked with 20 μL of the solution containing citalopram and desipramine, and then was subjected to the extraction procedure, and in paralel the other was subjected first to the extraction procedure and then spiked with 20 μL of the citalopram and desipramine solution.

In an analogous way, to evaluate the analyte's extraction efficiency using the hair extraction protocol, two levels of concentration were selected (162.2 and 973.2 ng/mL) with each concentration in triplicate. For each level, two aliquots of hair weighing between 1.9–3.2 mg were prepared. One aliquot was spiked with 20 μL of the spike solution containing citalopram and desipramine, and then was subjected to the extraction procedure. The other aliquot was first subjected to the extraction procedure and then was spiked with 20 μL of the citalopram and desipramine solution.

In both plasma and hair, 10 μL of sulfamethazine-D4 at 141.2 ng/mL were added, at the end of the procedure. The sulfamethazine-D4 was used to compensate injection losses as well as diferences in the system sensitivity, since extraction efficiency of desipramine was also evaluated.

Matrix effect

The assessment of matrix effect was performed at two concentration levels (162.2 and 973.2 ng/mL) each one in triplicate for both plasma and hair. Blank samples were subjected to the respective extraction procedure and then spiked with 20 μL of a solution with citalopram and desipramine. In parallel, pure solutions with citalopram and desipramine in mobile phase (2% ACN: 0.1% FA) at the equivalent levels of concentration, were also analysed.

Carry-over

To evaluate the carry-over phenomena, analytical blanks (2% ACN: 0.1% FA) were injected after a highest level of standard concentration (1622 ng/mL corresponding to 5 pmol/μL). This procedure was repeated in three different days.

Results

Method validation

Selectivity: The selectivity of the method was evaluated by analyzing blank samples of plasma and hair from six different animals to investigate the potential interferences at the peak region for citalopram and IS. The criteria for compound identification were the ones proposed by the World Anti-Doping Agency (WADA) [17]. In what concerns the chromatography, the acceptance criterion was the relative retention time (ratio between the RT of the interest compound and RT of the internal standard) that should not differ by more than ± 1% when compared with the relative retention time of the control sample of the same compound.

Regarding the mass spectrometry identification in the MRM mode, it is required to monitor at least two precursor-product ion transitions. The second criterion to monitor is the relative abundance of a diagnostic ion that is expressed as a percentage of the intensity of the most intense fragment (base peak). The relative intensities should not differ by more than a tolerated range (Table 2) from those generated by the control sample of the same compound. Moreover, the signal-to-noise ratio should be greater than 3. Citalopram and desipramine were successfully identified in all spiked samples (Tables A.1 and A.2). In the non-spiked samples, for both plasma (Figure 1) and hair (Figure 2), no chromatographic peaks of citalopram were detected. Even though spiked samples display small differences in the retention times, these differences fullfil the criteria based on the WADA guideline (Tables A.1 and A.2). Therefore, the two described methods were considered selective for the determination of citalopram.

Linearity: Calibration curves were obtained by plotting the peak area ratio between each analyte and the internal standard, against analyte concentration. One of the assumptions of linear regression is a constant variance of measured values between the limits of the working range [18]. A test of homogeneity of variances (F-test) [19] showed that there is a significant difference between those limits (since $F_{crit}= 715.2 > F_{crit}=6.54$, p-value=1.87×10^{-11}), corresponding to heteroscedastic data. Usually when a homogeneous variance is not verified, a weighted least squares regression is used [20-22]. Therefore, six weighting factors were evaluated ($1/\sqrt{x}, 1/x, 1/x^2, 1/\sqrt{y}, 1/y, 1/y^2$) and the one which originated the lowest sum of percentage of relative error (presenting simultaneously a mean r^2 value of at least 0.99) was selected taking into account the data relative to five independent calibration curves [23,24]. The chosen weighting factor for citalopram, given these criteria was $1/x$. The calibrators accuracy for the five days was within 80 and 120%. Calibration data is shown on Table 3. The LOQ was defined as the lowest concentration of analyte which could be measured reproducibly and accurately (%CV<20% and accuracy of 80–120%). The value of LOQ achieved was 33.8 ng/mL.

Precision and Accuracy: To calculate the concentration of the QCs, prepared in three concentration levels (40.5, 243.3 and 892.1 ng/mL) each one in triplicate, calibration curves on five different days were used. First it was performed a Cochran's test to detect if the assumption of constant variance is accomplished [25]. Then the results obtained for the different levels of concentration for each compound were analysed by ANOVA [26,27].

The intermediate precision and the repeatability were expressed in terms of %CV and the acceptance criteria was set at 15% for all the concentrations, except at the LOQ, for which 20% was accepted.

Relative abundance (% of base peak)	Maximum tolerance ranges (%)
>50	± 10 (absolute range)
25 to 50	± 20 (relative range)
5 to <25	± 5 (absolute range)
<5	± 50 (relative range)

Table 2: Maximum Tolerance ranges for relative ion intensities to ensure appropriate confidence in identification [17].

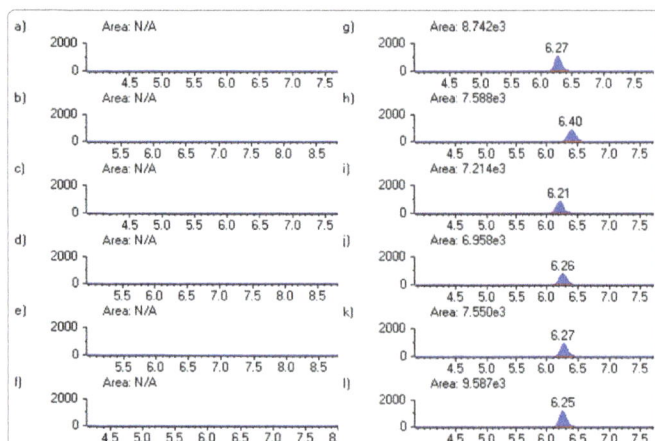

Figure 1: Chromatographic spectra of citalopram (325.3→109.0) for the selectivity analysis in plasma. (a-f) Six different blank samples not spiked with citalopram and desipramine (negative samples). (g-l) Six blank samples spiked with citalopram and desipramine (positive samples), at the final concentration of 162.2 ng/mL. Y-axis in counts per second (cps).

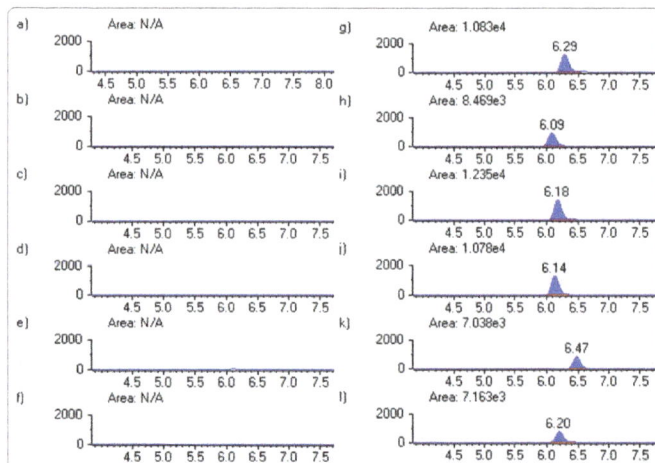

Figure 2: Chromatographic spectra of citalopram (325.3→109.0) for the selectivity analysis in hair. (a-f) Six different blank samples not spiked with citalopram and desipramine (negative samples). (g-l) Six blank samples spiked with citalopram and desipramine (positive samples), at the final concentration of 162.2 ng/mL. Y-axis in counts per second (cps).

Accuracy was evaluated in terms of percentage of mean relative error (%MRE) and as an acceptance criteria, the accuracy for each level of concentration should be within ± 15% of the nominal concentration [15,16]. The obtained %CVs were lower than 15.1% for LOQ concentration level and 12.5 for the others concentrations, with an accuracy lower than 9.3% (Table 4).

Extraction efficiency: The extraction efficiency was determined by comparing the representative peak areas of extracted blank samples spiked before extraction with the peak area of blank samples fortified after the extraction at equivalent concentrations. The experiment was conducted in triplicate at two levels of concentration (162.2 and 973.2

ng/mL). The results showed that the extraction efficiency values for citalopram were 90.9 and 101.5% in plasma, and 33.9 and 38.9% in hair (Table 5). The extraction efficiency for desipramine was 68.9 and 92.2% in plasma, and 36.6 and 36.3% in hair respectively.

The method provided good extraction efficiency for citalopram and desipramine in plasma at both concentrations. In hair, lower recoveries were achieved, however Causon et al. [14], states that it is unlikely that recoveries of 50% or less will compromise the integrity of the method if adequate detection can be attained, which is the case of this method.

Matrix effect: Matrix effects can be described as the difference between the response of an analyte in pure solution and the response of the same analyte in biological matrix. A normalized matrix factor can be measured by dividing the peak response ratio (analyte/IS) in the presence of matrix ions with the peak response ratio in the absence of matrix ions [28]. Hence, a normalized matrix factor of one represents no matrix effects, a value lower than one suggests ionization suppression and a value greater than one may be due to ionization enhancement [28].

The results for plasma and hair (Table 6) at higher concentration showed a normalized matrix factor of one, indicating no matrix effects. At lower concentrations, there is possibly some ion suppression for plasma and ion enhancement for hair samples. Nevertheless a value of one is not necessary for a reliable bio analytical assay, because coefficient of variation for the lowest concentration were lower than 15% demonstrating that results are reproducible [28]. Overall, coefficients of variation obtained for plasma were below 10.33%, and for hair these were slightly higher than the defined criteria (16.9% in the highest concentration). However, the normalized matrix factor for the highest hair concentration presents a value close to one, despite being a more complex matrix. The intra-day precision of the ratio of citalopram with desipramine in the presence of matrix ions and in the absence of matrix ions, presented %CV values lower than 15%. Moreover, the normalized matrix effect showed values close to one and also presented lower values of %CV. Thus, it can be concluded that the use of the internal standard has a compensatory effect on variations between injections. Therefore, these results showed that ion suppression or enhancement from the plasma and hair matrix was negligible under the current conditions.

Carry-over: Carry-over is the amount of analyte retained, in this case in the LC system, from a preceding sample that carries over into the next injected sample. This phenomena can be measured by the response of the blank sample after the injection of a preceding sample at high concentration [29]. The results showed that no carry-over was observed for citalopram. In any case, a blank injection between samples, and three blank injections were introduced between batches.

Application of the analytical method to the animal study

The developed analytical method was applied in plasma and hair samples that were collected from mice daily injected with a solution of citalopram (10 mg/kg) for 1, 2, 4, 8, 15 and 30 days (five independent replicates per day). The plasma concentration time profile of citalopram in mice shows values closer to zero, as these samples were collected 24 hours after the last administration (Figure 3). On the other hand, the concentration time profile of citalopram in hair increases over the days (Figure 4), showing a direct correlation, probably explained by the fact that there is more compound incorporated into the hair shaft, providing evidence of longer term exposure of drugs.

Conclusion

A method for plasma based on protein precipitation and for hair

Transition (m/z)	Weight	Linear range (ng/mL)	Linearity		r²	LOQ[b] (ng/mL)
			Slope[a]	Intercept[a]		
325.3→109.0	1/x	32.4–973.2	2.372 ± 0.657	0.020 ± 0.025	0.994 ± 0.003	32.4

[a]Mean values ± standard deviation
[b]Limit of quantification

Table 3: Analytical parameters of the calibration curve for citalopram.

Nominal concentration (ng/mL)	Repeatability (%CV)	Intermediate precision (%CV)	Accuracy(%MRE)	Estimated concentration (ng/mL)
40.5	10.6	15.1	0.04	40.5
243.2	5.8	11.7	8.5	263.6
891.8	5.6	12.5	9.3	973.0

%CV - percentage coefficient of variation.
%MRE- Percentage of mean relative error.

Table 4: Precision and accuracy in three concentration levels for citalopram.

Matrix	Compound	% Extraction efficiency (mean ± C.I.)	
		162.2 ng/mL	973.2 ng/mL
Plasma	Citalopram	90.9 ± 14.6	101.5 ± 23.7
	Desipramine	68.9 ± 6.6	92.2 ± 24.5
Hair	Citalopram	33.9 ± 10.4	38.9 ± 4.2
	Desipramine	36.6 ± 12.7	36.3 ± 7.1
C.I: Confidence Interval			

Table 5: Extraction efficiency, in percentage, of the extraction of the citalopram and desipramine in plasma and hair at two concentration levels.

Matrix	Nominal concentration (ng/mL)	Analyte with matrix ions peak area ratio[1]	Analyte pure solutions peak area ratio[1]	Normalized matrix factor[1]
Plasma	162.2	0.65 ± 1.43	0.76 ± 12.27	0.85 ± 1.43
	973.2	0.69 ± 10.33	0.68 ± 2.72	1.02 ± 10.33
Hair	162.2	1.81 ± 5.83	1.39 ± 5.56	1.31 ± 5.83
	973.2	1.53 ± 16.91	1.45 ± 7.46	1.06 ± 16.91
[1]Mean values ± % Coefficient of variation				

Table 6: Matrix effect of citalopram in plasma and hair at two concentration levels.

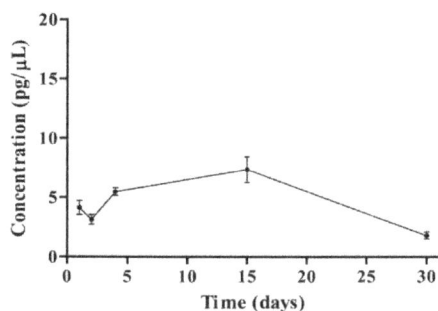

Figure 3: Mean plasma concentration time profile of citalopram in mice. Plasma was collected from mice daily injected with a solution of citalopram (10 mg/kg) for different periods of time: 1, 2, 4, 15 and 30 days (mean ± S.E.M, 4 ≤ n ≤ 5).

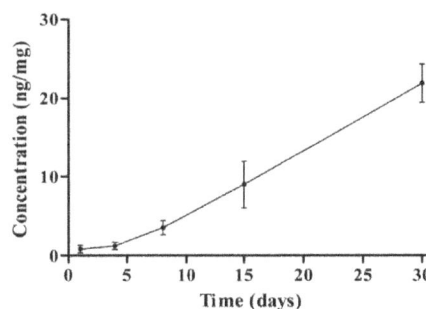

Figure 4: Mean hair concentration time profile of citalopram in mice. Hair was collected from mice daily injected with a solution of citalopram (10 mg/kg) for different periods of time: 1, 4, 8, 15 and 30 days (mean ± S.E.M, 3 ≤ n ≤ 4).

samples with MeOH incubation followed by solid phase extraction was developed and validated. In terms of plasma, this method uses smaller sample volumes in contrast to other published methods, which require sample volumes of more than 200 µL to achieve results with similar workload. Furthermore, the current methods in the literature for hair samples involve more extraction steps as well as a greater amount of hair (>10 mg) than the presented method.

The LC-MS/MS method developed for the determination of citalopram in mice plasma and hair fulfils the criteria generally required for bio analytical methods. The developed method was successfully applied for quantification of plasma and hair samples of mice injected daily with citalopram for 1, 2, 4, 8, 15 and 30 days.

Acknowledgment

This work is funded by FEDER funds through the Operational Programme Competitiveness Factors- COMPETE, the National Mass Spectrometry Network (RNEM) (REDE/1506/REM/2005) and national funds by FCT - Foundation for Science and Technology under the project (PTDC/SAU-NEU/103728/2008, Pest–C/SAU/LA0001/2013-2014) and strategic project UID/NEU/04539/2013.

Sandra Rocha performed the animal injections and tissue collection and was supported by PTDC/SAU-NEU/103728/2008.

References

1. Uckun Z, Süzen HS (2009) Quantitative Determination of Citalopram and its Metabolite Desmethycitalopram in Plasma by High Performance Liquid Chromatography. FABAD J Pharm Sci 34: 195-201.

2. Prahlow J (2010) Forensic Pathology for Police, Death Investigators, Attorneys, and Forensic Scientists. Springer: New York.

3. Sampedro MC, Unceta N, Gomez-Caballero A, Callado LF, Morentin B, et al. (2012) Screening and quantification of antipsychotic drugs in human brain tissue by liquid chromatography-tandem mass spectrometry: application to postmortem diagnostics of forensic interest. Forensic Sci Int 219: 172-178.

4. Pistos C, Panderi I, Atta-Politou J (2004) Liquid chromatography-positive ion electrospray mass spectrometry method for the quantification of citalopram in human plasma. J Chromatogr B Analyt Technol Biomed Life Sci 810: 235-244.

5. Singh SS, Shah H, Gupta S, Jain M, Sharma K, et al. (2004) Liquid chromatography-electrospray ionisation mass spectrometry method for the determination of escitalopram in human plasma and its application in bioequivalence study. J Chromatogr B 811: 209-215.

6. Rocha A, Marques MP, Coelho EB, Lanchote VL (2007) Enantioselective analysis of citalopram and demethylcitalopram in human and rat plasma by chiral LC-MS/MS: application to pharmacokinetics. Chirality 19: 793-801.

7. Jiang T, Rong Z, Peng L, Chen B, Xie Y, et al. (2010) Simultaneous determination of citalopram and its metabolite in human plasma by LC-MS/MS applied to pharmacokinetic study. J Chromatogr B 878: 615-619.

8. Lendoiro E, Quintela O, de Castro A, Cruz A, Lopez-Rivadulla M, et al. (2012) Target screening and confirmation of 35 licit and illicit drugs and metabolites in hair by LC-MSMS. Forensic Sci Int. 217: 207-215.

9. Muller C, Vogt S, Goerke R, Kordon A, Weinmann W (2000) Identification of selected psychopharmaceuticals and their metabolites in hair by LC/ESI-CID/MS and LC/MS/MS. Forensic Sci Int 113: 415-421.

10. Montesano C, Johansen SS, Nielsen MK (2014) Validation of a method for the targeted analysis of 96 drugs in hair by UPLC-MS/MS. J Pharm Biomed Anal 88: 295-306.

11. Fisichella M, Morini L, Sempio C, Groppi A (2014) Validation of a multi-analyte LC-MS/MS method for screening and quantification of 87 psychoactive drugs and their metabolites in hair. Anal Bioanal Chem. 406: 3497-3506.

12. International Conference on Harmonisation (2005) Validation of analytical procedures: Text and Methodology, Q2 (R1) ICH.

13. Wille SMR, Peters FT, Fazio V, Samyn N (2011) Practical aspects concerning validation and quality control for forensic and clinical bioanalytical quantitative methods. Accred Qual Assur 16: 279-292.

14. Causon R (1997) Validation of chromatographic methods in biomedical analysis. J Chromatogr B Biomed Sci Appl 689: 175-180.

15. United States Food and Drug Administration (2001) Guidance for industry-Bio analytical Method Validation FDA, USA

16. European Medicines Agency (2011) Guideline on bioanalytical method validation. EMA

17. World Anti-Doping Agency (2010) Identification criteria for qualitative assays incorporating column chromatography and mass spectrometry, WADA.

18. Peters FT, Maurer HH (2002) Bioanalytical method validation and its implications for forensic and clinical toxicology - A review. Accred Qual Assur 7: 441-449.

19. International Union of Pure and Applied chemistry (1999) Guidelines for calibration in analytical chemistry. IUPAC

20. Ermer J, Miller J (2006) Method Validation in Pharmaceutical Analysis. A Guide to Best Practice Wiley: Weinheim.

21. Miller JN, Miller JC (2005) Statistics and Chemometrics for Analytical Chemistry. Pearson&Prentice Hall: England.

22. International Union of Pure and Applied chemistry (1997) A statistical overview of standard (IUPAC and ACS) and new procedures for determining the limits of detection and quantification: Application to voltammetric and stripping techniques. IUPAC.

23. Almeida AM, Castel-Branco MM, Falcao AC (2002) Linear regression for calibration lines revisited: weighting schemes for bioanalytical methods. J Chromatogr B Analyt Technol Biomed Life Sci 774: 215-222.

24. Mansilhz C, Melo A, Rebelo H, Ferreira IM, Pinho O, et al. (2010) Quantification of endocrine disruptors and pesticides in water by gas chromatography-tandem mass spectrometry. Method validation using weighted linear regression schemes. J Chromatogr A 1217: 6681-6691.

25. Cheremisinoff NP, Ferrante L (1987) Practical Statistics for Engineers and Scientists. Taylor & Francis: Pennsylvania.

26. Maroto A, Boqué R, Riu J, Rius, FX (2001) Estimation of measurement uncertainty by using regression techniques and spiked samples. Analytica Chimica Act 446: 131-143.

27. Maroto A, Boqué R, Riu J, Rius FX (1999) Estimating uncertainties of analytical results using information from the validation process. Analytica Chimica Acta 391: 173-185.

28. Viswanathan CT, Bansal S, Booth B, DeStefano AJ, Rose MJ, et al. (2007) Workshop/Conference Report-Quantitative Bioanalytical Methods, Validation and Implementation: Best Practices for Chromatographic and Ligand Binding Assays. AAPS J 9: E30-E42.

29. Zeng W, Musson DG, Fisher AL, Wang AQ (2006) A new approach for evaluating carryover and its influence on quantitation in high-performance liquid chromatography and tandem mass spectrometry assay. Rapid Commun Mass Spectrom 20: 635-640.

Development and Validation of a Novel Stability Indicating UPLC Method for Dissolution Analysis of Bexarotene Capsules: An Anti Cancer Drug

Venkata Subba Rao D[1*], Raghuram P[1] and Harikrishna KA[2]

[1]ScieGen Pharmaceuticals INC, 89 Arkay Drive, Hauppauge, NY, USA

[2]Biological E Limited, ICICI Knowledge Park, Shamirpet, Hyderabad-500078, India

*Corresponding author: Venkata Subba Rao D, ScieGen Pharmaceuticals INC, 89 Arkay Drive, Hauppauge, NY, USA
E-mail: drdvsubbarao@yahoo.com

Abstract

A novel stability indicating liquid chromatographic method for dissolution analysis of Bexarotene capsules, 75 mg has been developed and validated. Efficient chromatographic separation was achieved on a C18 column (50 mm × 2.1 mm, 1.7-µm particles) with a simple isocratic mobile-phase at a flow rate of 1.0 mL min^{-1}. Quantification was achieved by use of ultraviolet detection at 260 nm. After the determination of the solubility the conditions selected were paddle at 50 RPM, with 900 mL of 0.5% HDTMA (hexadecyltrimethylammonium bromide) in 0.05 M phosphate buffer, pH adjusted to 7.5 with 1 N Sodium hydroxide at 37°C ± 0.5°C. Under these conditions the in vitro release profile of Bexarotene capsules, 75 mg shown good results. The drug release was evaluated by Reverse phase HPLC using mixture of Acetonitrile, water and trifluro acetic acid 70:30:0.1 (v/v/v). The method was validated for linearity, accuracy, precision, ruggedness, solution stability, mobile phase stability as per ICH guidelines to meet requirements for a global regulatory filing.

Keywords: Column liquid chromatography; UPLC; Bexarotene; Dissolution testing; Validation; Stability-indicating

Introduction

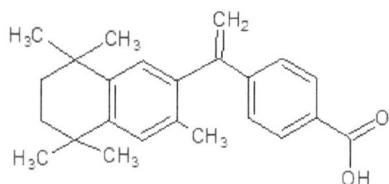

Figure 1a: The structures of Bexarotene and its impurities. Bexarotene. 4-[1-(3,5,5,8,8-pentamethyl-6,7-dihydronaphthalen-2-yl)ethenyl]benzoic acid.

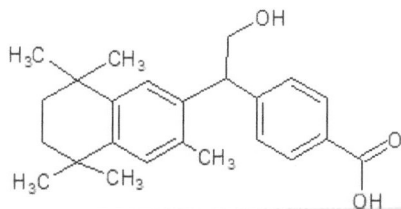

Figure 1b: The structures of Bexarotene and its impurities. 4-[1-(3,5,5,8,8-pentamethyl-6,7-dihydronaphthalen-2-yl)ethyl hydroxy]benzoic acid.

Bexarotene is an antineoplastic (anti-cancer) agent. Bexarotene is a retinoid specifically selective for retinoid X receptors, as opposed to the retinoic acid receptors and is chemically known as 4-[l-(3,5,5,8,8-pentamethyltetralin-2-yl)ethenyl] benzoic acid (Figure 1a and Figure 1b). Bexarotene (brand name: Targretin) is approved by the U.S.Food and Drug Administration (FDA) (in late 1999) and the European Medicines Agency (EMA) (early 2001) for use as a treatment for cutaneous T cell lymphoma (CTCL) [1]. It is a third-generation retinoid. Bexarotene is indicated for the treatment of cutaneous manifestations of cutaneous T-cell lymphoma in people who are refractory to at least one prior systemic therapy (oral) and for the topical treatment of cutaneous lesions in patients with CTCL who have refractory or persistent disease after other therapies or who have not tolerated other therapies.

Extensive literature survey revealed, few scientific articles have been published in which Bexarotene was quantified in human plasma [2] and rat plasma [3]. Despite this, a literature review reveals the lack of availability of a simple stability-indicating LC method for dissolution analysis of Bexarotene capsules.

The dissolution method conditions are considered from FDA Dissolution data base in which, 0.5% HDTMA (Hexadecyl trimethyl ammonium bromide) in 0.05 M phosphate buffer adjusted to 7.5 with 1 N NaOH/1 N HCl was given as Dissolution media. Since the dissolution medium is in basic side (pH 7.5) and Bexarotene is prone to degradation in basic medium, it is necessary to develop an analytical HPLC method for dissolution analysis which can separate the degradatnts arise due to basic hydrolysis of Bexarotene. As far as we are aware no stability-indicating LC method for dissolution Bexarotene drug product has been developed and validated.

A sensitive, simple, and selective chromatographic method was therefore sought for dissolution analysis of Bexarotene capsules.

Accordingly, the objective of this study was to establish the inherent stability of Bexarotene by use of stress studies under a variety of ICH recommended test conditions [4-6]. By development of a new analytical method for dissolution analysis of Bexarotene capsules [6-8], which could separate Bexarotene from all its related impurities and the various degradation products which could arise during stress testing.

Experimental

Chemicals and solutions

Bexarotene drug substance, standard, was supplied by Gensen Laboratories (Mumbai, India). Targretin capsules generic version was puchased. LC-grade acetonitrile and analytical reagent-grade tri-floro acetic acid were purchased from Merck (Darmstadt, Germany). High-purity water was prepared by use of a Millipore Milli-Q plus water-purification system.

Stock solutions of Bexarotene standard and sample (1.0 mg mL^{-1}) were prepared by dissolving appropriate amounts in ethanol (this solvent will subsequently be denoted 'diluent'). Working solutions of 83 μg mL^{-1} were prepared from this stock solution for analysis of bexarotene capsuels dissolution.

Chromatography

Chromatography was performed with a Waters Aquity UPLC quaternary pump, FTN auto sample manager and a photodiode-array detector (PDA). The output signal was monitored and processed by use of Empower software on a Pentium computer (Digital Equipment). Efficient separation was achieved on a C18 column (50 mm × 2.1 mm, 1.7-μm particles), BEH C18 column by use of a mobile phase isocratic prepared from a degassed 70:30:0.1 (v/v/v) mixture of Acetonitrile, water and trifluro acetic acid. The mobile phase flow rate was 1.0 mL min^{-1}, the column temperature was 30°C, and detection was at 260 nm. The injection volume was 5 μL.

Dissolution test conditions

The solubility study and percentage drug release was determined in 900 ml of 0.5% HDTMA (Hexadecyl trimethyl ammonium bromide) in 0.05 M phosphate buffer adjusted to 7.5 with 1 N NaOH/1 N HCl. Drug release tests were carried out with paddle method (USP apparatus II) at 50 rpm. The temperature of the cell was maintained at 37°C ± 0.5°C by using a thermostatic bath. Sampling aliquots of 10.0 ml were withdrawn at 15, 30, 45 and 60 min and replaced with an equal volume of the fresh medium to maintain a constant total volume. After the end of each test time, sample aliquots were filtered and quantified. The percentage content was calculated by validated stability indicating RP-HPLC method and these contents results were used to calculate the percentage release on each time of dissolution profile. The cumulative percentage of drug released was plotted against time in order to obtain the release profile.

Results and Discussion

Solubility determination and dissolution test condition

When dissolution test is not defined in the monograph of the dosage form, comparison of drug dissolution profiles is recommended on three different dissolution media, in the pH range of 1-7.5. The selection of a dissolution medium may be based on the solubility data and dosage range of the drug product. Hydrochloric acid, phosphate buffer and purified water are typical mediums used for dissolution test and these mediums were evaluated.

Bexarotene was insoluble in aqueous medium. For poorly soluble drugs, a percentage of surfactant can be used to enhance drug solubility and it is also recommended by USFDA. Then different concentrations of HDTMA (Hexadecyl trimethyl ammonium bromide) (0.25%, 0.5% and 1%) were prepared in in 0.05 M phosphate buffer adjusted to 7.5 with 1 N NaOH/1 N HCl and used for dissolution study.

At 25 rpm, the cumulative percentage drug release was considerably less than that at 50 rpm in above said dissolution medium. It was observed that less than 75% of drug was dissolved at 30 min in hydrochloric acid and phosphate buffer at a speed of both 25 rpm and 50 rpm.

In 0.1 N hydrochloric acid medium, pH 4.5 phosphate buffer and pH 7.5 buffer dissolution medium the drug was not completely soluble; some of the drug particles were detected at the bottom of the dissolution vessel. The cumulative percentage drug release obtained was low when compared to 0.5% HDTMA dissolution medium. At the end of 75 min the 0.5 % HDTMA medium showed 100% cumulative drug release at 50 rpm. In 0.25 %, HDTMA solutions showed more than 50% of drug release within 75 min. But in 0.5 %, HDTMA medium, % drug dissolved was nearly 100% within 75 min at stirring rate of 50 rpm. The cumulative percentages of drug in all the above said solutions were recorded. The analysis of variance showed no significant difference between the results obtained at 50 rpm and 75 rpm.

Analytical method development

To establish a selective and sensitive method the primary concern during development was to achieve symmetry of the Bexarotene peak and resolution between the peaks of Bexarotene and degradation peaks. Different types of buffer of different pH, for example phosphate buffer of pH 3-7, citrate buffer of pH 3-5, and ammonium acetate buffer of pH 4-6.5, were studied in combination with acetonitrile and methanol (30, 50, and 70%). The effects of acetonitrile, methanol, and their combinations on peak height and symmetry and on resolution between Bexarotene and degradation peaks were investigated. To improve peak height and peak symmetry, use of different ion-pairing reagents, for example trifluoroacetic acid, and triethylamine, was also investigated. The chromatographic data retention factor (k) and number of theoretical plates (N) were also recorded during these studies.

Solution of Bexarotene was prepared in diluent at a concentration of 100 ppm and the UV–visible spectra were acquired. The first UV absorption maxima of Bexarotene, was at approximately 260 nm, so detection at 260 nm was selected for method-development (Figure 2). Because Bexarotene and its possible degradation impurities are acidic in nature, because of the presence of carboxylic acid groups, a 80:20 (v/v) mixture of disodium hydrogen phosphate monohydrate buffer (0.01 m), pH 3.0, and acetonitrile was chosen for an initial trial with a 25 cm length, 4.6 mm i.d. column containing 5-μm particle size C18 stationary phase.

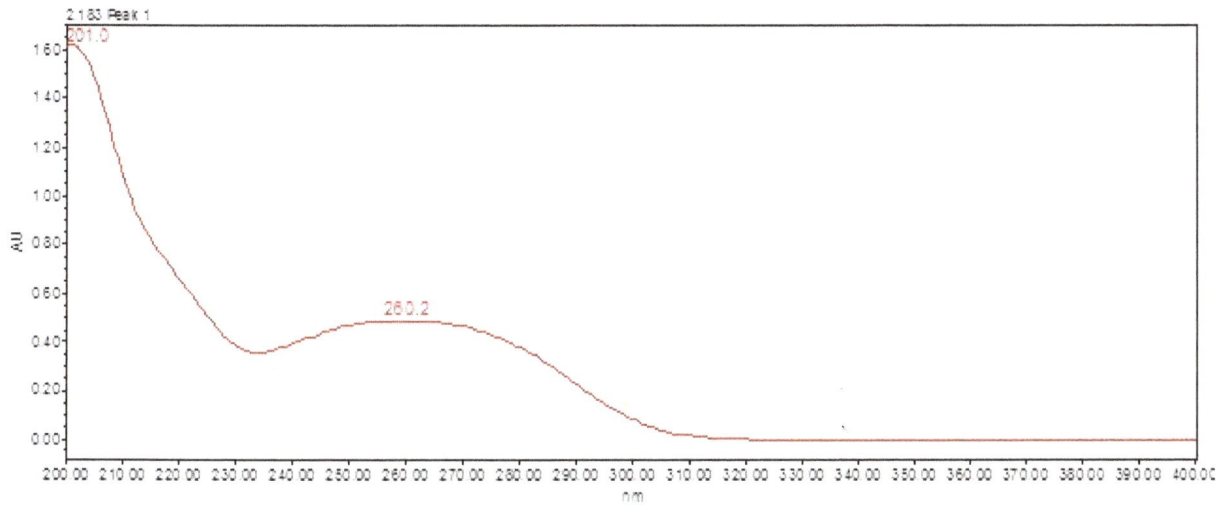

Figure 2: Typical UV Spectrums of Bexarotene and all impurities. Peak-1: imp-1, Peak-2: imp-2, Peak-3: Bexarotene, Peak-4: imp-3 and Peak-5: imp-4.

The flow rate was 1.0 mL min⁻¹. When Bexarotene sample spiked with all the impurities was injected under these conditions, Bexarotene and its impurities, peaks are not eluted even after 60 minutes of run time. To improve the retention time of Bexarotene, to decrease the interactions with stationary phase, acetonirile content has been increased to 40% i.e., buffer: acetonitrile 60:40 (v/v). Even with the increase acetonitrile content in mobile phase, Bexarotene peaks were not eluted before 60 minutes runt time.

To improve the retention times, shifted to UPLC system and submicron C18 column (50 mm × 2.1 mm, 1.7-µm particles). A 60:40 (v/v) mixture of disodium hydrogen phosphate monohydrate buffer (0.01 m), pH 7.5, and Acetonitrile was chosen for an initial trial on UPLC with BEH C18 column (50 mm × 2.1 mm, 1.7-µm particles) because of ionization of the carboxylic acid groups in Bexarotene and its degradation impurities, their polarity increased and interaction

with the column was reduced, so all degradation impurities and Bexarotene were co eluted rapidly (~1 min) (Figure 3a).

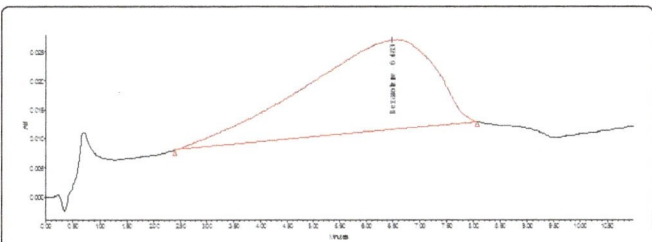

Figure 3a: Bexarotene in mobile phase (Buffer: ACN:60:40 V/V), pH 7.5.

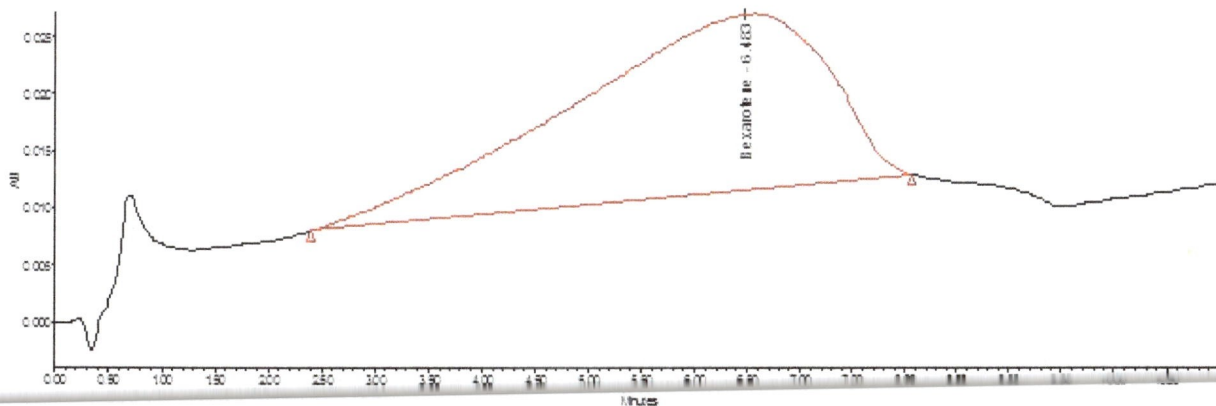

Figure 3b: Bexarotene in mobile phase (Buffer: ACN:80:20 V/V), pH 3.0.

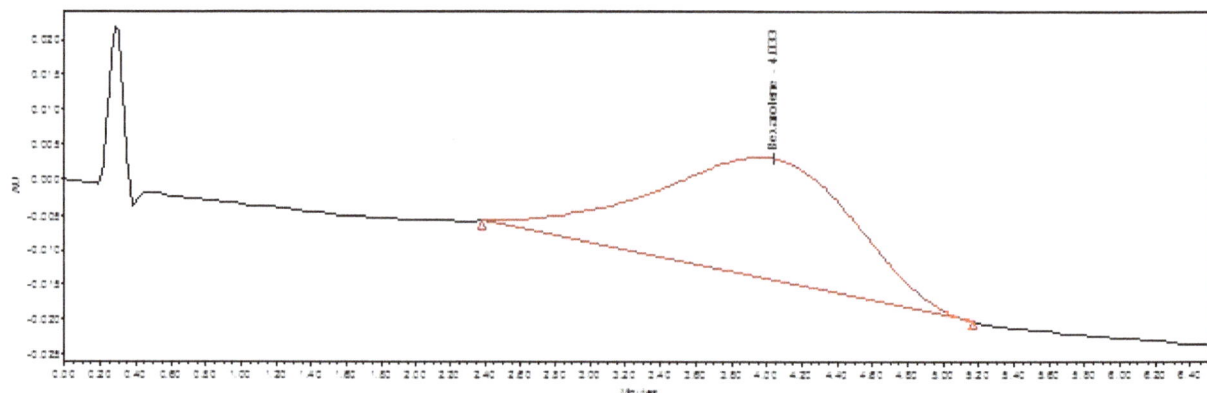

Figure 3c: Bexarotene in mobile phase (Buffer: ACN:70:30 V/V), pH 3.0.

To improve the retention time 80:20 (v/v) mixture of disodium hydrogen phosphate monohydrate buffer (0.01 m), pH 3.0, and acetonitrile was chosen, in acidic environment the ionization of Bexarotene decreases subsequently the polarity also decreased and the interactions with columns got increased, Bexarotene got eluted at about 6 minutes, but the peak shape of Bexarotene is too broad (Figure 3b). To further improve the retention, time a 70:30 (v/v) mixture of disodium hydrogen phosphate monohydrate buffer (0.01 m), pH 3.0, and acetonitrile was chosen Bexarotene got eluted at about 4 minutes but the peak shape of Bexarotene is too broad (Figure 3c).

To have to improve peak shape 0.25 mL trifluoroacetic acid was added to 1000 mL of 70:30 (v/v) mixture of DI water and acetonitrile, the peak shape of Bexarotene was improved, but still the trailing factor is greater than 2.0 observed (Figure 3d). To further improve the peak shape 1.0 mL trifluoroacetic acid was added to 1000 mL of mobile phase (700:300:1.0; water, Acetonitrile and Trifluoroacetic acid) with 1 mL trifluroacetic acid /1000 mL, the peak shape of Bexaroten was improved and the resolution between basic degradation peak and bexarotene is also improved (Figure 3e).

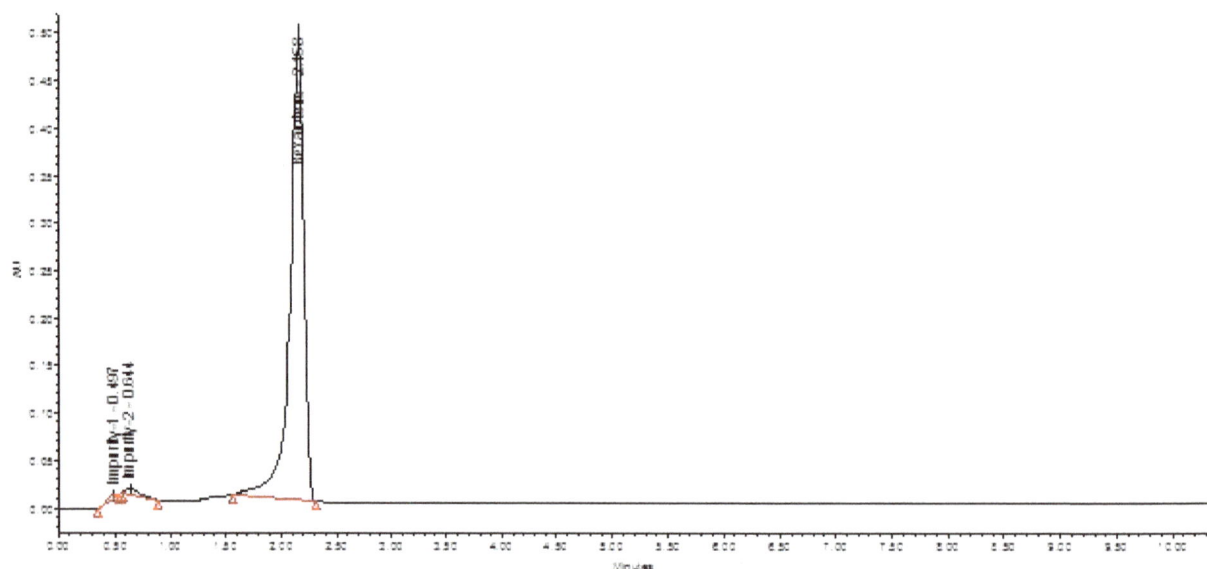

Figure 3d: Bexarotene in mobile phase (Water: AC:70:30 V/V), pH 3.0, 0.25 mL TFA.

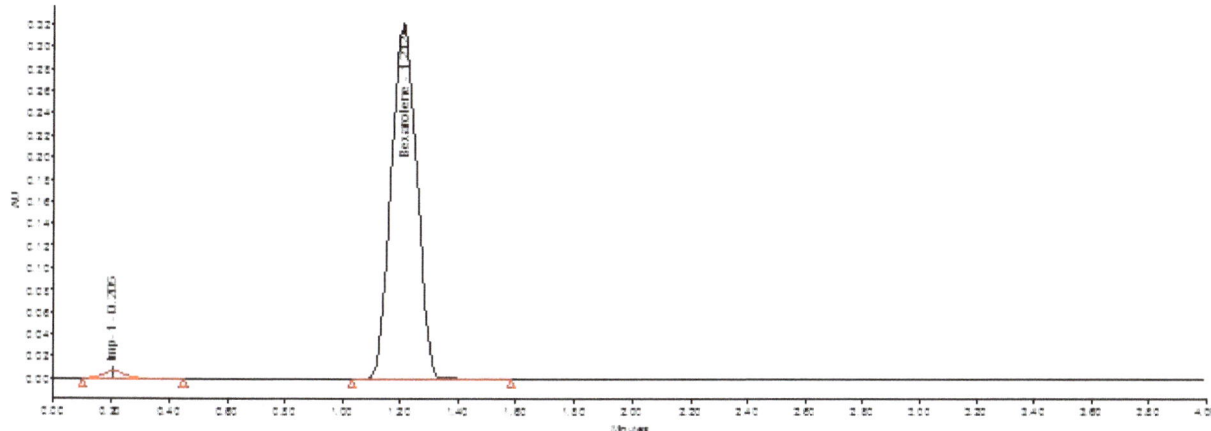

Figure 3e: Bexarotene in mobile phase (Water: ACN:70:30 V/V), pH 3.0, 1.0 mL TFA.

Buffer pH, amount of acetonitrile and amount of trifluro acetic acid all played a major role in achieving a symmetric Bexarotene peak and separation of Bexarotene from degradation impurities.

Method validation

The method was validated for specificity, linearity, range, precision, accuracy, sensitivity, robustness, and system suitability (Table 1).

Name of the parameter	Observed Value	Acceptance Criteria
%RSD of Bexarotene peak in five consecutive injections of standard preparation	0	Not more than 2.0
The tailing factor for Bexarotene peak in all injections of standard preparation	1.1 (mean of 5 injections value)	Not more than 2.0
The number of theoretical plates for Bexarotene peak in all injections of standard preparation	10078 (mean of 5 injections value)	Not less than 2000

Table 1: System suitability results.

Specificity

The dissolution test specificity was evaluated by preparing sample placebo of the commercial formulation of Bexarotene Capsules. These samples were transferred to separate vessels with 900 mL of dissolution medium at 37°C ± 0.5°C and stirred for 45 min at 100 rpm using a paddle (USP apparatus 2). Aliquots of these solutions were withdrawn, filtered through 0.45 μ membrane filters and analyzed by the RP-HPLC method using Bexarotene standard solution of 83 μg/mL.

The specificity [4,6,7] and stability-indicating nature of the method were assessed in the presence of the products obtained during forced degradation studies performed on Bexarotene [8,9]. Stress studies were conducted using a conventional reflux method.

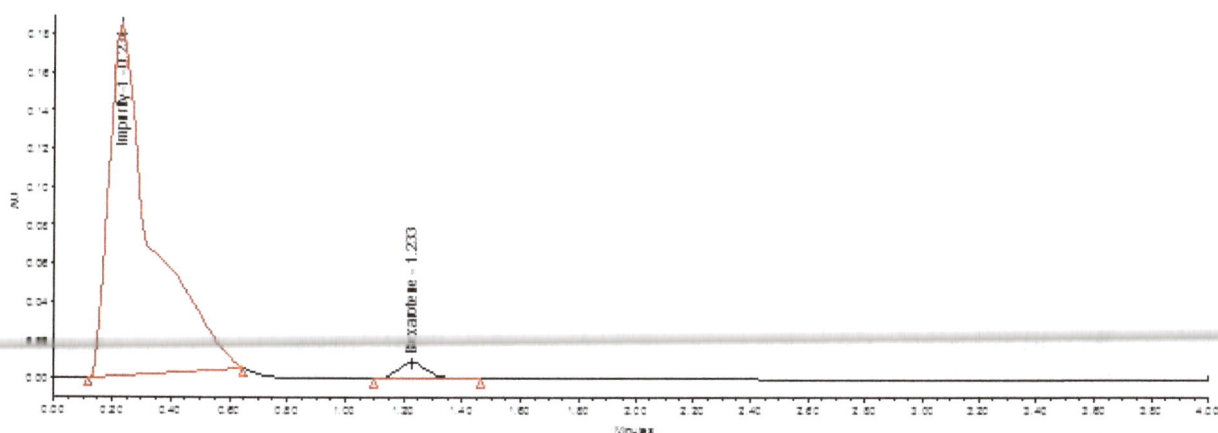

Figure 3f: Bexarotene capsules degradation with 1 m NaOH at Room temperature for 24 h.

All stress studies were performed at an initial drug concentration of 83 µg mL⁻¹. The stress conditions used for the degradation studies included light (exposure of the drug powder to overall illumination of 1.2 million lux hours at an integrated near ultraviolet energy of 200-watt hours/square meter (Whm⁻²) in a Sanyo (Leicestershire, UK) photostability chamber, conducted in accordance with ICH Q1B), heat (dry heat at 60°C, for 10 days, in a Mack Pharmatech (Hyderabad, India) dry air oven), acid hydrolysis (1 m HCl, at Room temperature, for 24 h), basic hydrolysis (1 m NaOH, at Room temperature, for 24 h) (Figure 3f), and oxidation (10% H₂O₂, at room temperature, for 24 h). For heat and light studies the study period was 7 days. Water baths equipped with MV controllers (Julabo, Seelabach, Germany) were used for hydrolytic studies. Stability studies were conducted in a laboratory humidity chamber (Thermo, India) (Figure 3g and Figure 3h).

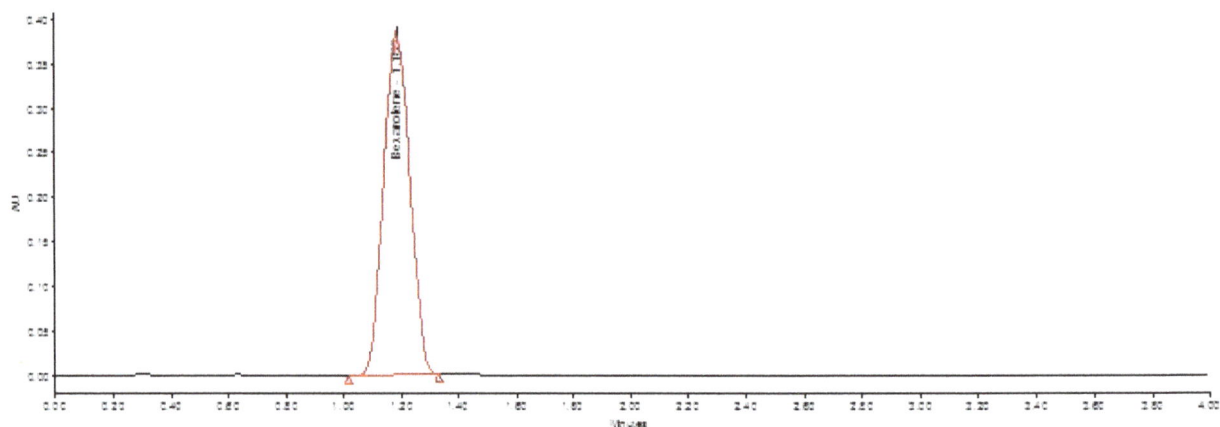

Figure 3g: Bexarotene standard chromatogram in final conditions.

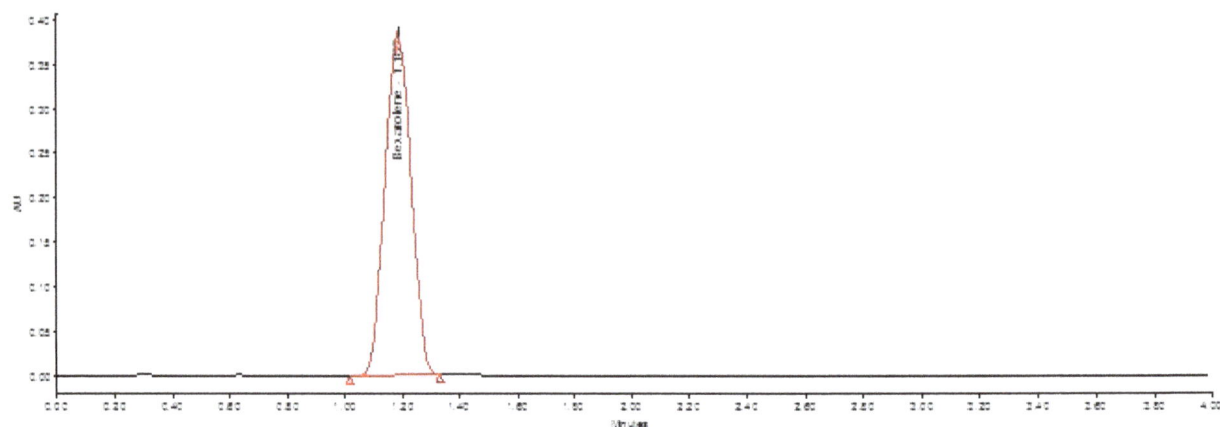

Figure 3h: Bexarotene sample chromatogram in final conditions.

Peak purity of Bexarotene from stressed samples was checked by use of the PDA. The purity angle within the purity threshold limit obtained for all stressed samples demonstrates analyte peak homogeneity. All stressed samples of Bexarotene were analysed for an extended run time of 100 min to check for late-eluting degradation products.

Precision

The precision of the dissolution by HPLC method was evaluated by performing six independent dissolutions of Bexarotene capsules against qualified reference standard. RSD (%) was calculated for the six assay values obtained.

Intermediate precision for the dissolution assay methods was evaluated by performing six independent dissolutions of Bexarotene capsules with the use of a different column (same packing but from a different batch), by a different analyst, using a different instrument (both HPLC and Dissolution), in the same laboratory.

Linear range

Linearity test solutions for the dissolution assay method at different concentrations from 25 to 150% of analyte concentration (i.e., 21, 41, 66, 00, 100, and 125 µg mL⁻¹) were prepared by dilution of the stock solution. Peak area and concentration data were treated by least-squares linear regression analysis and the correlation coefficients, slopes, and Y-intercepts of the calibration plots were calculated.

Accuracy

Accuracy for the dissolution method was studied by triplicate analysis at 50, 100 and 150% of the analyte concentration of 83 µg mL

[-1]. The accuracy was evaluated for the proposed method by adding known amount of bexarotene standard drug (50%, 100%, 150% level) to the tablet placebo powder, which were subjected to dissolution test conditions described above. Each solution was analysed in triplicate. The accuracy was calculated as the percentage of the drug recovered from the formulation matrix.

Robustness

Robustness was studied by making small but deliberate changes in the optimized method conditions and evaluating the effect on the results of Bexarotene capsules. With respect to chromatographic conditions, the mobile-phase flow rate was changed by 0.1 units from 1.0 mL min^{-1} to 0.9 and 1.1 mL min^{-1}. The Trifluro acetic acid was varied by ± 0.1mL units to 0.9 and 1.1 mL. The column temperature was changed from 30°C to 25°C and 35°C. When any changes were made all the other conditions, including the components of the mobile phase, were held constant.

With respect to dissolution conditions; dissolution Media Temperature was changed by ± 2°C from 37°C to 39° and 35°C. The RPM was varied by ± 2 from 50 RPM to 48 RPM and 52 RPM. Dissolution Media volume was changed ± 10 mL from 900 mL to 890 mL and 910 mL.

Solution stability and mobile phase stability

The solution stability of Bexarotene standard and Bexarotene capsules was assessed by leaving a test solution of the sample in a tightly capped volumetric flask at room temperature for 48 h and assay of the solution at 6-h intervals by comparison against freshly prepared standard solution. Stability in the mobile phase was assessed by assay of freshly prepared sample solutions against freshly prepared reference standard solutions at 6-h intervals up to 48 h. The prepared mobile phase was not changed during the study period. Assay was also checked for the test solutions under study.

Validation Results

Precision

RSD for assay of Bexarotene capsules during study of the precision of the dissolution method was within 2.0%. These results confirm the high precision of the method. In the study of intermediate precision the RSD of dissolution results was within 1.0% (Table 2) confirming the ruggedness of the method (Table 3).

Type of precision	Concentration (µg/mL)	%RSD Capsules	
Intra day precision	83	0.2	
Inter day Precision	83	0.4	
System precision	83	0.1	Acceptance criteria:NMT 2.0%

Table 2: Intra day and Inter day system precisions for Bexarotene Drug substance and Bexarotene Capsules.

Experimental parameters for robustness	Prep#	Peak area	% Drug release	% Mean Drug release	% Difference
Normal Conditions	1	1972867	97.65		
	2	1935733	95.81		
	3	2000993	99.04	97.5	N/A
Medium Temperature (-2°C) 35°C	1	2019084	100.11		
	2	1969414	97.65		
	3	2007052	99.51	99.1	1.6
Medium Temperature (+2°C) 39°C	1	2022086	100.26		1.8
	2	2014261	99.87		
	3	1972777	97.81	99.3	
RPM (-2) 48	1	1978283	98.09		
	2	1962521	97.3		
	3	1977095	98.03	97.8	0.3
RPM (+2) 52	1	2027156	100.41		
	2	2011941	99.66		
	3	2013872	99.75	100	2.5
Low HDTMA (-5%)	1	1964706	97.35		
	2	2005515	99.37		
	3	2002011	99.2	98.6	1.1
High HDTMA (+5%)	1	2010462	99.62		
	2	1991332	98.67		
	3	2007179	99.45	99.2	1.7
890 mL Media Volume (-1%)	1	2017873	99.98		
	2	2033031	100.86		
	3	1981251	98.17	99.7	2.2
910 mL Media Volume (+1%)	1	1929052	95.58		96	1.5

		2	196503 3	97.37		

Table 3: Data compilation of robustness study-part-b: sample solution. bexarotene capsules 75mg.

Linear Range

In the dissolution method a linear calibration plot was obtained over the range tested, i.e., 21–125 μg mL^{-1}; the correlation coefficient was>0.999. This result is indicative of excellent correlation between peak area and analyte concentration. The slope and Y-intercept of the calibration plot were 24623.7311 and -105573.0854, respectively (Table 4).

% level	Concentration in μg/mL	Response area	Predicted Response (Y)	Residuals
		(R)		(Y-R)
25	20.95	405310	410294.1	4984.08
50 40	41.1	901753	906462.3	4709.26
80	66.49	1547201	1531659	-15542.2
100	83.01	1934883	1938443	3559.83
125	100.74	2381390	2375022	-6368.42

150	124.92	2961766	2970423	8657.4
Correlation Coefficient				0.999
Slope(m)				24623.73
Intercept(C)				-105573
Residual Sum of Squares(RSS)				4.17E+08
% Y intercept				5.5
Limit value of residuals (±)				38697.66

Table 4: Actual concentrations and chromatographic data of bexarotene linearity solutions.

Accuracy

The individual recovery of Bexarotene is in between 95.7% and 98.9% and the grand recovery (grand mean) value is 97.21%. Over the entire range, 20% to 150% of Bexarotene concentration i.e., 20 μg/mL, the average recovery is 97.21%. This value is well within the validation criteria. Thus the method is valid for the proposed range of concentration. The % RSD of six preparations at different levels (20%-150%) is less than 5.0%, indicating the method is precise.

This study proves that the method is capable of recovering the active drug component Bexarotene from placebo matrix. Thus the method is accurate and precise for the proposed range of concentration (Table 5).

% level	Prep#	Peak area	Amount added (μg/mL)	Amount recovered (μg/mL)	% Recovery	%Mean recovery	SD	% RSD	95% CI
	1	391381	20.15412	19.97899	99.77				Lower
	2	391712	20.15412	19.98236	99.85				99.75
	3	391690	20.15412	19.98213	99.85				
	4	391106	20.15412	19.97619	99.7				Upper
	5	392773	20.15412	19.99314	100.12				100
25	6	391839	20.15412	19.98365	99.89	99.86	0.143	0.1	
	1	1967882	83.31627	83.00657	100.48				Lower
	2	1951794	83.31627	82.84301	99.66				99.83
	3	1955533	83.31627	82.88102	99.85				
	4	1976123	83.31627	83.09035	100.9				Upper
	5	1958333	83.31627	82.90949	99.99				100.6
100	6	1965505	83.31627	82.98241	100.36	100.2	0.4586	0.5	
	1	2947566	125.7886	124.9666	100.34				Lower
	2	2936331	125.7886	124.8524	99.95				100.23
	3	2967968	125.7886	125.174	101.03				
	4	2960372	125.7886	125.0968	100.77				Upper
150	5	2948908	125.7886	124.9802	100.38	100.58	0.4367	0.4	100.9

	6	2968689	125.7886	125.1813	101.05				
Minimum %Recovery					99.7				
Maximum %Recovery					101.1				
Grand Recovery					100.2				

Table 5: Chromatographic data and results of accuracy and precision study.

Robustness

Robustness was studied during the final phase of method development. When the chromatographic conditions flow rate, pH, and column temperature and dissolution conditions RPM, bath temperature, dissolution media volume were deliberately varied there is no changes in the results of Bexarotene capsules dissolution, illustrating the robustness of the method (Table 3).

Solution stability and mobile phase stability

The RSD obtained for assay of Bexarotene was within 1% during studies of stability in the diluent and in the mobile phase, and there was no significant change in Bexarotene content during the studies. These results confirm the compounds were stable in diluent and in the mobile phase used during dissolution assay analysis for a study period of up to 48 h.

Results from Forced Degradation Studies

Basic and acidic hydrolysis

When the drug was exposed to 1 m HCl, significant degradation was observed at RRT ~0.12.

When the drug was exposed to 1 m NaOH, significant degradation was observed at RRT ~0.12, after 24 hours at Room temperature, Bexarotene totally converted to imp^{-1}.

When the drug was exposed to aqueous hydrolysis no major degradation products were observed, so the drug was stable to aqueous hydrolysis.

Oxidizing conditions

When Bexarotene was exposed to 6% hydrogen peroxide at room temperature for 24 h no degradation was observed. The drug was stable to oxidation.

Photolysis

The drug was stable to the effect of photolysis and no degradation was observed.

Thermal degradation

The drug was stable to the effect of temperature and no degradation was observed. All the forced degradation samples were analyzed or assay, and peak purity was calculated.

Peak purity test results obtained by use of the PDA confirmed the Bexarotene and degradation product peaks obtained from analysis of all the stressed samples were homogeneous and pure. Because

dissolution assay of Bexarotene was unaffected by the presence of the degradation products (Table 6) the method can be regarded as stability-indicating.

Stress condition	Time	Degradation (%, approx.)
Acid hydrolysis	24 h	0.002
(1 m HCl, Room Temperature)		
Basic hydrolysis	24 h	0.97
(1 m NaOH, Room Temperature)		
Oxidation (6% H_2O_2, room temperature)	24 h	0
Thermal (60°C)	10 days	0
Light (photolytic degradation)	10 days	0

Table 6: Summary of results from forced degradation studies for Bexarotene Capsules.

Conclusion

A new, sensitive, and stability-indicating LC method has been successfully developed for quantitative analysis of Bexarotene capsules dissolution. The method was accurate and precise with good consistent recovery at all the levels studied. The method enables high resolution of degradation products from Bexarotene and from each other. The validated method may be used for routine analysis of Bexarotene capsules dissolution for analysis of other quality-control samples during product development.

Acknowledgements

The authors wish to thank the management of Gensen Laboratories for supporting this work.

References

1. Bohmeyer J, Stadler R, Kremer A, Nashan D, Muche M, et al. (2003) Bexarotene--an alternative therapy for progressive cutaneous T-cell lymphoma? First experiences J Dtsch Dermatol Ges 1: 785–789.

2. Van de Merbel NC, van Veen JH, Wilkens G, Loewen G (2002) Validated liquid chromatographic method for the determination of bexarotene in human plasma. J Chromatogr B 5: 189–195.

3. Lee, Jong L, Egah, Atheer, Kim, et al. (2017) Simple and sensitive HPLC-UV method for determination of bexarotene in rat plasma. J Chromatogr B 1040: 73-80.

4. International Conference on Harmonization (2003) Stability Testing of New Drug Substances and Products Q1A (R2). International Conference on Harmonization, IFPMA, Geneva.

5.   Singh S, Bakshi M (2000) Stress Tests to Determine Inherent Stability of Drugs. Pharm Tech On-line 24: 1–14.

6.   International Conference on Harmonization (1996) Photostability Testing of New Drug Substances and Products Q1B. International Conference on Harmonization, IFPMA, Geneva.

7.   Bakshi M, Singh S (2002) Development of validated stability-indicating assay methods--critical review. J Pharm Biomed Anal 28: 1011–1040.

8.   Singh S, Singh B, Bahuguna R, Wadhwa L, Saxena R (2006) Rapid simultaneous determination of telmisartan, amlodipine besylate and hydrochlorothiazide in a combined poly pill dosage form by stability-indicating ultra performance liquid chromatography. J Pharm Biomed Anal 41: 1037–1040.

9.   Carstensen JT, Rhodes CT. Drug stability principles and practices. Drugs and the Pharmaceutical Sciences.

3

A Simple Protein Precipitation-based Simultaneous Quantification of Lovastatin and Its Active Metabolite Lovastatin Acid in Human Plasma by Ultra-Performance Liquid Chromatography-Tandem Mass Spectrometry using Polarity Switching

Wujian J[1], Kuan-wei P[1], Sihyung Y[1], Huijing S[2], Mario S[2] and Wang MZ[1]*

[1]*Department of Pharmaceutical Chemistry, The University of Kansas, Lawrence, KS, USA*
[2]*Eshelman School of Pharmacy, The University of North Carolina at Chapel Hill, Chapel Hill, NC, USA*

Abstract

Lovastatin is an anti-cholesterol lactone drug indicated for the treatment of hyperlipidemia and to reduce the risk of coronary heart disease. It is converted to the β-hydroxy acid form (lovastatin acid) *in vivo*, which is the major pharmacologically active metabolite. Here, we describe the development and validation of an ultra-performance liquid chromatography-tandem mass spectrometry (UPLC-MS/MS)-based method utilizing polarity switching for the simultaneous quantification of lovastatin and lovastatin acid in human plasma. Simple protein precipitation extraction and direct injection of the extracted samples without drying/reconstitution showed good recoveries of both analytes (~70%). The developed method exhibited satisfactory intra-day and inter-day accuracy and precision. The interconversion between lovastatin and lovastatin acid during sample preparation and storage was minimal (< 1.9%). The lower limits of quantification were 0.5 and 0.2 nM (or 0.2 and 0.084 ng/mL) for lovastatin and lovastatin acid, respectively, using only 50 μL of plasma during extraction. The validated method was successfully applied to analyze plasma samples obtained from a healthy human subject who enrolled in a clinical drug interaction study involving lovastatin.

Keywords: Lovastatin; Lovastatin acid; UPLC-MS/MS; Polarity switching; Protein precipitation extraction; Pharmacokinetics

Abbreviations: Auc: Area Under The Plasma Concentration-Time Curve; Ce: Collision Energy; Cid: Collision-Induced Dissociation; C_{max}: Maximum Concentration; Cyp: Cytochrome P450; Cv: Coefficient Of Variation; Dmso: Dimethyl Sulfoxide; Hmgcr: 3-Hydroxy-3-Methylglutaryl Coenzyme A Reductase; Hplc-Uv: High-Performance Liquid Chromatography-Ultraviolet Detection; Is: Internal Standard; Lle: Liquid–Liquid Extraction; Lloq: Lower Limit Of Quantification; Lv: Lovastatin; Lva: Lovastatin Acid; Mrm: Multiple Reaction Monitoring; Ppe:protein Precipitation Extraction; Qc: Quality Control; S.d.: Standard Deviation; Spe: Solid-Phase Extraction; S/N: Signal-To-Noise Ratio; Sv: Simvastatin; Sva: Simvastatin Acid; $T_{1/2}$: Terminal Elimination Half-Life; T_{max}:time To Reach C_{max}; Uplc-Ms/Ms: Ultra-Performance Liquid Chromatography-Tandem Mass Spectrometry

Introduction

Lovastatin (LV) is a member of the class of cholesterol-lowering agents known as statins and is indicated for the treatment of dyslipidemia and the prevention of coronary heart disease. LV, a lactone, is readily hydrolyzed *in vivo* to form the pharmacologically active metabolite lovastatin β-hydroxy acid (LVA), which inhibits 3-hydroxy-3-methylglutaryl coenzyme A reductase (HMGCR), a key enzyme in the cholesterol biosynthetic pathway [1]. In human liver, LV also is metabolized by cytochrome P450 (CYP) 3A4 to form 6'-β-hydroxy- and 6'-exomethylene-LV, both of which also are HMGCR inhibitors [2,3]. However, neither metabolite is readily detected *in vivo*. Additional LV metabolites (*e.g.*, 3'-hydroxy-LV and conjugates formed after β-oxidation of the heptanoic acid moiety of LVA) can be detected *in vivo*, but do not inhibit HMGCR [2]. Therefore, LV and LVA are the major drug-derived molecules that can be used for therapeutic drug monitoring of LV treatment. Sensitive and specific analytical methods for simultaneous measurement of LV and LVA in human plasma are needed.

A number of studies have described the development and validation of analytical methods for LV. Pasha et al. [4] described an HPLC method coupled to UV detection (HPLC-UV), monitoring at a wavelength of 237 nm, to separate LV and four other statins. The lower limit of quantification (LLOQ) for LV was 0.1 μg/mL (~250 nM). This method is useful for pharmaceutical formulation analysis and *in vitro* metabolism studies, but lacks the sensitivity (sub-nM LLOQ) required for pharmacokinetic studies. A more sensitive HPLC-UV method, which achieved an LLOQ of 1.0 ng/mL (~2.5 nM) for LV, was described more recently [5]. However, it requires a relatively large volume (2 mL) of human plasma. Marked enhancement in sensitivity for LV has been achieved by employing tandem mass spectrometric (MS/MS) detection with the LLOQ ranging from 0.025 to 0.1 ng/mL (~0.06 to 0.25 nM) using 0.2-0.5 mL of human plasma [6-9]. However, all sample preparation methods were complicated and time-consuming, requiring liquid-liquid extraction (LLE) with *tert*-butyl methyl ether, ethyl acetate, ether-methylene chloride or *n*-hexane-methylene dichloride-isopropanol, followed by evaporation under nitrogen at an elevated temperature (40–50°C). In general, recovery following LLE was approximately 85%, but ethyl acetate gave a low of 55%.

Few reports have described the simultaneous measurement of LV

*Corresponding author: Wang MZ, Pharmaceutical Chemistry, The University of Kansas, 2095 Constant Ave, Lawrence, KS 66047, USA
E-mail: michael.wang@ku.edu

and LVA in human plasma. Sun et al. [10] described a solid-phase extraction (SPE)- and HPLC-MS/MS-based analytical method for LV and LVA in human plasma, applying the method for a comparative pharmacokinetic study of different LV dosage forms. However, the report did not provide details of the method (*e.g.*, plasma volume used, extraction recovery, ions monitored, matrix effect or stability) to allow cross-evaluation. In another report, Wu et al. [11] described an SPE- and HPLC-MS/MS-based analytical method for LV and LVA in mouse and rat plasma. This method, which used 0.1 mL of plasma, had an LLOQ of 0.5 ng/mL (~1.2 nM) and an extraction recovery of ~55% for LV. HPLC-UV methods [12,13] also have been reported for the simultaneous quantification of LV and LVA, exhibiting an LLOQ ≥ 25 ng/mL (~62 nM). However, in general, these methods lack the necessary sensitivity for pharmacokinetic studies. The purpose of the study described herein is to develop and validate a simple protein precipitation extraction (PPE), direct injection, UPLC-MS/MS-based analytical method for the simultaneous quantification of LV and LVA in human plasma.

Materials and Methods

Materials

LV was purchased from U.S. Pharmacopeia (Rockville, MD, USA). LVA, simvastatin (SV) and simvastatin acid (SVA) were obtained from Toronto Research Chemicals Inc. (North York, ON, Canada). SV and SVA were used as internal standards (IS) for the quantification of LV and LVA, respectively. Ammonium acetate and dimethylsulfoxide (DMSO) were purchased from Sigma-Aldrich (St. Louis, MO, USA). Optima-grade water, acetonitrile, methanol and acetic acid were obtained from Fisher Scientific (Pittsburgh, PA, USA). Blank human plasma (collected in K_2-EDTA tubes) was purchased from Innovative Research (Novi, MI, USA).

Instrument and chromatographic conditions

The UPLC–MS/MS system consisted of a Waters Acquity I-Class UPLC and a Xevo TQ-S triple quadrupole mass spectrometer equipped with an electrospray ionization (ESI) source (Waters Corporation, Milford, MA, USA). Chromatographic separation was achieved using an ACQUITY UPLC BEH C_{18} (1.7 μm, 2.1 x 50 mm) column, which was protected by an ACQUITY column in-line filter (0.2 μm). UPLC mobile phase (A) consisted of 100% water with 5 mM ammonium acetate (pH 4.5), while (B) consisted of acetonitrile-ammonium acetate (50 mM; pH 4.5) (90:10, v/v). The pH of each mobile phase (adjusted with acetic acid) was selected to minimize interconversion of LV and LVA, based on previous observations and recommendations for SV and SVA [14,15]. The gradient began with 40% B and was held for 0.5 min. Mobile phase B increased linearly to 70% B at 0.8 min, then to 80% B at 3.5 min. The column was washed with 100% B for 0.5 min before the system was re-equilibrated with 40% B for 1 min prior to the next injection. The total run time was 5 min with a flow rate of 0.4 mL/min. The autosampler was set at 6°C and the UPLC column heated at 50°C. The sample injection volume was 50 μL.

The mass spectrometer was operated under negative ion mode for the first 1.9 min to allow for the detection of LVA and SVA (as IS). Instrumental conditions during this time were: capillary voltage, 3.10 kV; cone voltage, -32 V; source offset, 50 V; desolvation temperature, 500°C; desolvation gas flow, 1000 L/h; cone gas flow, 150 L/h; nebulizer gas flow, 7.0 bar; and collision gas flow, 0.15 mL/min. Analyte-specific instrument parameters (*i.e.*, collision energy, capillary voltage and cone voltage) were optimized prior to analysis using the IntelliStart™ auto-tune with infusion of analyte standards

(500 nM in methanol at 0.010 mL/min infusion rate, supplemented with 50:50 (v/v) mobile phase A:mobile phase B at 0.4 mL/min). The specific MRM transitions used for quantification were m/z 421.4 → 319.3 for LVA and 435.4 → 319.3 for SVA. After 1.9 min, the instrument was switched to positive ion mode for the detection of LV and SV (as IS). The instrumental conditions were: capillary voltage, 1.10 kV; cone voltage, 30 V; source offset, 50 V; desolvation temperature, 500°C; desolvation gas flow, 1000 L/h; cone gas flow, 150 L/h; nebulizer gas flow, 7.0 bar; and collision gas flow, 0.15 mL/min. Similar to before, analyte-specific instrument parameters were optimized prior to analysis using the IntelliStart™ auto-tune with infusion of analyte standards (500 nM in methanol at 0.010 mL/min infusion rate, supplemented with 50:50 (v/v) mobile phase A:mobile phase B at 0.4 mL/min). The specific MRM transitions used for quantification were m/z 427.3 → 325.2 for LV as a sodium adduct and m/z 441.3 → 325.2 for SV as a sodium adduct. Peak area ratios of analyte *vs.* IS were used to generate calibration curves and calculate analyte concentrations in plasma samples.

Preparation of calibration standards and quality controls

Individual stock solutions of LV, LVA, SV, and SVA were prepared by dissolving an appropriate amount of each authentic standard in DMSO to give a final concentration of 10 mM. These stock solutions were stored at -20°C. To prepare working standard stocks, the individual stocks were first diluted to 10 μM with methanol and then further diluted with methanol to yield working concentrations that were 50-fold higher than the intended final concentrations in plasma. Working standard stocks were prepared fresh from the 10 μM stocks and were not stored after being used to prepare calibration standards and QCs. Calibration standards were generated by spiking 1.0 μL of working standard stock into 49 μL blank human plasma, yielding the following final concentrations: 0.1, 0.2, 0.5, 1, 2, 5, 10, 20, 50 and 100 nM for LV. Similarly, a separate set of calibration standards was prepared for LVA. LV and LVA quality controls (QCs) also were prepared separately from the individual stock solutions to give final concentrations of 2, 10, 50 and 100 nM LV or LVA in human plasma. Calibration standards were prepared in triplicate and QCs in quadruplicate. All were processed as plasma samples prior to UPLC-MS/MS analysis.

Sample preparation by protein precipitation extraction

Each plasma sample (50 μL), including calibration standards and QCs, was thawed at room temperature and immediately placed on ice. Fifty μL of ice-cold sodium acetate buffer (100 mM; pH 4.5 at room temperature) was added to minimize the interconversion of LV and LVA during sample preparation. Proteins in the mixture were precipitated and extracted with ice-cold methanol (200 μL) containing 0.1% (v/v) acetic acid and 3 nM each of SV and SVA as IS. After vortexing for 10 s, the samples were centrifuged at 2250 *g* for 15 min at 4°C. The supernatant (180 μL) was transferred to 96-well polypropylene plates, from which an aliquot (50 μL) of sample was injected for UPLC-MS/MS analysis.

Method validation

The specificity of the method was investigated by comparing UPLC-MS/MS extracted ion chromatograms of blank plasma to blank plasma spiked with working standard stocks at the LLOQ. The LLOQ was defined as the lowest concentration of calibration standards that had a signal-to-noise ratio (S/N) greater than 5, an accuracy of 80-120%, and an imprecision of ≤ 20%. The limit of detection (LOD) was defined as the lowest concentration of calibration standards that resulted in an S/N of at least 3.

The extent of interconversion between LV (lactone) and LVA (β-hydroxy acid) that would occur during sample preparation and storage was evaluated by quantifying both LV and LVA present in freshly made 50 and 100 nM LV-only and LVA-only QC samples. These QC samples were analyzed immediately after preparation and again after 24 h of storage in an autosampler at 6°C. Statistical analyses were performed using unpaired, two-tailed Student's t-tests (GraphPad Prism 5.04; GraphPad Software, Inc., La Jolla, CA, USA).

The intra-day accuracy and precision were determined by replicate analyses of the QCs on the same day. The inter-day accuracy and precision were determined by replicate analyses of the QCs on three separate days. The accuracy (%) was defined as the closeness of the average QC concentration determined by the current method to the true QC concentration. The precision (coefficient of variation [CV]; %) was defined as the spread of individual measures of multiple QC preparations.

The extraction recovery was determined by comparing peak area ratios of an analyte $vs.$ IS obtained from QC samples that were spiked with analyte prior to extraction to those obtained from blank plasma extracts that were spiked with known amounts of pure analyte following extraction (represents 100% recovery). The extraction recovery was examined at three QC concentrations (2, 10 and 50 nM for LV and LVA) and in quadruplicate at each concentration.

Matrix effects during the MS/MS analysis were evaluated by comparing peak areas of analyte from blank plasma extracts that were spiked with known amounts of pure analyte to peak areas from samples prepared in the neat solvent (water-100 mM sodium acetate [pH 4.5]-methanol containing 0.1% [v/v] acetic acid [50:50:200, v/v/v]) with known amounts of pure analyte. The matrix effects were examined at three QC concentrations (2, 10 and 50 nM for LV and LVA) and in quadruplicate at each concentration. Results were expressed as % signal remaining relative to standards in the neat solvent (represents 100% signal remaining or no suppression).

Freeze-thaw stability was evaluated by exposing QCs to three freeze (-20°C)-thaw (room temperature) cycles before sample preparation. Long-term storage stability was evaluated by storing QCs and sodium acetate-buffered QCs at -80°C for up to one month. The sodium acetate-buffered QCs were prepared by mixing QCs with an equal volume of 100 mM sodium acetate (pH 4.5). Freeze-thaw stability and long-term

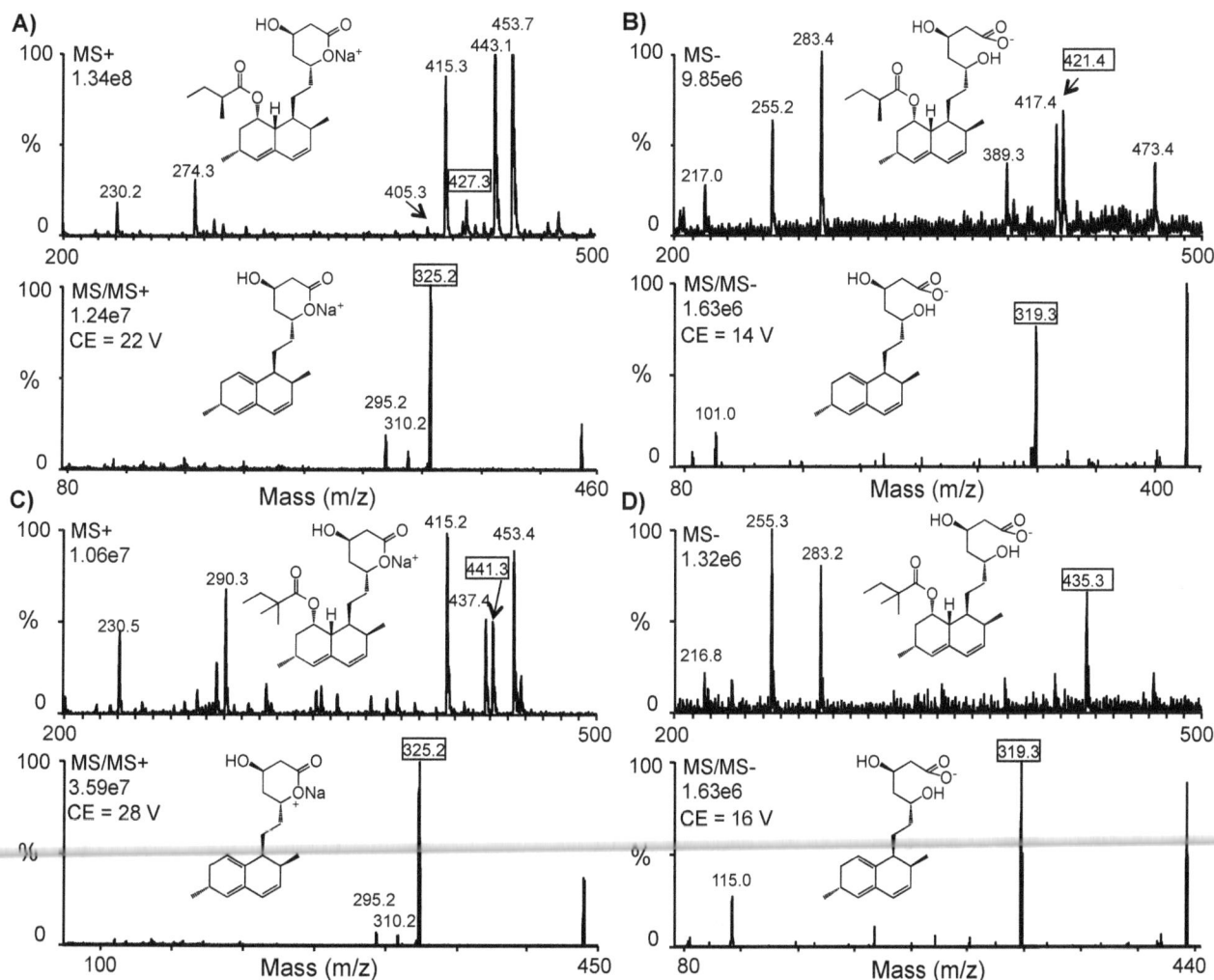

Figure 1: Q1 full-scan mass spectra (MS) and product ion mass spectra (MS/MS) of LV (A) and LVA (B), and the internal standards SV (C) and SVA (D)
The precursor and product ion pairs selected for MRM detection are shown in boxes. The postulated structure of each also is depicted. The signal intensity for the most intense peak is in the upper-left corner of each spectrum. CE, collision energy.

storage stability were examined at three QC concentrations (2, 10 and 50 nM for LV and LVA) and in quadruplicate at each concentration. Results were expressed as % signal remaining relative to freshly-made QCs (represents 100% signal remaining).

Analysis of human plasma samples from a clinical study

Blood samples were obtained from a healthy volunteer who participated in a drug-drug interaction clinical study involving lovastatin. Lovastatin was given once daily for 14 days. On the morning of the 7th day following an overnight fast, the volunteer received the 7th dose of lovastatin (40 mg) and a single dose of warfarin (10 mg) by mouth simultaneously. Venous blood (5 mL) was collected in K_2-EDTA Vacutainer* tubes (BD Biosciences, Franklin Lakes, NJ, USA) *via* an intravenous line at 0, 0.5, 1, 1.5, 2, 3, 4, 6, 8, 10, 12, and 24 h post-drug administration. Within 1 h after collection, blood samples were centrifuged (3000 rpm at 4°C for 10 min) and the resulting plasma samples transferred to pre-labeled cryotubes for storage at -80°C until analysis. The clinical study (ClinicalTrials.gov registry number: NCT01250535) was approved by the Institutional Review Board of the University of North Carolina at Chapel Hill (Chapel Hill, NC, USA). The area under the plasma concentration–time curve (AUC), terminal elimination half-life ($t_{1/2}$), maximum plasma drug concentration (C_{max}), and the time to reach C_{max} (T_{max}) were calculated using the trapezoidal rule–extrapolation method and non-compartmental analysis (Phoenix WinNonlin Version 6.3; Pharsight, Mountain View, CA, USA).

Results and Discussion

UPLC–MS/MS analysis

Apostolou et al. [16] previously reported that ESI was more sensitive for both SV and SVA than atmospheric pressure chemical ionization (APCI). Due to the structural similarity between LV/LVA and SV/SVA (Figure 1), ESI was selected for method development in this study. Under positive ion mode, LV and SV produced molecular ions at m/z 427.3 and 441.3, indicating the formation of sodium adducts with little protonated ions detected (Figures 1A and 1C). Upon collision-induced dissociation (CID) fragmentation, three major product ions were observed for both LV and SV (m/z 295.2, 310.2 and 325.2). The MRM transitions m/z 427.3→325.2 and 441.3→325.2 were selected for monitoring LV and SV, respectively. Under negative ion mode, LVA and SVA produced molecular ions at m/z 421.4 and 435.3, indicating deprotonated anions (Figures 1B and 1D). Upon CID fragmentation, two major product ions were observed for both LVA (m/z 319.3 and 101.0) and SVA (m/z 319.3 and 115.0). The MRM transitions m/z 421.4→319.3 and 435.3→319.3 were selected for monitoring LVA and SVA, respectively. These precursor and product ions for LV/LVA and SV/SVA also were observed in earlier studies [6,11], although protonated lactone cations and acetonitrile-sodium adducts of the lactones were predominant in other studies [16-18].

Representative extracted ion chromatograms of blank human plasma and calibration standards at LLOQ are shown in Figure 2. Since LV and LVA may undergo interconversion during MS analysis (*i.e.*,

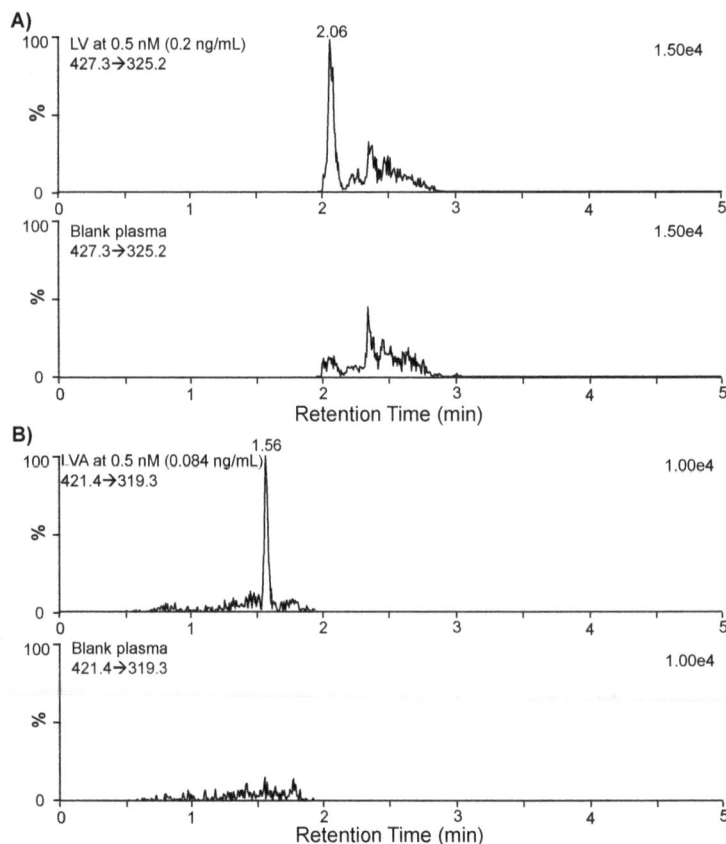

Figure 2: Representative extracted ion chromatograms of LV (A) and LVA (B) calibration standards at LLOQ in human plasma (upper panel) and blank human plasma (lower panel)
The monitored MRM transitions and signal intensity for the most intense peak are shown in the upper-left and upper-right corners of each chromatogram, respectively.

in-source conversion), chromatographic resolution of LV from LVA is required to detect and monitor the extent of such interconversion [19]. Chromatographic resolution of SV from SVA also is required for the same reason. Upon chromatographic separation, LV and LVA eluted at 2.06 and 1.56 min, while SV and SVA eluted at 2.34 and 1.69 min (Figures 2 and 3). All are well-resolved from each other. Retention times did not drift after repeat injections (n=3 ; S.D. ≤ 0.3 s) of the same QC samples separated by 30 injections (data not shown).

Calibration standards, accuracy and precision

A linear calibration curve (0.2-100 nM or 0.084-42 ng/mL) with 1/x weighting was obtained for LVA, while a quadratic calibration curve (0.5-100 nM or 0.2- 40 ng/mL) with 1/x weighting was obtained for LV. Coefficients of determination (r^2) were greater than 0.99 for the calibration curves. Standards with concentrations greater than 100 nM were not tested because LV and LVA plasma concentrations are not expected to exceed 100 nM following a therapeutic dose of LV (40 mg once daily in this study). The on-column sensitivity of the method was 25 and 10 fmol (or 2.8 and 1.2 pg) for LV and LVA, respectively. For QC samples at the low, medium, and high concentration, the intra-day accuracy were range from 97.7-104.9% for LV and 94.5-103.6% for LVA, and the inter-day accuracy were range from 98.2-105.3% for LV and 95.7-99.5% for LVA (Table 1). The precision was determined by calculating CV for QC. The intra-day precision and inter-day precision were all within 10% for LV and LVA. These values are well within the criteria of 15% bias and 15% CV stated in the US FDA bioanalytical method validation guidance [20].

Method validation

Interconversion of LV and LVA during sample preparation and storage was examined by monitoring both LV and LVA in single-component QC samples (50 and 100 nM). Immediately after sample preparation, a small amount of LVA was detected in the LV-only QC samples with the LVA concentration representing an average 1.7% of the LV concentration (Table 2). Similarly, a small amount of LV was detected in the LVA-only QC samples with the LV concentration representing an average 1.1% of the LVA concentration (Table 3). Upon storage in

Compound	QC Concentration (nM)	Intra-day		Inter-day	
		Accuracy (%)	CV (%)	Accuracy (%)	CV (%)
LV	2	104.9	2.9	105.3	6.8
	10	99.9	3.4	102.7	3.2
	50	97.7	7.8	98.2	0.4
LVA	2	103.6	4.8	98.5	3.7
	10	94.5	1.7	99.5	4.1
	50	94.6	9.8	95.7	2.9

CV, coefficient of variance

Table 1: Intra-day and inter-day accuracy and precision for LV and LVA.

QC Concentration (nM)	LVA		LVA after 24-h storage	
	Concentration (nM)	Relative to QC Concentration (%)	Concentration (nM)	Relative to QC Concentration (%)
50	1.3	2.6	1.1	2.2
50	1.1	2.2	1.2	2.4
50	0.8	1.6	1.6	3.2
100	0.9	0.9	1.2	1.2
100	1.6	1.6	1.6	1.6
100	1.2	1.2	0.6	0.6

Table 2: Conversion of LV to LVA during sample preparation and storage in autosampler.

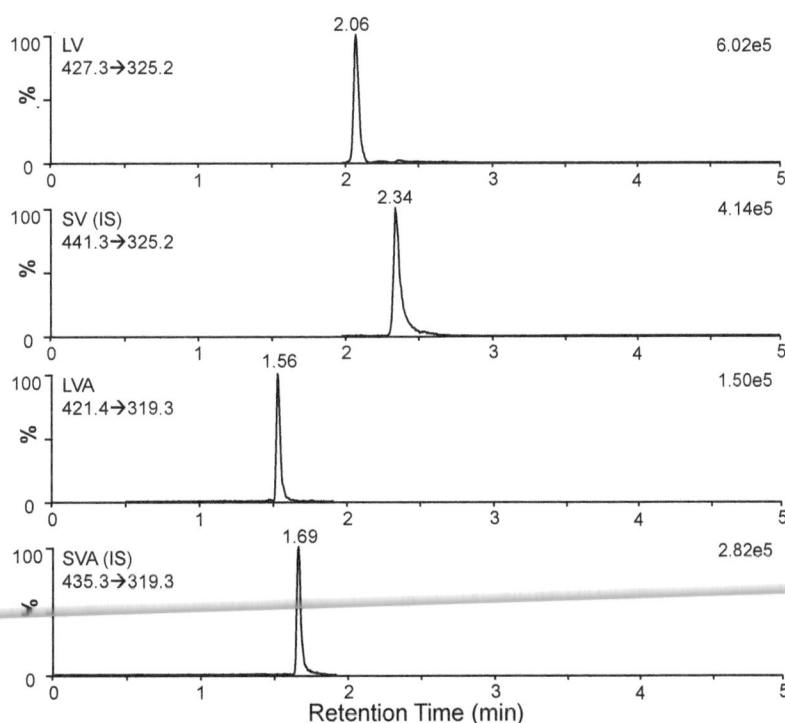

Figure 3: Representative extracted ion chromatograms of LV and LVA, as well as the internal standards SV and SVA, in a human plasma sample collected 4 h post-lovastatin administration
The monitored MRM transitions and signal intensity for the most intense peak are shown in the upper-left and upper-right corners of each chromatogram, respectively.

the autosampler at 6°C for 24 h, the same processed QC samples were analyzed again. The extent of LV-LVA conversion (<1.9%; Tables 2 and 3) did not exhibit statistically significant increases after storage.

The extraction recoveries of LV and LVA from human plasma following PPE were 67-74% for LV and 70-75% for LVA (Table 4), which

QC Concentration (nM)	LV		LV after 24-h storage	
	Concentration (nM)	Relative to QC Concentration (%)	Concentration (nM)	Relative to QC Concentration (%)
50	0.8	1.6	1.8	3.6
50	0.5	1.0	0.6	1.2
50	0.4	0.8	0.5	1.0
100	1.3	1.3	1.7	1.7
100	1.0	1.0	1.2	1.2
100	1.0	1.0	0.7	0.7

Table 3: Conversion of LVA to LV during sample preparation and storage in autosampler.

Compound	QC Concentration (nM)	PPE Extraction		Matrix Effect	
		Recovery (%)	S.D. (%)	Remaining Signal (%)	S.D. (%)
LV	2	73	2.1	155	8.1
	10	74	2.6	149	1.9
	50	67	4.0	134	1.4
LVA	2	75	9.5	102	9.5
	10	70	5.1	101	5.3
	50	71	8.5	95	4.7

S.D., standard deviation

Table 4: Extraction recovery and matrix effect for LV and LVA.

are comparable to recoveries previously reported for PPE extraction with human plasma (52% and 57%, respectively) [18]. Both higher and lower extraction recoveries have been reported for LV using an LLE method (54.8% [8]; ~86% [6,9]) or SPE method (55% [11]). For SV and SVA, higher extraction recoveries have been reported using an SPE method (78-89%) [17,21]. Little matrix effect was observed for LVA, whereas appreciable signal enhancement was observed for LV in the presence of human plasma matrix (Table 4). Since the LV molecular ions monitored in our MRM method were sodium adducts, the presence of plasma matrix could have favored the formation of sodium adduct ions, leading to the observed signal increase compared to the neat solvent.

Both LV and LVA were stable in the QC samples following three freeze-thaw cycles with ≤ 10% reduction in concentration (Table 5), although LVA appeared to be slightly less stable than LV (90-93.2% vs. 97.3-100.5% remaining). In addition, both LV and LVA were stable after long-term storage of plasma QC samples at -80°C (Table 5). No appreciable difference between plasma samples stored as is and sodium acetate-buffered plasma samples was seen (Table 5), indicating that immediate buffering of plasma samples after collection is not necessary.

Application for pharmacokinetic study in human

The validated simple PPE-based UPLC–MS/MS method was applied for the simultaneous quantification of LV and LVA in human plasma samples obtained from a clinical drug-drug interaction study involving LV. Representative extracted ion chromatograms of LV and LVA for a plasma sample collected 4 h post-lovastatin administration are shown in Figure 3. The plasma concentration-time profiles were plotted for a human subject, who received 40 mg lovastatin orally the morning of the study and was monitored for 24 h post-administration (Figure 4). Pharmacokinetic measurements

Compound	QC Concentration (nM)	Freeze-Thaw Stability		Remaining after one week (%)		Remaining after two week (%)		Remaining after one month (%)	
		Remaining (%)	S.D. (%)	Plasma	Buffered Plasma	Plasma	Buffered Plasma	Plasma	Buffered Plasma
LV	2	97.3	4.6	117	107	89	93	114	106
	10	98.1	3.0	101	97	87	87	103	89
	50	100.5	3.6	91	86	90	84	98	81
LVA	2	90.0	6.7	101	112	106	98	121	107
	10	93.2	4.8	86	86	84	88	97	86
	50	91.6	6.7	104	100	97	95	106	101

Table 5: Freeze-thaw and long-term storage stability of LV and LVA.

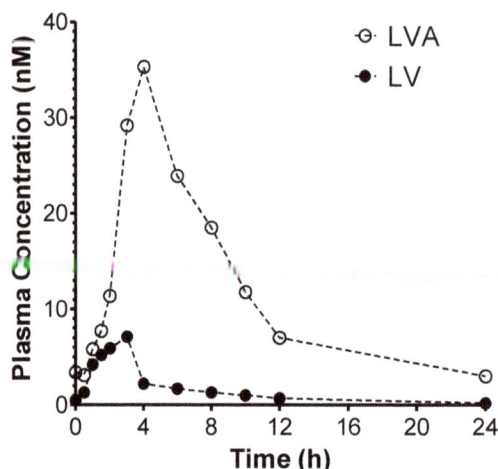

Figure 4: Plasma concentration-time profiles of LV and LVA for a human subject that received 40 mg of lovastatin orally.

Outcomes	Units	LV	LVA
C_{max}	nM (ng/mL)	7.1 (2.9)	35.3 (14.9)
T_{max}	h	3	4
AUC_{last}	nM•h (ng/mL•h)	33 (13.4)	270 (114)
$AUC_{0\rightarrow\infty}$	nM•h (ng/mL•h)	35 (14.2)	300 (127)
$t_{1/2}$	h	6.0	6.7

C_{max}, maximum concentration; T_{max}, time to reach C_{max}; AUC_{last}, area under the curve from time zero to the last measurable concentration; $AUC_{0\rightarrow\infty}$, area under the curve from time zero to infinite; $t_{1/2}$, terminal elimination half-life

Table 6: Pharmacokinetics of LV and LVA in a human subject administered 40 mg lovastatin orally.

are summarized in (Table 6). These values are in agreement with those previously reported for LV [10], although a greater exposure (*i.e.*, C_{max} and AUC) of LVA was observed in our study.

Conclusions

A simple PPE-based, direct injection, UPLC-MS/MS analytical method using polarity switching has been developed and validated for the simultaneous quantification of LV and LVA in human plasma samples. The interconversion of LV and LVA during sample preparation and storage was minimized by a lower handling temperature (on ice or ≤ 6°C) and the use of acidified plasma by addition of sodium acetate buffer. The validated method has been successfully applied for the quantification of LV and LVA in human plasma samples obtained from a clinical drug-drug interaction study involving LV.

Acknowledgements

This work was supported by the National Institutes of Health research grant R01-GM089994(MZW). MS was supported by the National Institute of General Medical Sciences training grant T32-GM086330.

References

1. Alberts AW, Chen J, Kuron G, Hunt V, Huff J, et al. (1980) Mevinolin: a highly potent competitive inhibitor of hydroxymethylglutaryl-coenzyme A reductase and a cholesterol-lowering agent. Proc Natl Acad Sci U S A 77: 3957-3961.

2. Vyas KP, Kari PH, Pitzenberger SM, Halpin RA, Ramjit HG, et al. (1990) Biotransformation of lovastatin. I. Structure elucidation of in vitro and in vivo metabolites in the rat and mouse. Drug Metab Dispos 18: 203-211.

3. Wang RW, Kari PH, Lu AY, Thomas PE, Guengerich FP, et al. (1991) Biotransformation of lovastatin. IV. Identification of cytochrome P450 3A proteins as the major enzymes responsible for the oxidative metabolism of lovastatin in rat and human liver microsomes. Arch Biochem Biophys 290: 355-361.

4. Pasha MK, Muzeeb S, Basha SJ, Shashikumar D, Mullangi R, et al. (2006) Analysis of five HMG-CoA reductase inhibitors-- atorvastatin, lovastatin, pravastatin, rosuvastatin and simvastatin: pharmacological, pharmacokinetic and analytical overview and development of a new method for use in pharmaceutical formulations analysis and in vitro metabolism studies. Biomed Chromatogr 20: 282-293.

5. Hamidi M, Zarei N, Shahbazi MA (2009) A simple and sensitive HPLC-UV method for quantitation of lovastatin in human plasma: application to a bioequivalence study. Biological & pharmaceutical bulletin 32: 1600-1603.

6. Wang D, Wang D, Qin F, Chen L, Li F (2008) Determination of lovastatin in human plasma by ultra-performance liquid chromatography/electrospray ionization tandem mass spectrometry. Biomed Chromatogr 22: 511-518.

7. Li L, Sun J, Sun YX, He ZG (2008) LC-ESI-MS determination of lovastatin in human plasma. Chromatographia 67: 621-625.

8. Pilli NR, Mullangi R, Inamadugu JK, Nallapati IK, Rao JV (2012) Simultaneous determination of simvastatin, lovastatin and niacin in human plasma by LC-MS/MS and its application to a human pharmacokinetic study. Biomed Chromatogr 26: 476-484.

9. Yuan H, Wang F, Tu J, Peng W, Li H (2008) Determination of lovastatin in human plasma by ultra-performance liquid chromatography-electrospray ionization tandem mass spectrometry and its application in a pharmacokinetic study. J Pharm Biomed Anal 46: 808-813.

10. Sun JX, Niecestro R, Phillips G, Shen J, Lukacsko P, et al. (2002) Comparative pharmacokinetics of lovastatin extended-release tablets and lovastatin immediate-release tablets in humans. J Clin Pharmacol 42: 198-204.

11. Wu Y, Zhao J, Henion J, Korfmacher WA, Lapiguera AP, et al. (1997) Microsample determination of lovastatin and its hydroxy acid metabolite in mouse and rat plasma by liquid chromatography/ionspray tandem mass spectrometry. J Mass Spectrom 32: 379-387.

12. Stubbs RJ, Schwartz M, Bayne WF (1986) Determination of mevinolin and mevinolinic acid in plasma and bile by reversed-phase high-performance liquid chromatography. J Chromatogr 383: 438-443.

13. Ye LY, Firby PS, Moore MJ (2000) Determination of lovastatin in human plasma using reverse-phase high-performance liquid chromatography with UV detection. Ther Drug Monit 22: 737-741.

14. Yang AY, Sun L, Musson DG, Zhao JJ (2005) Application of a novel ultra-low elution volume 96-well solid-phase extraction method to the LC/MS/MS determination of simvastatin and simvastatin acid in human plasma. J Pharm Biomed Anal 38: 521-527.

15. Zhang N, Yang A, Rogers JD, Zhao JJ (2004) Quantitative analysis of simvastatin and its beta-hydroxy acid in human plasma using automated liquid-liquid extraction based on 96-well plate format and liquid chromatography-tandem mass spectrometry. J Pharm Biomed Anal 34:175-187.

16. Apostolou C, Kousoulos C, Dotsikas Y, Soumelas GS, Kolocouri F, et al. (2008) An improved and fully validated LC-MS/MS method for the simultaneous quantification of simvastatin and simvastatin acid in human plasma. J Pharm Biomed Anal 46: 771-779.

17. Barrett B, Huclova J, Borek-Dohalsky V, Nemec B, Jelinek I (2006) Validated HPLC-MS/MS method for simultaneous determination of simvastatin and simvastatin hydroxy acid in human plasma. J Pharm Biomed Anal 41: 517-526.

18. Jemal M, Ouyang Z, Powell ML (2000) Direct-injection LC-MS-MS method for high-throughput simultaneous quantitation of simvastatin and simvastatin acid in human plasma. J Pharm Biomed Anal 23: 323-340.

19. Jemal M, Ouyang Z (2000) The need for chromatographic and mass resolution in liquid chromatography/tandem mass spectrometric methods used for quantitation of lactones and corresponding hydroxy acids in biological samples. Rapid Commun Mass Spectrom 14: 1757-1765.

20. (2001) Guidance for Industry: bioanalytical method validation. US Department of Health and Human Services, FDA, Center for Drug Evaluation and Research.

21. Zhao JJ, Xie IH, Yang AY, Roadcap BA, Rogers JD (2000) Quantitation of simvastatin and its beta-hydroxy acid in human plasma by liquid-liquid cartridge extraction and liquid chromatography/tandem mass spectrometry. J Mass Spectrom 35: 1133-1143.

4

Detection of Growth Hormone Releasing Peptides in Serum by a Competitive Receptor Binding Assay

Ferro P[1]*, Krotov G[2], Zvereva I[2], Pérez-Mana[3,4], Mateus JA[3] and Segura J[1,5]

[1]Bioanalysis Research Group, Neuroscience Research Program, IMIM (Hospital del Mar Medical Research Institute), Barcelona, Spain
[2]Anti-Doping Centre, Moscow, Russia
[3]Integrative Pharmacology and Systems Neuroscience Research Group, IMIM (Hospital del Mar Medical Research Institute), Barcelona, Spain
[4]Department of Pharmacology, Therapeutics and Toxicology, Autonomous University of Barcelona, Cerdanyola, Spain
[5]Department of Experimental and Health Sciences, Pompeu Fabra University (UPF), Barcelona Biomedical Research Park, Barcelona, Spain

Abstract

The use of small peptides for medical purposes has gained great importance in the last decade. Among these molecules are the growth hormone releasing peptides (GHRPs). These substances are highly sought by cheating athletes because of their ergogenic effects. Thus, GHRPs are listed in the World Antidoping Agency Prohibited List and several methods have been developed for their detection in urine samples. Serum is another useful matrix frequently used for anti-doping control. An experimental adaptation of a competitive methodology for GHRPs wide-group detection, previously developed for urine, is reported here using a serum matrix instead. The modifications have been mainly addressed the need to remove endogenous ghrelin (always present in serum) to avoid interference in the competitive binding assay. The modified test was first evaluated using serum spiked with different pure GHRPs (i.e., GHRP-2, GHRP-6 and Hexarelin), then using serum samples from pilot studies in which healthy subjects were treated intravenously with GHRP-2 and intranasally with GHRP-2 and GHRP-6. The results showed the reliability of the modified methodology for the detection of GHRPs in a serum matrix.

Keywords: Growth hormone secretagogue; Ghrelin; Radiocompetition assay; Doping substances; Doping abuse

Introduction

During the past decade, small molecules have been target of a wide range of applications in medicine and biotechnology. Therapeutic peptide research is also currently experiencing a renaissance for commercial interest [1], among others reasons. Growth hormone secretagogues (GHSs) are active small molecules with medical and sport performance applications. GHSs constitute a complex heterogeneous family of compounds that keeps growing in number and can be roughly divided into two main groups: 1) growth hormone releasing hormone and analogs such as Sermorelin and CJC-1295, and 2) growth hormone releasing peptides (GHRPs), which show a high stimulatory activity on growth hormone (GH) release [2].

GHRPs stimulate GH release and have been regarded as an alternative to support diagnostics and treat diseases related to GH deficiency [3,4]. Their effects have been described *in vivo* and *in vitro*, in both animals and humans, using various doses and administration routes [5-10]. However, only the peptide GHRP-2 (also known as pralmorelin) has been clinically tested and approved in Japan for diagnostic purposes [11], while other GHRPs have not passed clinical trials or are still in different stages of development.

One of the interesting features of GHRPs is the fact that they can be active by several administration routes, including oral administration. Besides, their ergogenic ability has been reported, so they are becoming very appreciated as substances of abuse by cheating athletes that seek new alternatives to traditional doping [12,13]. In fact, several of these GHRPs are available in the black market and can be obtained easily on the Internet [14,15]. Consequently, the World Anti-Doping Agency (WADA) incorporated GHRPs in its Prohibited List of doping substances since 2013, and must be detectable by anti-doping laboratories [16].

Anti-doping detection methods based in liquid chromatography–mass spectrometry (LC-MS) have been developed in recent years using urine as main matrix [17-21]. These methods are focused especially on those GHSs with known structure and metabolism which could be used more frequently in sport abuse. Nonetheless, considering that these molecules belong to a pharmacological family, the development of wide scope detection methods for all GHSs known as well as unknown, could be a highly desirable resource [22-24]. Such universal detection methods could be based on a common feature, e.g., their interaction with the GHS-R1a receptor, for which the endogenous natural ligand is ghrelin, a 3-octanoylated 28 amino acid peptide [25].

Traditionally, urine was the sample of choice for the screening and identification of unknown drugs due to high concentration of drugs in urine. Therefore, this matrix has been used or GHSs (no: or GHSs) detection in athlete samples with anti-doping purposes [21]; although the use of another matrices, as serum, could be highly desirable. However, the GHRPs studies performed so far with serum samples have been carried out only in a few clinical trials with healthy males and children administered with GHRP-2 and GHRP-6 respectively [26,27]. In these clinical trials a pharmacokinetic study for both GHRPs was performed for a maximum time of 12 hours. For sample analysis, a specific anti-GHRP-2 polyclonal antibody was used in a radioimmunoassay to detect GHRP-2 [26] and a LC-MS method was developed to detect GHRP-6 [27].

*Corresponding author: Ferro P, Bioanalysis Research Group, Neuroscience Research Program, IMIM (Hospital del Mar Medical Research Institute), Barcelona, Spain, E-mail: pedro.ferro@ibima.eu

Here we present an experimental study which was built on the previous screening method described by Pinyot et al. and Ferro et al. in urine samples [22,28,29] to develop an alternative method that uses serum as a new matrix. The serum is a valuable matrix used for analysis of some doping substances e.g., CERA [30], human chorionic gonadotrophin [31], and growth hormone [32,33]. A GHRPs screening method for anti-doping control in serum samples would be highly desirable. However, serum contains a substantial amount of the natural ligand ghrelin, which was expected to interfere with the determination of GHRPs by competitive binding to the corresponding membrane receptor (GHS-R1a) in the assay. This work assessed this potential interference and implemented steps to remove it. The removal of natural ghrelin from serum by plate immunocapture appeared as a suitable alternative. The performance of this new protocol was assessed in serum samples from pilot studies in which healthy subjects were administered intravenously with a bolus of 100 μg of GHRP-2, intranasally with 200 μg of GHRP-2, intranasally with 200 μg of GHRP-6, or as untreated controls (n=4). Clear competitive signal of radioactive ligand displaced from the GHS-R1a receptor was verified in all administration studies.

Methods and materials

Chemicals

For *in vitro* studies, growth hormone-releasing peptide 2 (GHRP-2, pralmorelin) was kindly supplied by Dr. S. Kageyama (Tokyo Laboratory, Anti-Doping Center, Mitsubishi Chemical Medience, Japan) whereas GHRP-6 and Hexarelin were kindly supplied by Giampiero Muccioli (University of Torino, Italy). For the intravenous excretion studies, GHRP KAKEN100® (pralmorelin dihydrochloride, 100 μg/vial) was purchased from Kaken Pharmaceutical Co., Ltd. For intranasal studies, GHRP-2 and GHRP-6 were synthesized by Thermo Fisher Scientific GmbH (Ulm, Germany). Radiolabeled (125I-His9) ghrelin (no: Rdiolabeled ghrelin (a125-His9) ghrelin) was purchased from Perkin Elmer (Waltham, MA, USA). All other chemicals were of the highest grade commercially available.

Cell culture, membrane preparations and competition binding assay

Cell culture, membrane preparations and competition binding assay parameters have previously been published by Pinyot et al. [22] and Ferro et al. [28] Data were analysed and plotted with GraphPad Prism 5 (San Diego, CA, USA) or Excel software. Statistical analysis was performed using Excel and SPSS programs. To contrast independent continuous variables, Student's t-test was used. Null hypothesis was rejected for values of $p \leq 0.05$.

Reference samples

Two reference samples in binding buffer were included in each competition binding experiment, one as blank or negative control (containing cells and all reagents but not spiked with GHRPs) as the maximum possible binding (100% of relative specific binding, RSB) and a second sample or positive control, spiked with 7.5 μM GHRP-2, as the minimal possible specific binding (0% RSB). All sample binding values were calculated relative to these limits.

GHRPs treatment in human subjects

Intravenous GHRP-2: An injectable solution of GHRP KAKEN100® (10 μg/mL in saline), containing 100 μg pralmorelin dihydrochloride (GHRP-2), was administered intravenously after overnight fasting. The protocol was approved by the Clinical Research Ethical Committee from Parc de Salut Mar (CEIC-Parc de Salut Mar), Barcelona (n°2014/5760). Eight healthy caucasian male volunteers were administered with the drug and two control subjects were left untreated. The subjects were 20 to 30 years old, mean weight 72.9 ± 8.7 and body mass index (BMI): 21.5 ± 3.5. All subjects were informed in advance of the study details and signed the corresponding informed consent letter. The volunteers refrained from ingesting alcohol and any medicines and did not perform any exercise during the studies. Urine and serum samples were collected prior to administration of the drug and several times during the 24 h period after administration. All samples were stored at -20°C until analysis.

Intranasal GHRP-2 and GHRP-6: A single nasal dose of one of the peptides GHRP-2 or GHRP-6 (200 μg) was administered to four volunteers (two for GHRP-2 and two for GHRP-6) alongside with two control subjects without any administration) by the Anti-Doping Centre in Moscow. Serum samples from each volunteer were collected prior to drug administration and several times during the 24 h period after administration. Samples were stored at −20°C until analyses.

Analytical aspects

Serum: During method development, serum samples were subjected to different treatments. Additional details appear in the results section.

Thermic treatment: 150 μl of serum samples were incubated at 37°C and 50°C for 1,2 and 4 h in a thermomixer and evaluated by the competition binding assay (see above).

Incubation at different pHs: 150 μl of serum samples were treated to reach pH 7.4 or 9 and were incubated at 37°C and 50°C in a thermomixer for 1 h. After treatment samples were evaluated by the competition binding assay (see above).

Immunocapture: Antibody anti-ghrelin coated microplates (Cayman Chemicals, Ann Arbor, MI, USA) were washed 5 times with 300 μl of wash buffer (phosphate-buffered saline (PBS) and 0.05% Tween-20). 150 μl of serum samples were thawed, deposited in a well and incubated at 4°C under agitation (orbital shaker) for 1, 2, 4 h or overnight. Serum samples recovered from the wells were evaluated by the competition binding assay (see above).

According to these developments and the results obtained (see below), the final chosen methodology involved 4 hours incubation followed by a second incubation of the same serum in another well in the same conditions for 2 h. Eventually, samples were recovered again from the second well and 40 μl were evaluated in triplicate by the competitive binding assay (see above).

Urine: Urine samples were processed following the protocol established by Pinyot et al. [29]. Briefly, urine samples were thawed and centrifuged for 15 min at 3500g at 4°C. After desalting of 2.5 ml of urine by loading into a 3-ml Oasis HLB solid phase extraction (SPE) cartridge (Waters, Mildford, USA), the eluted sample obtained with methanol was dried in an N$_2$-evaporator TurboVap LV (Caliper, Honkinton, USA). The dried residue was reconstituted in 1 ml fresh prepared binding buffer, sonicated for 5 min and GHRPs were purified from the sample by receptor affinity. Purified samples were reconstituted in 150 μl of binding buffer and analysed by the competitive binding assay (see above).

Results and Discussion

There were several important reasons for our interest in the development of a novel detection method for GHRPs in serum: 1) the increased and frequent use by cheating athletes of these compounds, which can be easily purchased in the black market or via Internet, 2) the relative complexity of the method for urine (the protocol takes three days to be completed), and 3) the fact that blood samples can be used for the detection of some doping substances that cannot be detected in urine [33].

The previously described radio competition assay methodology to detect GHRPs in urine [22,28,29] was modified and simplified for the detection of these compounds in serum samples. As was expected, however, the direct application of the competitive assay on "unprocessed" serum samples (without any denaturing or immunocapture pretreatment) from 10 non-treated subjects showed an interfering effect. An RSB with mean and SD of 82.97% ± 3.68, indicated the presence of background amounts of receptor binding compound(s) (presumably ghrelin) when compared with negative reference sample in binding buffer (100% RSB). Notably, serum samples from volunteers 8, 9 and 10 had RSB values of 70.98 ± 6.23, 66.74 ± 1.37 and 68.07 ± 0.93%, respectively, indicating a high interference background. As a consequence, different variables had to be studied to remove the endogenous interfering substance(s) from the serum, as mentioned below.

Temperature and pH treatment

Ghrelin (together with non-acylated ghrelin) concentrations in plasma or serum have been found around 100–200 fmol/ml and their levels are increased before meals to 300 fmol/ml [34-36]. This could cause interference in the assay with serum samples and consequently generate However, it is well known that the ghrelin protein is sensitive to specific pH and temperature conditions [37,38], with a partial conversion from acylated ghrelin to des-acyl ghrelin. Des-acyl ghrelin has much lower capacity to interact with the GHS-R1a receptor. To assess whether the temperature and pH have effects on the assay regarding the reduction of endogenous ghrelin concentrations, serum samples were incubated at 37°C and 50°C for several hours (1, 2 and 4 h). Results are displayed in Table 1. It was found an increase in the RSB values in samples treated at 37°C and 50 °C after two hours of incubation. Higher effects were observed in samples incubated at 50°C for 4 hours, with 92.67 ± 1.29% RSB, as compared to 74.63 ± 1.26% RSB for the non-incubated serum. Unfortunately, samples from volunteers having a lower baseline RSB showed improvement with temperature treatment, but it was not the case in those with baseline RSB greater than 77%.

Regarding the experiments involving incubation steps at neutral and basic pHs (pH 7.4 and pH 9.0, respectively) and different temperatures (37°C and 50°C) for 1 h, the results did not show significant differences before and after incubation (Table 1), therefore a different alternative would be needed in order to remove the endogenous ghrelin.

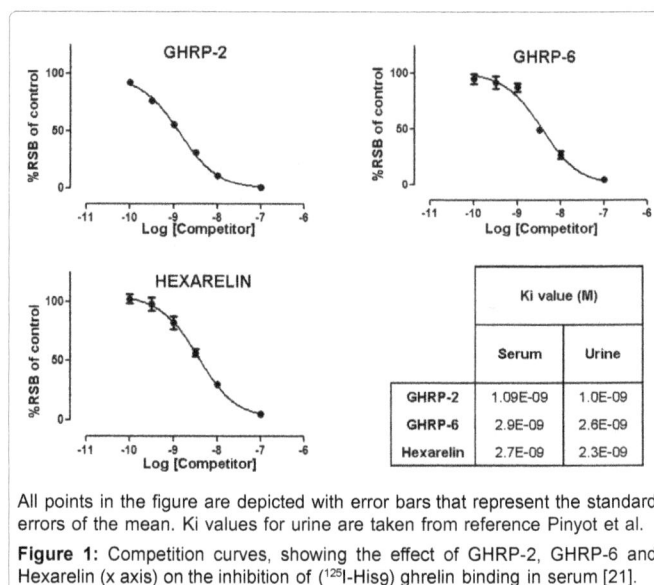

All points in the figure are depicted with error bars that represent the standard errors of the mean. Ki values for urine are taken from reference Pinyot et al.

Figure 1: Competition curves, showing the effect of GHRP-2, GHRP-6 and Hexarelin (x axis) on the inhibition of (^{125}I-His9) ghrelin binding in serum [21].

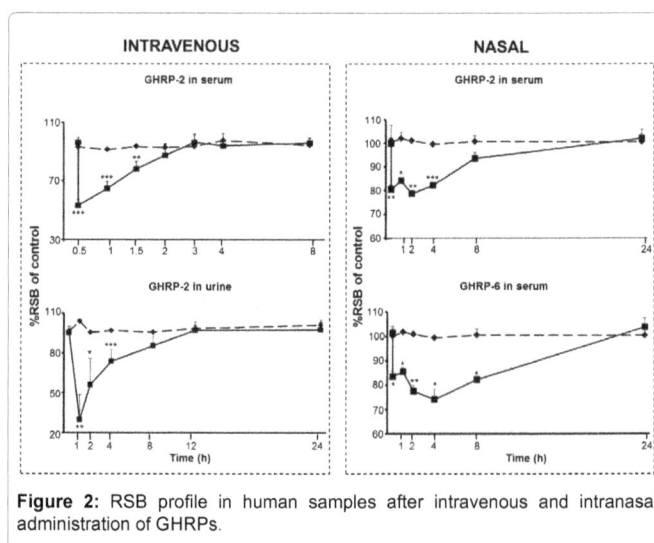

Figure 2: RSB profile in human samples after intravenous and intranasal administration of GHRPs.

Immunocapture

The most successful approach to remove endogenous ghrelin content in serum samples was based on antibody capture using microplates coated with anti-ghrelin antibodies. With this goal, 150 µl of serum were incubated in the antibody-coated wells for different time periods (1, 2, 4 h and overnight) and the serum recovered was assessed by radiocompetition assay. The results showed a RSB values of 86.71 ± 0.55%, 89.03 ± 0.68%, 92.67 ± 1.29% and 88.42 ± 5.03% for 1, 2, 4 h and overnight incubation, respectively, what indicated an increase ghrelin removal as compared to the non-incubated serum samples with mean RSB value of 74.33 ± 2.05%. However, a small background binding was still detected in the samples after a single immunocapture step.

Therefore, to remove all endogenous ghrelin, serum samples from 4 volunteers were subjected to two consecutive immunocapture steps: incubating first for 4 h in anti-ghrelin antibody-coated wells, and then incubating the recovered serum for 2 additional hours in distinct antibody-coated wells. After analysis by radiocompetitive assay, all samples showed RSB values close to 100%, with mean and SD between 96.02 ± 1.42% and 111.12 ± 1.74%, indicating the removal of most of

	pH			
	7.4		9	
Time (h)	37°C	50°C	37°C	50°C
1	71.73 ± 9.91	73.48 ± 8.69	74.04 ± 4.16	74.04 ± 1.71
2	81.29 ± 1.33	87.61 ± 8.16	--	--
4	90.85 ± 2.09	94.68 ± 3.77	--	--
NI	74.63 ± 1.26			

NI: Non incubated.

Table 1: Temperature and pH treatment in serum samples.

the interfering ghrelin.

Radiocompetition assay with different GHRPs

The assessment of the urine protocol with several different GHSs showed high robustness and sensitivity [22]. To optimize the radiocompetition assay to detect GHRPs in serum (ghrelin-depleted by immunocapture), the method was applied to serum samples spiked with different concentrations of three different GHRPs (GHRP-2, GHRP-6 and Hexarelin). The respective competition binding curves (Figure 1) showed serum Ki values similar to those obtained by Pinyot et al. for these GHRPs in urine samples [22]. These results confirmed that the newly implemented ghrelin immunocapture step in serum samples led to the development of a successful method for this matrix.

Detectability in serum and urine samples after GHRP-2 intravenous administration

The few alternative methods previously described to detect GHSs in serum samples are rather specific for one or few GHRPs based on radioimmunoassay [26] or LC-MS methods [27]. However, the new radiocompetition assay presented in this study could be the first method which can detect any known or unknown GHRP in this matrix. In order to verify the effectiveness and the detection time window of our protocol in serum samples, a pilot study in healthy male volunteers treated with GHRP-2 was performed. Serum and urine samples from a group of healthy male volunteers dosed intravenously with 100 μg of pralmorelin dihydrochloride (GHRP-2) were collected at different time points (0, 0.5, 1, 1.5, 2, 3, 4 and 8 h for serum and 0, 1, 2, 4, 6, 12, and 24 h for urine) post-administration. Samples from eight treated subjects and two untreated controls were assessed by the radiocompetition assay.

The results, presented in Figure 2-left, showed that all serum and urine samples collected immediately (and up to 2 h for serum and 6 h for urine) after the intravenous GHS injection displaced (125I-His9) ghrelin from the GHS-R1a receptor to a large extent. The basal state was subsequently recovered and remained unaltered until at least 24 h post-administration. The detection time window appeared shorter in the serum as the drugs can be detected for 2 hours after administration versus 6 hours in urine samples [29], although these data are similar to that obtained previously from GHRP-2 and GHRP-6 in serum samples with other methods [26,27].

GHRP-2 and GHRP-6 detection in serum samples after intranasal administration

Serum samples from healthy volunteers administrated with an intranasal dose of GHRP-2 or GHRP-6 were collected at different time points (0, 0.5, 1, 2, 4, 8 and 24 h for serum) post-administration. Samples from four treated subjects (two for GHRP-2 and two for GHRP-6) and two controls were assessed by the radiocompetition assay. The results, presented in Figure 2-right, showed that all GHRP-2 and GHRP-6 serum samples collected immediately and up to 8 h after intranasal administration, displaced (125I-His9) ghrelin from the GHS-R1a receptor to a large extent. The basal state was subsequently recovered and remained unaltered until at least 24 h post-administration.

The detection time window for GHRP-2 appeared shorter in serum as the drug can be detected for 8 h after administration in serum versus 18 h in urine samples [39]. For GHRP-6, the detection windows obtained here in serum and previously in urine [39] samples were very similar, so it also would be very interesting to study other GHRPs

Conclusion

A new wide-group screening method to detect GHRPs in serum samples has been developed and tested in human samples after administration of various GHRPs through different administration routes. This method showed a similar sensitivity and detection time window in serum in comparison with other protocols specified for a unique GHRP after intravenously administration. The detection window after intranasal administration of some GHRPs might be even similar regardless of the matrix used for the analysis (i.e., urine or serum). The approach described here has the additional advantages of being completed in a short time (only one day versus three days in the case of urine), easy to perform (needing only an initial immunocapture step with anti-ghrelin antibody to remove endogenous ghrelin) and requires a small volume of sample (40 μl), which is easy to obtain from any venipuncture protocol.

Acknowledgments

The authors thank Professor F. Casanueva and Professor R. Smith for HEK293 cells expressing GHS-R1a; Professor Rafael de la Torre and Professor Magi Farré for sharing the GHRP-2 study samples; and the Clinical Research Unit staff (Esther Menoyo, Marcos Wenceslao Pestarino, Cristina Llop, Clara Gibert, Soraya Martin and Marta Pérez) for their assistance. Clara Pérez-Mañá was funded by a Juan Rodes fellowship (ISC-III, JR15/00005).

References

1. Fosgerau K, Hoffmann T (2015) Peptide therapeutics: current status and future directions. Drug Discov Today 20: 122-128.

2. Arvat E, Di Vito L, Maccagno B, Broglio F, Boghen MF, et al. (1997) Effects of GHRP-2 and hexarelin, two synthetic GH- releasing peptides, on GH, prolactin, ACTH and cortisol levels in man. Comparison with the effects of GHRH, TRH and hCRH. Peptides 18: 885-891.

3. Bowers CY, Momany FA, Reynolds GA, Hong A (1984) On the in vitro and in vivo activity of a new synthetic hexapeptide that acts on the pituitary to specifically release growth hormone. Endocrinology 114: 1537-1545.

4. Bowers CY, Alster DK, Frentz JM (1992) The growth hormone-releasing activity of a synthetic hexapeptide in normal men and short statured children after oral administration. J Clin Endocrinol Metab 74: 292-298.

5. Esposito S, Deventer K, Geldof L, Peter VE (2015) In vitro models for metabolic studies of small peptide hormones in sport drug testing. J Pept Sci 21: 1-9.

6. Johansen PB (1998) Pharmacokinetic evaluation of ipamorelin and other peptidyl growth hormone secretagogues with emphasis on nasal absorption. Xenobiotica 28: 1083-1092.

7. Pihoker C, Badger TM, Reynolds GA, Bowers CY (1997)Treatment effects of intranasal growth hormone releasing peptide-2 in children with short stature. J Endocrinol 155: 79-86.

8. Hayashi S, Okimura Y, Yagi H, Toshiyuki U, Yashuhiro T (1991) Intranasal administration of His-D-Trp-Ala-Trp-D-Phe-LysNH2 (growth hormone releasing peptide) increased plasma growth hormone and insulin-like growth factor-I levels in normal men. Endocrinol Jpn 38: 15-21.

9. Ghigo E, Arvat E, Gianotti L, Imbimbo BP, Lenaerts V (1994) Growth hormone-releasing activity of hexarelin, a new synthetic hexapeptide, after intravenous, subcutaneous, intranasal, and oral administration in man. J Clin Endocrinol Metab 78: 693-698.

10. Okano M, Nishitani Y, Sato M, Ayaka I (2010) Influence of intravenous administration of growth hormone releasing peptide-2 (GHRP-2) on detection of growth hormone doping: growth hormone isoform profiles in Japanese male subjects. Drug Test Anal 2: 548-556.

11. Graul AI, Prous JR (2006) The Year's New Drugs: A Historical and Research Perspective on the 41 New Products that Reached their First Markets in 2005. Drug News & Perspectives (Prous Science) 19: 33.

12. Cox HD, Hughes CM, Eichner D (2015) Detection of GHRP-2 and GHRP-6 in urine samples from athletes. Drug Test Anal 7: 439-444.

13. Gaudiano MC, Valvo L, Borioni A (2014) Identification and quantification of the doping agent GHRP-2 in seized unlabelled vials by NMR and MS: a case-report. Drug Test Anal 6: 295-300.

14. Kohler M, Thomas A, Geyer H, Michael P, Wilhelm S, et al. (2010) Confiscated Black Market Products and Nutritional Supplements with Non-Approved Ingredients Analyzed in the Cologne Doping Control Laboratory 2009. Drug Test Anal 2: 533–537.

15. Thomas A, Kohler M, Mester J, Hans G, Wilhelm S, et al. (2010) Identification of the growth-hormone-releasing peptide-2 (GHRP-2) in a nutritional supplement. Drug Test Anal 2: 144-148.

16. WADA. The 2015 Prohibited.

17. Okano M, Sato M, Ikekita A, Shinji K (2010) Determination of growth hormone secretagogue pralmorelin (GHRP-2) and its metabolite in human urine by liquid chromatography/electrospray ionization tandem mass spectrometry. Rapid Commun. Mass Spectrom 24: 2046-2056.

18. Thomas A, Höppner S, Geyer H, Wilhem S, Michael P, et al. (2011) Determination of growth hormone releasing peptides (GHRP) and their major metabolites in human urine for doping controls by means of liquid chromatography mass spectrometry. Anal Bioanal Chem 401: 507-516.

19. Thevis M, Thomas A, Schänzer W (2011) Doping control analysis of selected peptide hormones using LC–MS(/MS). Forensic Sci Int 213: 35-41.

20. Thevis M, Thomas A, Pop V, Schanzer W (2013) Ultrahigh pressure liquid chromatography–(tandem) mass spectrometry in human sports drug testing: Possibilities and limitations. J Chromatogr A 1292: 38-50.

21. Semenistaya E, Zvereva I, Thomas A, Mario T, Grigory K, et al. (2015) Determination of growth hormone releasing peptides metabolites in human urine after nasal administration of GHRP-1, GHRP-2, GHRP-6, Hexarelin, and Ipamorelin. Drug Test Anal 7: 919-925.

22. Pinyot A, Nikolovski Z, Bosch J. (2010) On the use of cells or membranes for receptor binding: growth hormone secretagogues. Anal Biochem 399: 174-181.

23. Leyris JP, Roux T, Trinquet E (2011) Homogeneous time-resolved fluorescence-based assay to screen for ligands targeting the growth hormone secretagogue receptor type 1a. Anal Biochem 408: 253-262.

24. Liu Y, Shao XX, Zhang L, Ge Song, Ya LL, et al. (2015) Novel bioluminescent receptor-binding assays for peptide hormones: using ghrelin as a model. Amino Acids 47: 2237-2243.

25. Kojima M, Hosoda H, Date Y, Nakazato N, Matsua Y, et al. (1999) Ghrelin is a Growth- Hormone-Releasing Acylated Peptide from Stomach. Nature 402: 656–660.

26. Pihoker C, Kearns GL, French D, Cyris YB (1998) Pharmacokinetics and pharmacodynamics of growth hormone-releasing peptide-2: a phase I study in children. J Clin Endocrinol Metab 83: 1168-1172.

27. Cabrales A, Gil J, Fernández E, Valenzuela C, Hernandez F, et al. (2013) Pharmacokinetic study of Growth Hormone-Releasing Peptide 6 (GHRP-6) in nine male healthy volunteers. Eur J Pharm Sci 48: 40-46.

28. Ferro P, Gutiérrez-Gallego R, Bosch J, M farre (2015) Fit-for purpose radioreceptor assay for the determination of Growth Hormone Secretagogues in urine. J Biomol Screen 20: 1268-1276.

29. Pinyot A, Nikolovski Z, Bosch J, Gerard SS, Shinji K, et al. (2012) Growth hormone secretagogues: out of competition. Anal Bioanal Chem 402: 1101-1108.

30. Leuenberger N, Lamon S, Robinson N, Giraud S, Saugy M (2011) How to confirm C.E.R.A. doping in athletes' blood? Forensic Sci Int 213: 101-103.

31. Lund H, Snilsberg AH, Halvorsen TG, Peter H, Leon R (2013) Comparison of newly developed immuno-MS method with existing DELFIA (®) immunoassay for human chorionic gonadotropin determination in doping analysis. Bioanalysis 5: 623-630.

32. Holt RI, Böhning W, Guha N, Bartlett C, Cowan DA, et al. (2015) The development of decision limits for the GH-2000 detection methodology using additional insulin-like growth factor-I and amino-terminal pro-peptide of type III collagen assays. Drug Test Anal 7: 745-755.

33. Bidlingmaier M, Suhr J, Ernst A, Wu Z, Keller A, et al. (2009) High-sensitivity chemiluminescence immunoassays for detection of growth hormone doping in sports. Clin Chem 55: 445-453.

34. Cummings DE, Purnell JQ, Frayo RS, Schmidova K, Wise BE, et al. (2001) A preprandial rise in plasma ghrelin levels suggests a role in meal initiation in humans. Diabetes 50: 1714-1719.

35. Tschop M, Wawarta R, Riepl RL, Friedrich S, Bidlingmaier K, et al. (2001) Post-prandial decrease of circulating human ghrelin levels. J Endocrinol Invest 24: RC19-RC21.

36. Yoshimoto A, Mori K, Sugawara A, Mukoyama M, Yhaata K, et al. (2002) Plasma ghrelin and desacyl ghrelin concentrations in renal failure. J Am Soc Nephrol 13: 2748-2752.

37. Tvarijonaviciute A, Martínez-Subiela S, Ceron JJ (2013) Influence of different storage conditions and anticoagulants on the measurement of total and acylated ghrelin in dogs: a preliminary study. Vet Rec 172: 289.

38. Hosoda H, Doi K, Nagaya N, Okumura H, Nakagawa E, et al. (2004) Optimum collection and storage conditions for ghrelin measurements: octanoyl modification of ghrelin is rapidly hydrolyzed to desacyl ghrelin in blood samples. Clin Chem 50: 1077-1080.

39. Ferro P, Krotov G, Zvereva I, Rodchenkov G, Segura J, et al. (2016) Structure-activity relationship for peptídic growth hormone secretagogues. Drug Test Anal 9: 87-95.

5

Determination of Methadone in Human Urine Using Salting-Out Effect and Dispersive Liquid-Liquid Microextraction Followed by HPLC-UV

Reza Akramipour[1], Mitra Hemati[1], Simin Gheini[1], Nazir Fattahi[2]* and Hamid Reza Ghaffari[3,4]

[1]School of Medical, Kermanshah University of Medical Sciences, Kermanshah, Iran
[2]Research Center for Environmental Determinants of Health (RCEDH), Kermanshah University of Medical Sciences, Kermanshah, Iran
[3]Social Determinants in Health Promotion Research Center, Hormozgan University of Medical Sciences, Bandar Abbas, Iran
[4]Department of Environmental Health Engineering, School of Public Health, Tehran University of Medical Sciences, Tehran, Iran

Abstract

The salting-out effect combined with the dispersive liquid-liquid microextraction based on solidification of floating organic drop (DLLME-SFO) has been developed as a high preconcentration technique for the determination of drug in urine samples. Methadone was employed as model compound to assess the extraction procedure and were determined by high performance liquid chromatography-ultraviolet detection (HPLC-UV). In this method, initially, NaCl as a separation reagent is filled into a small column and a mixture of urine and acetonitrile is passed through the column. By passing the mixture, NaCl is dissolved and the fine droplets of acetonitrile are formed due to salting-out effect. The produced droplets go up through the remained mixture and collect as a separated layer. Then, the collected acetonitrile is removed with a syringe and mixed with 30.0 µL 1-undecanol (extraction solvent). In the second step, the 5.00 mL K_2CO_3 solution (2% w/v) is rapidly injected into the above mixture placed in a test tube for further DLLME-SFO. Under the optimum conditions, calibration curves are linear in the range of 2-2000 µg L^{-1} and limit of detection (LOD) is 0.7 µg L^{-1}. The extraction recovery and enrichment factor were 83% and 140%, respectively. Repeatability (intra-day) and reproducibility (inter-day) of method based on seven replicate measurements of 100 µg L^{-1} of methadone were 4.1% and 5.3%, respectively. The relative recoveries of urine samples spiked with methadone are 92%-106%.

Keywords: Salting-out effect; Dispersive liquid-liquid microextraction; Methadone; Urine analysis

Introduction

Methadone (MET) is a synthetic analgesic drug, which has been widely used for the treatment of opioid dependence since the mid-1960s [1]. The treatment has been controversial as it replaces a short-acting opioid (heroin) with a long acting one [2]. The mechanism of action by which MET can alleviate opioid dependence and diminishing symptoms in affected individuals, have been discussed thoroughly [3]. Many analytical methods have been applied to the quantitation of MET in biological fluids. However, isolation and preconcentration of the drug have been the main challenges in the analysis of MET due to the complex matrix of biological samples and the low concentration of residual MET.

Among LLE methods, dispersive liquid-liquid microextraction (DLLME) [4] garnered considerable attention from researchers due to its fast, simple, and efficient extraction procedure, and direct injection of the extracted phase to various instruments. Up to now, DLLME has been successfully applied to the extraction of various families of organic and inorganic compounds from different matrices [5-10], and several reviews have been written on this issue [11-13]. Despite the many benefits of DLLME, the high density of the extraction solvents is its main drawback.

DLLME integrated with the solidification of a floating organic drop (DLLME-SFO) [14] has overcome the abovementioned drawbacks. It is based on an organic extractant that has lower density than water, but solidifies at near ambient temperature. After centrifugation, the test tube is cooled by inserting into an ice bath for 5 min. The enriched analyte in the floated solid phase is collected simply by a spatula and melted at room temperature; then, it is finally determined by chromatography or spectrometry methods.

The performance of DLLME-SFO was illustrated by extraction of

different organic and inorganic compounds [15-18]. In the previous research, we applied DLLME-SFO for extraction and preconcentration of methadone in urine samples [19]. Despite many benefits of the DLLME-SFO, the pretreatment and dilution of urine samples is its main drawback. Because of decrease in matrix effect, urine samples should be pretreated and diluted before DLLME-SFO.

The aim of this work is the combination of salting-out effect and DLLME-SFO, as a sample-preparation method for high performance liquid chromatography (HPLC). Methadone was chosen as model analyte to investigate the feasibility of the improved salting-out-DLLME-SFO technique. To the best of our knowledge, for the first time, the salting-out-DLLME-SFO is developed and applied to the analysis of methadone in human urine without pretreatment and dilution of the samples.

Materials and Methods

Reagents

Standards of MET (with a certified purity >98%) was obtained from Cerilliant (Round Rock, TX, USA) as 1 mg mL^{-1} methanol solution. The methadone stock standard solution was prepared in methanol

*Corresponding author: Nazir Fattahi, School of Medical, Kermanshah University of Medical Sciences, Kermanshah, Iran
E-mail: nazirfatahi@yahoo.com

at the concentration levels of 1.00 mg L^{-1} for MET. Afterwards, they were stored in a freezer at -20°C. Working standard solutions were prepared daily by diluting the stock solution with methanol. The ultra-pure water (six times distilled) was purchased from Shahid Ghazi Company (Tabriz, Iran). Methanol (for spectroscopy), acetone (Suprasolv for gas chromatography), acetonitrile (Hyper grade for liquid chromatography), acetic acid, sodium dihydrogen phosphate, sodium dodecyl sulfate, sodium chloride, 1-undecanol, n-hexadecane, 2-dodecanol and 1-decanol were obtained from Merck (Darmstadt, Germany). Drug free urine sample (blank) collected from healthy volunteer in our lab was used for the study. Actual human urine samples taken from two young people who were suspicious to consumption of MET were stored at -20°C and analyzed within 48 h after collection without any previous treatment or filtration.

Instrumentation

Quantitative analysis of the MET was performed on a Knauer HPLC system (Berlin, Germany) equipped with aSmartline-1000 binary pumps and Smartline-UV-2500 detector variable wavelength programmable, an on-line solvent vacuum degasser and manual sample injector fitted with a 20 μL injection loop (model 7725i, Rheodyne, Cotati, CA, USA). Chromatographic separation was achieved on an ODS-3 column (25 cm × 4.0 mm, with 5 μm particle size) from Waters (Milford, MA, USA). The mobile phase consisted of water/methanol (50:50, v/v) adjusted to pH 2.7 with phosphate buffer (0.1 M). A mobile phase flow-rate of 1.0 mL min^{-1} was used in isocratic elution mode and the detection was performed at the wavelength of 210 nm. The Hettich Zentrifugen (EBA20, Tuttlingen, Germany) was used for centrifugations. Chromatographic data were recorded and analyzed using Chromgate software version 3.1.

Extraction procedure

In the first step, a 10 mL glass syringe barrel was cleaned with pure water and then a frit was placed in the bottom of the barrel and a stopcock was installed. Afterwards, 5 g of NaCl was poured into the barrel and slightly compressed with the syringe plunger. A 5.0 mL of urine sample (spiked or not with MET) was mixed with 1.5 mL acetonitrile and passed through the barrel at a flow rate of 0.5 mL min^{-1}. By passing the above homogenous solution through the barrel, fine droplets of acetonitrile were formed at the interface of solid (NaCl) and solution due to dissolution of salt into solution (salting-out effect). The produced droplets moved through the remained solution to top of the barrel and floated on the surface of solution as a separated layer due to lower density of acetonitrile with respect to water. During this step, the analytes were extracted into the fine droplets of acetonitrile. After passing all aqueous solution, the stopcock was closed. The volume of the acetonitrile (separated phase) on the top of remained NaCl solid was about 0.70 ± 0.03 mL. Subsequently, the organic phase obtained from the first step was transferred into a 10 mL glass test tube and 30.0 μL 1-undecanol (extraction solvent) was added to the test tube. Then, K$_2$CO$_3$ solution (2% w/v, 5.00 mL) were rapidly injected into a test tube using a 5.00 mL syringe (Gastight, Hamilton, Reno, NV, USA). A cloudy solution, resulting from the dispersion of the fine 1-undecanol droplets in the aqueous solution, was formed in the test tube and the mixtures were centrifuged for 5 min at 4000 g. Accordingly, the organic solvent droplet was floated on the surface of the aqueous solution due to its low density. The sample vial was there after put into an ice bath for 4 min; at this time, the floated solvent was solidified because of the low melting point (14°C). The solidified solvent was transferred into a conical glass sample cup where it was melted immediately. Finally, 25

μL of the extractant was collected with a syringe and injected onto the HPLC-UV.

Results and Discussion

In this research, the salting-out and DLLME-SFO conjunction was designed and employed for extraction of different drugs from urine samples. To reach a high extraction recovery and enrichment factor with the employment of salting-out-DLLME-SFO, the salting-out and DLLME-SFO conditions must be examined and optimized. Since the DLLME-SFO conditions had been optimized in our previous research [19], those results were used in this research. Only the salting-out conditions together with some notable parameters in the salting-out-DLLME-SFO combination were studied. The enrichment factor (EF), the extraction recovery (%ER) and the relative recovery (%RR) were calculated according to the equations described in our previous research [20].

Selection of extraction solvent in salting-out step

In salting-out-DLLME-SFO procedure, the extraction solvent in salting-out step should be able to play the role as a disperser solvent in the following DLLME-SFO step. For this purpose, acetone, acetonitrile, methanol, and tetrahydrofuran, displaying this ability, were selected. The obtained results showed that only acetonitrile formed a two-phase system, while other solvents could not be separated from the aqueous solution by passing through a syringe barrel filled with NaCl. Therefore, acetonitrile was selected as an extraction solvent for the further studies.

Selection of extraction solvent volume in salting-out step

For obtaining optimized volume of extraction solvent in salting-out step, various experiments were performed by using different volumes of acetonitrile (i.e., 0.50, 0.75, 1.00, 1.50, 2.00, and 2.50 mL) and all experiments were performed in triplicates (n=3). The results are shown in Figure 1. According to Figure 1 and considering the experimental errors on the data points, the enrichment factor of MET was found to increase by increasing volume of acetonitrile up to 1.50 mL; while, further increase in volume of acetonitrile caused a small decrease in the enrichment factor. This observation could be attributed to the fact that at lower acetonitrile volumes, the cloudy suspension of the 1-undecanol droplets was not formed well (in DLLME-SFO step), resulting in a decrease in the enrichment factor. By using more than 1.50 mL acetonitrile, the solubility of MET in aqueous phase increases and it causes a small decrease in the enrichment factor. Also, no collected phase was obtained in the case of 0.50 mL acetonitrile. Thus, according to the results, 1.50 mL of acetonitrile was chosen as the optimum volume of extraction solvent in salting-out step.

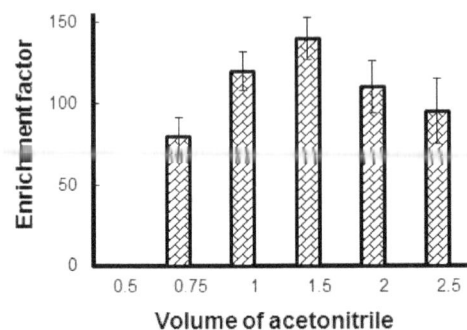

Figure 1: Effect of volume of extraction solvent in salting-out step on the enrichment factor of MET from urine sample.

Effect of the flow rate of the sample solution

The flow rate of the sample solution through the solid (NaCl) is an important factor because it controls the time of analysis and extraction recovery. The flow rate of the sample solution must below enough to perform an effective salting-out. On the other hand, it must be high enough not to waste time. The effect of the flow rate of sample solution was examined from 0.5 to 5 mL min⁻¹. As it is illustrated in Figure 2, the flow rates up to 2 mL min⁻¹ have no effect on enrichment factor of MET while, at higher speeds, the enrichment factor decreased. This behavior can be explained because the amount of dissolved salt in aqueous phase is decreased at high flow rates and leads to a decrease in efficiency. Thus, a flowrate of 2 mL min⁻¹ was selected for further studies.

Analytical performance of the method

The optimized salting-out-DLLME-SFO and HPLC-UV procedure was validated with respect to limit of detection (LOD), precision (intra-day and inter-day), linear range (LR), EF, and ER. Table 1 summarizes the analytical characteristics of the optimized method. The repeatability (intra-day) and reproducibility (inter-day) were studied by extracting the spiked urine samples (100 µg L⁻¹ for MET). The repeatability and reproducibility were calculated and were 4.1% and 5.3%, respectively. Determination coefficients (r^2) was 0.988. Good linearity 2-2000 µg L⁻¹ were obtained. The limits of detection, based on a signal-to-noise ratio (S/N) of 3, were 0.7 µg L⁻¹. Moreover, the EF and the ER of MET were 140 and 83%, respectively.

Analysis of MET in real samples

The proposed method was firstly applied to determination of the concentration of MET in human urine samples, provided by one male volunteer in our lab, who was not exposed to any drugs or MET for at least 10 months. The results from urine samples showed that they were free of MET. These samples were spiked with MET standards at different concentration levels to assess matrix effects. The results of relative recovery of urine samples are shown in Table 2. As seen, the relative recoveries for MET in spiked urine samples are between 92% and 106%. Four actual urine samples taken from male and female young persons who were suspicious to consumption of MET were also subjected to the proposed procedure. MET was detected in all of the actual urine samples in the range of 38.5-74.6 µg L⁻¹. The concentration of MET in different actual urine samples are listed in Table 2. The presences of MET in these samples were confirmed by spiking MET at the different concentration levels. The results of relative recoveries and concentrations obtained by analysis of spiked actual urine samples are also included in Table 2. Figure 3 shows the obtained chromatograms of blank urine sample (A), blank urine sample spiked with 0.7 µg L⁻¹ of MET (B), actual urine sample taken from a 22-year-old male (C) and corresponding spiked ones at concentration level of 100.0 µg L⁻¹ for MET (D). These results demonstrate that the matrices of the analyzed actual urine samples have little effect on salting-out-DLLME-SFO followed by HPLC-UV for determination of MET.

Conclusion

In this study, the salting-out effect was combined with the DLLME-SFO technique for the determination of MET in urine samples prior to analysis by HPLC-UV. This combination not only resulted in a high enrichment factor, but also it could be used in complex matrices (such as urine, fruit juice and highly saline solution) without any pretreatment or dilution. As compared with the other sample preparation methods, the analytical procedure offered numerous advantages such as simplicity, ease of operation, high preconcentration factor, low detection limit

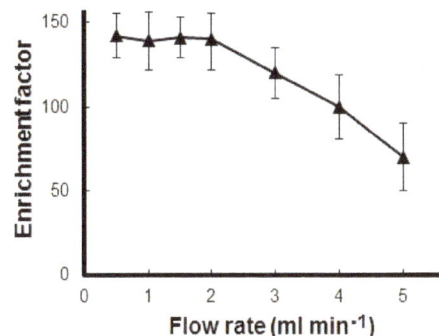

Figure 2: Effect of sample solution flow rate in salting-out step on MET enrichment factor from urine.

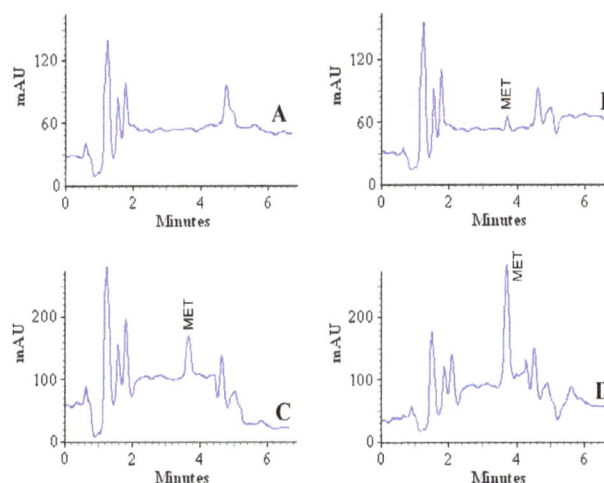

Figure 3: (A) Chromatograms of blank urine sample (B) Blank urine sample spiked with 0.7 µg L⁻¹ of MET (C) Urine sample from a person (22-year-old male) suspicious of MET consumption (D) and the same urine sample spiked with 100 µg L⁻¹ of MET.

Analyte	RSD[a]% (Intra-day, n=7)	RSD% (Inter-day, n=7)	EF[b]	ER[c]%	LR[d] (µg L⁻¹)	r[2e]	LOD[f] (µg L⁻¹)
MET	4.1	5.3	140	83	2–2000	0.988	0.7

[a]RSD% at a concentration of 100 µg L⁻¹ for MET; [b]EF: Enrichment factor; [c]ER: Extraction recovery; [d]LR: Linear range; [e]r²: Correlation coefficient; [f]LOD: Limit of detection for a S/N=3

Table 1: Figures of merit of salting-out DLLME-SFO in a MET-free urine sample.

Urine samples	Added (µg L⁻¹)	Found mean ± SD[a] (µg L⁻¹)	Relative recovery (%)
Taken from healthy volunteer (blank)	0	ND[b]	-
	100	97.5 ± 6.3	97.5
	200	198.4 ± 14.7	99.2
Taken from a 22-year-old male	0	38.5 ± 2.6	-
	100	135.0 ± 7.4	96.5
	200	247.8 ± 12.2	104.6
Taken from a 24-year-old female	0	74.6 ± 5.3	-
	50	120.6 ± 8.4	92.0
	100	180.6 ± 137.0	106.0

[a]SD: Standard deviation (n=3); [b]ND: Not detected

Table 2: Analysis of blank and actual urine samples for determination of MET.

and relatively short analysis time. Although the obtained results in this work are related to determination of MET, the system could be readily applied for the determination of other drugs from complex biological and pharmaceutical matrices, using different analytical instruments.

Acknowledgements

The authors thank the Deputy of Research and Technology, Kermanshah University of Medical Sciences, Kermanshah, Iran for financial support.

References

1. Dole VP, Nyswander ME (1976) Methadone Maintenance Treatment: A Ten-Year Perspective. J Amer Med Assoc 235: 2117-2119.

2. Cooper JR (1992) Ineffective use of psychoactive drugs methadone treatment is no exception. J Amer Med Assoc 267: 281-282.

3. Martin TJ, Kahn WR, Xiao R, Childers SR (2007) Differential regional effects of methadone maintenance compared to heroin dependence on μ-opioid receptor desensitization in rat brain. Synapse 61: 176-184.

4. Rezaee M, Assadi Y, Milani Hosseini MR, Aghaee E, Ahmadi F, et al. (2006) Determination of organic compounds in water using dispersive liquid–liquid microextraction. J Chromatogr A 1116: 1-9.

5. Fattahi N, Assadi Y, Hosseini MRM, Jahromi EZ (2007) Determination of chlorophenols in water samples using simultaneous dispersive liquid–liquid microextraction and derivatization followed by gas chromatography-electron-capture detection. J Chromatogr A 1157: 23-29.

6. Daneshfar A, Khezeli T, Lotfi HJ (2009) Determination of cholesterol in food samples using dispersive liquid–liquid microextraction followed by HPLC–UV. J Chromatogr B 877: 456-460.

7. Farajzadeh MA, Djozan DJ, Bakhtiyari RF (2010) Use of a capillary tube for collecting an extraction solvent lighter than water after dispersive liquid–liquid microextraction and its application in the determination of parabens in different samples by gas chromatography-Flame ionization detection. Talanta 81: 1360-1367.

8. Shamsipur M, Fattahi N, Assadi Y, Sadeghi M, Sharafi K (2014) Speciation of As(III) and As(V) in water samples by graphite furnace atomic absorption spectrometry after solid phase extraction combined with dispersiveliquid–liquid microextraction based on the solidification of floating organic drop. Talanta 130: 26-32.

9. Rezaee M, Yamini Y, Khanchi A, Faraji M, Saleh A (2010) A simple and rapid new dispersive liquid–liquid microextraction based on solidification of floating organic drop combined with inductively coupled plasma-optical emission spectrometry for preconcentration and determination of aluminium in water samples. J Hazard Mater 178: 766-770.

10. Shamsipur M, Ramezani M (2008) Selective determination of ultra trace amounts of gold by graphite furnace atomic absorption spectrometry after dispersive liquid–liquid microextraction. Talanta 75: 294-300.

11. Santaladchaiyakit Y, Srijaranai S, Burakham R (2012) Methodological aspects of sample preparation for the determination of carbamate residues: A review. J Sep Sci 35: 2373-2389.

12. Andruch V, Balogh IS, Kocurova L, Sandrejova J (2013) Five years of dispersive liquid–liquid microextraction. Appl Spectrosc Rev 48: 161-259.

13. Kokosa JM (2013) Advances in solvent-microextraction techniques. Trends Anal Chem 43: 2-13.

14. Leong MI, Huang SD (2008) Dispersive liquideliquid microextraction method based on solidification of floating organic drop combined with gas chromatography with electron-capture or mass spectrometry detection. J Chromatogr A 1211: 8-12.

15. Pirsaheb M, Fattahi N, Pourhaghighat S, Shamsipur M, Sharafi K (2015) Simultaneous determination of imidacloprid and diazinon in apple and pear samples using sonication and dispersive liquideliquid microextraction. LWT-Food Sci Technol 60: 825-831.

16. Pirsaheb M, Fattahi N, Shamsipur M (2013) Determination of organophosphorous pesticides in summer crops using ultrasound-assisted solvent extraction followed by dispersive liquideliquid microextraction based on the solidification of floating organic drop. Food Control 34: 378-385.

17. Ahmadi-Jouibari T, Fattahi N, Shamsipur M, Pirsaheb M (2013) Dispersive liquid–liquid microextraction followed by high-performance liquid chromatography–ultraviolet detection todetermination of opium alkaloids in human plasma. J Pharm Biomed Anal 85: 14-20.

18. Ataee M, Ahmadi-Jouibari T, Fattahi N (2016) Application of microwave-assisted dispersive liquid–liquid microextraction and graphite furnace atomic absorption spectrometry for ultra-trace determination of lead and cadmium in cereals and agricultural products. Int J Environ Anal Chem 96: 271-283.

19. Taheri S, Jalali F, Fattahi N, Jalili R, Bahrami G (2015) Sensitive determination of methadone in human serum and urine by dispersive liquid–liquid microextraction based on the solidification of a floating organic droplet followed by HPLC–UV. J Sep Sci 38: 3545-3551.

20. Ahmadi-Jouibari T, Fattahi N, Shamsipur M (2014) Rapid extraction and determination of amphetamines in human urinesamples using dispersive liquid–liquid microextraction andsolidification of floating organic drop followed by high performanceliquid chromatography. J Pharm Biomed Anal 94: 145-151.

6

Application of a Validated Stability-Indicating HPTLC Method for Simultaneous Estimation of Paracetamol and Aceclofenac and their Impurities

Ambekar Archana M and Kuchekar Bhanudas S

MAEER's Maharashtra Institute of Pharmacy, Department of Pharmaceutical Chemistry, Pune, Maharashtra, India

Abstract

A sensitive, accurate, precise and stability indicating HPTLC method has been developed and validated for the simultaneous determination of Paracetamol (PCT), Aceclofenac (ACF) and their impurities namely, 4-amino phenol (AP) and Diclofenac acid (DA), respectively, which are the hydrolytic degradation products and related substances of the drugs. The chromatographic separation was achieved using optimized mobile phase, ethyl acetate: methanol: glacial acetic acid (8.5:1.5:0.25 V/V) on pre-activated silica gel 60 F_{254} TLC plates with ultraviolet detection at 231 nm. Statistical analysis revealed linear relationships in the concentration ranges of 3900–9100, 1200–2800 for PCT, ACF respectively and 100-600 ng/band for AP and DA with correlation coefficient more than 0.99. The chromatographic conditions gave compact spots at R_f value (± SD) 0.60 (± 0.011), 0.21 (± 0.01), 0.83 (± 0.01) and 0.08 (± 0.015) for PCT, ACF, AP and DA respectively. The method was validated as per ICH guidelines, demonstrating to be accurate and precise within the corresponding linearity range of analytes. Inherent stability of these drugs was studied by exposing drug substances to stress conditions as per ICH guidelines namely oxidative, thermal, photolysis and hydrolytic conditions under different pH. Relevant degradation was found to take place under these conditions. PCT and ACF were found to undergo extensive acidic and basic hydrolysis while ACF was also susceptible to neutral hydrolysis; producing stated impurities as their major degradation products. The developed method resolved the drugs from degradation products generated under stress conditions; with the peak purity values greater than 0.999. Thus, a new simple, accurate, precise, sensitive and economic stability-indicating High Performance Thin Layer Chromatography (HPTLC)/densitometry method has been developed and validated for the simultaneous estimation of the titled analytes and their stated impurities in pharmaceutical dosage form. The high sensitivity and specificity make it valuable quantitative method suitable to be employed in the routine estimation of impurities and in stability studies.

Keywords: Paracetamol; Aceclofenac; HPTLC; Densitometry; Degradation; Related substances; Impurities; Forced degradation; Validation

Introduction

Paracetamol (PCT) is chemically N-(4-hydroxyphenyl) acetamide (Figure 1A). It is widely used as a minor analgesic, which is an alternative to aspirin without the side effects of salicylate on gastric mucosa [1]. 4-amino phenol (AP) (Figure 1C) has been found to be the primary hydrolytic degradation product of PCT. It is also reported impurity and related substance of PCT [2-4].

Aceclofenac (ACF) is chemically 2-[2-[2-[(2,6- dichlorophenyl) amino]phenyl]oxyacetic acid (Figure 1B). It is used as an effective non-steroidal anti-inflammatory drug (NSAID) derived from the phenyl acetic acid in the relief of pain and inflammation in rheumatoid arthritis, ankylosing spondylitis and osteoarthritis [5]. It has pronounced anti-inflammatory, analgesic and antipyretic properties. It undergoes extensive degradation in alkaline conditions, yielding its main degradation product as diclofenac acid (DA) (Figure 1D) derived from the phenyl acetic acid which is also related substance [6]. The combination of PCT and ACF is available in the market for its synergetic effect in pain management.

Literature survey reveals that numerous methods have been reported for analysis of PCT such as UV spectrophotometry [7], electro analytical method [8], stability indicating HPLC [9], HPLC applied to related Substances [10] and HPTLC [11] either alone or in combination with other drugs. For ACF methods reported are spectrophotometry [12,13], stability indicating HPLC [14], UPLC [15] and HPTLC [16,17] either alone or in combination with other drugs. It was further revealed that there are analytical methods reported for simultaneous estimation

of PCT and ACF like Spectrophotometry [18], HPLC [19-21], HPTLC [22], stability indicating HPLC [23] while few reported methods like HPLC [24], HPTLC [25] and stability indicating HPTLC [26] involve simultaneous estimation of PCT and ACF combined with one more additional drug. No study so far has been reported that involves studies related to simultaneous estimation of the titled analytes and their structurally related substances i.e., AP and DA that might co-exist as impurities in bulk drug.

Analytical monitoring of impurities is a key component of the recent guidelines issued by the ICH Q3A [27]. Nowadays, it is mandatory requirements in various pharmacopoeias to know the impurities present in drug substances [28]. The ICH Q1A guideline states that the validated stability indicating testing methods must be employed to monitor the characteristic features of drug substance which are susceptible to change during storage and are likely to affect the quality, safety and/or efficacy of the formulation [29]. Therefore, it was decided to develop and validate stability-indicating simultaneous

***Corresponding author:** Archana Ambekar, MAEER's Maharashtra Institute of Pharmacy, Department of Pharmaceutical Chemistry, Pune, Maharashtra, India, E-mail: archanacontact@yahoo.co.in*

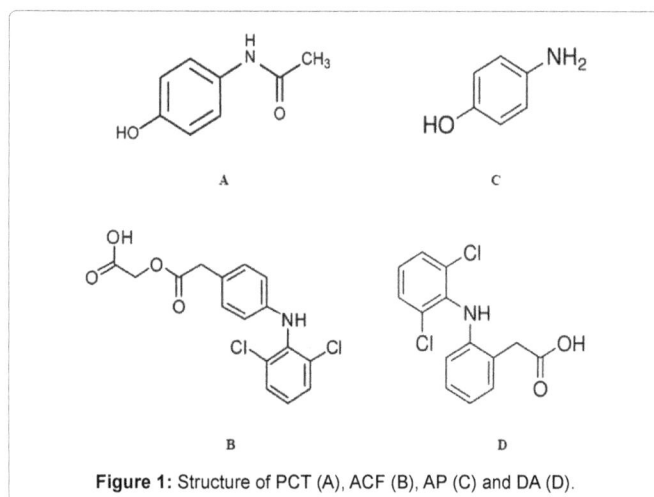

Figure 1: Structure of PCT (A), ACF (B), AP (C) and DA (D).

quantification HPTLC method for PCT and ACF and their related substances in bulk drug and in pharmaceutical dosage form. The HPTLC technique was chosen as it has an advantage of handling large number of samples at a time, it is sensitive enough to detect small amount of related substances/impurities, degradation products formed under stress conditions and at the same time it has low maintenance cost, short run time, less mobile phase consumption per sample; hence it can fulfill the demands of a routine analytical technique. It also facilitates automated application of sample and scanning of the plate [30]. Thus, the aim of the present study was to develop an accurate, precise, specific, sensitive, robust and stability indicating HPTLC method for simultaneous determination of PCT and ACF and their stated impurities in bulk drug and pharmaceutical dosage form. The analytes were purposely degraded by forced degradation to check the stability and to develop stability indicating assay method that can resolve degradation products from analytes. The developed analytical method was validated as per ICH guidelines.

Materials and Methods

Chemical and reagents

PCT and ACF standards were kindly provided as a gift sample by Zest Pharma, Indore and Vapi Care Pharma. Pvt. Ltd., Gujarat respectively, its impurities AP (99.52%) and DA (99.75%) were procured from Aarti Drugs Ltd., Thane. The marketed formulation used in this study was Acemiz Plus (Lupin Lab. Ltd., India) procured from the local market, labeled to contain 325 mg of PCT and 100 mg of ACF per tablet. The solvents and chemicals used in the development of method were of AR grade purchased from Merck Pvt. Ltd., Mumbai.

Instrumentation

Camag HPTLC system consisting Linomat V sample applicator, chromatogram development chambers (Twin Trough Chambers 20 cm × 10 cm), pre-coated silica gel 60 F$_{254}$ aluminium plates (20.0 cm × 10.0 cm, 250 µm thickness; Merck, Germany), TLC scanner III and winCATS version 1.4.4 Software, (Sr. No. 1508W015) was used for development of chromatographic separation.

Chromatographic conditions

TLC plates were pre-washed with methanol and activated at 110°C for 10 min prior to application. Suitable volumes of standard, sample and resolution solutions were applied to the HPTLC plates using CAMAG Linomat V sample applicator (Muttenz Switzerland)

equipped with 100 µl syringe (Hamilton, Reno, Nevada, USA) with spraying speed as 150 nl/s. Sample and standard zones were applied to the plates as bands (10 mm from the bottom and 8 mm from the side edges) with band length of 6 mm. The linear ascending chromatographic separation was carried using ethyl acetate: methanol: glacial acetic acid (8.5:1.5:0.25, by v/v) as mobile phase with chamber saturation time of 20 minutes and the migration distance of 90 mm. Before detection, the plates were dried to eliminate mobile phase and separation was achieved within 10 minutes. A common wavelength for the simultaneous determination of all the analytes was selected based on the overlaying spectra (Figure 2). The wavelength at 231 nm was chosen for detection as it resulted in better detection sensitivity for drug substances and their related impurities compared to the well accepted λmax of either drug substance. Densitometric scanning was performed using Camag TLC scanner III in the reflectance/absorbance mode at 231 nm with a slit dimension (4.00 mm × 0.45 mm, Micro), scanning speed 20 mm/s and data resolution 100 µm/step. Data were integrated using winCATS Software.

Preparation of solutions

Standard solution: Standard stock solutions of drug substances were prepared by dissolving 130 mg of PCT and 40 mg of ACF, in 10 ml of methanol separately to obtain a concentration of 13 mg/ml of PCT and 4 mg/ml of ACF (Solution A). An aliquot of 1 ml of solution A of each drug substance was diluted to 10 ml with methanol separately to obtain 1300 µg/ml of PCT and 400 µg/ml of ACF (Solution B). Similarly, the impurity stock solutions were prepared by dissolving 10 mg of AP and DA, in 10 ml of methanol separately to obtain a concentration of 1 mg/ml (Solution C). An aliquot of 1 ml of each impurity solution was further diluted to 1 ml with methanol separately, to obtain 100 µg/ml of each impurity. The ratio of drug substances was fixed, taking into account their proportion in tablet formulation.

Resolution solution: An aliquot of 1 ml of standard solution A of each drug substance and 1 ml of solution C of each impurity was mixed and diluted to 10 ml with methanol to get resolution solution containing 1300 µg/ml PCT, 400 µg/ml ACF, 100 µg/ml of each impurity.

Formulation assay

Twenty tablets were weighed and crushed to fine powder. An accurately weighed powder sample equivalent to 65 mg of the PCT (20 mg of ACF) was transferred to a 40 ml methanol, sonicated for about 10 min to ensure complete dissolution of drugs and the volume was made up to 50 ml with methanol, resulting into the concentration of 1300 µg/ml of PCT and 400 µg/ml of ACF. The extract was filtered through Watmann filter paper No 41. An aliquot (5ml) of sample

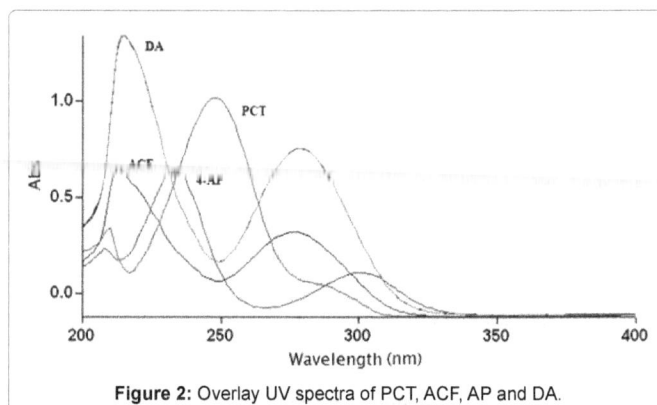

Figure 2: Overlay UV spectra of PCT, ACF, AP and DA.

solution was further diluted to 10 ml with methanol; 10μl of the resulting solution (containing 6500 ng/band of PCT and 2000 ng/band of ACF) was applied to TLC plate, developed as per optimized chromatographic conditions and scanned. The amount of each drug present was determined using calibration curves.

Forced degradation studies

The study was intended to ensure the optimum resolution of PCT and ACF and their degradation product peaks. Forced degradation studies were performed to evaluate the stability indicating properties and specificity of the method. The objective of stress study was to generate the degradation products under various forced degradation conditions and to verify that the degradation peaks are well resolved from the analyte peaks by the developed method [29,31]. Each drug substance was subjected to stress degradation, carried out in the dark, to avoid the possible degradation due to light except for photo stability studies. Forced degradation studies were carried out on the bulk drug separately and in combination. It was also applied to tablet formulation to study the presence of additional degradation products that may arise from drug excipient interaction.

Acid, base, oxidative and wet heat degradation

To 1 ml of working standard solution A of each analyte (i.e., 13 mg of PCT and 4 mg of ACF), 5 ml of each stressor i.e., 0.1N HCl / 0.1N NaOH/ 6% H_2O_2/ HPLC grade water was added and refluxed at 80°C for 2 h. The resultant solution was neutralized except for oxidative and wet heat degradation and diluted up to 10 ml with methanol. The suitable volume (5 μl containing 6500 ng/band of PCT and 2000 ng/band of ACF) was applied to a TLC plate.

Thermal degradation and photo-degradation

For solid state dry heat and photo degradation studies, PCT and ACF 50 mg each was spread as thin film in two separate petri dishes (50 mm diameter) in two sets. The first set of petri dish was heated in an oven at 70°C for 7 days and the second was exposed to UV radiation 200 Watt h/m² for 28 h in the UV chamber [32]. The solution (10 ml) was prepared using 13 mg of PCT and 4 mg of ACF of the exposed powder in methanol and suitable volume (5 μl) was applied to TLC plate.

Validation

The method was validated for parameters namely, linearity, range, accuracy, precision, sensitivity, specificity and robustness in accordance with ICH guidelines [28,33].

Linearity and range: The calibration curve was obtained in the range of 3900–9100 ng/band and 1200–2800 ng/band for PCT and ACF respectively and for impurities, in the range of 100–600 ng/band; The appropriate volume of resolution solution (3-7 ml) was diluted to 10 ml with methanol to get mixed standards in the stated concentration range and 10μl of the resulting solution was applied on TLC plate. The analytes were resolved under optimized chromatographic conditions and the peak area was observed. The whole procedure was repeated thrice starting from weighing of analytes to preparation of resolution solution as described under standard solution preparation. The average of peak area of three readings was plotted against concentration to obtain calibration curve. The slope and intercept were calculated. The linearity of the method was evaluated by linear regression analysis, using least square method, F test and the residual plot of relative response against concentration applied. The value of experimental Fischer ratio was compared against the critical value found in statistical tables, at the 95% confidence level while the trend was observed in residual plot [34].

Accuracy (recovery study): Recovery studies were conducted by the standard addition method where known amounts of the standard substances were spiked to the pre-analyzed sample at 80%, 100% and 120% level. The resulting samples were analyzed independently in triplicate starting from weighing. In short, the quantity of tablet blend equivalent to 32.5 mg of PCT was spiked with 26 mg, 32.5 mg and 39 mg of standard PCT powder separately. These blends were also spiked with 8 mg, 10 mg and 12 mg of standard ACF powder separately. The assay was carried out as per the procedure elaborated under formulation assay. For impurity recovery study, a standard powder blend containing 91 mg PCT and 10 mg AP was spiked with 8 mg, 10 mg and 12 mg of standard AP separately. Likewise, a standard powder blend containing 28 mg ACF and 10 mg DA was spiked with 8 mg, 10 mg and 12 mg of standard AP separately. These mixtures were suitably diluted and analyzed under optimized chromatographic conditions. Thus, the recovery studies were performed in order to find out how much of the drug substance is retrieved during the process of sample preparation from tablet triturate and how much of impurity is retrieved during the process of sample preparation from bulk drug.

Precision: The precision of proposed analytical method was demonstrated by repeatability and intermediate precision studies. Resolution solution containing 6500 ng/band of PCT, 2000 ng/band of ACF and 200 ng/band of each impurity was obtained after independent preparation of each solution starting from weighing were analyzed six times on the same day and six times by different analysts on different days. Briefly, 65 mg of PCT, 20 mg of ACF and 20 mg of each impurity was diluted suitably with methanol to obtain 650 μg/ml of PCT, 200 μg/ml of ACF and 20 μg/ml of each impurity. An aliquot of 10 μl was applied on TLC plate. Similarly, the assay of tablet triturate was performed six times independently, as per the procedure elaborated under formulation assay on different days. The % RSD value and mean recovery were calculated to demonstrate precision of the method.

Method sensitivity (limit of detection and limit of quantification): The LOD and LOQ of the developed method were calculated as per ICH guidelines. A series of standard preparations containing 3900–9100 ng/band and 1200–2800 ng/band of PCT and ACF respectively and 100–600 ng/band of impurities, were prepared over different levels. Calibration graphs were plotted for the obtained area under the curve of each level against the concentration. The LOD and LOQ were calculated using equations, LOD=3.3 × σ/S; LOQ=10 × σ/S, respectively where σ is the standard deviation of the response and S is the slope of the calibration curve. Thus, obtained LOD and LOQ values were further confirmed by applying different volumes of stock solution of each analytes in three replicate separately on a TLC plate and % RSD values were calculated.

Specificity: Specificity is the capacity of the method to measure the analyte response in the presence of components that may be expected to be present like impurities, degradants and formulation excipients. Therefore these studies were conducted in two parts, Part-I and Part-II. Part-I, involved demonstration of the ability of the developed method for the separation and resolution of PCT, ACF and their impurities without any interference by analytes and matrix. Peak purity of the analytes was measured to evaluate the specificity of the method. The sample and standard bands were scanned at three distinct levels, i.e., peak start (S), peak apex (M), and peak end (E) positions. The peak purity was determined by winCATS software. Part-II, involved forced degradation of drug substances under various stress conditions, Viz. Oxidative, photolytic, thermal and hydrolytic conditions under different pH values. Degraded samples were applied on the TLC plates, developed as mentioned previously. Developed densitograms were observed for separation of degraded product and spots were evaluated for peak purity as described before.

Robustness and stability of sample solution: The effect of small but deliberate variations in method parameters like the volume and composition of the mobile phase, time from spotting to development and development to scanning were evaluated in this study. One variable at a time was altered and the effect on the R_f and % recovery was checked. Similarly sample solution was applied at regular time interval for 8 h and the percentage assay was noted to study stability of the sample solution.

Results and Discussion

Development and optimization

For the selection of optimized mobile phase to achieve sufficient resolution of PCT, ACF, their impurities and degradation products several runs were made by using mobile phases containing solvents of varying polarity in different proportion like toluene, chloroform, n-butanol, methanol, glacial acetic acid, formic acid, ethyl acetate, ammonia, acetone. Several different compositions of the above solvents were tried at different concentration levels to achieve optimum resolution between analytes and their impurities. The mobile phase consisting of ethyl acetate: methanol: glacial acetic acid (8.5:1.5:0.25 V/V) was found to give the optimum resolution with sharp well defined peaks at R_f value 0.60, 0.21, 0.83 and 0.08 for PCT, ACF, AP and DA respectively. Densitogram obtained shows no interference between analyte and impurities peak and/or from other constituents originating from the tablet excipients (Figure 3). The peak purity of more than 0.999 indicates that there is not any interference in the separation and estimation of the analytes and the developed method is specific for

drug substances and their impurities (Figure 4).

Validation of the developed stability-indicating method

The developed method was validated as per ICH guidelines. The following parameters were evaluated [28,33].

Linearity and range: The calibration curve was plotted for drug substances and their impurities in the stated range. The linearity was evaluated by regression analysis, F-test and residual plots. The correlation coefficient was found greater than 0.99 for both the drug substances and their impurities. The calculated experimental F value was found less than the critical tabulated F value at 95% confidence level for titled analytes (Table 1). The plot for residuals showed random distribution without any trend/pattern in it (Figure 5). These results proved that an excellent correlation existed between peak areas and concentration of the drugs within the selected concentration range.

Method sensitivity: The low values of LOD and LOQ with % RSD values <10 (Table 1) for drug substances and impurities indicate that the developed method is sensitive enough to detect impurities in the drug substance and to be used as stability indicating and can be applied to dilute sample.

Accuracy: The % mean recoveries for PTC and ACF for each 80%, 100%, 120% level were in the range of 99.61% to 101.1% while for impurities % mean recoveries were found to be in the range of 98% to 100.6%. The overall % RSD was observed to be satisfactory thus, indicates accuracy of the developed method (Table 2).

Precision: The overall % RSD values for repeatability were found not more than 0.79% for drug substances while for impurities it was not more than 1.80%. The overall % RSD values for intermediate precision were found not more than 1.14% for drug substances and it was not

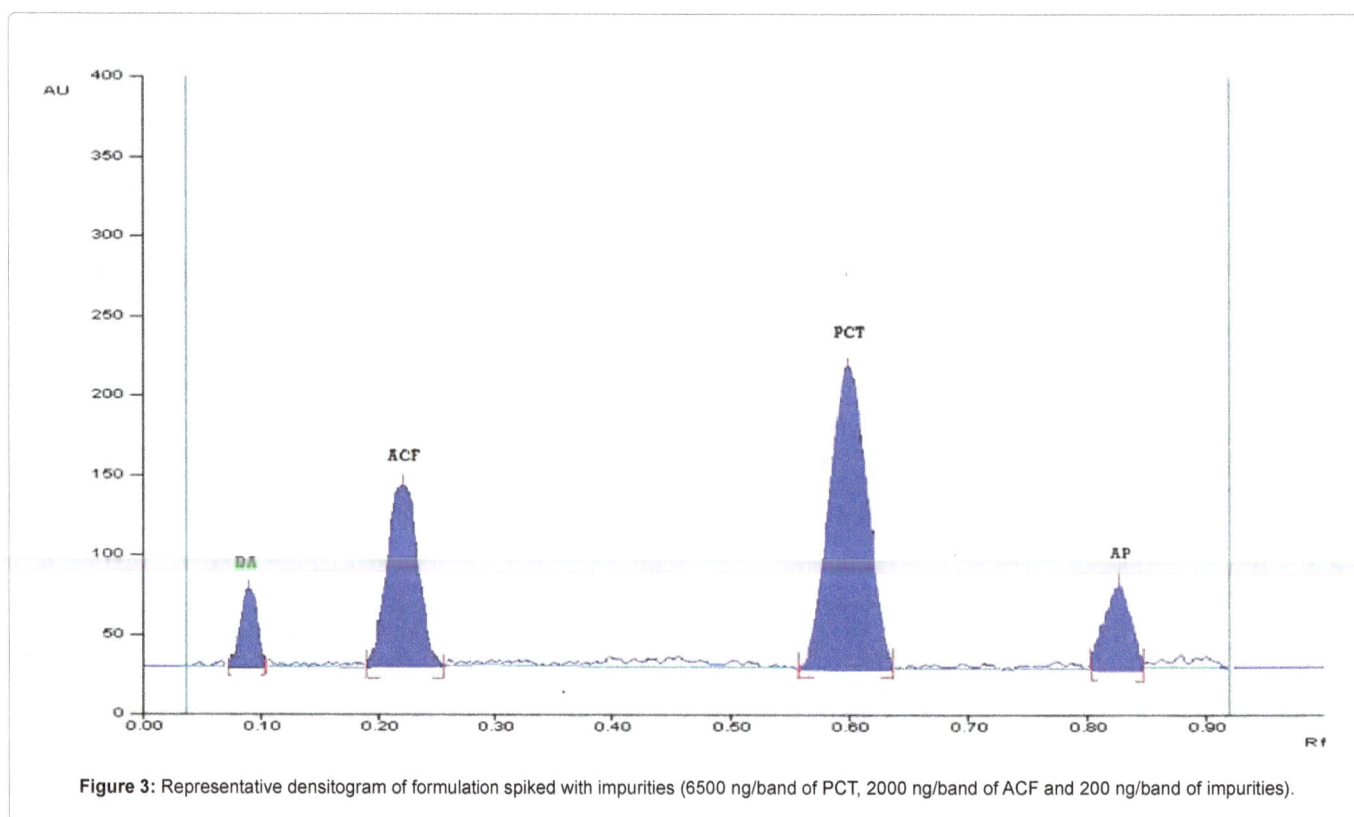

Figure 3: Representative densitogram of formulation spiked with impurities (6500 ng/band of PCT, 2000 ng/band of ACF and 200 ng/band of impurities).

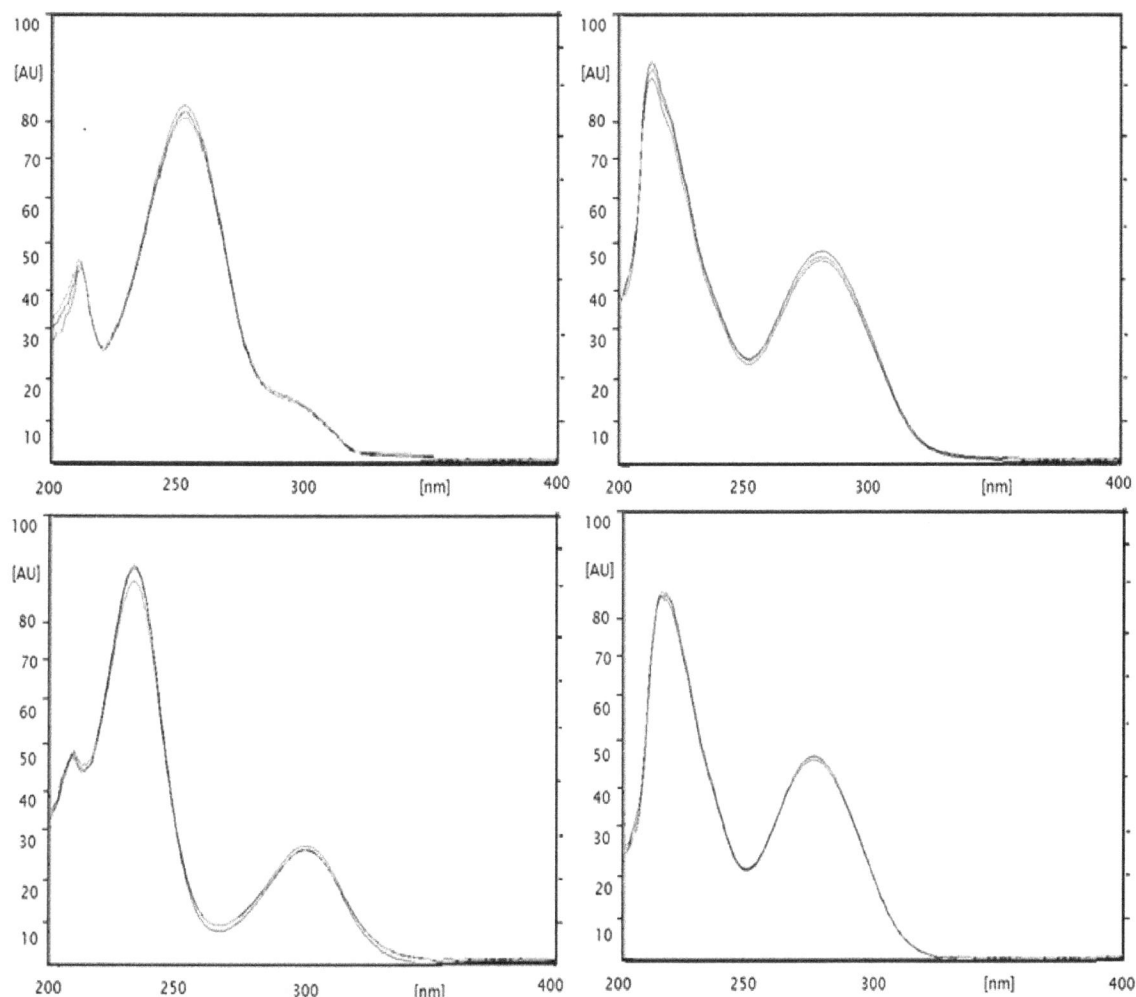

Figure 4: Peak purity spectra of Drug substances: PCT (A), ACF (B) and Impurities: AP (C) DA (D) at 231 nm.

Parameter↓ / Analytes →	PCT	ACF	AP	DA
Concentration range (ng/band)	3900–9100	1200–2800	100–600	100–600
Rf	0.60	0.21	0.83	0.08
Linearity (n=3)				
Regression equation	y=2.5741x+6234.4	y=4.1713x+117.87	y=3.7625x+335.06	y=5.3694x+29.978
Correlation coefficient (R^2)	0.998	0.998	0.998	0.998
Experimetal Variance σ^2_e	45296.77	5867.11	1229.83	2011.33
Lack of fit variance σ^2_{Lof}	116239.99	12963.45	3720.36	6197.10
Experimental Fischer variance ratio (F)	2.566	2.209	3.025	3.081
Tabulated Fischer variance ratio (F)	3.259	3.259	3.259	3.259
Mean % Assay (% RSD) (n=6)	100.73 (0.72)	99.92 (0.32)		
Precision (n=6)				
Repeatability Mean % Assay (% RSD)	99.97 (0.79)	100.14 (0.71)	98.92 (1.73)	99.81 (1.80)
Intermediate Mean % Assay (% RSD)	99.45 (1.03)	100.02 (1.14)	99.80 (2.15)	100.00 (2.01)
Method sensitivity (n=3)				
LOD (ng/band) (% RSD)	128.82 (6.17)	122.44 (7.60)	28.14 (8.03)	24.50 (9.05)
LOQ (ng/band) (% RSD)	387.34 (4.86)	371.03 (5.79)	85.29 (6.12)	74.26 (7.64)
Specificity (Typical Peak purity data)				
r(S,M)	0.9997	0.9999	0.9996	0.9998
r(S,E)	0.9996	0.9996	0.9997	0.9996

% RSD=Relative standard deviation; n=Number of repetitions

Table 1: Result of Linearity, range, precision, method sensitivity and specificity study.

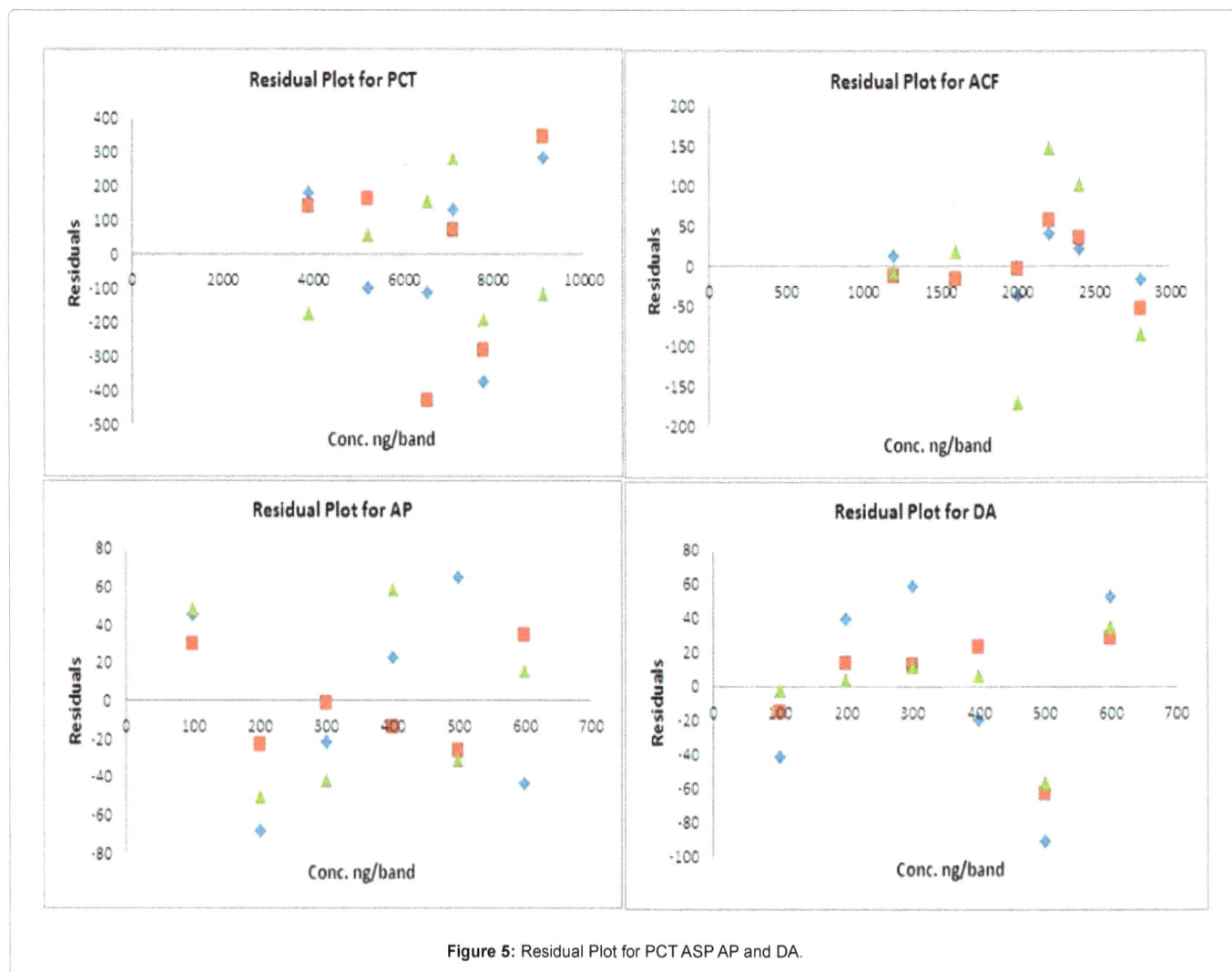

Figure 5: Residual Plot for PCT ASP AP and DA.

Recovery level	PCT		ACF		AP		DA	
	%MR	%RSD	%MR	%RSD	%MR	%RSD	%MR	%RSD
80%	100.77	0.75	99.61	0.4	99.2	1.8	98.76	2.01
100%	100.51	0.92	100.41	0.55	98.49	1.73	100.56	1.52
120%	101.1	0.94	100.92	0.36	100.44	1.95	98.58	1.37

n=Number of repetitions; MR-Mean Recovery

Table 2: Results of Accuracy (Recovery) study by standard-addition method (n=3).

more than 2.15% for impurities. The low % RSD indicates sensitivity and repeatability of the proposed method which proves the proposed method is precise (Table 1).

Robustness: Small deliberate changes in optimized chromatographic conditions did not affect the assay and R_f value significantly for the investigated compounds; low % RSD indicates that the developed method is robust enough to minor changes in the experimental conditions during routine usage (Table 3). The assay values were found to be above 99% and % RSD less than 1.80 indicates stability of sample solution over a period of 8 h.

Specificity: The specificity of the method was verified by peak purity studies. In the specificity part I study, the densitogram showed well resolved peaks without any overlapping (Figures 3 and 4) indicating separation without any interference by each other or formulation

excipients (Table 1). While the specificity part II study, showed peak purity values greater than 0.999 for drug substances and impurities indicating resolution without any interference by degradation products or matrix (Table 4). Thus, the presented HPTLC method in this study is selective for PCT, ACF and their related substances AP and DA which might co-exist as impurities.

Forced degradation studies: The results of forced degradation studies of PCT and ACF using the developed method are summarized in Table 4. Degradation products of both analytes were well resolved from the drug substances under the optimized chromatographic conditions; which were further confirmed by peak purity study. The peak purity r (s,m) and r (s,e) values were above 0.999; indicating homogeneous peaks. Thus developed method is demonstrated to be specific. The percentage degradation of PCT and ACF was calculated by area normalization technique [29]. PCT and ACF were found to undergo considerable degradation under acid hydrolysis (Figure 6A). ACF showed DA as major hydrolytic degradation product with R_f value 0.08 (densitogram 1 about 20.5%) while the PCT showed AP as major hydrolytic degradation product with R_f value 0.82 (densitogram 6 about 9.6%). Both the analytes showed additional negligible degradation peaks. Base induced degradation studies were performed with and

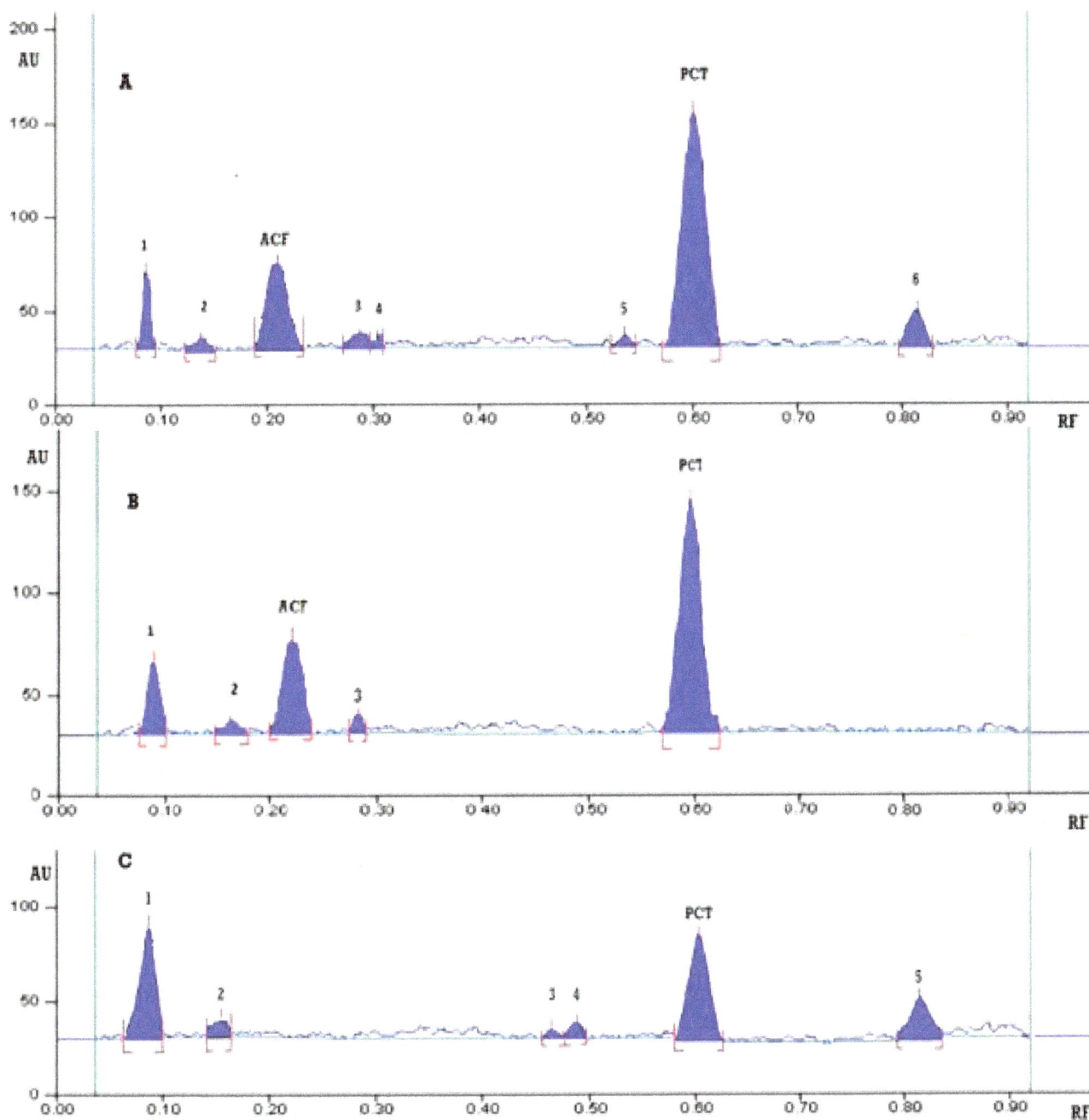

Figure 6: Representative Densitogram showing degradation under A: Acid induced hydrolysis, B: Base induced hydrolysis (without reflux), C: Base induced hydrolysis (with reflux).

without reflux. Under base hydrolysis without reflux, ACF formed DA as the major degradation product with R_f value of 0.08 (densitogram 1, about 28.56%) while PCT did not show any degradation and peak area remained almost constant (Figure 6B). Extensive degradation of both the drug substances was found to occur under base hydrolysis with reflux condition. The ACF was extremely susceptible to reflux condition and found to be completely degraded, producing DA as the major degradation product with R_f value 0.08 (densitogram 1, about 94.28%) and PCT showed AP as major degradation product at R_f value

0.82 (densitogram 5, about 38%) (Figure 6C). Additional negligible degradation peaks were observed. Thermal degradation studies involved wet and dry heat induced degradation. ACF was found to be susceptible to wet degradation and showed additional peak of DA at R_f value 0.08 (densitogram 1, about 20%) while PCT showed negligible degradation with an additional peak at R_f value 0.82 (densitogram 2, about 1.03%) and drug peak area remained almost constant (Figure 7A). The analytes were found to be stable with no additional peaks under dry heat degradation (Figure 7B). PCT and ACF under oxidative and photo degradation produced negligible degradation products

Condition	PCT %Assay,% RSD	PCT R_f	ACF %Assay, %RSD	ACF R_f	AP %MR, %RSD	AP R_f	DA %MR, %RSD	DA R_f
Mobile phase Composition (± 0.1ml)	99.51, 0.89	0.69	99.60, 1.26	0.29	99.47, 1.92	0.88	98.92, 1.52	0.15
	100.2, 1.37	0.68	100.54, 1.18	0.28	100.69, 2.36	0.88	99.26, 1.79	0.15
Chamber Saturation Time (± 5min)	100.87, 1.22	0.67	101.18, 0.82	0.29	99.91, 1.93	0.88	100.99, 2.08	0.15
	101.05, 1.00	0.68	101.84, 0.36	0.28	99.73, 2.30	0.89	100.08, 2.13	0.15
Time from spotting to development (± 10 min.)	100.81, 0.91	0.69	100.34, 1.18	0.28	100.27, 1.60	0.88	98.48, 1.82	0.15
	99.33, 0.90	0.70	100.58, 1.09	0.28	99.94, 2.68	0.88	99.39, 2.02	0.15
Time from development To scanning (± 10 min.)	100.31, 1.02	0.69	100.61, 0.97	0.29	99.47, 2.34	0.89	100.82, 1.96	0.15
	101.03, 0.64	0.68	100.06, 1.20	0.28	98.05, 1.72	0.89	99.49, 2.01	0.15

MR-Mean Recovery; n- Number of repetitions

Table 3: Robustness study for the developed method (n=3).

Analyte→ Stress condition ↓	ACF R_f of Degraded Product	ACF % Degradation	ACF % Recovery	ACF Peak Purity r(S,M), r(S,E)	PCT R_f of Degraded Product	PCT % Degradation	PCT % Recovery	PCT Peak Purity r(S,M), r(S,E)
Acid (0.1 N HCL, reflux 80°C 2 h.)	0.08,0.14,0.29, 0.31	20.5,1.30, 1.41,1.05	71.28	0.9997, 0.9999	0.82, 0.54	9.6,1.14	89.26	0.9997, 0.9998
Base (0.1 N NaOH, R.T. 4 h)	0.08, 0.16, 0.28	28.56, 1.26, 1.81	68.37	0.9997, 0.9996	—	—	97.82	0.9999, 0.9996
Base (0.1 N NaOH, reflux, 80°C, 2 h)	0.08, 0.15	94.28, 5.72	—	0.9996, 0.9998	0.82, 0.46, 0.49	38,1.05, 1.8	59.15	0.9998, 0.9993
Wet Heat (reflux, 80°C, 2 h)	0.08	20.0	80.0	0.9997, 0.9998	0.82	1.03	98.07	0.9999, 0.9989
Dry Heat (80°C, 5 h)	—	—	99.60	0.9997, 0.9995	—	—	100.02	0.9995, 0.9999
Hydrogen peroxide 6% (reflux, 60°C, 2 h)	0.17	1.35	98.65	0.9996, 0.9999	0.30, 0.39, 0.49	1.01, 2.41, 2.10	95.49	0.9997, 0.9995
Photolysis	0.27	1.5	98.5	0.9995, 0.9996	—	—	100.21	0.9996, 0.9994

Table 4: Summary of Forced Degradation Study for ACF and PCT.

(Figure 8A and 8B). Thus, overall, it was found that PCT and ACF were susceptible to acid and base hydrolysis and formed the impurities i.e., AP and DA respectively as their major degradation products.

Conclusion

Monitoring and controlling of impurities assures the quality and safety of a drug. Thus, analytical activities concerning impurities in drugs are among the most important issues in modern pharmaceutical analysis. Forced degradation studies provide data to support identification of possible degradants; degradation pathways and intrinsic stability of the drug molecule. Therefore an attempt has been made in the proposed research work, to develop stability indicating method for simultaneous estimation of the titled analytes along with their stated impurities. The present research involves quantification of the related substances with vital information about the forced degradation under different stress conditions. The development and validation studies, give satisfactory % mean recovery, lower % RSD and optimum peak purity values; prove that the proposed method can resolve the drug substances, their stated impurities and degradants in short run time with optimum resolution and desirable sensitivity to be utilized in impurity quantification. The proposed method found robust to minor changes in the experimental conditions and can give accurate and precise results during routine analysis. The method has been successfully applied for the analysis of tablet formulation without any interference from excipients. The stability study reveals that, both drugs are more sensitive towards acid and base hydrolysis producing their related substance as major degradants. The published stability indicating HPLC method for titled analytes has high run time and more LOD and LOQ values indicating lower sensitivity in comparison to the proposed method [23]. Extensive literature revealed that there is stability indicating HPTLC method available for simultaneous estimation of the titled analytes along with one more drug substance, in bulk and pharmaceutical dosage form [26]. Although this method estimate the analytes simultaneously and is sensitive with short run time but is not applicable for simultaneous estimation of stated impurities and hence cannot be utilized in impurity quantification. Thus, the proposed stability indicating HPTLC method is simple, reproducible and economic with the appropriate level of sensitivity. It

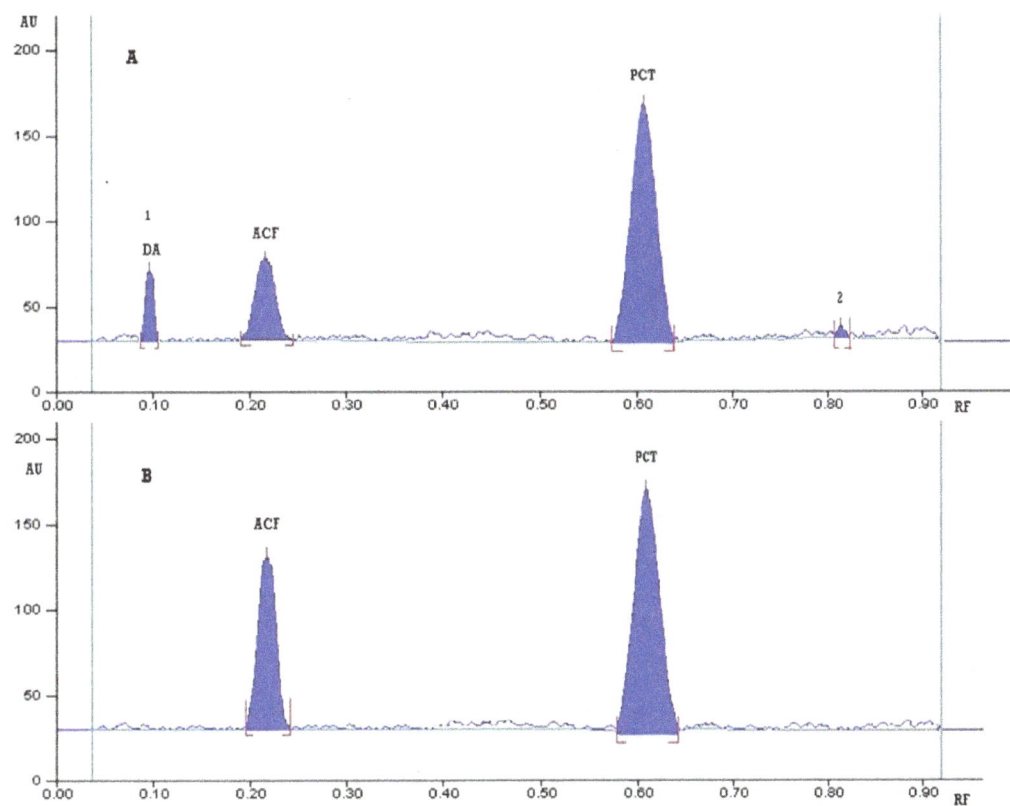

Figure 7: Representative Densitogram showing degradation under thermal condition A: Wet heat B: Dry heat.

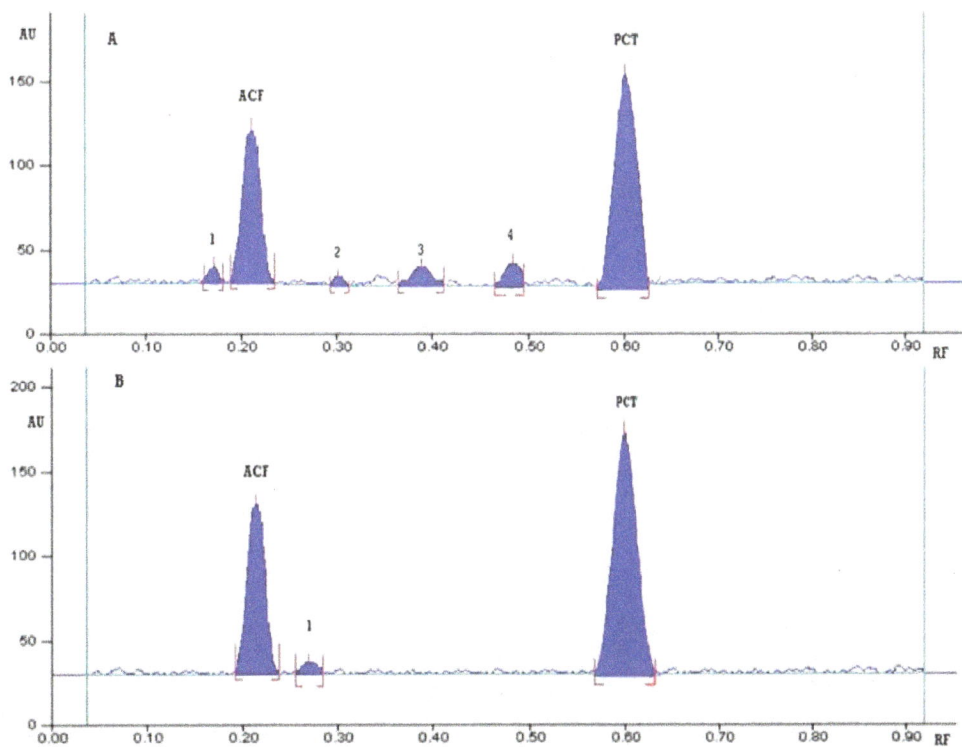

Figure 8: Representative Densitogram showing degradation under A: oxidative degradation, B: photo degradation.

can be utilized for simultaneous determination of PCT, ACF and their stated impurities in impurity estimation. The developed densitometric method is suitable not only for separation and determination of related impurities for monitoring of synthetic reaction but also for stability studies of drug substances and in routine analysis.

Acknowledgements

Authors are thankful to Anchrom Enterprises Ltd. Mulund, Mumbai for providing lab facility for carrying out HPTLC work.

References

1. United States Pharmacopoeia 32 National Formulary 27 (2009) Asian edition, United States Pharmacopoeia Convention, Rockville, MD, USA 2: 1388.

2. Bloomfield MS (2002) A sensitive and rapid assay for 4-aminophenol in paracetamol drug and tablet formulation, by flow injection analysis with spectrophotometric detection. Talanta 58: 1301-1310.

3. Mostafa NM (2010) Stability indicating method for the determination of paracetamol in its pharmaceutical preparations by TLC densitometric method. Saudi J Chem Soc 14: 341-344.

4. Lock EA, Cross TJ, Schnellmann RG (1993) Studies on the mechanism of 4-aminophenol-induced toxicity to renal proximal tubules. Hum Exp Toxicol 12: 383-388.

5. Indian Pharmacopoeia (2010) Ministry of Health and Family Welfare. Government of India, the Indian Pharmacopoeial Commission. Ghaziabad 2: 681-683.

6. Saraf S, Garg G, Singh D, Saraf S (2006) Aceclofenac: A Potent Non-Steroidal Anti-Inflammatory Drug. Pharmainfo.net 4: 3.

7. Behera S, Ghanty S, Ahmad F, Santra S, Banerjee S (2012) UV-Visible Spectrophotometric Method Development and Validation of Assay of Paracetamol Tablet. J Anal Bioanal Tech 3: 1-6.

8. Omran M, Mohammad AZ (2014) Kinetic Study of Paracetamol Hydrolysis: Evaluating of the Voltammetric Data. Anal Bioanal Electrochem 6: 450-460.

9. Kulandaivelu K, Gurusamy N, Elango KP (2014) A validated stability-indicating RP-HPLC method for paracetamol and lornoxicam: application to pharmaceutical dosage forms. Chem Ind Chem Eng Q 20: 109-114.

10. Karbhari PA, Joshi SJ, Bhoir SI (2014) RP-LC Gradient Elution Method For Simultaneous Determination Of Related Substances Of Zaltoprofen And Paracetamol And Application For Drug Excipient Compatibility Study. Int J Pharm Pharm Sci 6: 698-703.

11. Abdelaleem EA, Abdelwahab NS (2013) Stability-indicating TLC-densitometric method for simultaneous determination of paracetamol and chlorzoxazone and their toxic impurities. J Chromatogr Sci 51: 187-191.

12. Bose A, Dash P, Sahoo M (2010) Simple spectrophotometric methods for estimation of aceclofenac from bulk and formulations. Pharm Methods 1: 57-60.

13. el-Saharty YS, Refaat M, el-Khateeb SZ (2002) Stability-indicating spectrophotometric and densitometric methods for determination of aceclofenac. Drug Dev Ind Pharm 28: 571-582.

14. Bhinge JR, Kumar RV, Sinha VR (2008) A simple and sensitive stability-indicating RP-HPLC assay method for the determination of aceclofenac. J Chromatogr Sci 46: 440-444.

15. Balan P, Kannappan N (2014) Development and Validation of Stability-Indicating Rp-Uplc Method for Simultaneous Estimation of Thiocolchicoside and Aceclofenac in Combined Dosage Form. Int Current Pharma J 3: 296-300.

16. William H, Parimala DB, Jose K, Valsalakumari PK, Mohandas CK (2014) Validated High Performance Thin Layer Chromatographic Estimation of Aceclofenac in Bulk and Pharmaceutical Dosage Forms. Int J Sci Res, 3: 2613-2617.

17. Gandhi SP, Dewani MG, Borole TC, Damle MC (2012) Development and Validation of Stability Indicating HPTLC Method for Determination of Diacerein and Aceclofenac as Bulk Drug and in Tablet Dosage Form. E J Chem 9: 2023-2028.

18. Pawar VT, Pishawikar SA, More HN (2011) Spectrophotometric estimation of paracetamol and aceclofenac from tablet dosage form. Int Res J Pharm 2: 93-98.

19. Patil PP, Deshpande MM, Kasture V (2010) Simultaneous Estimation of Aceclofenac and Paracetamol in Bulk and Combined Tablet Dosage Form and From Biological Fluid by Planar Chromatography. Research J Pharm and Tech 3: 1194-1199.

20. Gopinath R, Rajan S, Meyyanathan SN, Krishnaveni N, Suresh B (2007) A RP-HPLC method for simultaneous estimation of paracetamol and aceclofenac in tablets. Indian J Pharm Sci 69: 137-140.

21. Momin MY, Yeole PG, Puranik MP, Wadher SJ (2006) Reversed phase HPLC method for determination of Aceclofenac and Paracetamol in tablet dosage form. Indian J Pharm Sci 68: 387-389.

22. Gandhi SV, Barhate NS, Patel BR, Panchal D, Bothara KG (2008) A validated densitometric method for analysis of Aceclofenac and Paracetamol as a bulk drugs and combined tablet dosage form. Acta Chromatographica 20: 175-182.

23. Jamil S, Talegaonkar S, Khar KR, Kohli K (2008) Development And Validation of Stability-Indicating LC Method for Simultaneous analysis of Aceclofenac and Paracetamol in conventional tablet and microsphere Formulations. Chromatographia 68: 557-565.

24. Shaikh KA, Devkhile AB (2008) Simultaneous determination of aceclofenac, paracetamol, and chlorzoxazone by RP-HPLC in pharmaceutical dosage form. J Chromatogr Sci 46: 649-652.

25. Apshingekar PP, Mahadik MV, Dhaneshwar SR (2010) Validated HPTLC Method for simultaneous quantitation of Paracetamol, Tramadol and Aceclofenac in tablet formulation. Der Pharmacia Letter 2: 28-36.

26. Mallikarjuna RN (2011) Development and Validation of Stability Indicating HPTLC method for Simultaneous Estimation of Paracetamol, Aceclofenac and Rabeprazole in Combined Tablet Dosage Formulation. Int J Pharm Tech Res 3: 909-918.

27. ICH guidance (2006) Impurities in New Drug Substances. Q3A (R2).

28. Jain D, Basniwal PK (2013) Forced degradation and impurity profiling: recent trends in analytical perspectives. J Pharm Biomed Anal 86: 11-35.

29. ICH guidance (2003) Stability Testing of New Drug Substances and Products Q1A (R2).

30. Sethi PD (2013) HPTLC: Quantitative Analysis of Pharmaceutical Formulations. 1st edn. CBS Publishers & Distributors, New Delhi, pp: 15-60.

31. Bakshi M, Singh S (2002) Development of validated stability-indicating assay methods--critical review. J Pharm Biomed Anal 28: 1011-1040.

32. ICH guidance (1998) Photostability Testing of New Active Substances and Medicinal Products Substances and Medicinal Products. Q1B.

33. ICH guidance (2005) Validation of analytical procedures, text and methodology Q2 (R1). International conference on harmonization of technical requirements for registration of pharmaceuticals for human use.

34. Ferenczi-Fodor K, Vegh Z, Nagy-Turak A, Renger B, Zeller M (2001) Validation and quality assurance of planar chromatographic procedures in pharmaceutical analysis. J AOAC Int 84: 1265-1276.

7

Development and Validation of RP-UPLC Method for Determination of Related Substances in Risperdal® Consta®

Cheng-Tung Chen*, Cheng-Wei Cheng and Yu-Fang Hu

Pharmaceutical Development Center, TTY Biopharm Company Limited, Taipei, Taiwan

Abstract

In this work, a validated stability-indicating ultra-high performance liquid chromatographic (UPLC) method has been developed for quantitative determination of related substances in Risperdal® Consta®. The chromatographic separation was achieved on a Waters BEH C18 column (2.1 × 100 mm, 1.7 µm) in an isocratic elution mode. The limit of quantification is 0.1 µg/mL, and the method is linear over the concentration range of 0.1-1.5 µg/mL for risperidone. The PLGA polymers have been dissolved in organic phase (acetonitrile) and precipitated in aqueous phase (0.01 N HCl). This simple step avoids the clogging of column and makes the method more robust. The mean recovery of extracted risperidone from PLGA microspheres is 99.26 ± 1.22%. Herein, this method was proved to be highly reliable and applicable for measuring impurities in PLGA microspheres.

Keywords: RP-UPLC; Validation; Risperdal® Consta®; PLGA; Relative substances

Introduction

Risperidone, which is a novel antipsychotic medicine mainly worked for treating schizophrenia and also the psychotic, affective, or behavioral symptoms associated with other disorders [1]. The main pharmacological effects of risperidone include dopamine D2 antagonism and serotonin 5-HT2 receptor blockade [2]. The molecular formula and IUPAC name of risperidone are $C_{23}H_{27}FN_4O_2$ and 3-[2-[4-(6-fluoro-1,2-benzisoxazol-3-yl)-1-piperidinyl]ethyl]-6,7,8,9-tetrahydro-2-methyl-4H-pyrido[1,2-a] pyrimidin-4-one respectively. The chemical structure of risperidone was depicted in Figure 1.

In recent years, a sustained-release API (active pharmaceutical ingredient) delivery system has benefit patients, and embedding these drugs inside. US Food and Drug Administration (FDA)-approved carriers such as liposomes [3], polyethylene glycol (PEG) [4], or poly (lactide-co-glycolide) PLGA [5,6] were wildly developed by researchers. PLGA microspheres have been presented to be a controlled release system for various drugs in demand. It degrades and to form water-soluble and non-toxic monomers under physiologic conditions. In April 2007, the US FDA approved Risperdal® Consta® (risperidone long-acting injection; Johnson & Johnson) for the treatment of schizophrenia. It overcomes the major issue with the current oral dosage is the difficulty of making patients take the oral medication on a daily basis.

The determination of the relative substances could be a critical item to monitor the goods quality. The assay techniques for determining the impurities content for several dosage forms including oral solution, tablets and orally disintegrating tablets have been published in pharmacopoeias (e.g., USP, JP). In general, high-performance liquid chromatography (HPLC) equipped with UV detector has been utilized for the impurities assay. However, the unsuitable column selection (e.g., the production process/control by different manufacturers) and the total analysis time consuming are the major concerns. Especially, isolating and identifying the impurities/degradants is not negligible in the least amount of time for the process control monitoring of pharmaceutical industry. Hence, more rapid and simple method should be developed at the early drug discovery stage.

Ultra performance liquid chromatography (UPLC) has been a proven technique in the past decade. Due to packing with sub-2 µm particles, analytes retain less time on the stationary phase which not only increase the resolution power but also improve the sensitivity at the same time. Besides, reduced solvent consumption is another benefit compared to conventional HPLC. Bindu et al. have reported a UPLC method for quantitating six impurities of paliperidone (9-hydroxy resperidone) in 8 min [7]. Nejedly et al. developed a UHPLC method for discrimination the main degradation impurities from risperidone tablet [8]. The new UHPLC method has reduced 4 times of total analysis time comparing with original pharmacopoeial HPLC method. A comparison for the determination of related substances in risperidone tablet using HPLC and UPLC system has been evaluated by Yan et al. [9]. UPLC revealed higher efficiency for determining the generated impurities from the force degraded samples.

However, there were no reported chromatographic methods for determining the related substances in Risperdal® Consta®. Completely extracting drugs or impurity substances from PLGA microspheres and removing the high molecular weight PLGA polymer were the major challenges. Herein, the aim of this study was to develop and validate a simple and rapid UPLC method to resolve the mentioned issue.

Materials and Methods

Chemical

USP reference standard Risperidone and Risperidone Related Compounds Mixture were purchased from USP convention, Inc. (Rockville, MD). Poly (D,L-lactide-co-glycolide) 75:25 was obtained from Lakeshore Biomaterials (Birmingham, Al). Distilled water was produced by Millipore water purification system. Risperdal® Consta® 25 mg long-acting injection was purchased from Janssen

*Corresponding author: Cheng-Tung Chen, Associate scientist, Pharmaceutical Development Center, TTY Biopharm Company Limited, 3F, No. 124, Xingshan Rd, Neihu Dist, Taipei City-11469, Taiwan E-mail: Eden_Chen@tty.com.tw

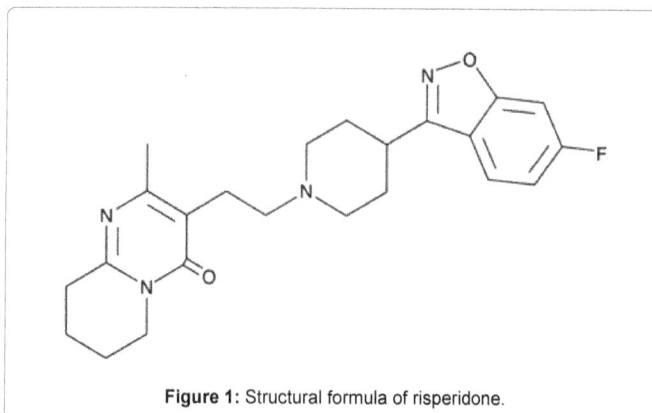

Figure 1: Structural formula of risperidone.

Pharmaceuticals (Titusville, NJ). The PLGA microspheres placebo was produced by R&D division of TTY Biopharm Company Ltd.

Instrumental and chromatographic conditions

For HPLC analysis, the study was employed using a BDS HYPERSIL C18 column (3 μm, 4.6 × 100 mm). The injection volume was 10 μL. An isocratic elution with a mixture of acetonitrile and 50 mM ammonium acetate solution (22:78, v/v). The flow rate was set at 1.5 mL/min.

For UPLC analysis, the test was performed using an ACQUITY UPLC BEH C18 column (1.7 μm, 2.1 × 100 mm) coupled with a VanGuard pre-column (1.7 μm, 2.1 × 5 mm). The injection volume was 3 μL. UPLC separation of related substances used isocratic elution with a mixture of acetonitrile and 10 mM ammonium acetate solution (26:74, v/v). The flow rate of the mobile phase was 0.3 mL/min.

The instrument used was a Waters ACQUITY H-Class UPLC system (Milford, MA) coupled with a PDA (Photo Diode Array) detector. The column temperature for both separation methods was maintained at 30°C and the eluted analytes were monitored at the wavelength of 275 nm. All mobile phase was filtered through 0.22 μm PVDF membrane filters and degassed by an ultrasonicator at least 30 min. The output signal was processed using Empower® 3 software.

Standard solutions preparation

Accurately weighed risperidone reference standard (10 mg) was transferred to a 50 mL volumetric flask, dissolved in and diluted to the mark with 0.01 N HCl. Standard solutions were prepared by diluting the stock solution to obtain suitable concentration for chromatographic measurements. All solutions were filtered by a 0.22 μm PTFE syringe filter before analysis.

The solution for system suitability

Ten milligram of USP Risperidone Related Compounds Mixture RS was transferred to a 10 mL volumetric flask, dissolved in and diluted to the mark with diluent (10 mM ammonium acetate: water: methanol=1:9:10, v/v/v). The chemicals contained in it are risperidone cis N-oxide, bicyclorisperidone, z-oxime and risperidone. All solutions were filtered by a 0.22 μm PTFE syringe filter before analysis.

Sample preparation

Fifty milligram of Risperdal® Consta® microspheres were accurately weighed and transferred to a 100 mL volumetric flask. 10 mL of acetonitrile was added into the flask, followed by 10 min sonication to dissolve PLGA polymer. 90 mL of 0.01 N HCl was slowly added and resulted in the precipitation of PLGA polymer. After well-mixed the

suspension, 2 mL of sample solution was centrifuged at 8,000 × g for 10 min to remove most of the polymer. For the remaining polymer depletion, the supernatant was filtered through a Microcon-YM10 filter with a 10,000-Da molecular mass cutoff. All solutions were filtered by a 0.22 μm PTFE syringe filter before analysis. Store under refrigeration at 4°C, the sample solution containing drug substance or impurities is stable over one week.

The similar procedure was utilized to investigate the PLGA precipitation study. After dissolving the 50 mg of Risperdal® Consta® microspheres in the 10 mL of acetonitrile, various composition of ACN-HCl (0.01N) solutions have been prepared. 1 mL of the sample was transferred to the cuvette and the transmission at 600 nm was recorded using a UV-vis spectrophotometer (DU-640, Beckman).

Stress degradation studies

To prove the specificity of the stability-indicating method, the stress degradation studies were conducted to check the possible degradation of the active pharmaceutical ingredient or PLGA under various states such as alkaline, heating, and oxidative conditions. The forced degradation conditions have been referred to the USP 36th official monograph of the risperidone tablet (peak identification solution). Briefly, 20 mg of Risperdal® Consta®/PLGA microspheres or 10 mg of risperidone API were suspended in the 100 mL volumetric flask with 10 mL of 0.1 N NaOH solution (pH 8.5) at 90°C for 24 h. Subsequently, 10 mL of 0.1% aqueous hydrogen peroxide was added to the flask and kept at 90°C for additional 2 hr. For the reaction work-up, the solution was cooled to room temperature and diluted with methanol to volume. The degraded sample solutions were filtered by a 0.22 μm PTFE syringe filter before analysis.

Validation procedures

Validation of the newly developed method was constructed of linearity, range, precision, accuracy, and robustness, limit of quantitation (LOQ) and limit of detection (LOD) as per ICH guidelines [10]. Since the lack of the qualified purity impurities, the diluted risperidone solution was used for all studies in terms of quantitation e.g., linearity, sensitivity, precision and accuracy. The risperidone related compounds mixture solution was subjected to prove system suitability. The resolution between bicyclorisperidone and z-oxime under various conditions should be evaluated and greater than 1.5. The USP tailing factor of a peak should less than 2.0. For linearity, the peak response for risperidone was carried out at five concentration levels from LOQ to 150% of the specification (0.10%). The calibration curves were generated and regression parameters were calculated by Excel. The precision of the method was determined by checking the intra-day and inter-day precision for five independent preparations. The %RSD should not more than 5%. Accuracy of the method was evaluated by calculating the percentage recoveries by adding risperidone reference standard to the placebo. LOD and LOQ were estimated at a signal-to-noise ratio was 3:1, 10:1, respectively. To determine the robustness of the developed method, different experimental conditions including mobile phase composition (± 2%), column temperature (± 5°C) and flow rate (± 0.02 mL/min) were altered. All specificity of the UPLC assay method was evaluated by stress test. Risperdal® Consta® and placebo were subjected to various forced degradation conditions. The peak purity was calculated by Empower® 3 to ensure the analyte chromatographic peak is not attributable to more than one component.

Results and Discussion

Method development

The developed and validated method was aimed to establish chromatographic conditions, capable of qualitative and quantitative of risperidone impurities. The chromatographic condition was initially based on the pharmacopoeial provided method. Figure 2a shows the conventional HPLC chromatogram for risperidone impurities assay. Although the resolution of bicyclorisperidone and z-oxime is qualified (Rs: 3.5>1.5), but the suggested flow rate (1.5 mL/min) often caused the relatively high backpressure to the system and resulted in poor peak shape. Based on our past experience, the tailing factor got worse after replicate injections. In order to overcome this problem, we transferred the method from HPLC to UPLC and varied the conditions until met the acceptance criteria. Various mobile phase compositions and flow rates were adjusted. An isocratic RP-UPLC method was developed for screening all contents with UV detection at 275 nm. In Figure 2b, a smooth and flat baseline and a symmetry peak of risperidone at 5.4 min were observed in the UV chromatogram. The compared results of UPLC and HPLC method are given in Table 1. The improved USP tailing factor (from 1.75 to 1.01) and reduced analyzing time demonstrated that the UPLC-based method is more efficient than HPLC. The system suitability results are given in Table 2 and developed UPLC method was found to be a well-separated approach for risperidone and its main impurities.

In general, for determination of the encapsulation efficiency or drug-loading, the sustained released PLGA microspheres would be dissolved in organic solvent and directly subjected for HPLC analysis [11,12]. However, for the measurement of the impurities content, a large amount of drug products should be injected into the column to satisfy the limit of quantitation with trace substances. One of the key issues in sample preparation is to avoid injecting PLGA polymer into the UPLC column. It may easily lead to physical clogging of the column even though the pre-column has been installed. To determine whether the PLGA precipitation depend on organic solvent composition, the transmission of the PLGA solution under different ACN concentrations were measured (Figure 3). At high volume of organic phase (>90% ACN), a clear sample solution and high transmission were observed. However, at the high volume of aqueous phase, measured low transmission means large amount of PLGA polymer has been precipitated. By these features, PLGA polymer could be depleted by organic-aqueous phase transfer. Meanwhile, the contained drug substances were dissolved and recovered under both solution systems for further analysis.

Specificity

The diluent sample, stress degraded Risperdal® Consta® and placebo (without API) were analyzed under the proposed chromatographic condition to prove the selectivity. Figure 4a displays the UV spectrum of stress degraded Risperdal® Consta®. Figure 4b, c shows the UV spectrum obtained from direct analysis of sample diluent (10% ACN in 0.01 N HCl) and forced degradation placebo. No signal of solvent or hydrolyzed PLGA monomers was presented in the spectrum. Fortunately, all generated impurity peaks do not overlap with the remaining signal of risperidone. The purity angle of risperidone is less than purity threshold which proves the specificity of this method was verified.

Linearity and range

Linearity was conducted by preparing diluted standard solutions at five concentration levels including from 0.1, 0.25, 0.5, 1.0 and 1.5 µg/mL. The regression equation of the linearity plot of concentration

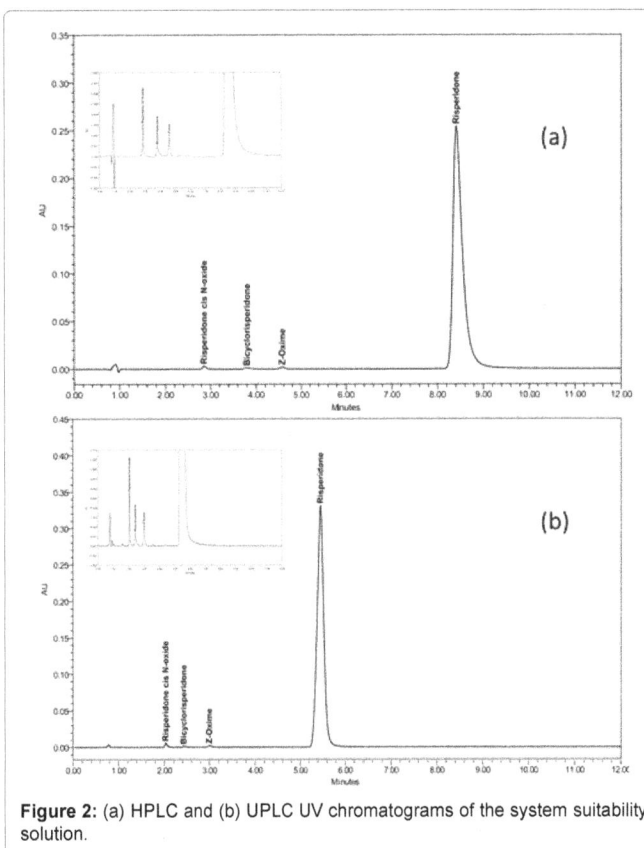

Figure 2: (a) HPLC and (b) UPLC UV chromatograms of the system suitability solution.

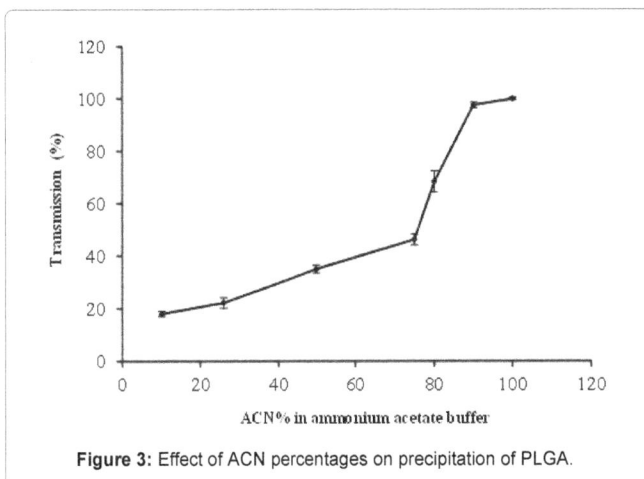

Figure 3: Effect of ACN percentages on precipitation of PLGA.

of risperidone over its peak area was found to be peak area=35380 × (conc.)−323.8. The correlation coefficient was found to be least R^2=0.9993. It indicates that the anayte gives a linear response for all ranges of concentrations.

Sensitivity

LOQ and LOD are expressed as a known concentration of analyte at a specified signal-to-noise ratio, usually 10:1 for LOQ or 3:1 for LOD. It is the lowest concentration of analyte in a sample can be quantitated or detected under the stated UPLC method. The LOQ (mean S/N=11.5) and LOD (mean S/N=4.9) of risperidone were 0.1 and 0.05 µg/mL, respectively.

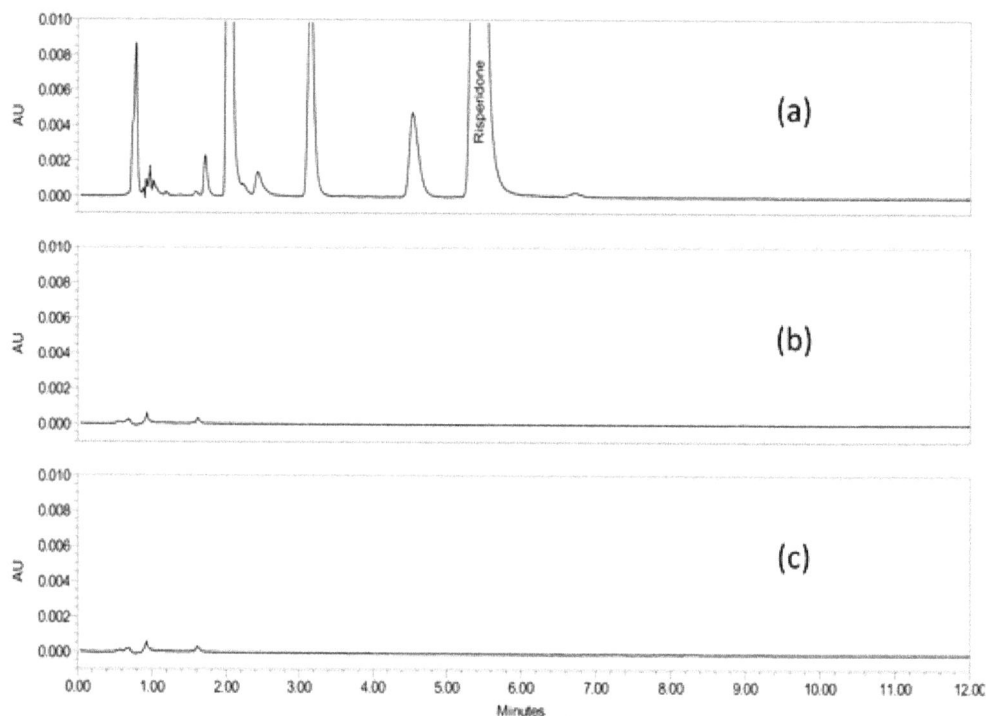

Figure 4: UPLC-UV chromatograms of (a) stress degraded Risperdal® Consta®, (b) diluent, and (c) PLGA microsphere placebo.

	K'	RT	N	T
UPLC	4.46	5.459	5974	1.01
HPLC	7.41	8.412	8257	1.75

Table 1: Comparison of developed UPLC and HPLC method. K': K prime; RT: retention time; N: theoretical plate; T: USP tailing.

	Risperidone cis N-oxide	Byclorisperidone	Z-Oxime	Risperidone
RT(min)	2.03	2.42	2.99	5.46
RRT	0.37	0.44	0.55	-
R_s	-	3.17	3.82	11.91
T	1.51	1.11	1.57	1.01
N	7006	3936	5855	5974

Table 2: System suitability report. RT: retention time; RRT: relative retention time; R_s: resolution; T: USP tailing; N: theoretical plate.

Nominal Conc of Risperidone(μg/mL)	Intra-day (n=5)%RSD	Inter-day (n=5) %RSD
0.5 μg/mL	0.92%	4.79%
1.0 μg/mL	0.58%	1.86%
1.5 μg/mL	0.14%	2.14%

Table 3: Intra-day and inter-day precision data for the method.

Level	Spiked Conc. (μg/mL)	Found Conc. (mean ± SD, μg/mL)	Recovery (%)
50%	0.5	0.49 ± 0.02	97.93%
100%	1	1.00 ± 0.01	100.33%
150%	1.5	1.49 ± 0.03	99.51%

Table 4: Extraction recovery of risperidone in PLGA microspheres (n=3).

Precision

The results of intra-day and inter-day precision are reported in Table 3. %RSD values were not more than 0.92% for intra-day precision and not more than 4.79% for inter-day precision. The data demonstrated that the values met the acceptance criteria.

Accuracy

Accuracy measurements are obtained by comparison of the results with the analysis of the target concentration (1 μg/mL of risperidone). The experiments are repeated for three times at three different concentration levels. The total average recovery is 99.26% with %RSD=1.23%. The detailed results are list in Table 4. Furthermore, stress degraded risperidone API was spiked into the PLGA microsphere placebo to estimate the extraction recovery (Figure S1). The total recovery was determined by calculating the sum of peak area. The mean recovery is 98.72 ± 2.43%. It means not only drug substance could be successfully extracted but also all impurities were easily recovered using this approach.

Robustness

The robustness of the method was studied by small but deliberate changes the acetonitrile content, column temperature and flow rate. The data provided in Table 5 indicates that the analytical method was robust, as small but deliberate changes in the method parameters have no detrimental effect on the method performance. As expected, a proportionate increase in retention time ratio of analytes with increasing acetonitrile content and vice versa with decreasing acetonitrile content was observed. A similar result has been obtained while deliberating the flow rate. Elevated flow rate for eluting analytes resulted in raised retention time ratio as well. Moreover, with the exception of light drift in retention time, the chromatographic properties of the method remained constant when different column temperatures were adjusted.

Retaintion time ratio of impurities to Risperidone					
Parameter altered	Variation	Risperidone cis N-oxide	Bicyclorisperidone	Z-Oxime	Resolution between Bicyclorisperidone and Z-Oxime	USP tailing factor of Risperidone
ACN Conc	24%	0.34	0.41	0.53	4.29	0.97
	26.00%	0.36	0.43	0.54	3.76	1.01
	28%	0.4	0.46	0.57	3.44	0.89
Column Temp	25°C	0.37	0.44	0.55	3.51	0.93
	30°C	0.36	0.43	0.54	3.76	1.01
	35°C	0.36	0.43	0.54	4.12	0.91
Flow rate (mL/min)	32.00%	0.41	0.45	0.54	3.54	0.97
	30.00%	0.36	0.43	0.54	3.76	1.01
	28.00%	0.32	0.4	0.52	3.86	1.13

Table 5: Summarized robustness data.

According to the resolution item, all results are still compliance with acceptance criterion.

Conclusion

A validated rapid, specific and reproducible method was developed for analyzing the relative substances in Risperdal® Consta®. The simple phase-to-phase precipitation strategy for the extraction recovery of risperidone from PLGA microspheres was to be found excellent and reproducible. The process impurities generated from forced degradation conditions are well-separated, showing satisfactory data for all the ICH guidance requirements. Hence, the proposed means can be employed in routine analysis for product or stability samples.

Acknowledgments

Authors would like to thank the TTY Biopharm Company Limited for supporting this research financially.

References

1. Bhat M, Shenoy SD, Udupa N, Srinivas CR (1995) Optimization of delivery of betamethasone-dipropionate from skin preparation. Indian Drugs 32: 211.

2. Bindu KH, Reddy IU, Anjaneyulu Y, Suryanarayana M (2012) A Stability-Indicating Ultra-Performance Liquid Chromatographic Method for Estimation of Related Substances and Degradants in Paliperidone Active Pharmaceutical Ingredient and its Pharmaceutical Dosage Forms. J Chromatogr Sci 50: 368-372.

3. International Conference on Harmonization (2005) Validation of analytical procedures: text and methodology. ICH guideline Q2 (R1).

4. Komossa K, Rummel-Kluge C, Schwarz S, Schmid F, Hunger H, et al. (2011) Risperidone versus other atypical antipsychotics for schizophrenia. Cochrane Database Syst Rev: CD006626.

5. Megens AA, Awouters FH, Schotte A, Meert TF, Dugovic C, et al. (1994) Survey on the pharmacodynamics of the new antipsychotic risperidone. Psychopharmacology (Berl) 114: 9-23.

6. Musumeci T, Ventura CA, Giannone I, Ruozi B, Montenegro L, et al. (2006) PLA/PLGA nanoparticles for sustained release of docetaxel. Int J Pharm 325: 172-179.

7. Nejedly T, Pilarova P, Kastner P, Blazkova Z, Klimes J (2014) Development and validation of rapid UHPLC method for determination of risperidone and its impurities in bulk powder and tablets. IJRPC 4: 261-266.

8. Selmin F, Blasi P, DeLuca PP (2012) Accelerated polymer biodegradation of risperidone poly(D, L-lactide-co-glycolide) microspheres. AAPS PharmSciTech 13: 1465-1472.

9. Song J, Xie J, Li C, Lu JH, Meng QF, et al. (2014) Near infrared spectroscopic (NIRS) analysis of drug-loading rate and particle size of risperidone microspheres by improved chemometric model. Int J Pharm 472: 296-303.

10. Stebbing J, Gaya A (2002) Pegylated liposomal doxorubicin (Caelyx) in recurrent ovarian cancer. Cancer Treat Rev 28: 121-125.

11. Yan XD, Gao M, Cai ZH (2012) A Comparison of determination of related substances in risperidone tablets by HPLC and UPLC. Anhui Medical Pharm J 11: 19.

12. Ye M, Kim S, Park K (2010) Issues in long-term protein delivery using biodegradable microparticles. J Control Release 146: 241-260.

8

Determination of Allopurinol and Oxypurinol in Dogs Plasma by High-Performance Liquid Chromatography with an Ultraviolet Detector: Application for Pharmacokinetic Studies

Godoy ALPC[1,2], De Jesus C[2], De Oliveira ML[3], Rocha A[3], Pereira MPM[3], Larangeira DF[2,4], Lanchote VL[3], Barrouin-Melo SM[2,4]

[1]Department of Clinical and Toxicological Analyzes, School of Pharmacy, Federal University of Bahia (UFBA) , Rua Barão de Jeremoabo, 147, Salvador, BA, Brazil
[2]Laboratory of Veterinary Infectology, Hospital-School of Veterinary Medicine, Federal University of Bahia, Av. Adhemar de Barros, 500, Salvador, BA, Brazil
[3]Department of Clinical, Toxicological and Bromatological Analyzes, Faculty of Pharmaceutical Sciences of Ribeirão Preto, University of São Paulo, Avenida do Café, S / N, Ribeirão Preto, SP, Brazil
[4]Department of Anatomy, Pathology and Clinics, School of Veterinary Medicine and Animal Science, UFBA, Salvador, BA, Brazil

Abstract

High performance liquid chromatography with ultraviolet detection (HPLC-UV) was developed and validated for quantification of allopurinol and its active metabolite oxypurinol; in dog plasma with naturally acquired zoonotic visceral leishmaniasis. Allopurinol is a potent inhibitor of the xanthine oxidase enzyme; an enzyme that catalyzes the conversion of hypoxanthine to xanthine and from xanthine to uric acid. In veterinary medicine allopurinol is indicated in the treatment of canine visceral leishmaniasis. Allopurinol is utilized to inhibit the synthesis of Leishmania RNA. The conditions defined for development of the chromatographic analysis of dog plasma samples by utilizing the mobile phase of the HPLC-UV consist of a mixture of 0.1% water, 88% formic acid and 0.25% acetonitrile. Allopurinol and oxypurinol were separated on a LiChroCART® 125-4 LiChrospher® 100 RP-8 (5 μm) column utilizing a flow rate of 0.7 mL min^{-1} and detector operation was at a wavelength of 254 nm. Acyclovir was utilized as an internal standard. The validation of the HPLC-UV method was determined by limits of detection and quantification, linearity, reproducibility and repeatability. The method had a lower quantification limit of 0.1 μg.mL^{-1} for both allopurinol and oxypurinol. The analysis was linear at concentrations of 0.1-20.0 μg.mL^{-1} for both allopurinol and its oxypurinol. The precision value intracurrent and intercurrent for all concentrations presented had a coefficient variation lower than 15%. The confidence limits of HPLC-UV for analysis of allopurinol and oxypurinol in dog plasma indicates that the method is applicable to the multiple dose pharmacokinetic study of allopurinol in dogs. It is efficient, accurate and sensitive. This study of Pharmacokinetic research of allopurinol and oxypurinol in dogs with Stage I and II visceral leishmaniasis resulted in similar research outcomes that correlated with the healthy dog investigation.

Keywords: HPLC-UV; Allopurinol; Oxypurinol; Pharmacokinetic; Canine visceral leishmaniasis

Introduction

Allopurinol (1H-pyrazolo (3,4-d) pyrimidin-4-ol) is an oxypurine base. It was first synthesized in the 1950s by George Hitchings and Gertrude Elion in New York with the aim to increase the efficacy of antineoplastic drugs [1,2]. However, it was later discovered that allopurinol was a strong inhibitor of the enzyme xanthine oxidase [1]. Currently allopurinol is utilized in the treatment of humans diagnosed with hyperuricemia [2,3].

In veterinary medicine, this drug has been used to dissolve uroliths formed by ammonia uric acid calcifications in dogs. Dalmatian genetics elicit a predisposition to developing high concentrations of uric acid [4,5]. Its use also extends to the treatment of canine visceral leishmaniasis (LV) therapy. In leishmaniostatic treatment is utilized either as monotherapy [6,7] or associated with leishmanicidal drugs [8-10].

Allopurinol is an analog of hypoxanthine and is converted by xanthine oxidase into its active metabolite oxypurinol (alloxanthin) (Figure 1). Both allopurinol and oxypurinol bind competitively to xanthine oxidase promoting its inhibition. They inhibit xanthine oxidase which is the enzyme that catalyzes the conversion of hypoxanthine to xanthine and xanthine into uric acid in both humans and dogs [3,11,12]. Peak plasma concentrations of allopurinol are obtained within 30 minutes to 1 hour after ingestion and about 60-70% of the drug is converted rapidly by xanthine oxidase to the active metabolite oxypurinol [3,13].

In humans, allopurinol and oxypurinol are excreted primarily via the kidneys. A dose of 100 mg allopurinol orally generates 90 mg of oxypurinol [11]. Allopurinol is excreted in the urine in its unchanged form at a rate of 10%; in the feces at a rate of 20%. The metabolite oxypurinol is eliminated in the urine at a concentration of 70% [12-14]. Limited research has been conducted on the pharmacokinetics of allopurinol in dogs [5,15,16]. The pharmacokinetics of allopurinol has been demonstrated in healthy Dalmatians where maximum concentration (C_{max}) was reached at 6.43 ± 0.18 μg.mL^{-1} at a time of (T_{max}) 1.9 hours and its apparent volume of distribution was (Vd / F) 1.17 ± 0.07 L.kg^{-1}. This illustrates it was well distributed within the tissues [5].

The allopurinol pharmacokinetic research by Ling et al. in healthy Dalmatians was also performed utilizing an HPLC-UV operating at 254 nm wavelength and had a limit of detection of 2.0 μg.mL^{-1} [5].

*Corresponding author: Ana Leonor Pardo Campos Godoy, Laboratory of Veterinary Infectology, Hospital-School of Veterinary Medicine, Federal University of Bahia (UFBA), Av. Adhemar de Barros, 500, Salvador, BA, 40170-110, Brazil, E-mail: leonor.godoy@ufba.br

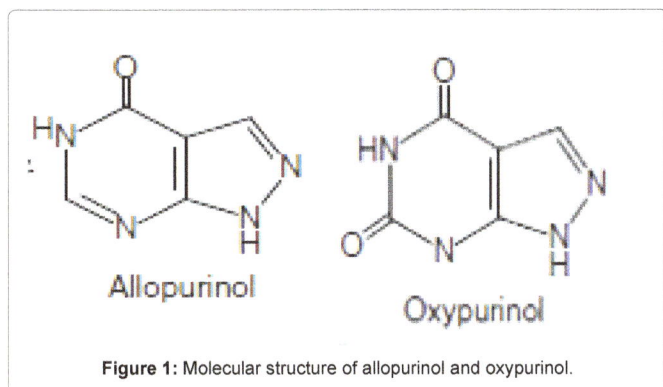

Figure 1: Molecular structure of allopurinol and oxypurinol.

In human analysis, allopurinol and oxypurinol were evaluated utilizing high performance liquid chromatography (HPLC) with UV-detection. The percentage recovery and intra-assay coefficient of variation (CV%) for allopurinol was 97.4-101.3% and 0.66-5.13% respectively in the concentration range of 0.5 to 5.0 µg.mL^{-1}. For oxypurinol the percentage recovery and the intra-assay coefficient of variation (CV%) was 93.2-98.1% and 0.88-5.62% respectively in the concentration range of 0.4 to 20.0 µg.mL^{-1} [17].

Reinders et al. also evaluated allopurinol and oxypurinol in human plasma. Validation results for allopurinol lower and upper limits of quantification were 0.5 and 10.0 mg.L^{-1} and for oxypurinol 1.0 and 40.0 mg.L^{-1}. The assay was linear over the concentration range of 0.5-10.0 mg.L^{-1} (allopurinol) and 1.0 to 40.0 mg.L^{-1} (oxypurinol). Intra- and inter-day precision showed coefficients of variation <15% over the complete concentration range. Accuracy was within 5% for allopurinol and oxypurinol. Endogenous purine-like compounds were separated from allopurinol, oxypurinol and acyclovir with a factor of >1.5% [18].

In the present study, both allopurinol and oxypurinol in dog plasma were tested utilizing HPLC-UV. The development of an accurate, sensitive and reproducible method of plasma analysis is a fundamental foundation in the research of the kinetic arrangement and metabolism of allopurinol and oxypurinol utilizing multiple dose administration in dogs with visceral leishmaniasis.

Materials and Methods

Chemicals and reagents

Standard research concentrates (European Pharmacopoeia Reference Standard) of allopurinol 99.8% and oxypurinol 98.4% was obtained from Sigma (St. Louis, MO). Acyclovir was donated by Cristália, Produtos Químicos e Farmacêuticos Ltda. (Brazil). Standard stock solutions of allopurinol and oxypurinol were prepared in NaOH (1M) at a concentration of 400 µg/mL.

Chromatography grade acetonitrile and formic acid were purchased from Merck (Darmstadt, Germany). The analytical grade reagent sodium hydroxide was purchased from Mallinckrodt Baker (Paris, Kentucky). Water was purified using a Milli-Q Plus System.

Chromatography

Plasma sample analysis: High performance liquid chromatography with ultraviolet detection (HPLC-UV) was the method used for the quantitative analysis of allopurinol and oxypurinol in dog plasma. This system consisted of a LC20AD high-pressure pump and a CTO-10ASvp column-conditioning oven from Shimadzu (Tokyo, Japan).

Allopurinol and oxypurinol were separated on a LiChroCART®

125-4 LiChrospher® 100 RP-8 (5 µm) column (Merck, Darmstadt, Germany) and using a pre-column LiChroCART® 4-4 LiChrospher® 100 RP-8 (5 Mm) (Merck, Darmstadt, Germany). The oven temperature was 25°C and the auto sampler was 12°C. The mobile phase consisted of a mixture of purified water, 0.1% formic acid (88%) and 0.25% acetonitrile. The HPLC-UV system operated at a flow rate of 0.7 mL.min^{-1}. Ultraviolet detection was performed at a wavelength of 254 nm.

Sample preparation solution: Plasma samples were obtained from a random sample of healthy dogs owned by the staff of the Teaching Hospital of the School of Veterinary Medicine and Zootechny, University Federal of Bahia (HOSPMEV-UFBA), Brazil. Aliquots of 100 µL of dog plasma were supplemented with 25 µL of IS solution (acyclovir). Allopurinol and oxypurinol were extracted from plasma samples with 200 µL of acetonitrile by shaking in a mechanical shaker (220 ± 10 cycles min^{-1}) for 1 minute. After 10 minutes at room temperature the samples were centrifuged at 21.500 × gravity for 10 minutes at 5°C to separate the organic phase. The organic phase was collected, evaporated and dried in an evaporator vacuum system set at 25°C (models RCT90 and RC10.22; Jouan AS, St. Herblain, France). The residues were dissolved in a 200 µL mixture of water plus 0.1% of formic acid and 0.25% of acetonitrile (mobile phase) and vortexed for 10 seconds. Then 60 µL was injected into the analytical column.

Method validation of plasma sample: The calibration curve was prepared from 100 µL of drug free plasma enriched with 25 µL of the solution for each concentration standard and the internal standard. The samples were prepared in triplicates resulting in concentrations of 0.1 to 20.0 µg.mL^{-1} of each plasma solution (allopurinol and oxypurinol).

Recovery of allopurinol and oxypurinol was evaluated five times at plasma concentrations of 0.25 µg.mL^{-1} and 16.0 µg.mL^{-1}. The internal standard acyclovir was present in the concentration at 40.0 µg.mL^{-1} (n=10). The peak concentration areas from the extracted samples were compared with the standard areas obtained in the analysis of standard solutions × 100.

The quantification limit was determined as the lowest quantified concentration of allopurinol and oxypurinol with precision and accuracy of less than 20%. Plasma samples were evaluated five separate times. Each plasma sample was enriched with allopurinol and oxypurinol at a concentration of 0.1 µg.mL^{-1}.

Linearity was verified at concentrations of 0.1 to 20.0 µg.mL^{-1}. Plasma samples for both allopurinol and oxypurinol were tested in triplicate at eight different concentrations.

Precision and accuracy were evaluated five times at concentrations of 0.1, 0.25, 10.0 and 16.0 µg.mL^{-1} for allopurinol and oxypurinol. In the analysis of the intra-crossover accuracy, four preparations of four different concentrations were evaluated. Five aliquots of each sample were used by means of a single calibration curve. The coefficient variation of the data achieved must be less than 15% to substantiate methodology accuracy. For precision analysis, the four independent concentration preparations were evaluated using five aliquots of each sample for three consecutive days. The precision evaluation was performed using the coefficient variation of the acquired values.

For the analysis of the intra and inter current accuracy the experimental results obtained in the precision research were utilized. The accuracy is expressed by the standard error at a rate of less than 15%. The long-term stability assessment (3 cycles for 24 hours) of five plasma samples of allopurinol and oxypurinol in plasma concentrations of 0.25 and 16.0 µg.mL^{-1} was evaluated. To evaluate the stability of freezing and thawing; the five samples were frozen at -20°C and after 24 hours were thawed and then frozen again for 12 hours. This cycle was repeated two more times and then the drug and metabolite tested.

The short-term stability assessment was also performed five times at the same concentrations of 0.25 and 16.0 µg.mL^{-1} of both allopurinol and oxypurinol in plasma samples. In this analysis, these samples were maintained at room temperature (25°C) for four hours.

To evaluate the post-processing stability assessment the samples were extracted and maintained at 12°C for twenty-four hours. The completed sample preparations were maintained until completion of analysis.

Experimental sampling

Animals and ethics: This investigation was conducted on male dogs consisting of various weights, ages and species. They were all carriers of zoonotic visceral leishmaniasis. These animals were patients at the HOSPMEV-UFBA in the Small Animal Clinic sector. The experiments were carried out in accordance with the rules and recommendations of The Ethics Committee Use of Animals, University Federal of Bahia.

Dosing and sample collection: The standard method was applied for pharmacokinetic studies of allopurinol and oxypurinol in dogs with naturally acquired visceral leishmaniasis. After 8 h fasting a blood serum sample was obtained from the dogs at a zero time; being in a steady state ($5 \times t_{1/2}$). An oral dose of 10 mg.kg^{-1} of allopurinol was then administered. Blood samples were obtained utilizing standard medical procedures at time intervals of: 15 minutes, 30 minutes, 1 h, 1.5 h, 2 h, 2.5 h, 3 h, 4 h, 5 h, 7 h, 10 h and 12 h. Blood samples were centrifuged and the plasma separated and stored at -20°C until analysis on high performance liquid chromatography (HPLC-UV).

Pharmacokinetics: The pharmacokinetic parameters were calculated on the basis of the plasma concentration *versus* time curves using WinNonlin software version 4.0 (Pharsight Corp, Mountain View, CA). No compartmental model was used in this study.

Statistical analysis: The median, mean and relative standard error with 95% confidence interval were calculated using Excel 2016.

Results and Discussion

Several preliminary experiments were evaluated to discriminate the allopurinol and oxypurinol analysis in plasma samples. The present study reports for the first time the simultaneous analysis of allopurinol and oxypurinol by HPLC-UV in dogs. This pharmacokinetic veterinary application was than applied to this study in male dogs with visceral leishmaniasis receiving multiple oral doses of allopurinol (10 mg.Kg^{-1}).

The optimal chromatographic separation of allopurinol and oxypurinol (Figure 2) was conducted with the LiChroCART® 125-4 LiChrospher® 100 RP-8 (5 µm) The mobile phase utilized a mixture of water plus 0.1% of formic acid and 0.25% of acetonitrile. Figure 2 show the chromatographic analysis of plasma blank (drug-free), plasma blank (drug-free) enriched with solutions of allopurinol, oxypurinol, acyclovir (internal standard - IS) and the plasma from one dog with leishmaniasis treated with allopurinol (10 mg.Kg^{-1}). It was observed that the retention times of allopurinol, oxypurinol and acyclovir were 6.0 min; 7.8 min and 9.9 min, respectively.

The calibration curve constructed for the analysis of allopurinol and

oxypurinol (0.1-20 µg.mL^{-1} plasma) showed a coefficient correlation higher than 0.99 (Table 1).

The methodology was validated utilizing assessments of recovery, limits of quantification, linearity, precision, accuracy and stability. Coefficients of variation and relative errors of less than 15% are within acceptable range. The limit of quantification values was extended to 20%.

Recovery rates obtained were ~90% for allopurinol and oxypurinol extracted from plasma with acetonitrile (Table 1). Reinders et al. utilized dichloromethane as an extractor solvent and these recovery rates were 65% for allopurinol and 75% for oxypurinol [18]. Reddy and Reddy utilized an extractor solvent of ethyl acetate for their plasma samples and elicited recovery rates of less than 50% [19].

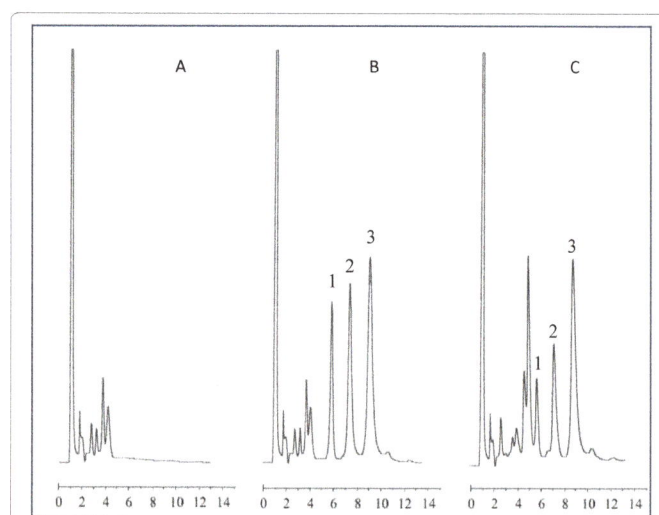

Figure 2: Chromatograms for analysis of allopurinol and oxypurinol in canine plasma: A) White plasma; B) White plasma+enriched with standard solution of 5 µg/mL plasma; C) Plasma sample collected within 2 h after oral administration of 10 mg/kg allopurinol. 1=Oxypurinol; 2=Allopurinol; 3=Aciclovir (Internal Standard).

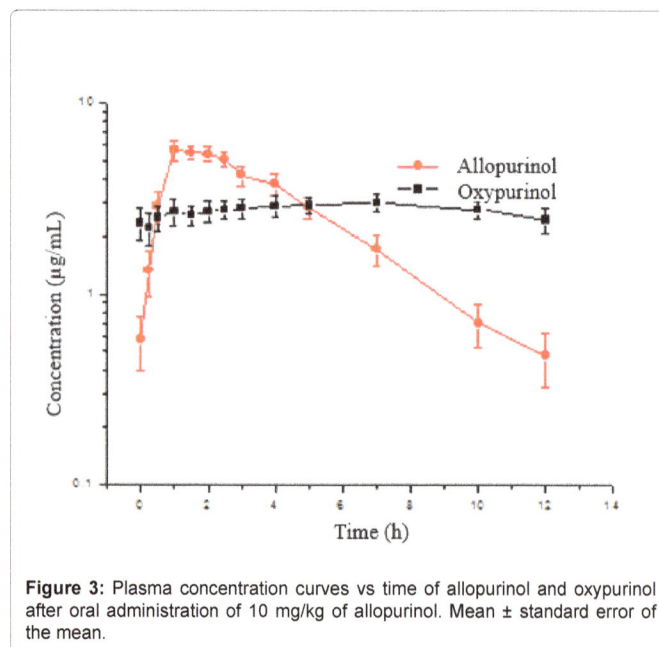

Figure 3: Plasma concentration curves vs time of allopurinol and oxypurinol after oral administration of 10 mg/kg of allopurinol. Mean ± standard error of the mean.

Quantification limits of 0.1 µg.mL^{-1} dog plasma were obtained for both allopurinol and oxypurinol, using 100 µL plasma sample and the HPLC-UV system. Ling et al. reported in his investigation quantification limits of 2 µg.L^{-1} for allopurinol when testing dog plasma samples [5]. Quantification limit research performed by Reinders et al. utilized 100 µL of human plasma samples [18]. It elicited quantification limits of 0.5 mg.L^{-1} and 10 mg.L^{-1} for allopurinol and oxypurinol, respectively.

Analysis of precision and accuracy showed coefficients of variation and percent of error of less than 15%. This indicates that the method is precise, duplicable and accurate (Table 1). The evaluation of stability after 3 days of freeze-thaw cycles and after storage at room temperature for 4 h showed deviations of less than 15% for both allopurinol and oxypurinol. This guarantees the stability of samples stored at -20°C.

This research study of kinetic disposition of allopurinol and oxypurinol as multiple doses of 10 mg.kg^{-1} to male dogs with naturally acquired visceral leishmaniasis was validated by recovery, limits of quantification, linearity, precision, accuracy and stability. In Figure 3 it is illustrated the curve of plasma concentrations versus the time of allopurinol and oxypurinol concentrations in a steady state.

Canine visceral leishmaniasis is classified in four stages: Stage I (mild diseases), Stage II (moderate disease), Stage III (severe disease) and Stage IV (very severe disease); diagnosis is based upon clinical signs, laboratory findings, serological status, and types of therapy and prognosis for each patient [20]. In this investigation, the studied dogs were in Stage I and II.

The present pharmacokinetic study of allopurinol and oxypurinol in dogs with Stage I and II of visceral leishmaniasis resulted in similar conclusions as the study of Ling et al. when they studied these drugs in healthy Dalmatians [5]. The following parameters of drug kinetic disposition found in our study were equivalent to the study of Ling et al.: distribution volume (Vd.F^{-1}.Kg^{-1}), time to reach the maximum concentration (T$_{max}$), maximum concentration (C$_{max}$), half-life (t$_{1/2}$) and constant of elimination (Kel), as illustrated in Table 2. Even though we have studied infected dogs, the results we have found using HPLC-UV method for the pharmacokinetic analyses of allopurinol and oxipurinol in canine plasma were comparable to the results in a study of the same parameters in healthy dogs reported in the literature [5].

Conclusion

In the present study, it has been shown that allopurinol and oxypurinol can be accurately and precisely measured in dog plasma samples in less than 10 minutes using HPLC-UV. This method proved to be simple, accurate, reproducible and sensitive when applied for the pharmacokinetic analyses of those drugs using plasma from dogs with visceral leishmaniasis treated with multiple doses of 10 mg.kg^{-1}

	Allopurinol			Oxypurinol		
Linearity	0.1-20.0 µg/mL plasma			0.1-20.0 µg/mL plasma		
Equation of the line	y=0.15257x-0.02218			y=0.13523x-0.0166		
(r)	0.99962			0.99966		
(r²)	0.99924			0.99933		
* Absolute Recovery (%) (n=10)	87%			86%		
Aciclovir-PI (n=10)	89%					
Precision and Accuracy						
Intracorridas	Mean	CV%	EPR%	Mean	CV%	EPR%
LIQ (0.1 µg/mL, n=5)	0.10	5.64	-0.20	0.10	3.96	-4.40
CQB (0.25 µg/mL, n=5)	0.23	3.74	-7.68	0.22	3.53	-11.68
CQM (10.0 µg/mL, n=5)	10.53	1.84	5.32	10.12	1.87	1.18
CQA (16.0 µg/mL, n=5)	16.45	1.53	2.81	15.76	1.49	-1.49
Intercorridas	Mean	CV%	EPR%	Mean	CV%	EPR%
LIQ (0.1 µg/mL, n=5)	0.10	9.22	1.40	0.10	6.90	-3.60
CQB (0.25 µg/mL, n=5)	0.24	3.01	-3.68	0.24	4.55	-5.97
CQM (10.0 µg/mL, n=5)	10.19	4.81	1.90	10.01	4.82	0.07
CQA (16.0 µg/mL, n=5)	15.14	5.14	-5.40	14.79	5.03	-7.53
Stability						
Freezing and Defrosting	Mean	CV%	EPR%	Mean	CV%	EPR%
CQB (0.25 µg/mL, n=5)	0.24	3.24	-3.36	0.22	2.38	-12.00
CQA (16.0 µg/mL, n=5)	16.73	3.18	4.56	17.46	3.05	9.13
Post Processing (12°C-24 h)	Mean	CV%	EPR%	Mean	CV%	EPR%
CQB (0.25 µg/mL, n=5)	0.23	9.88	-8.16	0.22	9.96	-12.48
CQA (16.0 µg/mL, n=5)	14.22	1.38	-11.3	14.45	1.47	-9.70
Short term (25°C-4 h)	Mean	CV%	EPR%	Mean	CV%	EPR%
CQB (0.25 µg/mL, n=5)	0.26	0.52	3.04	15.65	3.68	-2.20
CQA (16 µg/mL, n=5)	0.27	5.38	9.78	16.40	3.44	2.49

r: Linear Correlation coefficient; r²: determination coefficient; CV: coefficient of variation, EPR: Relative standard error; LIQ: Lower Limit of Quantification; CQB: Low Concentration Quality Control; CQM: Quality Control of Medium Concentration; CQA: High Concentration Quality Control'; *Recovery=Mean determined by the ratio of analyte processed/analyte in solution × 100, Concentrations of allopurinol and oxypurinol of 0.25 µg/mL (CQB, n=5) e 16.0 µg/mL (CQA, n=5), and acyclovir de 40.0 µg/mL solution (n=10)

Table 1: Confidence limits of the analysis method of allopurinol and oxypurinol in dog plasma. Linearity, recovery, precision, accuracy, and stability values for allopurinol and oxypurinol.

Parameters	Allopurinol	Allopurinol
	Mean ± SD LV dogs	Men ± SD Healthy dogs
Cl/F/Kg (L/h/Kg)	0.35 ± 0.11	-
Vd/F/Kg (L/Kg)	1.37 ± 0.83	1.17 ± 0.07
Tmax (hr)	1.50 ± 0.67	1.9 ± 0.01
Cmax (µg/mL)	6.00 ± 1.78	6.43 ± 0.18
$t_{1/2}$ (hr)	2.81 ± 1.29	2.69 ± 0.14
Kel (1/hr)	0.31 ± 0.17	0.26 ± 0.01
AUC 0-12 h (hr* µg/mL)	29.47 ± 10.58	-

Cl_F: Apparent clearance; Vd: Volume of Distribution; T_{max}: Time to reach C_{max}; C_{max}: Maximum concentration; $t_{1/2}$: Elimination half-life; Kel: Elimination constant; AUC 0-12: Area under the curve in the dose interval (0-12 h)

Table 2: Mean values of the pharmacokinetic parameters corresponding to the allopurinol and oxypurinol analysis in plasma obtained after allopurinol administration in multiple dosage of 10 mg/kg every 12 hours in dogs with naturally acquired visceral leishmaniasis.

of allopurinol. Thus, HPLC-UV is a suitable method to evaluate the pharmacokinetics of allopurinol and oxypurinol in naturally infected dogs under treatment for visceral leishmaniasis.

Acknowledgements

The present study has been financed by the Brazilian agency for research and development "Bahia Research Foundation" - FAPESB (Grant n. PRONEM: 498/2011 - PNE 0002/2011), and by the "Brazilian Coordination for Improvement of Higher Education Personnel" - CAPES (MSc degree scholarship for C. de Jesus). The authors thank Rafaela de Sousa Gonçalves and Gabriela Porfírio Passos for technical assistance.

References

1. Day RO, Birkett DJ, Hicks M, Miners JO, Rahm GG, et al. (1994) New uses for allopurinol. Drugs 48: 339-344.

2. Graham S, Day RO, Wong H, Mclachlan AJ, Bergendal L, et al. (1996) Pharmacodynamics of oxypurinol after administration of allopurinol to healthy subjects. Br J Clin Pharmacol 41: 209-304.

3. Guerra P, Frias J, Ruiz B, Soto A, Carcas A (2001) Bioequivalence of allopurinol and its metabolite oxypurinol in two tablet formulations. J Clin Pharm Ther 26: 113-119.

4. Andrade SF (2008) Manual de Terapêutica Veterinária. terceira edição, Roca, São Paulo, p: 347.

5. Ling GV, Case LC, Nelson H, Harrold DR, Johnson DL, et al. (1997) Pharmacokinetics of allopurinol in Dalmatian dogs. J Vet Pharmacol Therap 20: 134-138.

6. Denerolle P, Bourdoiseau G (1999) Combination Allopurinol and Antimony Treatment versus Antimony Alone and Allopurinol Alone in the Treatment of Canine Leishmaniasis (96 Cases). J Vet Intern Med 13: 413-415.

7. Plevraki K, Koutinas AF, Kaldrymidou H, Roumpies N, Papazoglou LG, et al. (2006) Effects of Allopurinol Treatment on the Progression of Chronic Nephritis in Canine Leishmaniosis (Leishmania infantum). J Vet Inter Med 20: 228-233.

8. Amusategui I (1998) Treatment of canine leishmaniasis: evaluation, characterization and comparison of response to different protocols based on meglumine antimoniate (associated or not with allopurinol).

9. Ginel PJ, Lucena R, López R, Molleda JM (1998) Use of allopurinol for maintenance of remission in dogs with leishmaniasis. J Small Anim Pract 39: 271-274.

10. Nery G (2015) Prospective study of response to a therapeutic protocol for the treatment of canine visceral leishmaniasis: clinical evolution, serology, parasitism and vector infectivity. PhD Thesis in Animal Health. Post-graduate Program in Animal Science in the Tropics, Federal University of Bahia, Brazil, p: 110.

11. Day RO, Graham GG, Hicks M, Mclachlan AJ, Stocker SL, et al. (2007) Clinical pharmacokinetics and pharmacodynamics of allopurinol and oxypurinol. Clin Pharmacokinet 46: 623 -644.

12. Murrel GA, Rapeport WG (1986) Clinical Pharmacokinetics of allopurinol Clin Pharmacokinet. 11: 343-353.

13. Alppelbaum SJ, Mayersohn M, Dorr T, Perrier D (1982) Allopurinol Kinects and Bioavailability. Cancer Chemother Pharmacol 8: 93-98.

14. Sandoz (2014) Sandoz do Brasil Industria Farmacêutica Ltd. Allopurinol. Available from: http://www.anvisa.gov.br/datavisa/fila_bula/frmVisualizarBula.asp?pNuTransacao=7935722014&pIdAnexo=2216412/

15. Bartges JW, Osborne CA, Felice LJ, Koehler LA, Ulrich LK, et al. (1997) Bioavailability and pharmacokinetics of intravenously and orally administered allopurinol in healthy beagles. Am J Vet Res 58: 504-510.

16. Bartges JW, Osborne CA, Felice LJ, Koehler LA, Ulrich LK, et al. (1997) Influence of two diets on pharmacokinetic parameters of allopurinol and oxypurinol in healthy beagles. Am J Vet Res 58: 511-515.

17. Tada H, Fujisaki A, Itoh K, Suzuki T (2003) Facile and rapid high-performance liquid chromatography method for simultaneous determination of allopurinol and oxypurinol in human serum. J Clin Pharm Ther 28: 229-234.

18. Reinders MK, Nijdam LC, Van Roon EN, Movig KL, Jansen TL, et al. (2007) A simple method for quantification of allopurinol and oxypurinol in human serum by high-performance liquid chromatography with UV- detection. J Pharm Biomed Anal 45: 312-317.

19. Rami Reddy VV, Hussain Reddy K (2015) Rapid and sensitive LC-MS/MS method for the analysis of allopurinol in human plasma. Inter J Pharma Bio Sci 6: 1-11.

20. Solano-Gallego L, Koutinas A, Miró G, Cardoso L, Pennisi MG, et al. (2009) Directions for the diagnosis, clinical staging, treatment and prevention of canine leishmaniosis. Vet Parasitol 165: 1-18.

9

Development and Validation of High Performance Liquid Chromatography Tandem Mass Spectrometry (HPLC-MS/MS) Method for Determination of Tenofovir in Small Volumes of Human Plasma

Mulubwa M[1], Rheeders M[1], Du Plessis L[2], Grobler A[2] and Viljoen M[1]*

[1]Centre of Excellence for Pharmaceutical Sciences (Pharmacen), Division of Pharmacology, North-West University, Potchefstroom 2520, South Africa
[2]DST/NWU Preclinical Drug Development Platform (PCDDP), North-West University, Potchefstroom 2520, South Africa

Abstract

Determination of plasma tenofovir (TFV) concentrations in small plasma volumes and in a short time period is most desirable in a clinical setting. The HPLC-MS/MS method was developed and validated for the determination of TFV. Plasma sample volumes of 10 μL were extracted by protein precipitation. Cimetidine, used as internal standard (ISTD) and TFV were separated on a C18 (Phenomenex Kinetex ™ 30 mm × 2.1 mm, 2.6 μm) reversed phase column with a pre-column. The gradient mobile phase consisted of 10 mmol/L ammonium acetate in water and acetonitrile/methanol (50:50, v/v). TFV and ISTD retention times were 0.27 and 0.90 minutes, respectively, with a run time of 2 minutes. Transition of the parent to the product ion of TFV (m/z 288.072→176.038) and ISTD (m/z 253.13→158.987) were monitored in positive ionization mode. Calibration curves were linear (average r^2, 0.9958) over a TFV concentration range of 12.5-600 ng/mL. The mean recovery was 96.9%. Accuracy, inter and intra-assay of three quality controls and lower limit of quantification (12.5 ng/mL) were within the acceptable limit of <15% relative standard error. TFV was stable at studied conditions and neither matrix effect nor significant carry-over was observed. A distinct, reliable and robust method for determination of TFV in a small volume (10 μL) of human plasma was developed, validated and incorporated to determine the plasma TFV levels in 30 HIV-infected women on antiretroviral treatment.

Keywords: Tenofovir plasma levels; Small plasma volumes; Clinical application; Antiretroviral treatment

Introduction

Tenofovir disoproxil fumarate (TDF) is a prodrug of tenofovir diphosphate (TFV), a nucleotide analogue of adenosine monophosphate with potent activity against human immunodeficiency virus (HIV-1 and HIV-2) [1,2]. The absorption of TDF following an oral dose of 300 mg/day is described by first-order kinetics with an absorption rate constant of 1.03 h^{-1}. Its oral bioavailability is approximately 25% in fasting state and is enhanced by a high-fat meal [3,4]. The volume of distribution of TFV in adults is 0.813 L/kg and is best described by a two-compartment model. The concentrations of TFV in the cerebro-spinal fluid are only 5% of plasma concentrations [3,5]. It is minimally bound to human plasma or serum proteins *in vitro*. The drug is not a substrate of the cytochrome P-450 enzyme system but is mainly excreted unchanged in urine. After intracellular uptake, TFV inhibits the activity of HIV-1 reverse transcriptase by competing with the natural substrate deoxyadenosine 5'-triphosphate for incorporation into DNA [1,2]. A dose of 300 mg/day TDF is used in combination with other antiretroviral drugs to manage HIV-1 infection in adults with proven improved efficacy [6]. The maximum plasma TFV concentration ranges from 195.6-306.3 ng/mL and the trough concentrations from 33.5-62.1 ng/mL [7]. Elevated plasma trough TFV concentration (>90 ng/mL) is associated with renal impairment and bone mineral loss [8,9] and this suggests the need of monitoring plasma TFV as a way of managing adverse effects experienced by patients. The monitoring of TFV in plasma, not only does it confirm patients' compliance but also ensure acceptable plasma drug exposure.

To this effect, several LC-MS/MS analytical methods to determine TFV as a single drug or in combination with other antiretroviral drugs in plasma have been published [10-17,20,21,24-26]. The analytical methods differ with the steps involved in sample preparation, the cost of the method and the duration or run time of the assay.

The aim of the present study was to develop and validate HPLC-MS/MS method for the determination of TFV in human plasma.

Sample preparation was improved by using smaller volumes of plasma and run time was reduced while maintaining the desired sensitivity and stability for clinical application.

Materials and Methods

Chemicals and reagents

Tenofovir monohydrate USP ($C_9H_{14}N_5O_4P.H_2O$) reference standard (Figure 1) was obtained from Industrial Analytical (Pty) Ltd, South Africa (CAS Number 206184-49-8). Cimetidine ($C_{10}H_{16}N_6S$) as internal standard (ISTD) (Figure 1) was sourced from Sigma-Aldrich, USA (CAS Number 51481-61-9). HPLC water was obtained from a Milli-Q water purification system (Millipore SAS 67120 Moisheim, France). Ammonium acetate, LC/MS/MS hyper grade acetonitrile and methanol were sourced from Merck (Pty) Ltd, South Africa. Drug-free plasma containing sodium citrate anticoagulant was kindly donated from healthy volunteers and stored at -23°C.

Instrumentation

An Agilent 1290 Infinity HPLC System consisting of binary pump with two identical high pressure (1200 bar) pumps; a two-channel solvent degasser and four-channel inlet solvent selection valve and

---

*Corresponding author: Michelle Viljoen, Centre of Excellence for Pharmaceutical Sciences (Pharmacen), Division of Pharmacology, North-West University, Private Bag X6001, Potchefstroom 2520, South Africa
E-mail: Michelle.Viljoen@nwu.ac.za

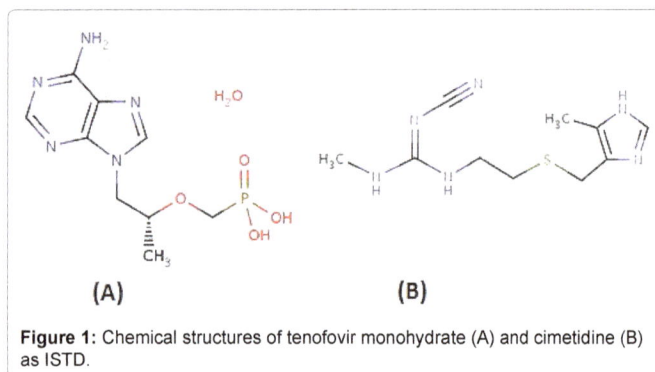

Figure 1: Chemical structures of tenofovir monohydrate (A) and cimetidine (B) as ISTD.

CTC PAL HTx-xt auto sampler with a 20 μL sample loop were used for analysis. Mass spectrometric detection was performed on an AB SCIEX 4000 QTRAP° MS/MS system. Quantitation was performed in multiple reactions monitoring mode (MRM) and Analyst° 1.6 software was used to manage and execute analysis. MultiQuant 3.0 was used to quantify results and create reports.

Chromatographic conditions

Chromatographic separation was carried out on C18 (Phenomenex Kinetex™ 30 mm × 2.1 mm, 2.6 μm) reversed phase column with a pre-column (UHPLC C18, 2.1 mm ID). Column and auto sampler tray temperatures were maintained at 20°C and 19°C, respectively. A gradient was used to elute TFV and ISTD as shown in Table 1. The total chromatographic run time was 2 minutes. Mobile phase A consisted of 10 mmol/L ammonium acetate in water and phase B consisted of methanol: acetonitrile (50:50, v/v).

Mass spectrometric conditions

Optimization of the signal was performed by constant injection of high concentration of TFV and ISTD. The transition of the parent to product ion was studied with the use of a turbo spray ionization source, operating in the positive ionization mode. The transition of parent → production (m/z) for TFV was 288.072→176.038. Dwell time (msec), declustering potential (DP), collision energy (CE) and collision exit potential (CXP) were optimized at 200 msec, 106, 35 and 8 volts, respectively. The parent → product ion (m/z) for ISTD was 253.13→158.987. The dwell time was 200 msec. DP, CE and CXP was 61, 21 and 12 volts, respectively.

Preparation of stock and internal standard solutions

The stock solution was prepared by dissolving 8.5 mg of TFV reference standard in 50 mL of Milli-Q water to a final concentration of 170 μg/mL. Stock quality control (QC) solution was prepared by dissolving 9.1 mg of TFV reference standard in 50 mL of Milli-Q water to a concentration of 182 μg/mL. This solution was then further diluted with Milli-Q water to a final QC concentration of 170 μg/mL. 10.08 mg of ISTD was dissolved in 25 mL of methanol: water (40:60, v/v) to a final concentration of 403.2 μg/mL. This solution was diluted to 10 ng/ml with Milli-Q water on the day of analysis. All solutions were clearly labeled and stored in Eppendorf tubes at -23°C.

Preparation of calibration standards and quality control samples

Calibration standards were prepared by spiking 4235 μL of drug-free plasma with 15 μL of TFV stock solution which was then diluted with drug-free plasma. The nine calibration standard concentrations were 12.5, 25, 50, 100, 200, 300, 400, 500 and 600 ng/mL.

QC samples were prepared at 30, 150 and 450 ng/mL for low quality control (LQC), medium quality control (MQC) and high quality control (HQC), respectively by spiking 4235 μL of drug-free plasma with 15 μL of stock QC solution and then diluting appropriately with drug-free plasma. Calibration standards and quality control samples were stored in Eppendorf tubes at -23°C in 100 μL aliquots. Prior to analysis, all frozen samples from participants, calibration standards and quality control were left to thaw unassisted until equilibrated with room temperature.

Sample pre-treatment

Nine calibration standards and QC plasma samples in Eppendorf tubes were heated in a water bath at 57°C for 30 minutes and were left to equilibrate to room temperature. 25 μL (10 ng/mL) of ISTD stock solution was added to 10 μL of the plasma sample. To precipitate plasma proteins, 100 μL of methanol: acetonitrile (50:50 v/v) was added to the sample. The sample was then vortex mixed for 10 seconds, left in a freezer at -23°C for 10 minutes, vortex mixed for a second time and then centrifuged for 4 minutes at 6000 rpm. 50 μL of the supernatant was transferred into a clean auto sampler vial, 120 μL of Milli-Q water was added and vortex mixed. 20 μL of this solution was injected onto the column for analysis.

The same pre-treatment procedure was performed for all human plasma samples. Heating of plasma samples in a water bath ensured inactivation of HIV.

Method validation

The analytical method was validated for linearity, sensitivity, accuracy, precision, selectivity, carry-over, recovery, matrix effect and stability according to the European Medicines Agency and US Food and Drug Administration guidelines for bio analytical method validation [18,19]. Each analytical validation run comprised nine spiked standards, QC samples, zero blank (with ISTD, without TFV) and blank (without ISTD or TFV) samples.

Clinical application

This method was developed to determine plasma TFV concentrations of 30 HIV-1-infected women in a pilot cross-sectional sub-study within the Prospective Urban and Rural Epidemiology-South Africa study (PURE-SA). Women were all on the first line antiretroviral regimen, taking a TDF dose of 300 mg *nocte*. This investigation was approved by Human Research Ethics Committee (HREC) of North-West University in Potchefstroom (NWU-00016-10-A1 on 12/04/2013). Participants signed the informed consent to have their blood drawn for analysis. Blood was collected in tubes containing sodium citrate anticoagulant between 11-14 hours post TDF dose and plasma was stored at -80°C until analyzed.

Results

Method development

TFV and ISTD were successfully separated on C18 (Phenomenex Kinetex™ 30 mm × 2.1 mm, 2.6 μm) reversed phase column with a

Run time (min)	Flow rate (μL/min)	A (%)	B (%)
0.00	250	95.0	5.0
0.10	250	95.0	5.0
0.60	250	5.0	95.0
1.30	250	5.0	95.0
1.40	250	90.0	10.0
2.00	250	90.0	10.0

Table 1: Chromatography gradient elution.

pre-column. Retention times for TFV and ISTD were 0.27 and 0.90 minutes, respectively. Representative chromatographic profiles of the upper limit of quantification (600 ng/mL) and zero blank samples are presented in Figure 2.

In order to achieve optimal separation of TFV and ISTD, sharp peak shape and shortest possible run-time was used as criteria and we performed chromatographic analysis via gradient elution. Ultra C18 column, 20% carbon load (100 Å pores, 100 mm × 2.1 mm, 5 μm, Restek, Bellefonte) and C18 (Phenomenex Kinetex™ 30 mm × 2.1 mm, 2.6 μm) columns were employed in order to obtain the best separation efficiency. The C18 (Phenomenex Kinetex™ 30 mm × 2.1 mm, 2.6 μm) column provided better peak shape and shorter run-time. The mobile phases tested to provide the best resolution were: acetonitrile/ ammonium (150 mM) formate in water and ammonium acetate (10 mmol/L) in water/ methanol: acetonitrile (50:50, v/v). The latter mobile phase was found more suitable for eluting TFV and ISTD with better resolution at a flow rate of 250 μL/min.

Linearity and sensitivity

Linearity and sensitivity were determined by a nine-point TFV calibration standard curve on three separate days. The curve was constructed (Y-axis) using peak area ratios of chromatograms (TFV peak area/ISTD peak area) versus (X-axis) nominal concentration of TFV over the concentration range of 12.5-600 ng/mL. A linear regression weighting factor of 1/x best described the linearity of the calibration curve with regression coefficient (r^2) ranging from 0.9942-0.9970 with an average of 0.9958. The lower limit of quantification (LLOQ) was 12.5 ng/mL and the signal was more than 5 times the signal of the blank sample (Figure 3) using the FDA/ EMA guidelines [18,19]. Back calculated mean concentrations with percentage relative standard deviation and accuracy are presented in Table 2.

Accuracy and precision

Accuracy and precision for the analyte were determined by analyzing QCs including the lower limit of quantification samples (HQC, MQC, LQC and LLOQ) with the standard calibration curve. Sets of samples were analyzed within a single analytical run (within-run precision) and between different runs (between-run precision) on different days. Table 3 summarizes the accuracy and precision results.

Selectivity and carry-over

Selectivity was demonstrated by analyzing six drug-free plasma samples obtained from six random individual donors in sodium citrate tubes spiked at LLOQ. Negligible interference in one donor's plasma was observed at the TFV retention time (0.27 minutes); blank and TFV peak area counts were 96.909 and 988.236 (9.8%), respectively [18,19]. Chromatograms of the blank and spiked samples from five donors showed no endogenous peaks at TFV retention time or internal standard retention time (0.90 minutes). Hence, the method was selective and unaffected by interference from endogenous components in the matrix.

Carry-over was assessed by injecting blank samples after the upper limit of quantification (600 ng/mL) of the calibration standard in each analytical run. TFV carry-over for three runs was negligible (0.00, 0.00 and 0.286 ng/mL) with average of 0.095 ng/mL (0.76% of LLOQ). This was within the acceptable range of not greater than 20% of the LLOQ. There was no observed carry-over in the ISTD.

Recovery

Extraction recovery of TFV from sodium citrate plasma was determined at 4 concentrations (HQC, MQC, LQC and LLOQ) in triplicate by comparing analyte-internal standard area ratio of samples spiked in plasma with those of spiked in Milli-Q water. The mean recovery of TFV at HQC, MQC, LQC and LLOQ was 99.4 ± 11.0,

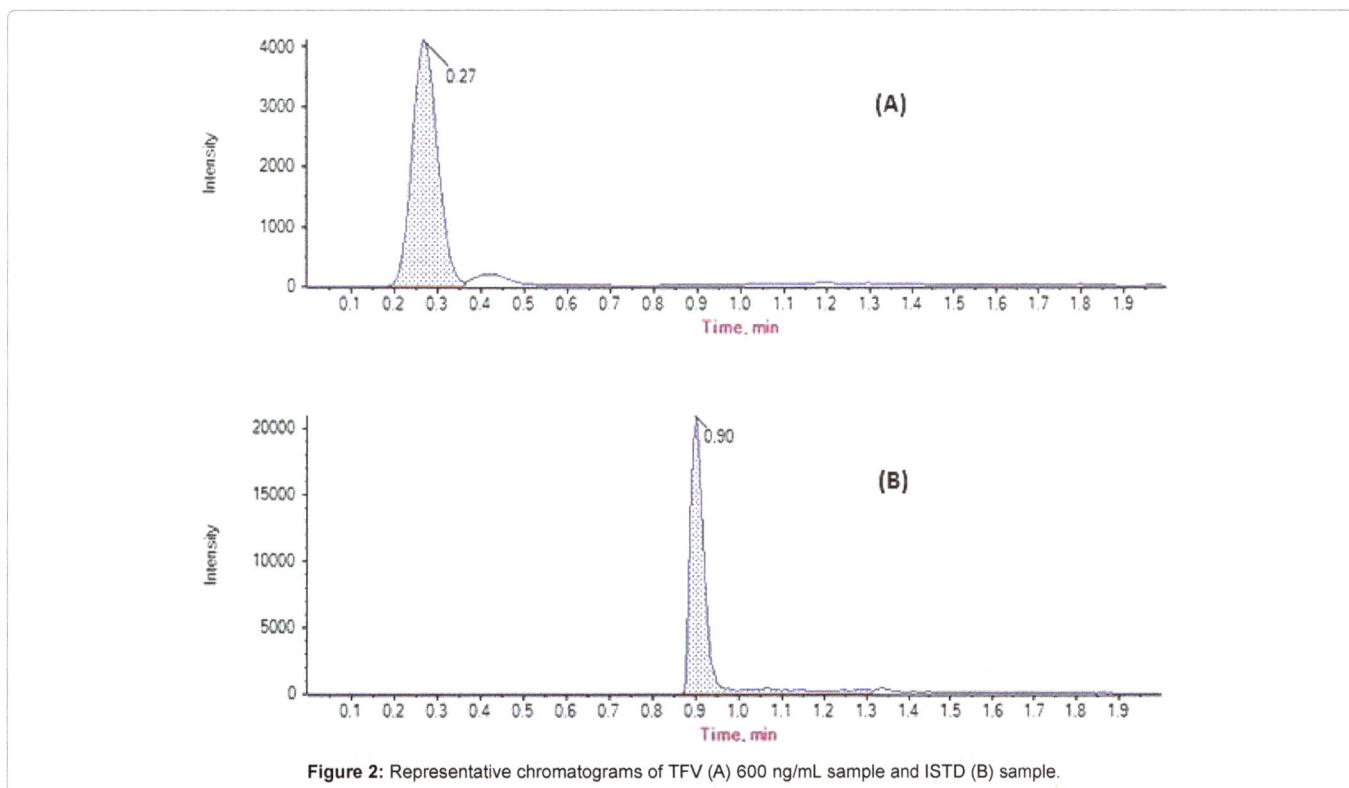

Figure 2: Representative chromatograms of TFV (A) 600 ng/mL sample and ISTD (B) sample.

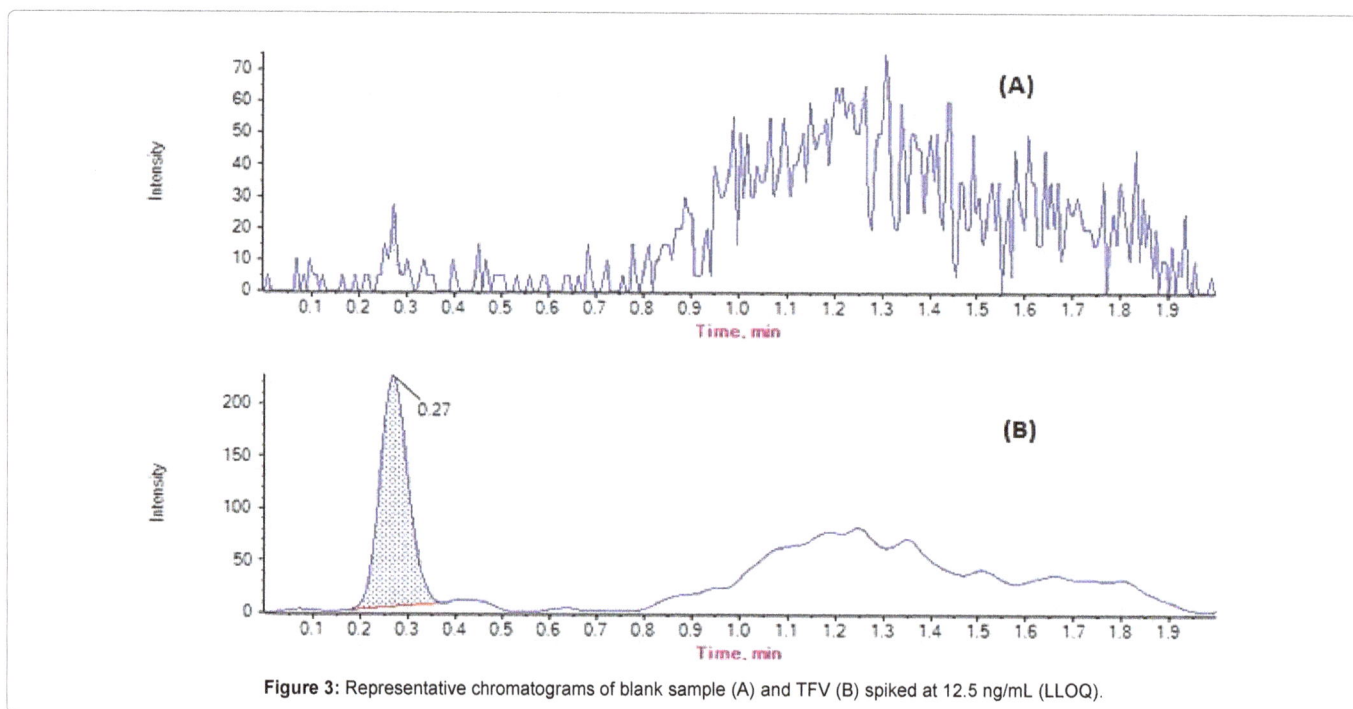

Figure 3: Representative chromatograms of blank sample (A) and TFV (B) spiked at 12.5 ng/mL (LLOQ).

Nominal Conc. (ng/mL)	[a]Calculated Mean Conc. (ng/mL)	[b]%RSD	Accuracy (%)
12.5	13.2	2.0	105.5
25	25.8	4.7	103.4
50	47.4	6.1	94.8
100	94.9	5.8	94.9
200	199.7	13.3	99.9
300	308.0	3.6	102.7
400	388.5	2.9	97.1
500	503.1	6.2	100.6
600	606.9	6.2	101.1

[a]Mean of 3 injections on three separate days; [b]Percent relative standard deviation.

Table 2: Summary of standard calibration curve concentrations.

Nominal conc. (ng/mL)	Calculated conc. (ng/mL)	Accuracy (%)	Precision (%RSD)
Within-run assay			
12.5[a]	12.2	98.2	3.6
30[b]	30.2	100.6	5.9
150[b]	146.4	97.6	7.9
450[b]	453.1	100.7	11.2
Between-run assay			
12.5[c]	11.5	92.0	3.6
30[d]	28.0	93.2	8.8
150[d]	139.0	92.5	14.9
450[d]	399.2	88.7	5.9

[a]n=5, [b]n=6, [c]n=3, [d]n=4.

Table 3: Within-run (intra) and between-run (inter) assay performance for TFV in plasma.

100.1 ± 2.3, 92.8 ± 3.7 and 95.3 ± 2.0%, respectively. The overall mean extraction recovery (96.9%) was high and acceptable.

Matrix effect

Co-elution of some endogenous compounds from the matrix can affect sensitivity, precision and accuracy of the bio-assay. The matrix effect was thus determined by assessing variability in instrument response from six lots of plasma of individual donors. Each plasma lot was spiked with TFV at 30, 150 and 450 ng/mL and analyzed in 6 runs. The overall coefficient of variation (CV) for the calculated concentrations at three levels was less than 15% as presented in Table 4. No significant differences were observed in peak areas of the ISTD.

Stability

Stability of TFV was investigated in plasma spiked at LQC, MQC and HQC. Samples were heated in a water bath at 57°C for 30 minutes and analyses were performed in quadruplicate. The coefficient of variation ranged between 2.6-12.8% and accuracy was between 93.3-99.1%. After a second freeze-thaw cycle at -23°C of the same QC plasma samples, analyses were performed in triplicate and accuracy ranged between 90.3-93% with the coefficient of variation between 5.1-10.0%. The mean concentration of freeze-thaw samples compared to heated samples decreased by 8.0, 2.6 and 5.9% for LQC, MQC and HQC, respectively. Hence, TFV was stable after heating in a water bath and after two freeze-thaw cycles.

Clinical application

This validated assay was successfully applied in measuring TFV in plasma samples of 30 HIV-infected women and investigated the association of TFV plasma levels and its toxicity on renal function and bone metabolism. TFV could not be quantified in 5 participant's plasma samples as these levels were below the LLOQ. The mean concentration in 25 participants was found to be 114.0 ± 78.9 ng/mL. The lowest and highest concentration was 17.2 and 434.2 ng/mL, respectively corresponding to mid-dose concentration [9]. The concentrations were all in the range of 12.5-600 ng/mL. The typical chromatograms of high and low TFV plasma concentrations from two participants are presented in Figure 4.

Discussion

In several published analytical methods where TFV was analyzed in the presence of other antiretroviral drugs [11,12,14-16,20-22,23,24] or alone [10,13,17,25,26], fairly large plasma volumes were used:

	LQC (30 ng/mL)		MQC (150 ng/mL)		HQC (450 ng/mL)	
	Calculated conc. (ng/mL)	Accuracy (%)	Calculated conc. (ng/mL)	Accuracy (%)	Calculated conc. (ng/mL)	Accuracy (%)
1	26.64	88.8	149.31	99.5	466.43	103.7
2	33.16	110.5	133	88.7	405	90.0
3	28.9	96.3	130.52	87.0	403	89.6
4	24.1	80.3	117.73	78.5	331	73.6
5	29.65	98.8	146.4	97.6	444.02	98.7
6	28.42	94.7	121.8	81.2	362.2	80.5
Mean		94.9		88.7		89.3
%CV		10.7		9.6		12.5

Table 4: Matrix effect at LQC, MQC and HQC in six lots of plasma.

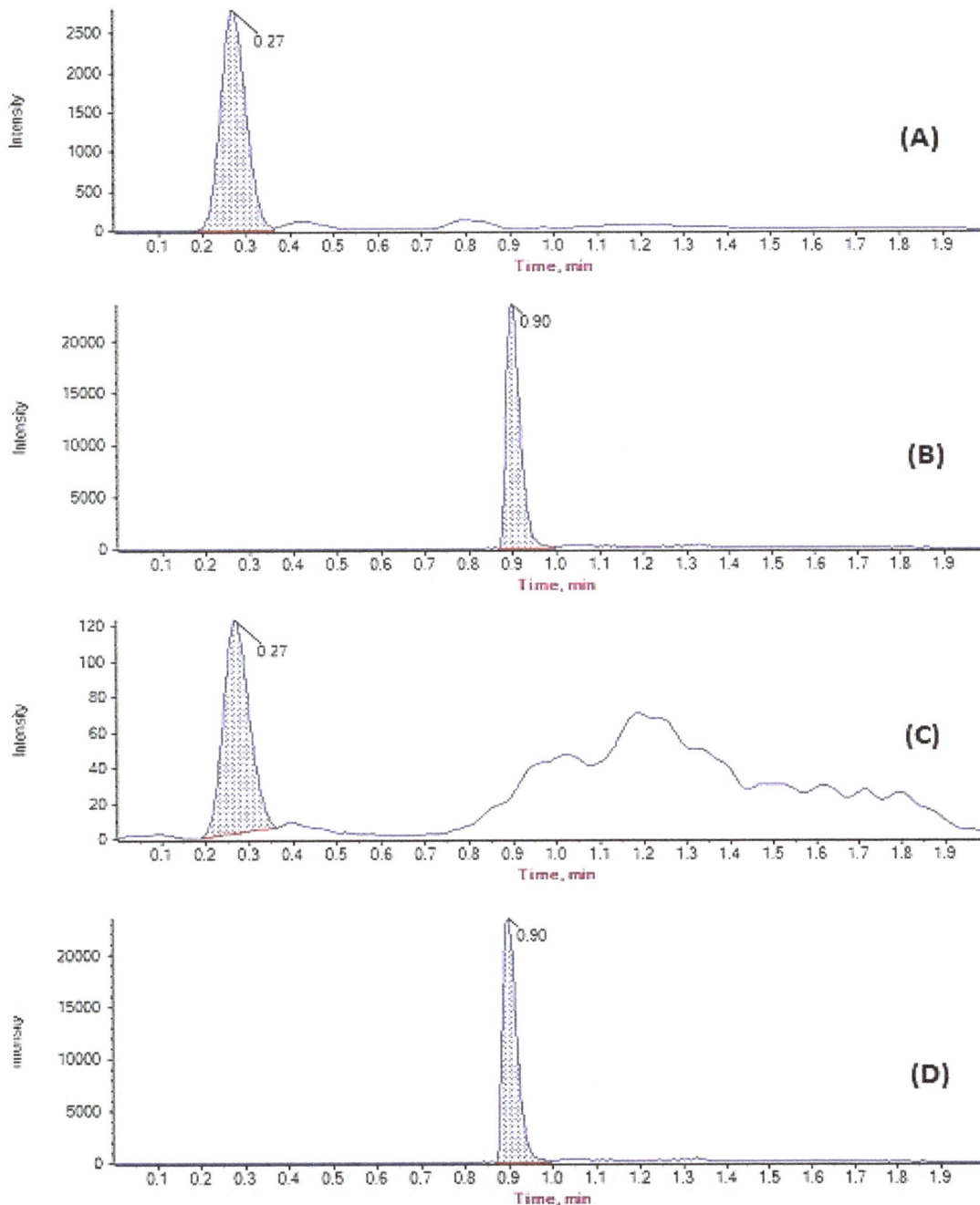

Figure 4: Typical chromatograms of plasma TFV at 434.2 ng/mL (A) and 17.2 ng/mL (C) with their respective ISTD chromatograms (B) and (D) at 10 ng/mL.

volumes between 100-200 µL and 200-1000 µL, respectively for protein precipitation or solid phase extraction were documented. This is the first method, to our knowledge, where only 10 µL of plasma was used for extraction of TFV by protein precipitation, making it preferable and suitable for clinical application. Furthermore, this method is convenient for the emerging interest in therapeutic drug monitoring of TDF in HIV-infected patients [3,8]. A method [13] with a similar run time (1.8 minutes) to this method (2 minutes) was reported in the literature. However, the former required a solid phase extraction procedure which is expensive and more time consuming in sample pre-treatment than the current method.

The method developed is robust, fast, accurate, and sensitive with high extraction recovery and it covered the whole range of expected plasma concentration [9]. A mixture of acetonitrile and methanol (50:50, v/v) made a better precipitating solvent than acetonitrile or methanol alone. Protein precipitation was further enhanced by refrigeration of samples in addition to vortex mixing. Dilution of supernatant with water at neutral pH produced a higher TFV intensity chromatogram compared to acidic or basic water. Similarly, the intensity of TFV chromatograms decreased with decreasing auto sampler temperatures but was optimal at 19°C.

TFV was stable after at least 2 freeze-thaw cycles, which was in agreement with previously published methods [23,24]. Additionally, this method showed that plasma TFV was stable when exposed to a temperature of 57°C in a water bath for 30 minutes.

Conclusion

A simple, fast, reliable and robust HPLC-MS/MS method has been developed and validated for determination of TFV in human plasma. This developed method is suitable for clinical application and was applied for analysis of small volumes (10 µL) of plasma TFV in HIV-1-infected women with desired sensitivity and accuracy.

Acknowledgements

The authors wish to thank all female study participants from PURE-SA cohort and field workers for their co-operation and trust during this sub-study investigation. The Ethics approval of the clinical part of this study within the PURE-SA cohort was approved by North-West University Ethics committee NWU-00016-10-A1 (12 April 2013) and the North West Department of Health (Policy, Planning, Research, Monitoring and Evaluation), Mahikeng on 08 August 2013.

Conflict of Interest

The analytical part of this study was funded by Pharmacen and the PCDDP, North-West University, Potchefstroom Campus, South Africa. The authors declare no conflict of interest.

References

1. Fung HB, Stone EA, Piacenti FJ (2002) Tenofovir disoproxil fumarate: a nucleotide reverse transcriptase inhibitor for the treatment of HIV infection. Clin Ther 24: 1515-1548.

2. Kearney BP, Flaherty JF, Shah J (2004) Tenofovir disoproxil fumarate: clinical pharmacology and pharmacokinetics. Clin Pharmacokinet 43: 595-612.

3. Baheti G, Kiser JJ, Havens PL, Fletcher CV (2011) Plasma and intracellular population pharmacokinetic analysis of tenofovir in HIV-1 infected patients. Antimicrob Agents Chemother 55: 5294-5299.

4. Gagnieu M, Barkil ME, Livrozet J, Cotte L, Miailhes P, et al. (2008) Population pharmacokinetics of tenofovir in AIDS patients. J Clin Pharmacol 48: 1282-1288.

5. Best BM, Letendre SL, Koopmans P, Rossi SS, Clifford DB, et al. (2012) Low cerebrospinal fluid concentrations of the nucleotide HIV reverse transcriptase inhibitor, tenofovir. J Acquir Immune Defic Syndr 59: 376-381.

6. Esser S, Haberl A, Mulcahy F, Gölz J, Lazzarin A, et al. (2011) Efficacy, adherence and tolerability of once daily tenofovir DF-containing antiretroviral therapy in former injecting drug users with HIV-1 receiving opiate treatment: results of a 48-week open-label study. Eur J Med Res 16: 427-436.

7. Pruvost A, Negredo E, Théodoro F, Puig J, Levi M, et al. (2009) Pilot pharmacokinetic study of human immunodeficiency virus-infected patients receiving tenofovir disoproxil fumarate (TDF): investigation of systemic and intracellular interactions between TDF and abacavir, lamivudine, or lopinavir-ritonavir. Antimicrob Agents Chemother 53: 1937-1943.

8. Poizot-Martin I, Solas C, Allemand J, Obry-Roguet V, Pradel V, et al. (2013) Renal impairment in patients receiving a tenofovir-cART regimen: impact of tenofovir trough concentration. J Acquir Immune Defic Syndr 62: 375-380.

9. Rodríguez-Nóvoa S, Labarga P, D'Avolio A, Barreiro P, Albalate M, et al. (2010) Impairment in kidney tubular function in patients receiving tenofovir is associated with higher tenofovir plasma concentrations. AIDS 24: 1064-1066.

10. Bennetto-Hood C, Long MC, Acosta EP (2007) Development of a sensitive and specific liquid chromatography/mass spectrometry method for the determination of tenofovir in human plasma. Rapid Commun Mass Spectrom 21: 2087-2094.

11. Rezk NL, Crutchley RD, Kashuba AD (2005) Simultaneous quantification of emtricitabine and tenofovir in human plasma using high-performance liquid chromatography after solid phase extraction. J Chromatogr B 822: 201-208.

12. Nirogi R, Bhyrapuneni G, Kandikere V, Mudigonda K, Komarneni P, et al. (2009) Simultaneous quantification of a non-nucleoside reverse transcriptase inhibitor efavirenz, a nucleoside reverse transcriptase inhibitor emtricitabine and a nucleotide reverse transcriptase inhibitor tenofovir in plasma by liquid chromatography positive ion electrospray tandem mass spectrometry. Biomed Chromatogr 23: 371-381.

13. Yadav M, Mishra T, Singhal P, Goswami S, Shrivastav PS (2009) Rapid and specific liquid chromatographic tandem mass spectrometric determination of tenofovir in human plasma and its fragmentation study. J Chromatogr Sci 47: 140-148.

14. Le Saux T, Chhun S, Rey E, Launay O, Weiss L, et al. (2008) Quantification of seven nucleoside/nucleotide reverse transcriptase inhibitors in human plasma by high-performance liquid chromatography with tandem mass-spectrometry. J Chromatogr B 865: 81-90.

15. Delahunty T, Bushman L, Robbins B, Fletcher CV (2009) The simultaneous assay of tenofovir and emtricitabine in plasma using LC/MS/MS and isotopically labeled internal standards. J Chromatogr B 877: 1907-1914.

16. Gehrig AK, Mikus G, Haefeli WE, Burhenne J (2007) Electrospray tandem mass spectroscopic characterisation of 18 antiretroviral drugs and simultaneous quantification of 12 antiretrovirals in plasma. Rapid Commun Mass Spectrom 21: 2704-2716.

17. Takahashi M, Kudaka Y, Okumura N, Hirano A, Banno K, et al. (2007) Determination of plasma tenofovir concentrations using a conventional LC-MS method. Biol Pharm Bull 30: 1784-1786.

18. FDA. Guidance for industry bioanalytical method validation.

19. European Medicines Agency. Guideline on bioanalytical method validation.

20. Valluru RK, Reddy B, Sumanth SK, Kumar VP, Kilaru NB (2013) High throughput LC-MS/MS method for simultaneous determination of tenofovir, lamivudine and nevirapine in human plasma. J Chromatogr B 931: 117-126.

21. Gomes NA, Vaidya VV, Pudage A, Joshi SS, Parekh SA (2008) Liquid chromatography–tandem mass spectrometry (LC–MS/MS) method for simultaneous determination of tenofovir and emtricitabine in human plasma and its application to a bioequivalence study. J Pharm Biomed Anal 48: 918-926.

22. Kromdijk W, Pereira S, Rosing H, Mulder J, Beijnen J, et al. (2013) Development and validation of an assay for the simultaneous determination of zidovudine, abacavir, emtricitabine, lamivudine, tenofovir and ribavirin in human plasma using liquid chromatography–tandem mass spectrometry. J Chromatogr B 919: 43-51.

23. Tan R, Yan B, Shang J, Yang J, Huang W, et al. (2012) New solid phase extraction reversed phase high performance liquid chromatography ultraviolet (RP-HPLC-UV) method for simultaneous determination of tenofovir and emtricitabine in Chinese population. Afr J Pharm Pharmacol 6: 1890-1900.

24. Matta MK, Burugula L, Pilli NR, Inamadugu JK, Rao JVLNS (2012) A novel LC-MS/MS method for simultaneous quantification of tenofovir and lamivudine in human plasma and its application to a pharmacokinetic study. Biomed Chromatogr 26: 1202-1209.

25. Delahunty T, Bushman L, Fletcher CV (2006) Sensitive assay for determining plasma tenofovir concentrations by LC/MS/MS. J Chromatogr B 830: 6-12.

26. Barkil ME, Gagnieu M, Guitton J (2007) Relevance of a combined UV and single mass spectrometry detection for the determination of tenofovir in human plasma by HPLC in therapeutic drug monitoring. J Chromatogr B 854: 192-197.

10

A Targeted Metabolomics Assay to Measure Eight Purines in the Diet of Common Bottlenose Dolphins, *Tursiops truncatus*

Ardente AJ[1], Garrett TJ[2], Wells RS[1,3], Walsh M[1], Smith CR[4], Colee J[5] and Hill RC*

[1]College of Veterinary Medicine, University of Florida, Gainesville, FL, USA
[2]Southeast Center for Integrated Metabolomics, Mass Spectrometry Core Laboratory, University of Florida, Gainesville, FL, USA
[3]Chicago Zoological Society's Sarasota Dolphin Research Program, C/O Mote Marine Laboratory, Sarasota, FL, USA
[4]National Marine Mammal Foundation, San Diego, CA, USA
[5]Department of Statistics, Institute of Food and Agricultural Sciences, University of Florida, Gainesville, FL, USA

Abstract

Bottlenose dolphins managed under human care, human beings and Dalmatian dogs are prone to forming urate uroliths. Limiting dietary purine intake limits urate urolith formation in people and dogs because purines are metabolized to uric acid, which is excreted in urine. Managed dolphins develop ammonium urate nephroliths, whereas free-ranging dolphins do not. Free-ranging dolphins consume live fish, whereas managed dolphins consume different species that have been stored frozen and thawed. Differences in the purine content of fish consumed by dolphins under human care versus in the wild may be responsible for the difference in urolith prevalence. Commercially available purine assays measure only four purines, but reported changes in purines during frozen storage suggest that a wider range of metabolites should be measured when comparing fresh and stored fish. A method using high performance liquid chromatography with tandem mass spectrometry was developed to quantify eight purine metabolites in whole fish and squid commonly consumed by dolphins. The coefficient of variation within and among days was sometimes high for purines present in small amounts but was acceptable (≤ 25%) for guanine, hypoxanthine, and inosine, which were present in high concentrations. This expanded assay identified a total purine content up to 2.5 times greater than the total that would be quantified if only four purines were measured. Assuming additional purines are absorbed, these results suggest that additional purine metabolites should be measured to better understand the associated risk when fish or other purine-rich foods are consumed by people or animals prone to developing uroliths.

Keywords: Kidney stones; LC-MS/MS; IMP; Inosine; Hypoxanthine; Urate; Uric acid; Fish; Purines

Abbreviations

DNA: Deoxyribonucleic acid; RNA: Ribonucleic acid; ATP: Adenine triphosphate; GTP: Guanine triphosphate; AMP: Adenine 5'-monophosphate; IMP: Inosine 5'-monophosphate; LC: Liquid chromatography; HPLC: High performance liquid chromatography; MS/MS: Tandem mass spectrometry; HESI: Heated electrospray ionization; NH_4: Ammonium; NH_4OH: Ammonium hydroxide; MS 222: Tricaine methanesulfonate; TP4: Total purine four metabolite assay; TP8: Total purine eight metabolite assay.

Introduction

Purines contribute to the structure of deoxyribonucleic acid (DNA) and ribonucleic acid (RNA), adenine triphosphate (ATP) and guanine triphosphate (GTP). Purines are made by the body, salvaged and recycled, or absorbed from food. They are metabolized to uric acid, which is excreted in the urine (Figure 1) [1-4].

Dietary purines are found in high concentrations in organ meat and seafood [1,5]. Consumption of a purine-rich diet increases the production and excretion of uric acid. When the concentration of uric acid in the urine reaches a threshold, uric acid precipitates out of solution and can aggregate to form urate-based stones in the urinary tract. These urate uroliths can be composed of uric acid, ammonium urate, monosodium urate, sodium calcium urate, or potassium urate. Human beings primarily develop uric acid uroliths, and the prevalence of these uroliths has increased internationally as a consequence of increased consumption of meat [6,7]. Individuals living in Taiwan, for example, experience a high occurrence of uric acid stones as a consequence of consuming purine-rich diets that include seafood

and specifically grass shrimp [8]. Some Dalmatian dogs have a genetic predisposition to develop ammonium urate uroliths because of altered purine metabolism. For both species, consumption of a purine-restricted diet minimizes uric acid excretion and the risk of forming urate uroliths.

Foods are currently classified according to their total purine content, defined as the sum of the four nucleobases, adenine, guanine, hypoxanthine, and xanthine [1,8]. However, individual metabolites have different propensities for causing hyperuricosuria and individual metabolite concentrations may vary widely among purine-rich foods [9-12]. Human beings, for example, fed a diet supplemented with adenine and hypoxanthine excreted greater urinary uric acid concentrations than when fed a diet supplemented with guanine and xanthine [13]. Thus, it may be more important to describe the individual purine metabolite composition of food rather than total purine content, when determining a food's uricogenic potential.

It also may not be sufficient to describe just the nucleobase content of foods. Additional metabolites, including nucleotides

*Corresponding author: Hill RC, Small Animal Clinical Sciences, College of Veterinary Medicine, University of Florida, 2015 SW 16th Ave., Gainesville, FL 32608, USA, E-mail: hillr@ufl.edu

adenine 5'-monophosphate (AMP), inosine 5'-monophosphate (IMP) and the nucleoside inosine (Figure 1) are absorbed by the mammalian gastrointestinal tract and may also increase urinary uric acid concentrations [13-15]. Furthermore, the concentrations of purine metabolites are affected by food source and storage conditions [8,16,17]. For example, hypoxanthine concentrations progressively increase over time in pork stored at 2°C, but this effect is delayed at lower storage temperatures [14]. Also, IMP, hypoxanthine, xanthine, and inosine concentrations in fish fileted for human consumption have been demonstrated to vary with water temperature where fish were caught, cooking methods, storage methods and duration, and among species [8,18-20]. Thus, it would seem important to measure more purine metabolites than the four nucleobases that are usually measured and to report individual concentrations, when considering a food items' uricogenic potential.

This is particularly relevant when considering purines in whole fish and squid commonly consumed by dolphins. Common bottlenose dolphins (*Tursiops truncatus*) managed under human care develop ammonium urate kidney stones, but these stones rarely occur in free-ranging dolphins [21,22]. Why ammonium urate nephroliths form in managed dolphins is unclear, but the purine-rich whole fish diet may contribute to nephrolith development because the diets of free-ranging and managed bottlenose dolphins differ [5,23].

The total purine content of the fish species consumed by dolphins has not been previously reported. The purine content of the fillets of some fish species has been measured using high performance liquid chromatography (HPLC) with ultraviolet detection, but the authors are aware of only one study that has measured purines in a whole fish (gilted sea bream) [24]. Fish organs, including the liver and skin, are reported to be rich in purines; therefore, purines measured only in the muscle of fish may underestimate total purine content of whole fish [5,25].

We sought, therefore, to develop an expanded assay using HPLC with tandem mass spectrometry (MS/MS) that would accurately identify and quantify a broader range of purine metabolites in both fresh frozen fish commonly consumed by free-ranging dolphins and in stored frozen and thawed fish and squid commonly consumed by managed dolphins. We also hypothesized that it would not be possible to predict the total purine content obtained with the new assay using an assay that only measures four nucleobases. In particular, we suspected that the concentration of some of the additional metabolites would be lower and hypoxanthine concentrations would be higher for the frozen stored and thawed species fed to managed dolphins.

Materials and Methods

Chemicals

The stock solutions used for the mobile phase and for purine standards were prepared freshly at least monthly. The water used for mixing all solutions was HPLC-grade. These stock solutions included: 0.1% acetic acid (glacial) in water, 10% formic acid in water, 10 mM ammonium (NH_4) acetate in water, and 2 mM ammonium hydroxide (NH_4OH) in water. All chemicals were purchased from Fisher Scientific Company, LLC (Suwanee, GA 30024).

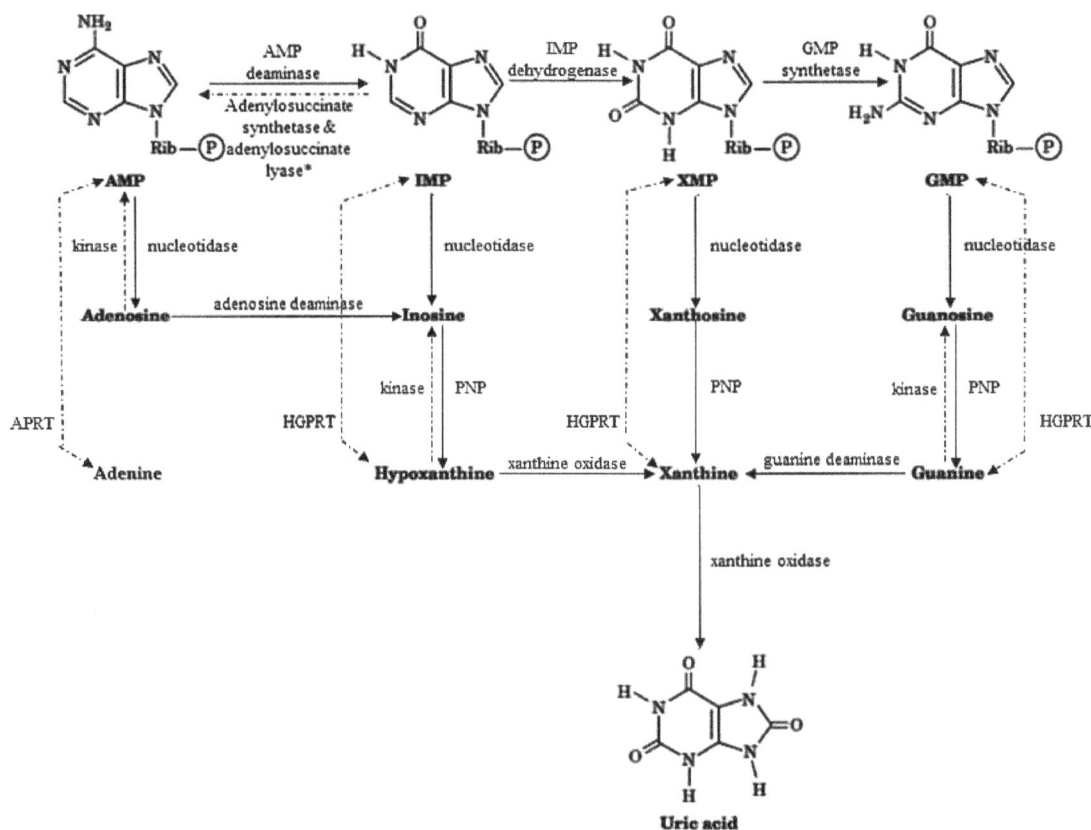

Figure 1: Synthetic degradation and salvage pathways of purines compiled from several sources. Dotted lines represent salvage pathways.

AMP (Adenine monophosphate); IMP (Inosine monophosphate); XMP (xanthine monophosphate); GMP (Guanine monophosphate); APRT (Adenosine phosphoribosyl transferase); HGPRT (Hypoxanthine-guanine phosphoribosyl transferase); PNP (Purine nucleoside phosphorylase)

Individual standard solutions were freshly prepared before each run. Purine metabolite solubility is highly pH dependent. Thus, AMP, IMP, adenine, and inosine were dissolved in 10 mM $NH_4CH_3CO_2$ (pH 6.5), whereas uric acid and xanthine were dissolved individually and hypoxanthine and guanine were dissolved together in 2 mM NH_4OH (pH 12, Sigma-Aldrich, St. Louis, MO). Each solution was then sonicated in a heated (35°C) water bath for 10-15 minutes to achieve complete dissolution. The individual standard purine solutions in $NH_4CH_3CO_2$ were combined to generate one mixed standard solution with final concentrations of 0.01 mM for adenine, 0.02 mM for IMP, 0.05 mM for AMP, and 0.15 mM for inosine. Individual standard solutions dissolved in NH_4OH were also combined to generate another mixed standard solution with final concentrations of hypoxanthine and guanine at 0.23 mM, xanthine at 0.22 mM, and uric acid at 0.01 mM. An internal standard, $^{15}N_2$-xanthine (Cambridge Isotopes Laboratory, Inc., Tewksbury, MA 01876), was mixed with 2 mM NH_4OH to achieve a concentration of 1.3 mM. The final concentrations of the individual metabolites in the combined solutions were half their original starting concentration.

Fish sampling and processing

Fish samples were collected by the Chicago Zoological Society's Sarasota Dolphin Research Program under the Mote Marine Laboratory and University of Florida (UF) Institutional Animal Care and Use Committee approvals.

Eight fish species commonly consumed by free-ranging bottlenose dolphins and six fish species and one squid species commonly fed to dolphins under human care were analyzed for purine metabolite content (Table 1) [26-28]. Free-ranging fish samples were caught during the months of May to September 2013 from the waters off the FL west coast by local fisherman using a rod and reel, crab trap, or cast net, or with a purse-seine net during fish surveys conducted by the SDRP team in Sarasota Bay. To mimic the rapid death of fish consumed by dolphins in the wild as closely as possible, fish were euthanized humanely by immersion in 500 ppm tricaine methanesulfonate (MS-222, Western Chemical, Ferndale, WA 98248) in sea water [29,30]. Death was confirmed by cessation of opercula movement for 10 minutes, and then fish were weighed, length was measured, and samples of fish were individually bagged, placed into a cooler of dry ice and transported to the UF nutrition laboratory where fish were stored at -80°C until further processing.

Boxes containing six fish species and one species of squid ('managed species') commonly fed to dolphins were supplied by two facilities where bottlenose dolphins are managed under human care (Table 2). Fish and squid were caught during one commercial fishing season, frozen stored at -18°C for 6 to 9 months, and then shipped overnight on dry ice from the dolphin management facilities to UF. Upon arrival, fish were stored at -20°C until further processing.

Five separate samples of each species were analyzed. To provide sufficient material to perform all the analyses on every sample, a minimum of 2 individual fish (or squid) were included in each sample; however, the number of individual fish (or squid) included in each sample varied depending on the size of the species so that each sample of smaller species contained more individuals than samples of large species. The five samples of each species were individually ground using commercial meat grinders with 4.5 and 10 mm plates (Biro 6642, Marblehead, OH 43440, and 1.5 HP, LEM Products, West Chester, OH 45011).

Free-ranging fish species were thawed the minimum amount needed to allow grinding, whereas managed diet fish species were thawed more completely using the standard operating procedure of one dolphin management facility. Free-ranging fish species were air thawed in a temperature controlled cold room (11-12°C) for approximately 1 hr., until fish thawed to a firm, slightly malleable texture. Managed diet fish were removed from the cardboard boxes, maintained wrapped in plastic, and air thawed in the cold room for approximately 20 hrs. Fish were then removed from the plastic and rinsed with cold water (approximately 16°C). Both minimally and well-thawed fish were then transported to the grinder in a cooler containing

Table 1: Fish and squid species commonly consumed by free-ranging and managed bottlenose dolphins.

Free-ranging diet species [26,27]	Managed diet species [28]
Pinfish (*Lagodon rhomboids*)	Pacific herring (*Clupea pallasii*)
Striped mullet (*Mugil cephalus*)	Atlantic herring (*Clupea harengus*)
Sheepshead (*Archosargus probatocephalus*)	Icelandic capelin (*Mallotus villosus*)
Ladyfish (*Elops saurus*)	Canadian capelin (*Mallotus villosus*)
Pigfish (*Orthopristis chrysoptera*)	Pacific mackerel (*Scomber japonicus*)
Spot croaker (*Leiostomus xanthurus*)	Pacific sardine (*Sardinops sagax*)
Spotted sea trout (*Cynoscion nebulosus*)	West coast Loligo squid (*Loligo opalescens*)
Gulf toadfish (*Opsanus beta*)	

Table 2: Columns tested for separation of purines and observed results.

Column (Manufacturer)	Chromatographic results
Acquity UPLC HSS T3 1.8 µm, 150 × 2.1 mm, with HSS T3 1.8 µm Vanguard (Waters Corporation, Milford, MA 01757)	Failed after ~ 800 injections
Symmetry C18 3.5 µm, 150 × 2.1 mm (Waters Corporation)	Poor retention
Excel C18-PFP 2 µm, 100 × 2.1 mm (ACE, Aberdeen, Scotland)	Poor separation
Excel C18-amide 2 µm 100 × 2.1 mm (ACE)	Poor elution AMP*, IMP*
Excel Ultra Core Super Phenyl Hexyl 2.5 µm, 75 × 2.1 mm (ACE)	Poor metabolite retention
HALO-PFP 2.6 µm, 150 × 2.1 mm (ACE)	Poor elution adenine
HALO-HILIC 2.6 µm, 150 × 2.1 mm (ACE)	Poor elution AMP*, IMP*
Prodigy ODS 3 µm, 100 × 2 mm (Phenomenex, Torrance, CA 90501)	Low sensitivity, except for AMP*, IMP*
Luna PFP(2) 3 µm, 150 × 3.0 mm† (Phenomenex)	Sufficient retention and separation, good sensitivity

*AMP (Adenine 5'-monophosphate); IMP (Inosine 5'-monophosphate); †Column selected for purine metabolite separation

ice. Grinder equipment was thoroughly rinsed between each sample to prevent contamination between samples. Ground samples were homogenized by hand and stored in sample bags (Whirl-Pak® bags, Nasco, Fort Atkinson, WI 53538) at -80°C until shipped overnight on dry ice to each laboratory for analysis.

Fish extraction

For each fish purine extraction, a 25 g sample bag was removed from the -80°C freezer where it had been stored and immersed in water at room temperature for 30-60 minutes until thawed. The aqueous extraction method published by Clariana et al. was followed, but required several modifications [17]. Thawed fish tissue was hand mixed by massaging the sample bag, and 2 g of mixed sample was transferred into a beaker with 20 mL of ultra-pure water and 500 µL of the internal standard solution. Following homogenization, sonication, heating, and cooling, as described by Clariana, et al., the extract supernatant was filtered through Whatman #1 filter paper (Whatman Inc, Clifton, NJ 07014) with a glass funnel. In all subsequent steps, extract and solvent volumes were doubled to provide a greater final volume for standard additions to be performed. Thus, a 10 mL aliquot of the filtrate was combined with 10 mL of HPLC-grade hexanes (Fisher) and centrifuged at 6,500 rpm for 7 minutes at 20°C. A 4 mL aliquot from the bottom layer of fluid was removed and combined with 4 mL of HPLC-grade methanol (Fisher), 4 mL of HPLC-grade acetone (Fisher), and 80 µL of 10% formic acid in water. Samples were centrifuged at 18,100 rpm for 17 minutes at 15°C. Five aliquots of 1,500 µL each were pipetted into 5 mL conical snap-top tubes (Eppendorf, Fisher).

Standard additions were used to quantify individual purine concentrations in the fish extract. Thus, for each fish sample, the first 1,500 µL aliquot was considered the blank fish extract sample. To the remaining 4 aliquots, the $NH_4CH_3CO_2$ mixed purine standard solution and the NH_4OH mixed purine standard solution were added in increasing quantities to provide standard additions of 2x, 4x, 6x, and 8x purine concentrations. Conical tubes were centrifuged at 3,000 rpm for 5 minutes at room temperature and then dried while heating to 35°C under a stream of nitrogen gas (MULTIVAP, Organomation, Berlin, MA 01503) for approximately 5-6 hrs.

The dried samples were reconstituted with 500 µL of 10 mM $NH_4CH_3CO_2$. Conical tubes were centrifuged at 4,000 rpm for 15 minutes. Following centrifugation, if the extract had a distinct pellet at the bottom of the conical tube, supernatant was transferred to a 2 mL LC vial (Thermo Scientific, Waltham, MA 02451) for HPLC analysis. If the sample had floating particulate matter, the extract was re-mixed, transferred to a 2 mL microcentrifuge 0.22 µm nylon filter tube (Corning Incorporated, Corning, NY 14831), and centrifuged at 11,884 rpm for 5 minutes. The supernatant was then transferred to a 2 mL LC vial for HPLC-MS/MS analysis.

Chromatographic conditions

Purine metabolite separation was carried out with an auto-sampler (Accela Open, Thermo) and HPLC system (Accela 1250, Thermo). Several HPLC columns were tested (Table 2) under various conditions, manipulating column and sample stack temperature and using several mobile phase solvent combinations, including 10 mM $NH_4CH_3CO_2$ in water (pH adjusted with acetic acid from 4-6), 10 mM $NH_4CH_3CO_2$ in 90:10 acetonitrile in water, 100% acetonitrile, 90:10 acetonitrile in water, 0.1% acetonitrile in water, 0.1% formic acid in water, 0.1% acetic acid in water, 0.1% NH_4OH in water, and 100% methanol.

The column and compartment were maintained at ambient temperature. The injection volume was 10 µL with a flow rate of 500 µL/min. The elution conditions for separation of AMP, IMP, inosine, adenine, guanine, hypoxanthine, xanthine, and uric acid began with 97% A (0.1% acetic acid, pH 3.5) and 3% B (methanol) for 2 minutes, gradually decreased to 90% A and increased to 10% B from 2 to 4 minutes, 70% A and 30% B from 4 to 5.5 minutes, 50% A and 50% B from 6 to 8 minutes, with a final rapid return back to the original conditions at 8.5 minutes. The column was maintained under these conditions for re-equilibration for a further 6.5 minutes. In addition to testing different columns, several mobile phases were tested to determine which combination of solvents and timing of transitions achieved the best separation of the metabolites with similar masses. The maximum number of injections per day, including duplicate sample injections and blanks, was 84; therefore, the maximum daily run time was approximately 22.5 hrs. To optimize the lifespan of the column, particularly considering the potential for particulate build-up on the column, the column was flushed at the end of every long run with methanol at 500 µL/min for approximately 15-20 minutes. Additionally, the guard column was changed after approximately 400 injections. For column storage between runs, the column was flushed at 500 µL/min with 65% acetonitrile and 35% water for at least 10 column volumes.

Mass spectrometry analysis

Purine metabolites were identified by mass and retention time with a triple quadrupole mass spectrometer (TSQ Quantum Access Max, Thermo) with heated electrospray ionization (HESI). The HESI source was operated in negative-ion mode with the following settings: spray voltage 4000 V, vaporizer temperature 250°C, sheath gas pressure 50 arb, ion sweep gas pressure 0.0 arb, auxillary gas pressure 10 arb, and capillary temperature 300°C. Compounds were identified by comparing the retention time and selected reaction monitoring (SRM) pairs (Table 3) during one scan event with a 20 msec scan time, as well as through the addition of purine standards. Data was processed using computer software (Xcalibur Quan Browser, Thermo).

Purine standard stock solution stability

The stability of purine standard stock solutions was assessed by first preparing combined stock solutions as described above and spiking the solutions into 1000 µL of 10 mM $NH_4CH_3CO_2$ to achieve equivalent 2x,

Table 3: Purine metabolite mass spectrometer identification parameters.

Metabolite	Precursor mass (g/mol)	Product mass (g/mol)	Collision energy (eV*)	Tube lens (V)
Adenine	134.00	107.00	24	74
Hypoxanthine	135.05	65.34	29	74
		95.22	18	74
Guanine	150.00	108.00	18	74
Xanthine	151.02	80.28	26	78
		108.16	18	78
$^{15}N_2$ - Xanthine	152.95	80.95	32	66
		109.20	18	66
Uric acid	167.00	124.24	16	65
Inosine	267.02	108.21	41	100
		135.15	25	100
AMP*	345.99	79.25	42	98
		134.23	39	98
IMP*	346.96	79.31	37	90
		135.13	34	90

*eV (Electron volts); AMP (Adenine 5'-monophosphate); IMP (Inosine 5'-monophosphate)

6x, and 8x standard addition concentrations. Samples were analyzed in triplicate immediately then stored at -80°C for 24 hrs. Samples were then permitted to thaw for at least 1 hr. until approximately ambient temperature, vortexed, and re-analyzed in triplicate alongside freshly prepared standard solutions. This procedure was repeated once more for a total of 2 freeze/thaw cycles at 24 and 48 hrs. In addition, stability over 24 hr while at ambient temperature was assessed for standard stock solutions and fresh prepared Pacific herring and mullet samples, prepared using the method described above.

Method validation

Repeatability was assessed on pooled samples of Pacific herring and striped mullet that were divided into separate bags and frozen at -80°C. Purines were analyzed in these samples over 4 separate days to determine between-day variability. Additionally, 4 extracts were prepared from one sample bag for each fish species and analyzed on the same day to determine within-day variability.

Purine concentration quantification

For each sample, the ratio of peak area relative to internal standard was regressed against the concentration of purine metabolite after standard additions. The purine metabolite concentration (mmol/L) in the fish extract was obtained from the absolute value of the x-intercept of the regression line. Total purine content (mmol/L) was calculated either as the sum of adenine, guanine, hypoxanthine, and xanthine, representing the commercially available four-metabolite assay (TP4), or as the sum of those four metabolites plus uric acid, AMP, IMP, and inosine, representing the total obtained using this expanded eight-metabolite assay (TP8).

Statistical analysis

Statistical comparisons were performed using statistical software (SAS° for Windows, version 9.4, Cary, NC, 27513). The coefficient of variation for within and between days (n=4) was calculated for individual samples of Pacific herring and striped mullet. The ratios of TP4 to TP8 within species were assessed visually and were found to be normally distributed using the Shapiro-Wilk test. The correlations between TP4 and TP8, and between the logarithms of TP4 and TP8, were compared between groups of species using a general linear model procedure (SAS proc glm). The ratio of TP8:TP4 and the ratio of inosine to hypoxanthine were compared among fish species' nested within species group using a general linear model (SAS glimmix) and post-hoc comparisons of least square means were performed with a Tukey-Kramer correction applied.

The primary endpoint was to determine whether there is a 50% increase in the concentration of hypoxanthine and other purines during frozen storage. Based on previous reports of the variability in concentrations of hypoxanthine and other purines in filleted fish during storage, comparing five samples of each species gave an 80% power to detect a 50% change in hypoxanthine concentration with a type I error of 0.05 [8,31,32].

Results

Purine metabolites were separated by mass and retention time using this HPLC/MS-MS method (Figure 2). Several of the metabolites were of similar mass: adenine and hypoxanthine, guanine and xanthine, and AMP and IMP. The reverse phase 100 Å, 150 mm × 3.0 mm column (Luna 5 μm PFP(2), Phenomenex, Torrance, CA 90501) with a guard column (SecurityGuard for PFP HPLC, Phenomenex) provided the

Figure 2: Chromatographic separation of eight purine metabolites in striped mullet with the addition of the 8x concentration standard. Time (minutes) and relative abundance are represented along the x-axis and y-axis, respectively. Each metabolite is labeled with purine name, precursor mass, and product mass(es). The retention time (RT) and the area under the curve (AA) are noted above each metabolite peak. AMP (Adenine 5'-monophosphate); IMP (Inosine 5'-monophosphate)

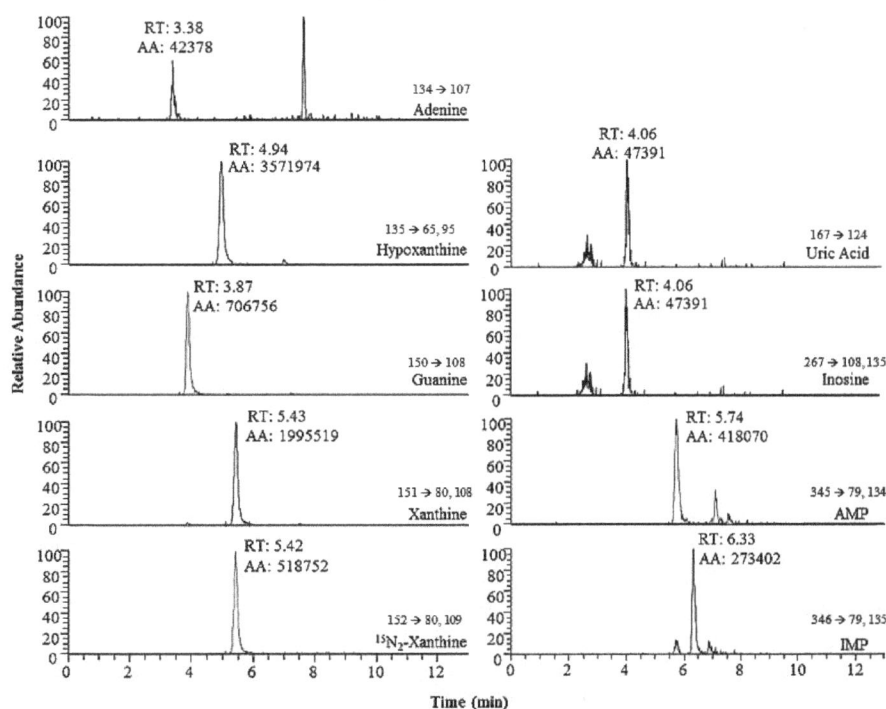

best metabolite separation and most stable retention times when used with the method conditions described above. This column also had sufficient lifespan (~ 800 injections) for fish tissue purine identification. All other columns did not achieve desired results (Table 2). Standard additions verified the location of each metabolite, particularly when concentrations in the fish extract were low. Each metabolite concentration in the fish extract was quantified as the x-intercept of the regression of concentration after standard additions with area ratio relative to that of the internal standard (Figure 3).

There was a 15-30% decrease in signal for all metabolites after the first 24 hr. freeze and thaw cycle. There was no signal for adenine, AMP, and IMP following the second freeze/thaw cycle and the signal for the other metabolites decreased by an average of 25%. During a single 24 hr. run, the signal for neat standards decreased by up to 90% for adenine, between 20-40% for AMP, IMP, and uric acid, and between 5-15% for all other metabolites. The decrease in signal for metabolites in the fish extract was less than 20% over a 24 hr. run.

The coefficient of variation, measured both within and between days, for the Pacific herring and mullet samples were much greater for metabolites with minimal measured concentrations, like AMP, adenine, IMP, but was mostly below 20% for metabolites with greater measured concentrations, like guanine and hypoxanthine (Table 4). The ratio of TP8:TP4 differed among fish species and between fresh frozen free-ranging diet species and managed species which had been frozen stored for several months and then thawed ($p \leq 0.0001$; Figure 4). The slope of linear correlations of TP8 to TP4 was steeper for fresh frozen fish commonly fed to free-ranging dolphins than in stored and thawed species commonly fed to managed dolphins (p=0.01). Regression of the logarithm of TP8 with the logarithm of TP4 improved the fit, and the logTP8:logTP4 regression slopes were almost identical for both groups of fish (0.963 and 0.969, respectively, Figure 5). Values for TP8 were on average 24% higher relative to $TP4^{0.966}$ in free-ranging diet species than in managed diet species but variation was substantial. The difference between TP8 and TP4 was largely due to a change in the ratio of inosine to hypoxanthine, which was on average greater for the free-ranging species (3.70) than for managed species (1.67; $p \leq 0.0001$).

Discussion

This study describes a new method that quantified eight purine metabolites contained in whole fish and a squid commonly consumed by bottlenose dolphins. The four metabolites in greatest concentration were guanine, hypoxanthine, xanthine, and inosine, and the method provided satisfactory, repeatable results for measuring the concentrations of those purines. Inosine is not measured by the four-nucleobase commercial assay, so the whole fish total purine content would be underestimated if only the usual four nucleobases, adenine, guanine, hypoxanthine, and xanthine, were measured.

The total purine content from eight metabolites cannot be reliably predicted by measuring just four metabolites because the ratio of the total obtained from eight versus four metabolites varied widely among species. Furthermore, the ratio of the total purine content measured by eight versus four metabolites differed between free-ranging and managed diet species possibly because of inherent species and handling differences. For example, inosine and IMP degrade to hypoxanthine during frozen storage and storage conditions can markedly affect purine degradation rates [16,33]. In this study, fresh frozen fish had a greater inosine to hypoxanthine ratio than frozen, stored, and thawed fish. This likely represents the degradation of inosine to hypoxanthine during frozen storage for managed diet species, whereas degradation

was probably minimal for the free-ranging fish species that were quickly frozen at -80°C.

These findings confirm our initial hypothesis that it is important to measure more than just four purine metabolites when comparing the purine intake of managed and free-ranging bottlenose dolphins. All of the metabolites measured are absorbed through the mammalian intestinal tract; however, the uricogenic importance may not be similar for all metabolites. For example, the uricogenic potential of inosine is not well-described. In rats, it is suspected that intestinal absorption of inosine is saturable, but the threshold for saturation has not been established [34]. This may be important because even though fish concentrations of inosine are large when compared to other purine metabolites, limited absorption may diminish its impact on urine uric acid excretion. Nevertheless, very little is known about purine metabolism in dolphins. This is the first attempt to quantify the purine content of the diet of dolphins, and the relationship between diet and uric acid excretion has not been established [21,22]. At this time, therefore, all purine metabolites should be considered equally important in assessing the dolphin diet.

Measuring more purines with this assay may also be relevant for human medicine. Purine-rich foods are classified based on the total nucleobase content. Human beings susceptible to developing gout or uric acid stones are advised to avoid foods containing increased purines. This study shows that foods may be misclassified as to total purine content if the additional purines that are known to be absorbed by the mammalian intestinal tract are not measured. In particular, this study suggests that the total purine content may be underrepresented by the traditional assay especially in fresh foods that are not frozen and stored for long periods.

This method of sample preparation and analysis has some limitations. First, fish were pooled and ground for analysis, and there was inherent sample heterogeneity of some species because the diverse tissues of whole fish respond differently to grinding. This was particularly true for some of the smaller, more bony species

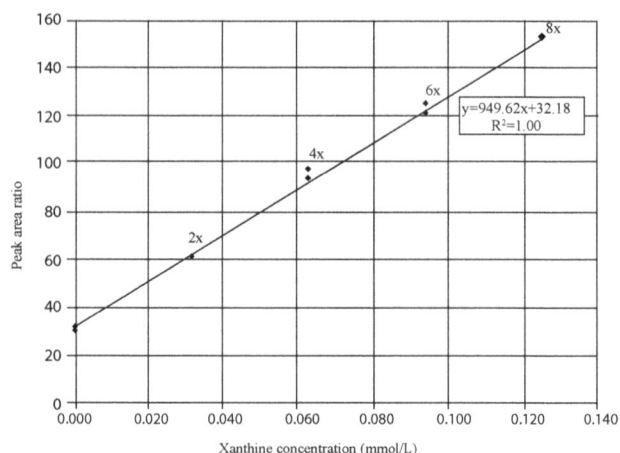

Figure 3: Example of a standard addition curve that shows the relationship between concentration and the ratio of peak area of xanthine to that of the internal standard that was used to establish the xanthine concentration in extract from a ground sample of striped mullet. Values are plotted either when no external standard was added or when increasing concentrations (2x, 4x, 6x, and 8x) of xanthine external standard were added to the fish extract. The xanthine concentration in the fish extract was calculated as the absolute value of the x-intercept of the linear regression line. Thus, for this curve x=(y-32.18)/949.62 (R²=1.00) and xanthine concentration in the extract was determined to be |-32.18/949.62|=0.0339 mmol/L.

Table 4: Purine metabolite concentrations and assay variability for two whole ground fish samples.

Metabolites	Within day		Between day	
	Mean ± SD* (µmol/L)	CV (%)†	Mean ± SD* (µmol/L)	CV (%)†
Pacific herring (*Clupea pallasii*)				
AMP*	5.46 ± 3.51	64	2.97 ± 3.21	108
Adenine	0.32 ± 0.09	27	0.14 ± 0.17	122
Guanine	201 ± 0.02	0	241 ± 40	17
IMP*	0.25 ± 0.36	147	0.10 ± 0.23	225
Inosine	228 ± 21	9	262 ± 63	24
Hypoxanthine	156 ± 24	15	166 ± 30	18
Xanthine	31.1 ± 4.8	15	34.4 ± 13.6	39
Uric acid	0.00 ± 0.00	N/A*	0.04 ± 0.13	316
Striped mullet (*Mugil cephalus*)				
AMP*	0.70 ± 0.09	13	0.77 ± 0.28	36
Adenine	1.34 ± 0.19	14	1.17 ± 0.50	43
Guanine	152 ± 17	11	165 ± 21	13
IMP*	0.95 ± 1.41	147	1.84 ± 2.16	117
Inosine	180 ± 15	8	191 ± 18	9
Hypoxanthine	35.95 ± 2.53	7	41.78 ± 6.65	16
Xanthine	12.18 ± 1.64	14	11.2 ± 2.5	22
Uric acid	0.00 ± 0.00	N/A	0.00 ± 0.00	N/A

*SD (Sandard deviation); AMP (Adenine 5'-monophosphate); IMP (Inosine 5'-monophosphate); N/A (Not applicable because concentration is zero); †Purine metabolite mean variability was calculated within day (n=4) and between days (n=4)

Figure 4: Ratios of the sum of concentrations of eight purine metabolites versus four purine metabolites in extracts of fish and squid samples of species commonly consumed by free-ranging dolphins and dolphins under human care. Boxes represent the range, the solid line represents the median, and the dotted line is the mean. Free-ranging diet species overall have a greater ratio than managed diet species (p<0.0001), and ratios of individual species with different letter superscripts are significantly different (p ≤ 0.0001; n=5).

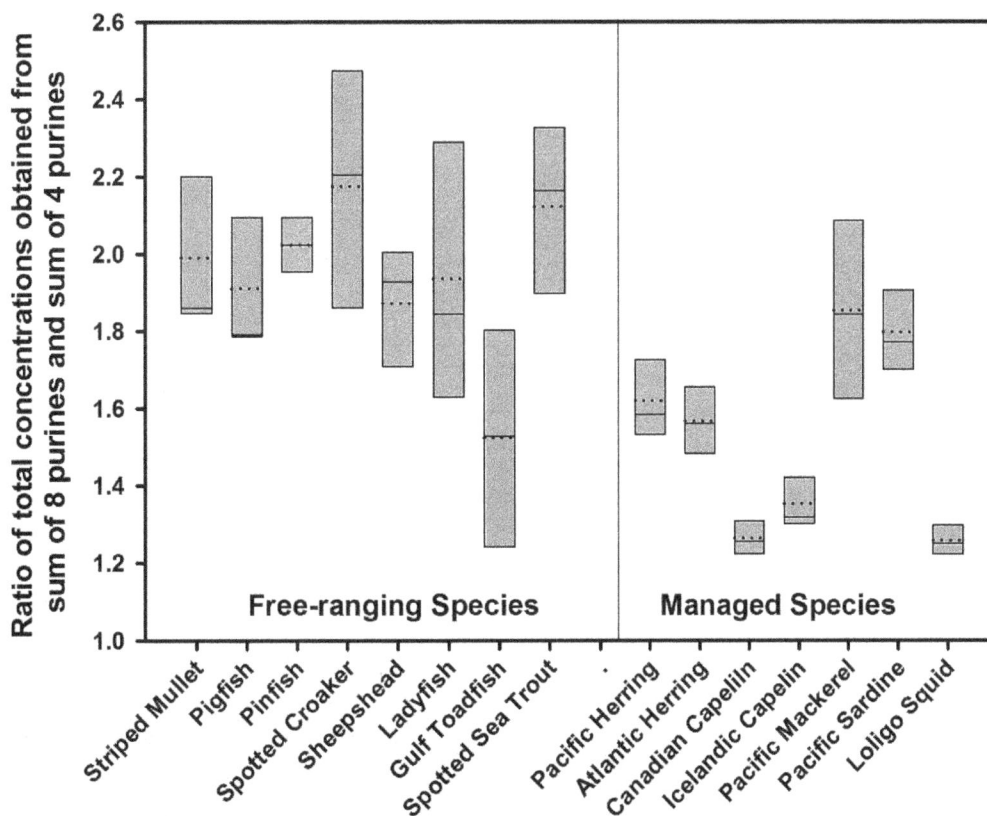

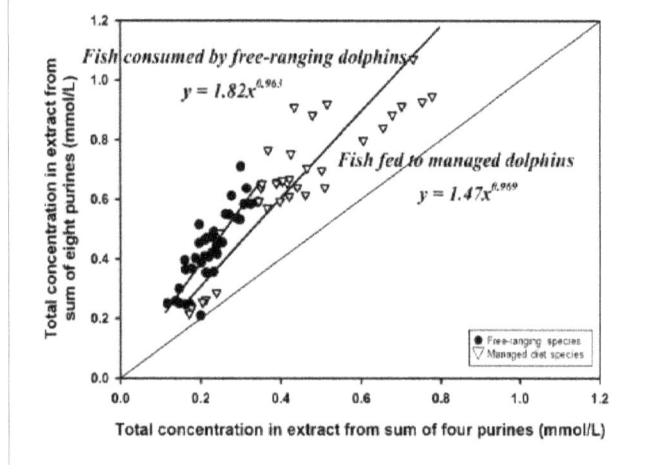

Figure 5: Comparison of the total concentration of eight purine metabolites (TP8) versus the total concentrations of four purine metabolites (TP4) among fish fed to free-ranging dolphins and species consumed by managed dolphins. Regression lines obtained by regressing the logarithm of TP8 with the logarithm of TP4 are shown. The exponents of the regression lines were approximately the same for both groups of fish but values for TP8 were on average 24% higher than for TP4.

like pinfish, where it was challenging to ensure ground sample was well homogenized. Additionally, the aqueous purine extraction method used was time-consuming, whereas acid-hydrolysis is a more commonly used and more time-efficient method for purine extraction. Nevertheless, acid-hydrolysis extraction of Pacific herring used with this HPLC-MS/MS method resulted in lower MS/MS signals for some purine metabolites, including guanine and inosine, when compared to the aqueous extraction method. There were also some noticeable differences in the appearance of the aqueous extract depending on the species. For example, Pacific herring aqueous extract had a more opaque, pink color and a more viscous consistency than striped mullet aqueous extract, which may account for the greater variability observed in Pacific herring purine measurements compared with mullet.

The fish matrix seemed to have a protective factor, stabilizing purine metabolites in the extracted sample. Nevertheless, lower MS signals for neat standards following the frozen storage and thawing experiments meant that purine standard stock solutions had to be prepared fresh for each experiment and added to experimental samples within 24 hrs. Thus, each analytical run had to be completed within 24 hrs. It was also necessary to use standard additions to quantify purine concentrations, which increased the number of runs necessary to analyze each sample and thus reduced the number of samples that could be analyzed in any one day. Standard additions were used for two important reasons. First, purines were soluble at different pH: AMP, IMP, inosine, and adenine dissolved only at a pH of approximately 7 ($NH_4CH_3CO_2$ solvent), whereas guanine, hypoxanthine, xanthine, and uric acid dissolved at a pH of approximately 10 (NH_4OH solvent). Thus, the aqueous extraction conditions of this method did not provide soluble conditions for all purine standards. Additionally, the retention times of neat individual purine standards prepared in $NH_4CH_3CO_2$ and NH_4OH solutions differed from retention times of purines found in the whole fish tissue; therefore, the matrix effect of the fish tissue did not permit accurate use of a standard calibration curve and isotopically labeled internal standards for each compound were not available. Thus, standard additions of four increasing concentrations were used to verify and quantify metabolites present in the fish.

This is an expanded assay in comparison to the commercially

available assay but is not inclusive of all metabolites in the purine degradation pathway that may lead to uric acid formation. The number of purine metabolites measured by this assay was limited by the variable solubility of individual metabolites, which was affected by the extraction solution volume and the pH of extraction and chromatographic methods. Thus, metabolites had to be prioritized for inclusion in the expanded assay, and this decision was based on their reported abundance in whole fish carcasses, ability to be intestinally absorbed in mammals, and/or uricogenic potential [1,13,34-36]. For example, adenosine 5'-diphosphate (ADP) and adenosine triphosphate (ATP) were initially included in the assay because moderate concentrations are reported in fish tissue; however, these metabolites were removed from the analysis because measured concentrations in whole fish were below the assay's limit of detection (<7.2 μmol/L) [37]. Rapid post-mortem degradation or lack of solubility relating to phosphorylation and pH may have resulted in the low concentrations of ADP and ATP measured by this HPLC method. We also elected to measure guanine rather than guanosine because of its abundance in the metallic scales of fish, despite reports of the metabolite being less urigogenic than other purines in human beings [13,25]. To the authors' knowledge, dolphins appear to digest fish scales, having no evidence of undigested scales in their feces, but it is unknown whether guanine is absorbed through the dolphins' intestinal tract. It is plausible, however, that guanine may be an important purine metabolite for the dolphin when compared to human beings because of their whole fish diet. In order to more accurately determine the total purine content of the dolphin diet, future work will involve development of two different extraction and chromatographic methods in order to accommodate the neutral- versus alkaline-soluble metabolites and permit a greater number of purines to be included in the assay.

In conclusion, this current expanded method of purine analysis enables precise and reproducible identification and quantification of the individual purine metabolites contained in whole fish. Furthermore, the inclusion of additional metabolites better quantifies the total purine content of a food item and represents a foods' uricogenic potential. Assessing the differences in the purine content of diets consumed by managed and free-ranging bottlenose dolphins may help explain why managed dolphins are predisposed to forming ammonium urate nephroliths.

Acknowledgements

The University of Florida (UF) Aquatic Animal Health Program, the US Navy Marine Mammal Program (MMP), Sea World Parks and Entertainment, Inc., and the Southeast Center for Integrated Metabolomics (NIH grant U24DK097209) provided funding to support this research. The MMP and Sea World Orlando also provided the managed diet fish samples analyzed, and the Chicago Zoological Society's Sarasota Dolphin Research Program provided the opportunity to collect the free-ranging diet fish species. Karen Scott, PhD, Brian Vagt, and the UF Meat Sciences Department assisted with sample processing.

References

1. Clifford AJ, Story DL (1976) Levels of purines in foods and their metabolic effects in rats. Journal of Nutrition 106: 435-442.

2. Choi HK, Liu SM, Curhan G (2005) Intake of purine-rich foods, protein, and dairy products and relationship to serum levels of uric acid - the third national health and nutrition examination survey. Arthritis and Rheumatism 52: 283-289.

3. Richette P, Bardin T (2010) Gout. The Lancet 375: 318-328.

4. Ho CY, Miller KV, Savaiano DA, Crane RT, Ericson KA, et al. (1979) Absorption and metabolism of orally administered purines in fed and fasted rats. The Journal of Nutrition 109: 1377-1382.

5. Choi H, Atkinson K, Karlson E, Willett W, Curhan G (2004) Purine-rich foods, dairy and protein intake, and the risk of gout in men. The New England Journal of Medicine 350: 1093-1103.

6. Serio A, Fraioli A (1999) Epidemiology of nephrolithiasis. Nephron 81: 26-30.

7. Sakhaee K, Maalouf NM (2008) Metabolic syndrome and uric acid nephrolithiasis. Seminars in Nephrology 28: 174-180.

8. Lou SN (1998) Purine content in grass shrimp during storage as related to freshness. Journal of Food Science 63: 442-444.

9. Bartges JW, Osborne CA, Felice LJ, Allen TA, Brown C, et al. (1995) Diet effect on activity product ratios of uric-acid, sodium urate, and ammonium urate in urine formed by healthy beagles. American Journal of Veterinary Research 56: 329-333.

10. Hardy R, Klausner J (1983) Urate calculi associated with poral vascular anomalies, in Current veterinary therapy. In: Kirk R. WB Saunders: Philadelphia, pp: 1073-1076.

11. Brockis JG, Levitt AJ, Cruthers SM (1982) The effects of vegetable and animal protein diets on calcium, urate and oxalate excretion. British Journal of Urology 54: 590-593.

12. Sarwar G, Brule D (1991) Assessment of the uricogenic potential of processed foods based on the nature and quantity of dietary purines. Progress in Food and Nutrition Science 15: 159-181.

13. Clifford AJ, Riumallo JA, Young VR, Scrimshaw NS (1976) Effect of oral purines on serum and urinary uric acid of normal, hyperuricemic and gouty humans. The Journal of Nutrition 106: 428-450.

14. Spann WK, Grobner W, Zollner N (1980) Effect of hypoxanthine in meat on serum uric acid and urinary uric acid excretion. Purine metabolism in man-III 122A: 215-219.

15. Brule D, Sarwar G, Savoie L (1992) Changes in serum and urinary uric acid levels in normal human subjects fed purine-rich foods containing different amounts of adenine and hypoxanthine. Journal of the American College of Nutrition 11: 353-358.

16. Fraser D, Dingle J, Hines J, Nowlan S, Dyer W (1967) Nucleotide degredation, monitored by thin-layer chromatography, and associated postmortem changes in relaxed cod muscle. Journal of the Fisheries Research Board of Canada 24: 1837-1841.

17. Clariana M, Gratacos-Cubarsi M, Hortos M, Garcia-Regueiro JA, Castellari M (2010) Analysis of seven purines and pyrimidines in pork meat products by ultra high performance liquid chromatography–tandem mass spectrometry. Journal of Chromatography A 1217: 4294-4299.

18. Aubourg SP, Piñeiro C, Gallardo JM, Barros-Velazquez J (2005) Biochemical changes and quality loss during chilled storage of farmed turbot (psetta maxima). Food Chemistry 90: 445-452.

19. Lou SN, Lin CD, Benkmann R (2001) Changes in purine content of tilapia mossambica during storage, heating, and drying. Food Science and Agricultural Chemistry 3: 23-29.

20. Bennour M, El Marrakchi A, Bouchriti N, Hamama A, El Ouadaa M (1991) Chemical and microbiological assessments of mackerel (scomber scombrus) stored in ice. Journal of Food Protection 54: 789-792.

21. Venn-Watson S, Townsend F, Daniels R, Sweeney J, McBain J, et al. (2010) Hypocitraturia in common bottlenose dolphins (tursiops truncatus): Assessing a potential risk factor for urate nephrolithiasis. Comparative Medicine 60: 149-153.

22. Venn-Watson S, Smith C, Johnson S, Daniels R, Townsend F (2010) Clinical relevance of urate nephrolithiasis in bottlenose dolphins, tursiops truncatus. Diseases of Aquatic Organisms 89: 167-177.

23. Bernard J, Allen M (2002) Feeding captive piscivorous animals: Nutritional aspects of fish as food. Nutrition Advisory Group Handbook. In: Baer D, Crissey S, Ullrey D. Nutrition Advisory Group, p: 12.

24. Huidobro A, Pastor A, Tejada M (2001) Adenosine triphosphate and derivatives as freshness indicators of gilthead sea bream (sparus aurata). Food Science and Technology International 7: 23-30.

25. Levy-Lior A, Pokroy B, Levavi-Sivan B, Leiserowitz L, Weiner S, et al. (2008) Biogenic guanine crystals from the skin of fish may be designed to enhance light reflectance. Crystal growth and design 8: 507-511.

26. McCabe E, Gannon D, Barros N, Wells R (2010) Prey selection by resident common bottlenose dolphins (tursiops truncatus) in sarasota bay, florida. Marine Biology 157: 931-942.

27. Wells RS, Rhinehart HL, Hansen LJ, Sweeney JC, Townsend FI, et al. (2004) Bottlenose dolphins as marine ecosystem sentinels: Developing a health monitoring system. EcoHealth 1: 246-254.

28. Smith CR, Poindexter JR, Meegan JM, Bobulescu IA, Jensen ED, et al. (2014) Pathophysiological and physicochemical basis of ammonium urate stone formation in dolphins. The Journal of Urology 192: 260-266.

29. Jacobs J (2012) NOAA/NCCOS/Oxford lab - tricaine methanesulfonate (ms-222) effect on tissue mineral concentrations. A. Ardente Oxford.

30. Posner L (2012) North carolina state university college of veterinary medicine, anesthesiologist - tricaine methanesulfonate (ms-222) effect on tissue mineral concentrations.

31. Piñeiro-Sotelo M, Rodríguez-Bernaldo de Quirós A, López-Hernández J, Simal-Lozano J (2002) Determination of purine bases in sea urchin (paracentortus lividus) gonads by high-performance liquid chromatography. Food Chemistry 79: 113-117.

32. Kabacoff R (2012) Power analysis. Quick R - Accessing the power of R.

33. Aubourg SP, Quitral V, Angélica Larraín M, Rodríguez A, Gómez J, et al. (2007) Autolytic degradation and microbiological activity in farmed coho salmon (oncorhynchus kisutch) during chilled storage. Food Chemistry 104: 369-375.

34. Salati LM, Gross CJ, Henderson LM, Savaiano DA (1984) Absorption and metabolism of adenine, adenosine-5'-monophosphate, adenosine, and hypoxanthine by the isolated vascularly perfused rat small intestine. The Journal of Nutrition 114: 753-760.

35. Yamamoto S, Inoue K, Murata T, Kamigaso S, Yasujima T, et al. (2010) Identification and functional characterization of the first nucleobase transporter in mammals: Implication in the species difference in the intestinal absorption mechanism of nucleobases and their analogs between higher primates and other mammals. The Journal of Biological Chemistry 285: 6522-6531.

36. Caulfield MJ, Munroe PB, O'Neill D, Witkowska K, Charchar FJ, et al. (2008) Slc2a9 is a high-capacity urate transporter in humans. Plos Medicine 5: 1509-1523.

37. Hattula T (1997) Adenosine triphosphate breakdown products as a freshness indicator of some fish species and fish products, in Biochemistry. University of Helsinki: VTT Publications, Technical Research Centre of Finland, p: 47.

11

Analysis of Rituximab, A Therapeutic Monoclonal Antibody by Capillary Zone Electrophoresis

Andrasi M[1,2], Gyemant G[1] and Gaspar A[1,2]*

[1]Department of Inorganic and Analytical Chemistry, University of Debrecen, 4032 Debrecen, Egyetem tér 1, Hungary
[2]Cetox Ltd., 4032 Debrecen, Egyetem tér 1, Hungary

Abstract

This paper focuses on the applicability of capillary zone electrophoresis (CZE) using uncoated fused silica capillaries for the determination of heterogeneity of rituximab (MabThera, Roche) and for the study of its solution and thermal stability. The best resolution for the charge variants of the main component was obtained with a buffer electrolyte containing 800 mM 6-amino caproic acid, 2 mM triethylene tetramine and 0.05% hydroxypropyl methylcellulose at pH = 5.2. It was found that the pH and the components of the buffer used for the electrophoretic separation and for the dilution of the sample prior to the analysis are important to the stability of the rituximab. We demonstrated the rituximab is stable in the pharmaceutical product MabThera due to the stabilizing additives, but the dilution of the MabThera caused a slow formation of acidic variants, while the amount of the basic variants did not change. After incubation of the diluted rituximab at higher temperature several charge variants could be determined by CZE.

Keywords: Monoclonal antibody; Charge heterogeneity; Capillary zone electrophoresis

Introduction

The biotechnologically engineered monoclonal antibody (mAb) contains four polypeptides linked via disulfide bonds: two light and two heavy chains. It is produced in mammalian cell culture. The mAbs can undergo post-translational modification (deamidation, oxidation, glycosylation, lysine truncation, aggregation) resulting in molecular heterogeneity, charge and size variants. Since these varieties of the mAb have effects on its pharmaceutical properties (antigen binding) and stability, it is important to investigate its molecular heterogeneity. Therefore numerous works were published about the analysis of mAb and the characterization of the molecular heterogeneity [1]. The rituximab (Rtx) is a monoclonal antibody used to treat cancers of blood system such as B cells leukemia (non Hodgkin's lymphoma) and some autoimmune diseases (rheumatoid arthritis) [2].

While the size heterogeneity of the monoclonal antibodies was studied by size-exclusion chromatography [3,4], peptide mapping [4], LC-MS [5] or gel electrophoresis [6-8], the charge heterogeneity was mostly studied by cation-exchange chromatography [3,4]. The different methods of capillary electrophoresis (CZE, CGE, CIEF) are useful in the complex analysis of the mAbs. The capillary zone electrophoresis (CZE) can reveal the charge variants of the mAbs. High separation efficiency could be achieved when the interaction between protein and the inner wall of the capillary was eliminated. In order to overcome this problem several strategies, such as appropriate choice of the pH and different additives of the background electrolyte and use of capillary coatings were applied [9]. The simplest way to reduce these interactions is if very low or high pH values are used. In this case the protein and the inner surface have the same charge and thus the proteins do not tend to adsorb onto the surface. However, the large difference between the pH of the buffer and the pI of the protein can cause structural changes in the protein resulting in low recovery [10]. This can often suggest modifying the inner wall of the capillary for preventing protein adsorption. Using bare fused silica capillaries different additives like 6-amino caproic acid (EACA) [11], triethylene tetramine (TETA) [12] or hydroxypropyl methylcellulose (HPMC) [11] are given into the background electrolyte. Shi et al. evaluated the separation of charge variants of mAb in fused silica capillary using a CZE running buffer containing polyethylene oxide (PEO) and TETA [13]. Five different commercially marketed mAb products were investigated by CZE in eCAP coated neutral capillary using EACA and polysorbate 20 in the running buffer [14]. Several mAbs, among them the rituximab, were analyzed in coated capillary with polyacrylamide-based hydrophilic surface [15]. Gassner et al. tested various statically coating (polyvinylalcohol, polybrene, dextran sulphate, polyethylenimine) for the analysis of rituximab by CZE [16].

This paper focuses on the applicability of CZE using uncoated fused silica capillaries for the determination of heterogeneity of rituximab (MabThera, Roche) and for the study of its solution and thermal stability.

Material and Methods

Instrumentation

The capillary electrophoresis instrument was a 7100 model (Agilent, Waldbronn, Germany). In the case of CZE measurements hydrodynamic sample introduction (30 mbar, 2 s or 30 mbar, 5 s) was used for injecting samples. The sample solutions were introduced at the capillary end closer to detection and the anodic end of the capillary as well. Separations were performed using a fused-silica capillary of 64.5 cm × 50 μm i.d. (Polymicro, Phoenix, USA). The applied voltage was +10-15 kV. The detection was carried out by on-column diode array photometric measurement at 200 nm. The electropherograms were recorded and processed by ChemStation computer program of B.04.02 version (Agilent).

***Corresponding author:** Gaspar A, Department of Inorganic and Analytical Chemistry, University of Debrecen, 4032 Debrecen, Egyetem tér 1, Hungary
E-mail: gaspar@science.unideb.hu

Chemicals

Reagents of analytical grade were obtained from various distributors. Disodium hydrogen phosphate, HCl, NaOH, EACA, HPMC and TETA for preparing buffer electrolytes were purchased from Sigma-Aldrich (St. Louis, USA). The 1 mg/mL Rtx sample solutions were prepared shortly before CE separation by dissolving the rituximab (10 mg/mL MabThera, Roche) in water. The CGE separations in sieving matrix gel buffer (Beckman Coulter SDS-MW, USA) were performed. In this case the 1 mg/ml sample sample solution is made by diluting the rituximab with SDS-MW sample buffer solution (Beckman Coulter). The post conditioning was carried out with 1 M NaOH or 1 M HCl (in case of using of EACA-TETA-HPMC) for 5 min and with background electrolyte for 6 min.

Results and Discussion

CZE separation of rituximab in uncoated capillary

The most simple approach to analyze Rtx with CZE if uncoated fused silica capillary is used. A phosphate buffer of pH= 9.3 (a higher pH than the pI value of the Rtx (pI=7.3 - 9.1 [13,16]) was selected to minimize interactions of Rtx with the deprotonated silanol groups of the surface. At pH 9.3 the surface became negatively charged and therefore the predominantly negatively charged Rtx was repelled by the surface (the estimated negative charge of the Rtx at pH= 9.3 is 40 according to the Protein Calculator version 3.4). In this condition a sharp peak was obtained for the Rtx (less than 0.08 min peak width at 16.6 min migration times), which indicates that only minimal adsorption occurred (Figure 1a). However, the electrophoresis of Rtx

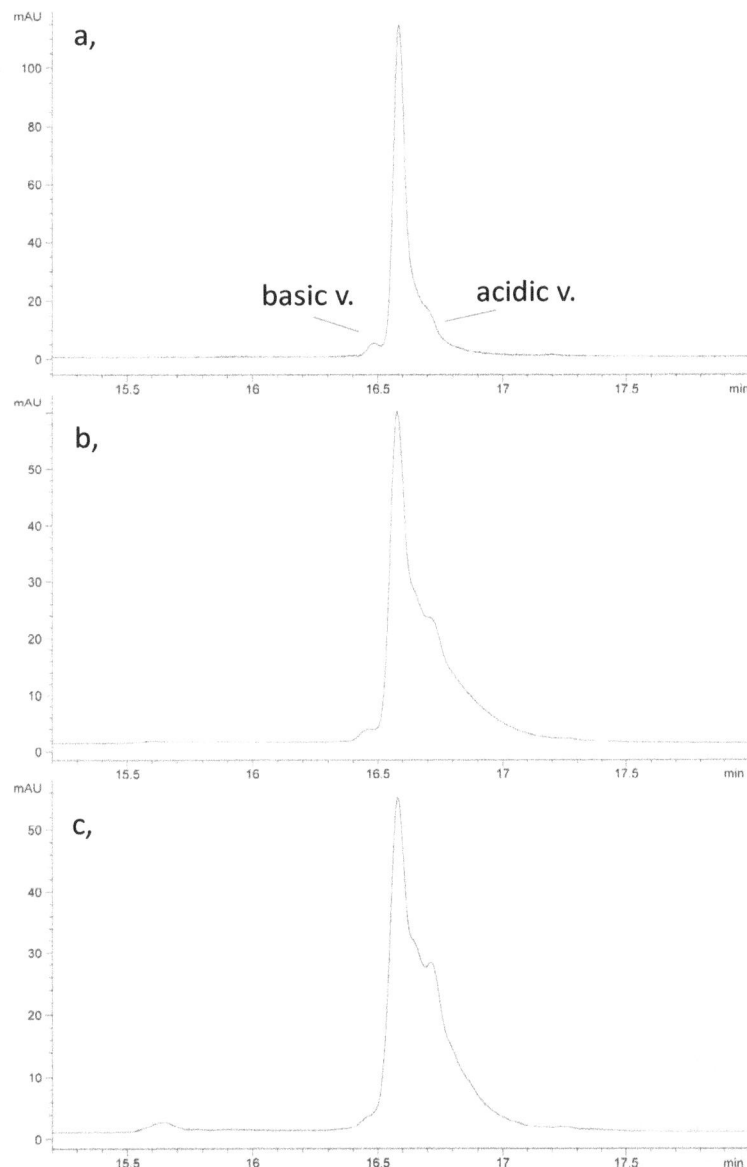

Figure 1: CZE separations of rituximab. The 10 mg/ml pharmaceutical product MabThera was tenfold diluted with water and the samples were injected within 5 min (a,), in 3 days (b,) and 7 days (c,) after the dilution.
Conditions: 100 mM phosphate buffer pH 9.3, capillary: 64.5 cm x 50 μm i.d., l_{eff}: 56.5 cm, 60 mbar s, +10 kV, λ= 200 nm.

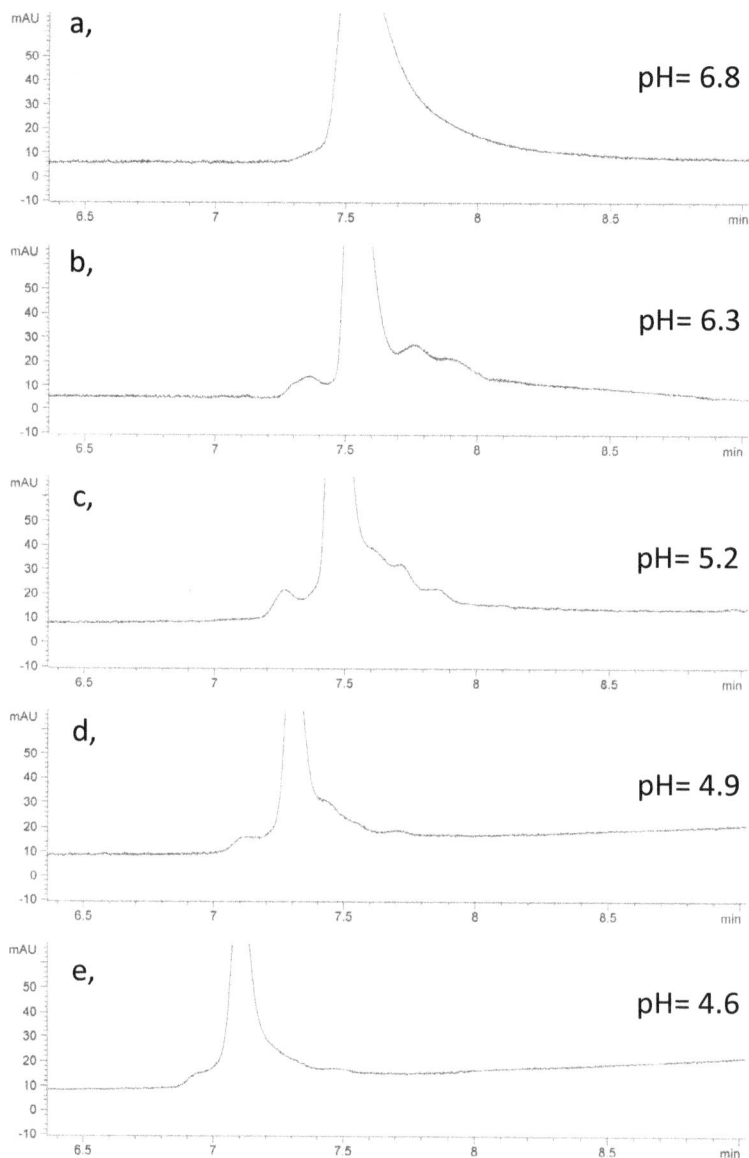

Figure 2: CZE separations of rituximab using running buffer of different pH. Conditions: 400 mM EACA, 2 mM TETA, 0.05% HPMC, capillary: 64.5 cm x 50 µm i.d., l_{eff}: 8.5 cm, samples were injected by 150 mbars at the anodic end of the capillary, closer to the detection window ("short end injection"), -15 kV, λ= 200 nm. The 10 mg/ml pharmaceutical product MabThera was tenfold diluted with water.

at pH lower than 8 resulted in a very broad signal if any. Although the CZE separations of proteins are generally carried out in coated (PVA, µSIL, PB, PEI, etc. [9,10]) capillaries, we did not get considerably better peak shape for Rtx using µSIL capillary. The CZE separation above pH= 10 could not be carried out because the Rtx degradated.

The 10 mg/ml pharmaceutical product was tenfold diluted with water and the samples were injected immediately, in 3 days and in 7 days after the dilution. In the obtained electropherograms of (Figure 1b) the basic and acidic variants can be observed at the sides of the main component, Rtx. In general, the acidic variants are formed by glycosylation with sialic acid or uronic acid, or by deamidation of asparagine or glutamine, or by pyroglutamate formation from glutamine and glutamate, whereas the basic variants are formed by formation of Fc-1 lysine or Fc-2 lysine, or by the noncyclization of N-terminal glutamine, or by succinimide formation from aspartic acid

or by C-terminal proline amidation [14]. In the electropherograms the basic variant is probably 1-lysine variant (~2.5 %), which is relatively well separated from the main component (0-lysine). The shoulder of the main component contains one or more acidic variants. The dilution of the pharmaceutical product, which contains stabilizing components (neutral detergent) caused a slow formation of acidic variants, while the amount of the basic variants did not change. Mainly the acidic variants in the drug product could have impact on biological activity, so the fast and reliable determination of the amounts of these variants is required [17]. The electrophoretic patterns presented in (Figure 1 c) are in good agreement with the electropherogram obtained in neutral coated capillary [16].

Since, we did not reach proper resolution for acidic variants using simple phosphate buffer in uncoated silica capillary, the application of different dynamic coatings was tested. Several components (detergents, polyamins, polymers) added to the running electrolyte have been investigated as dynamic coatings on the surface of the capillary [11-16]. Often a mixture of different additives with complex mechanism are applied [11-16]. Below pH= 8 the Rtx can be analyzed by CZE only in a (dynamically or statically) coated capillary. For the CZE of mAbs the mixture of a nonionic, hydrophilic polymer (HPMC), a compound forming ion pairs with peptides (EACA) and a polyamine (TETA) was suggested by Shi et al. in order to dynamically coat the capillary [13]. In our work this suggested mixture was used for the separation of Rtx. The nonionic linear polymer HPMC tends to adsorb to the neutral surfaces suppressing protein-wall interactions. Prior to conditioning the capillary with the running electrolyte short rinsing with 1 M HCl was used to protonate the silanol groups (and to remove the adsorbed proteins, as well). Using a higher polymer concentration a sieving effect can also be in play. The EACA on one side can interact with the residual silanol groups of the capillary surface that are not covered by HPMC, and on the other hand it can form ion pairs on the protein reducing its

net charge. The addition of TETA to the buffer can also form a dynamic coating on the inner wall of the capillary reducing the adsorption of proteins. Additionally, it competes with the protein to interact with the silanol groups. However, the excess of TETA can modify the charge heterogeneity of mAb caused by the interaction of amines of TETA with the carboxyl groups of the mAb. The CZE separations of Rtx shown in Figure 2 were performed using running buffer containing 400 mM EACA, 2 mM TETA and 0.05% HPMC (suggested concentrations from [12] for several mAbs but not Rtx) of different pH values. At pH= 6.8 (close to the pI of Rtx) the adsorption of the Rtx to the inner wall is still considerable, which resulted in peak tailing (Figure 2a,b). However, in the pH range of 4.9-6.3 the peaks at the bottom of the main peak Rtx were resolved to two-three acidic variants and one basic variant. At pH below 4.6 the separation of the variants from the main component became worse. The best resolution for the acidic variants were obtained at pH= 5.2 (Figure 2c,d,e). Similar patterns of peaks of basic/acidic

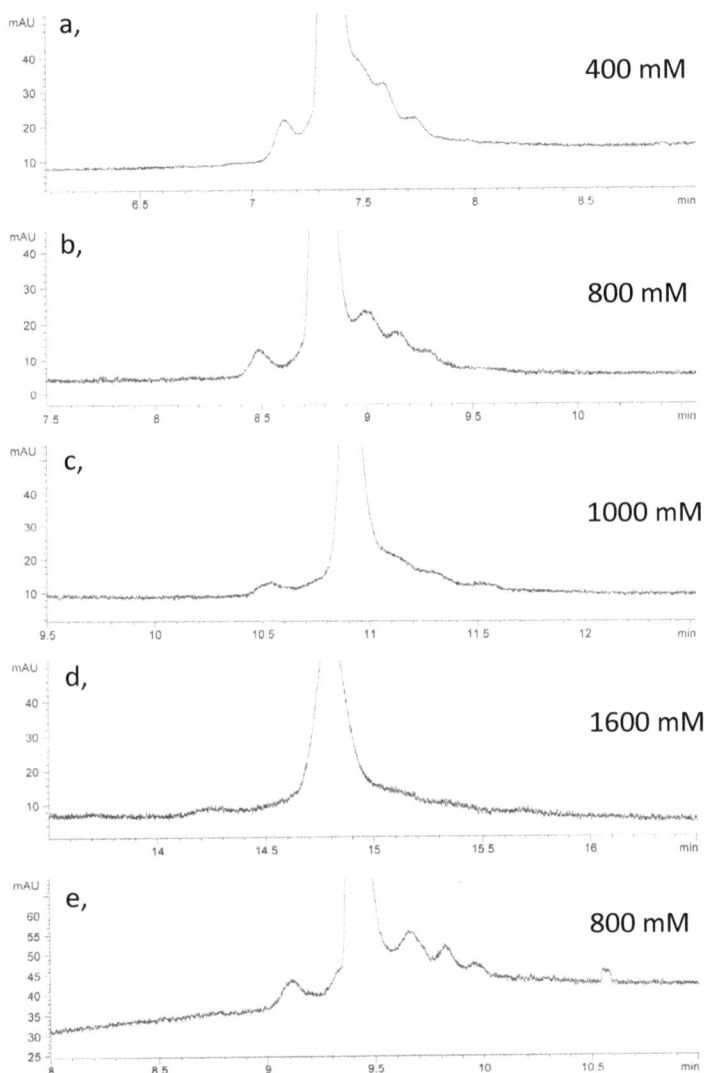

Figure 3: CZE separations of rituximab using running buffer of pH=5.2 with different concentration of EACA.
Conditions: 400-1600 mM EACA ((a,): 400 mM, (b,): 800 mM, (c,): 1000 mM, (d,): 1600 mM, (e,): 800 mM), 2 mM TETA, 0.05% HPMC, capillary: 64.5 cm x 50 µm i.d., l_{eff}: 8.5 cm, sample was injected by 150 mbars at the anodic end of the capillary, closer to the detection window ("short end injection"), -15 kV, λ= 200 nm. The 10 mg/ml pharmaceutical product MabThera was tenfold diluted with water. in case of (e,): capillary: 30 cm x 50 µm i.d., l_{eff}: 21.5 cm, +15 kV, sample was injected further to the detection window (normal mode).

variants were reported by He et al. for other mAbs [12,17], using a bit higher optimal pH (pH= 5.7).

The electropherograms of (Figure 3a,b,c,d) illustrate the CZE separations of Rtx using running buffer of pH= 5.2 including the three additives (EACA, TETA and HPMC) with 400-1600 mM concentration of EACA. When the concentration of EACA was doubled from 400 mM to 800 mM the improvement in the separation of acidic variants was quite significant, but the further increase in the concentration of the EACA led to a gradual disappearance of the variants. The considerable increase of the EACA in the buffer accompanied a slower migration of the components due to the suppressed of (increased ionic strength, smaller surface charge of the capillary). The phenomenon of the gradual disappearance of the variants above 1000 mM EACA could not be explained. A little better resolution was obtained for a slightly larger

electric field and longer separation length (Figure 3e), but the further increasement of the length of the capillary deteriorated the efficiency of the separation. Interestingly, the increase of the EACA concentration in the running buffer improved the resolution only for the basic variants for other mAbs [12].

Study of solution and thermal stability of rituximab

The stability of the mAb is critical in terms of its applicability and pharmacological properties. The characterization of the stability of the therapeutic monoclonal antibodies is important since they can go through several chemical and physical degradation/transformation processes, thus their proteic natures can alter. During the biotechnological preparation and the storage the mAbs are exposed to a lot of stress factors. The chemical stability relates to the integrity

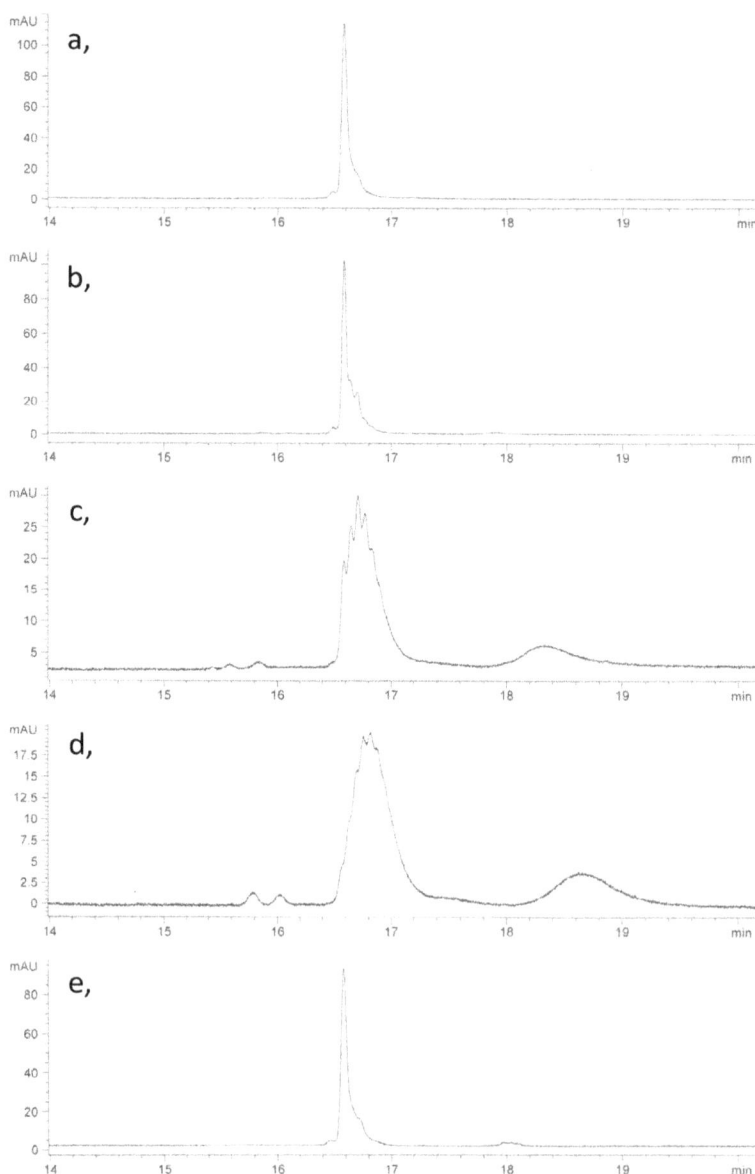

Figure 4: Thermal stability (55°C for 1.5 h (b,), 15 h (c, and e,) and 40 h (d,)) studies of rituximab.
(b,) - (d,): the samples (MabThera) were diluted prior the thermal treatment,
(a,) and (e,): the samples were diluted just before their injections. (a): the sample was kept at room temperature.
Conditions: 100 mM phosphate buffer pH 9.3, capillary: 64.5 cm x 50 µm i.d., 60 mbars, +10 kV, λ= 200 nm, the samples are diluted with water.

of covalent linkages, while the secondary bonds (hydrogen bond, electrostatic interaction) are responsible for the physical stability. The Rtx is formulated for parenteral administration in sodium citrate dihydrate buffer (pH 6.5) containing nonionic surfactant (Tween 80) to stabilize the aqueous formulation [4].

As it was stated in the previous section the acidic variants are slowly formed by diluting the pharmaceutical product by water, while the amount of the basic variants did not change. At higher temperature the acidic variants are formed with a much higher rate than at room temperature. Incubating the mAb samples at 55°C for 1.5 hours the degradation of the main component was around 10% and two peaks (acidic variants) appeared in the shoulder of the main peak (Figure 4a,b). The acidic variants are generally formed due to deamidations or pyroglutamate formation. In accelerated stability testing the deamidation

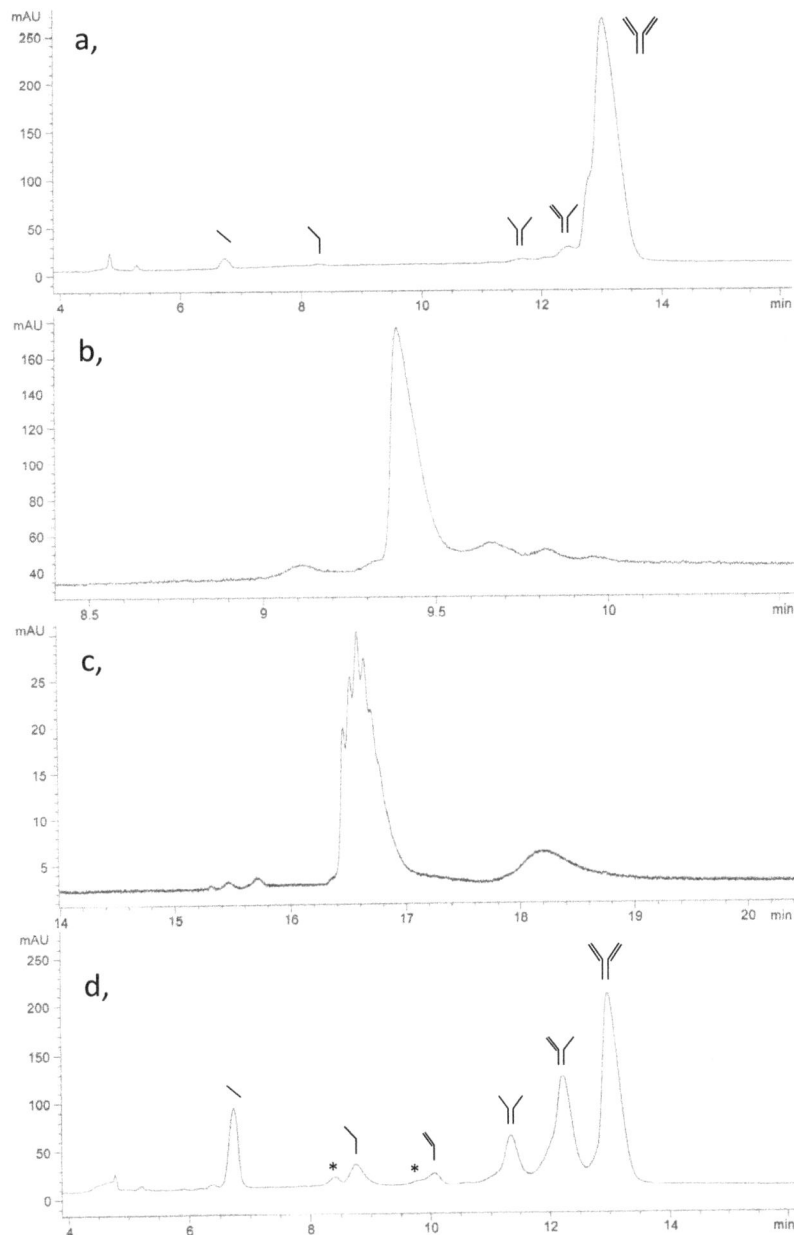

Figure 5: Comparison of electropherograms of rituximab obtained by CGE using sieving matrix (a, d), CZE using EACA-TETA-HPMC additives in the buffer (b,) and CZE using phosphate buffer (c,), (b,).
Conditions:
(a, d): CGE separations, Beckman Coulter SDS-MW gel buffer sieving matrix, electrokinetic (short-end) injection: +7.5 kV for 50 sec, capillary: 30 cm x 50 µm i.d., U= +15 kV, λ= 200 nm.
(b,): CZE separation, 800 mM EACA, 2 mM TETA, 0.05% HPMC, pH= 5.2, capillary: 30 cm x 50 µm i.d., 150 mbar s, +15 kV, λ= 200 nm.
(c,): CZE separation. The sample was stressed by 55°C for 15 h, 100 mM phosphate buffer pH 9.3, capillary: 64.5 cm x 50 µm i.d., 60 mbar s, +10 kV, λ= 200 nm
(d,): CGE separation. The sample was stressed by 55°C for 15 h, 100 mM phosphate buffer pH 9.
1 mg/ml pharmaceutical product MabThera samples were prepared by dilution with Beckman Coulter SDS-MW sample buffer (a, d,), water (b,) and with the buffer solution (c).

and the cyclization of N-terminal glutamate to pyroglutamate were verified by MALDI-TOF MS [5]. Incubating the mAb samples for 15 and 40 hours 5-6 intensive but overlapped peaks of acidic variants could be observed (Figure 4c,d). These results correspond with the statements of others, that is, at high temperature and at high pH the deamidation (the most common covalent modifications for protein pharmaceuticals) becomes faster [3]. A similar increase in acidic variants after stress (0,5 M phosphate, pH 8.0 at 40°C for 8 day) was obtained by Han et al. using conventional and microchip CZE [14]. When the mAb samples were incubated at 55°C for 15 and 40 hours (Figure 4c,d) two peaks could be observed between 15.5-16.5 min. According to the suggestion of ref. [18] these peaks can be assigned to the components formed by the oxidation of methionine.

The Rtx in the formulated product MabThera without dilution shows very good stability. (The dilution of the formulated product MabThera reduces the stabilizer effect of the nonionic surfactant and citrate buffer additives. While the citrate moderates the deamidation by changing the conformation of the protein, the phosphate ions may catalyze the deamidation [3].) Incubating the undiluted MabThera at 55°C for 15 hours it practically did not undergo any transformation (Figure 4e). Based on these obtained and reported results it can be stated that the roles of the pH and the buffer species are important in the stability of Rtx, the formulated product mAbs should not dilute with buffer (phosphate, pH=9.3), or if it is necessary it should be done just prior to the sample injection in order not to accelerate the degradation of the Rtx.

Comparison of CZE and CGE using SDS-sieving matrix for heterogeneity of rituximab

The SDS-protein complexes can be separated by CGE in the sieving matrix according to their molecular weight and shape. Earlier the CGE was also applied to separate the size variants (fragments, aggregates) of mAb [6-8]. In our experiments the Rtx was diluted with the sample matrix buffer just before the injection. Little peaks of four low molecular weight (LMW) fragments (light chain, heavy chain, two heavy chains and two heavy chains with one light chain) can be seen before the main component (monomer) (Figure 5a). These fragments observed on the CGE electropherogram are in a good agreement with the electrophoretic pattern obtained by Hunt [6]. Interestingly, in the CZE electropherogram of the same sample (Figure 5b) also four components could be observed next to the main component, however, the four components separated by CZE and CGE are probably not identical. The charge variants of the main component separated by CZE (Figure 5b) appeared as one peak (monomer with a given molecular size) in CGE electropherogram (Figure 5a). In this sample stored at room temperature mainly the peak of the light chain could be found.

The CZE separation of the components of the thermally degraded sample revealed a broad peak after the main component (monomer) (Figure 5c). Although this broad peak at 18.2 min seems to be a possible dimer (the increase of the temperature may alter the secondary, tertiary and quaternary structures of the protein and led to aggregation), the CGE measurement clearly shows the lack of aggregates of mAbs (Figure 5d). It is more likely that the effect of heating caused the rupture of the Y-shape in the hinge region resulting in formation of heavy chains and half antibodies (heavy chain + light chain) (Figure 5d). These two components could not be observed at room temperature. The amounts of the LMW components were increased by thermal stress (Figure 5d). Since the release of a light chain is easier than that of the heavy chains (which are held together by two disulfide bonds in hinge region), the light chain is predominant LMW fragment in the sample. The shoulders

of the peaks of the heavy chain and half antibody (marked with *) are probably their non-glycosylated forms (as suggested by [8]) or the results of incomplete SDS binding [6] and/or unfolding of the mAb [6]. With the CGE the LMW components could be well separated from each other and from the monomer. For the monomer the precision of migration time and peak area were 0.27% and 6.16%, respectively. The limit of detection (LOD, 6s) was 0.69 µg/ml using UV detection at 200 nm.

According to the obtained results it can be stated that the CZE analysis could reveal the charge variants of Rtx, while the CGE with sieving matrix gave information about the size heterogeneity. There is no chance to determine either the charge or the size variants by the same separation method.

Conclusions

In this work the applicability of CZE for the determination of the heterogeneity of rituximab was studied. A buffer of pH=5.2 with several additives was used to suppress the interactions between the Rtx and the capillary, thus even an uncoated fused silica capillary was applicable for the efficient separations of several charge variants and the main component. In order to support some conclusions obtained by CZE, CGE in sieving matrix separation was used. For instance, the charge variants of the main component that can be separated by CZE provide the same molecular size.

It was found that the dilution of the pharmaceutical product, which contains stabilizing components (neutral detergent) caused a slow formation of acidic variants. Incubating the Rtx at higher temperature or basic condition, charge variants and fragments of the Rtx were formed by deamidation and fragmentation. The CZE analysis could reveal the charge variants of Rtx, while the CGE with sieving matrix gave information about the size heterogeneity.

Acknowledgement

The authors gratefully acknowledge financial support for this research by grants from the University of Debrecen (RH/885/2013), the National Scientific Research Fund, Hungary (OTKA K111932, PD83071) and TÁMOP 4.2.2.A-11/KONV-2012-0043 project.

References

1. Cheung MC, Haynes AE, Meyer RM, Stevens A, Imrie KR (2007) Rituximab in lymphoma: A systematic review and consensus practice guidline from Cancer Care Ontario. Cancer Treat Rev 33: 161-176.

2. Beck A, Wagner-Rousset E, Ayoub D, Van Dorsselaer A, Sanglier-Cianférani S (2013) Characterization of therapeutic antibodies and related products. Anal Chem 85: 715-736.

3. Zheng JY, Janis LJ (2006) Influence of pH, buffer species and storage temperature on physicochemical stability of a humanized monoclonal antibody LA298. Int J Pharmac 308: 46-51.

4. Paul M, Vieillard V, Jaccoulet E, Astier A (2012) Long term stability of diluted solutions of the monoclonal antibody rituximab. Int J Pharmac 436: 282-290.

5. Liu H, Gaza-Bulseco G, Sun J (2006) Characterization of the stability of a fully human monoclonal IgG after prolonged incubation at elevated temperature. J Chromatogr B Analyt Technol Biomed Life Sci 837: 35-43.

6. Hunt G, Nashabeh W (1999) Capillary electrophoresis sodium dodecyl sulfate nongel sieving analysis of therapeutic recombinant monoclonal antibody: A biotechnology perspective. Anal Chem 71: 2390-2397.

7. Salas-Solano O, Tomlinson B, Du S, Parker M, Strahan A, et al. (2006) Optimization and validation of quantitative capillary electrophoresis sodium dodecyl sulfate method for quality control and stability monitoring of monoclonal antibodies. Anal Chem 78: 6583-6594.

8. Kamoda S, Ishikawa R, Kakehi K (2006) Capillary electrophoresis with laser-induced fluorescence detection for detailed studies on N-linked oligosaccharide profile of therapeutic recombinant antibodies. J Chromatogr A 1133: 332-339.

9. Hempe JM (2008) Protein analysis by capillary electrophoresis, in: Landers JP (Edn), Capillary and microchip electrophoresis and associated microtechniques, Taylor and Francis Group, 75-107.

10. Girard M, Lacunza I, Diez-Masa JC and de Frutos M (2008) Glycoprotein analysis by capillary electrophoresis, in: Landers JP (Edn), Capillary and microchip electrophoresis and associated microtechniques, Taylor and Francis Group, 631-705.

11. He Y, Lacher NA, Hou W, Wang Q, Isele C, et al. (2010) Analysis of identity, charge variants and disulfide isomers of monoclonal antibodies with capillary zone electrophoresis in an uncoated capillary column. Anal Chem 82: 3222-3230.

12. He Y, Isele C, Hou W, Ruesch M (2011) Rapid analysis of charge variants of monoclonal antidodies with capillary zone electrophoresis in dynamically coated fused-silica capillary. J Sep Sci 34: 548-555.

13. Shi Y, LI Z, Qiao Y, Lin J (2012) Development and validation of rapid capillary zone electrophoresis method for determining charge variants of mAb. J Chromatogr B Analyt Technol Biomed Life Sci 906: 63-68.

14. Han H, Livingston E, Chen X (2011) High throughput profiling of charge heterogeneity in antibodies by microchip electrophoresis. Anal Chem 83: 8184-8191.

15. Espinosa-de la Garza CE, Perdomo-Abúndez FC, Padilla-Calderón J, Uribe-Wiechers JM, Pérez NO, et al. (2013) Analysis of recombinant monoclonal antibodies by capillary zone electrophoresis. Electrophoresis 34: 1133-1140.

16. Gassner AL, Rudaz S, Schappler J (2013) Static coatings for the analysis of intact monoclonal antibody drugs by capillary zone electrophoresis. Electrophoresis 34: 2718-2724.

17. Gandhi S, Ren D, Xiao G, Bondarenko P, Sloey C, et al. (2012) Elucidation of degradants in acidic peak of cation exchange chromatography in an IgG1 monoclonal antibody formed on long-term storage in a liquid formulation. Pharm Res 29: 209-224.

18. Chumsae C, Gaza-Bulseco G, Sun J, Liu H (2007) Comparison of methionine oxidation in thermal stability and chemically stressed samples of fully human monoclonal antibody. J Chromatogr B Analyt Technol Biomed Life Sci 850: 285-294.

12

Development and Validation of HPLC Method for the Determination of Alcaftadine in Bulk Drug and its Ophthalmic Solution

Mishra PR[1], Satone D[2] and Meshram DB[1]*

[1]Department of Quality Assurance, Pioneer Pharmacy Degree College, Sayajipura, Vadodara, Gujarat, India
[2]Department of Pharmaceutical Sciences, RTM Nagpur University, Nagpur, Maharashtra, India

Abstract

A simple, precise and accurate stability indicating RP-HPLC method was developed for the estimation of Alcaftadine in bulk drug and its ophthalmic dosage form. Chromatographic separation of target analyte and degradation products was achieved using Enable HPLC ODS C_{18} G (250 × 4.6 mm, 5 μm) column and mobile phase comprising of methanol: water, 50:50% v/v at a flow rate of 1.2 ml/min with UV detection at 282 nm. The correlation coefficient (r^2) was found to be 0.999 in the concentration range of 1-16 μg/ml. The retention time was found to be 3.15 min. The limit of detection and limit of quantitation was found to be 0.25 μg/ml and 0.75 μg/ml respectively. Stress degradation of Alcaftadine was carried out under stress conditions like acid and base hydrolysis, oxidation, thermal and photolytic stress. The degradation products generated as a result of stress did not show any interference to the detection of Alcaftadine. The developed method was validated, and selectivity of the method was demonstrated by its ability to quantify the target analyte in presence of its degradation products.

Keywords: Alcaftadine; Lastacaft; HPLC; Stability indicating method

Introduction

The IUPAC name of Alcaftadine is 6,11-dihydro-11-(1-methyl-4-piperidinylidene)-5H-imidazo [2, 1-b][3] benzazepine-3-carboxaldehyde (Figure 1). Alcaftadine is an ophthalmic dual-acting H_1-antihistamine and mast cell stabilizer approved for the prevention of itching associated with allergic conjunctivitis. The drug was approved by USFDA in July 2010. It is commercially marketed under the name LASTACAFT. Lastacaft is a sterile, topically administered H_1 receptor antagonist containing Alcaftadine for ophthalmic use [1-2].

Alcaftadine is not official in any Pharmacopoeia. A literature survey on Alcaftadine revealed that, until now no analytical method was reported for its determination in bulk drug and its ophthalmic dosage form. However, a clinical pharmacological review report found during the survey showed that liquid chromatography with tandem mass spectrometry (LC/MS/MS) was used to quantitate concentrations of Alcaftadine and R90692 (active metabolite) in K3 EDTA human plasma [3]. It was also found that the metabolic fate of 14 C- Alcaftadine was determined by high performance liquid chromatography-based separation of parent compound from metabolites [4]. Since literature did not cite any method for determination of this drug from bulk drug as well as its formulation it was planned to develop a RP-HPLC method to determine alcftadine in presence of its degradation products.

Materials and Methods

Instrumentation

Shimadzu Prominence UFLC LC-20AD with UV detector and Enable C_{18} G RP column (250 × 4.6 mm, 5 μm) was used was chromatographic resolution of target analyte and degradation products. The acquisition and integration of data was performed using LC solution software.

Materials and reagents

Alcaftadine pure drug sample was obtained from JHP Pharmaceuticals LLC, Rochester, Michigan, USA. Lastacaft, Alcaftadine ophthalmic solution, was procured from MediPrime Pharmacy, Dubai. HPLC grade methanol and HPLC water was procured from Merck.

Solution preparation

Standard stock solution: Accurately 10 mg of drug was weighed and transferred to a 10 ml volumetric flask. It was dissolved using methanol and volume was made up to obtain a concentration of 1000 μg/ml. After sufficient dilutions, final stock having concentration equivalent to about 50 μg/ml of drug was prepared.

Working standard solution: The standard stock solution was used to prepare working standard solutions of concentrations 1, 4, 8, 12 and 16 μg/ml. Solution having drug concentration of 8 μg/ml was used as a working standard for stress degradation studies. Both the standard and sample solution of 8 μg/ml were estimated at 282 nm and the chromatograms were recorded (Figure 2).

Sample solution

From the ophthalmic solution (alcaftadine 2.5 mg/ml), one ml

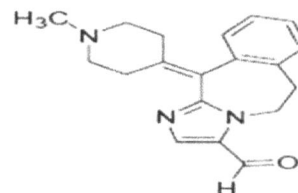

Figure 1: Structure of alcaftadine.

*Corresponding author: Dhananjay B Meshram, Department of Quality Assurance, Pioneer Pharmacy Degree College, Near Ajwa Cross Road, NH 8, Ajwa Nimeta Road, At and Post Sayajipura, Vadodara-390 019, Gujarat, India
E-mail: dbmeshram@yahoo.com

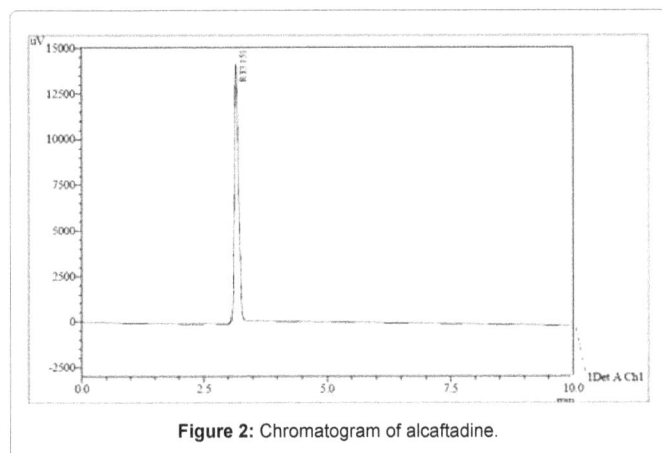

Figure 2: Chromatogram of alcaftadine.

was withdrawn and transferred to a 10 ml volumetric flask, dissolved and diluted with methanol to get the concentration of 2500 µg/ml. Further dilutions were made to get the final stock having concentration equivalent to 8 µg/ml.

Optimization of chromatographic conditions

Various trials were made for the selection of chromatographic conditions like mobile phase, its ratio and flow rate. Finally, the one giving the best results were optimized. The chromatographic estimation of Alcaftadine and its separation from degradation products was achieved using Enable C_{18} G HPLC column using mobile phase Methanol: Water 50:50% v/v at a flow rate of 1.2 ml/min. The UV detection was done at 282 nm. The optimized chromatographic conditions are summarized in Table 1.

Validation of proposed method

The method was validated for the parameters like specificity, linearity, precision, accuracy and robustness as per ICH guidelines [5,6].

System suitability testing

Six replicates of drug concentration of 8 µg/ml were injected and the chromatograms were recorded to check with the system suitability parameters [7].

Forced degradation studies

The forced degradation studies were carried out as per ICH Q1A (R2) for hydrolysis, oxidation and thermal stress conditions and ICH Q1 B for photo stability. The stress conditions employed were 1N HCl for acid hydrolysis, 1 N NaOH for base hydrolysis, 3% H_2O_2 for oxidative hydrolysis. The drug samples were also subjected to thermal stress conditions as well as subjected to UV light to test photo stability [8-15].

Acid and base hydrolysis: Acid and base induced forced degradation was performed by adding an aliquot of stock solution (1 mg/ml) of Alcaftadine to 1 N HCl and 1N NaOH respectively. These solutions were subjected to stress condition of 50°C for 24 hours. The resulting solutions were sufficiently diluted to obtain the concentration of 8 µg/ml.

Oxidation stress: To study the effects of oxidative conditions, aliquot of stock solution (1 mg/ml) was added to 3% H_2O_2 solution. It was then kept at 50°C for 24 hours. The resulting solution was sufficiently diluted to obtain the concentration of 8 µg/ml.

Thermal stress: The effect of temperature on Alcaftadine solution was studied by subjecting aliquot of stock solution (1 mg/ml) to hot air oven at 80°C temperature for two hours. The final solution of 8 µg/ml was prepared by adequate dilutions.

Photolytic stress: The aliquot of standard stock solution (1 mg/ml) was subjected to the UV exposure in UV chamber for 30 minutes. The resulting solution was adequately diluted to obtain the final concentration of 8 µg/ml.

Results and Discussion

Selection of solvent

Since Alcaftadine was soluble in methanol, it was used as a solvent. For dilution purpose also methanol was only used because on further dilutions with water resulted in splitting of peak.

Optimization of mobile phase

Final mobile phase Methanol: water was selected since trials done using ACN: Water caused peak splitting with different ratios. The ratio 50:50% v/v was optimized because altering the ratio to 60:40 or 70:30 resulted in asymmetrical peak. The flow rate 1.2 ml/min was optimized since at lower flow rates, i.e., at 0.8 ml/min or 1 ml/min, peak tailing of more than 2 was observed.

Optimization of chromatographic conditions

The aim was to develop a method so that it can specifically identify the analyte peak in presence of its degradants. The chromatographic separation of Alcaftadine from its degradants was achieved using Enable C_{18} G RP HPLC column with mobile phase Methanol: Water (50:50% v/v) at a flow rate of 1.2 ml/min and detection wavelength of 282 nm.

System suitability testing

The results of the system suitability parameters are shown in Table 2. The results were found within the acceptable limits. Hence the system was suitable for the proposed method.

Validation

Linearity: The linearity was observed in the concentration range of 1-16 µg/ml for Alcaftadine. The regression line equation was plotted between concentration and peak area. The regression equation y=1639x+50431 was obtained from the linearity data. The correlation coefficient was found to be 0.999.

Chromatographic parameters	Optimized conditions
Stationary phase	Enable HPLC C18 G 120A Column 250 × 4.6 mm, 5 µ
Mobile phase	Methanol: water, 50:50 (%v/v)
Flow rate	1.2 ml/min
Injection volume	20 µl
Detection wavelength	282 nm
Run time	10 minutes

Table 1: Optimized Chromatographic Conditions for RP-HPLC.

S No	Parameters	Observed results (n=6)	Acceptance criteria	Remarks
1	Theoretical plates (N)	3088.474	>2000	Method passes the system suitability test
2	Tailing factor (T)	1.36	T ≤ 1.5	
3	Repeatability (% RSD)	0.28	%RSD<2	
4	Capacity factor (k')	2.58	K'>2	

Table 2: Results for System Suitability Test by RP-HPLC.

Specificity: The results of specificity are shown in Figure 3. It showed no interference of the excipients present in drug formulation.

Precision: The results of precision studies are summarized in Table 3. The % RSD was found within the acceptable limit, i.e., <2.

Recovery studies: The % recovery was observed within the acceptable limits, i.e., 98.71%, 99.06% and 99.39% at the levels of 80%, 100% and 120% respectively. The results are summarized in Table 3.

LOD and LOQ: The results of LOD and LOQ were found to be 0.25 μg/ml and 0.75 μg/ml respectively. The results are summarized in Table 3.

Robustness: Robustness of the method was studied by making variations in the parameters like flow rate (± 0.2 ml/min), mobile phase composition (± 0.2) and detection wavelength (± 2 nm). The deliberate changes made in the flow rate, mobile phase composition and wavelength did not show major impact on retention time, assay value and peak area.

Forced degradation studies

Both the drug (API) as well as formulation was subjected to various

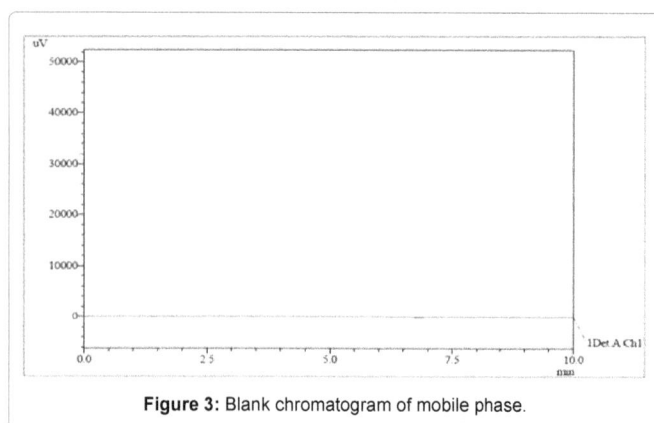

Figure 3: Blank chromatogram of mobile phase.

Parameters	Results	
Calibration curve range	1-16 μg/ml	
Regression line equation	Y=1639x+50431	
Slope	1639	
Intercept	50431	
Correlation coefficient (r^2)	0.999	
Summary of Validation		
Accuracy	% Recovery	% RSD
80%	98.87	0.495
100%	98.95	1.340
120%	99.46	0.330
Precision	Concentration (μg/ml)	% RSD
Repeatability (n=6)	8	1.205
Intraday precision (n=3)	4	0.986
	8	0.920
	12	1.08
Interday precision (n=3)	4	0.982
	8	0.582
	12	0.780
LOD	0.25 μg/ml	
LOQ	0.75 μg/ml	
Robustness	No significant changes observed with deliberate changes in the method parameters	

Table 3: Summary of the developed and validated HPLC Method.

Stress conditions	% Degradation	
	API	Formulation
Acid hydrolysis (1N HCl) 50°C, 24 hours	14.88%	14.32%
Base hydrolysis (1N NaOH) 50°C, 24 hours	15.23%	14.28%
Oxidation (3% H_2O_2) 50°C, 24 hours	6.89%	6.35%
Thermal (80°C, 2 hours)	15.13%	16.52%
Photostability (UV chamber, 30 mins)	26.32%	24.02%

Table 4: Results of Forced Degradation Studies.

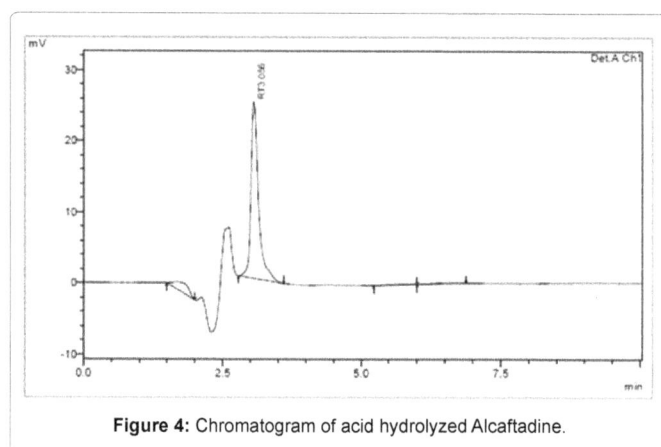

Figure 4: Chromatogram of acid hydrolyzed Alcaftadine.

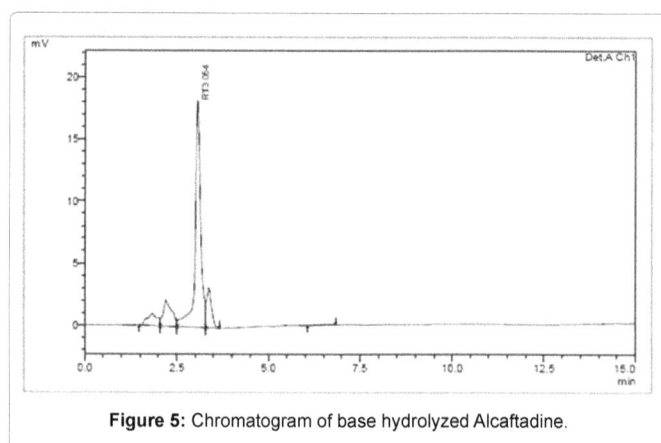

Figure 5: Chromatogram of base hydrolyzed Alcaftadine.

stress conditions like acid/ base hydrolysis, oxidation, thermal and photolytic stress conditions as per ICH guidelines. The results are tabulated in Table 4.

Acid and base hydrolysis: The drug degraded in the applied stress conditions of the acid and base hydrolysis i.e., 14.88% and 15.23% respectively for API and 14.32% and 14.28% respectively for formulation. The chromatograms showed the presence of degraded products (Figures 4 and 5).

Oxidation degradation: The drug was found to be quite stable in oxidation stress condition. It showed 6.89% and 6.35% for API and formulation respectively (Figure 6).

Thermal degradation: Adequate thermal degradation was achieved under thermal stress i.e., 15.13% and 16.52% for API and formulation respectively (Figure 7).

Photolytic degradation: The rate of photolytic degradation was higher as compared to others. The drug was degraded sufficiently i.e., 26.32% and 24.02% for API and formulation respectively (Figure 8).

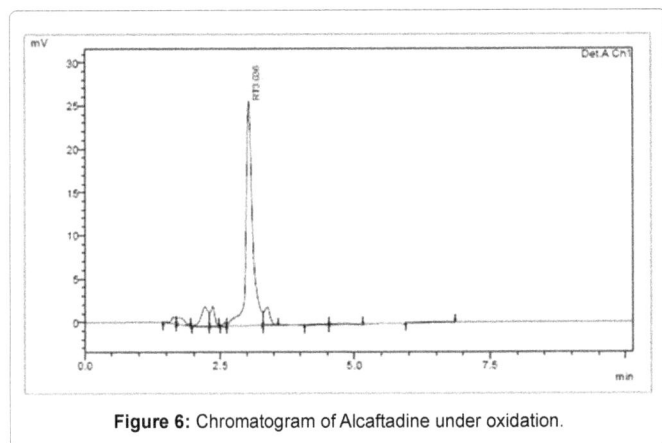

Figure 6: Chromatogram of Alcaftadine under oxidation.

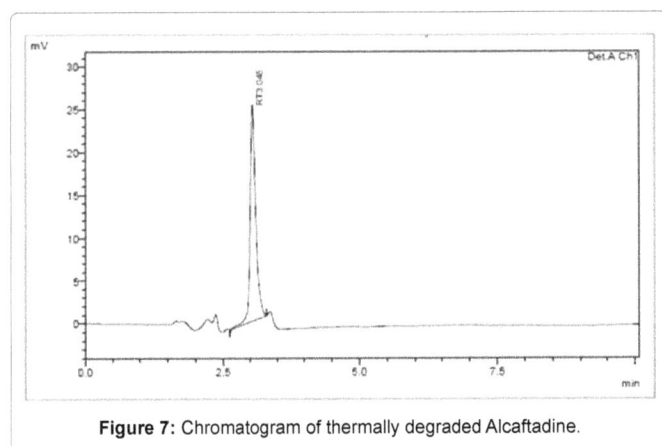

Figure 7: Chromatogram of thermally degraded Alcaftadine.

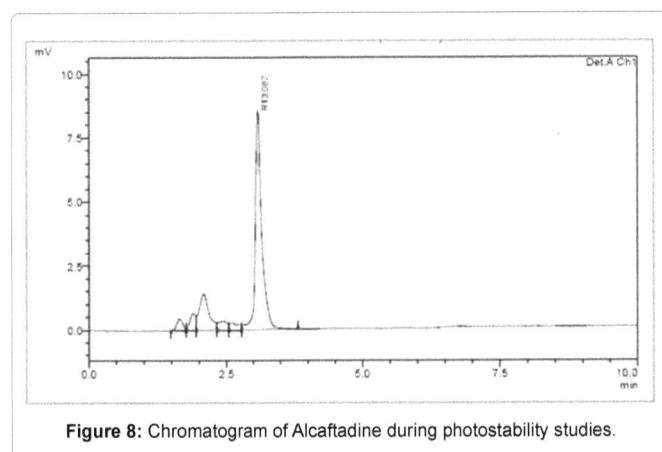

Figure 8: Chromatogram of Alcaftadine during photostability studies.

Formulation	Labeled amount ml/ml	Amount found mg/ml	% Label claim ± SD
LASTACAFT (ophthalmic solution)	2.5	2.49	99.90 ± 0.698

Table 5: Results for the Assay of Ophthalmic Solution of Alcaftadine.

Analysis of the marketed formulation

The proposed method was applied to the marketed formulation of Alcaftadine (LASTACAFT ophthalmic solution, 2.5 mg/ml). The assay results were found to be within acceptable limits, i.e., 99.10-100.39%. The results are shown in Table 5.

The developed RP-HPLC method for determination of alcaftadine from ophthalmic preparations is specific, reliable and accurate. The observed results also showed the value of relative standard deviations below 2, which indicate the accuracy and precision of the developed method. The proposed method is capable of identifying the target analyte in presence of its degradants. The degradation products generated during stress conditions were not further characterized.

Acknowledgements

The authors are heartily indebted to Shri D D Patel, President, Om Gayatri Education and Charitable Trust and Pioneer Pharmacy Degree College for their valuable guidance and support. Authors are grateful to the JHP pharmaceuticals, LLC, Rochester, Michigan, USA for providing the drug sample and the Management of Pioneer Pharmacy Degree College for providing necessary research facilities.

References

1. National PBM Drug Monograph (2015) Alcaftadine 0.25% Ophthalmic Solution (Lastacaft). VA Pharmacy Benefits Management Services. Medical Advisory Panel and VISN Pharmacist Executives.

2. (2015) US Food and Drug Administration Protecting and Promoting Your Health.

3. (2015) Office of Clinical Pharmacology Review.

4. Bohets H, McGowan C, Mannens G, Schroeder N, Edwards-Swanson K, et al. (2011) Clinical pharmacology of alcaftadine, a novel antihistamine for the prevention of allergic conjunctivitis. J Ocul Pharmacol Ther 27: 187-195.

5. ICH, Q2A, Hamonised Tripartite Guideline, Test on Validation of Analytical Procedures, IFPMA. In: Proceedings of the International Conference on Harmonization, Geneva. March, 1994.

6. ICH, Q2B, Hamonised Tripartite Guideline, Validation of Analytical Procedure: Methodology. IFPMA, In: Proceedings of the International Conference on Harmonization, Geneva. March, 1996.

7. Nash AR, Wachter AH (2005) Pharmaceutical process validation. 3rd edn. New York: Marcel Dekker Inc.

8. ICH, Q1A (R2): Stability Testing of New Drug Substances and Products (Revision 2), International Conference on Harmonization.

9. ICH, Q1B: Photostability Testing of New Drug Substances and Product, International Conference on Harmonization.

10. Biradar SP, Kalyankar TM, Wadher SJ, Moon RS, Dange SS (2014) Stability indicating HPLC method development: Review. Asian J Med and Anal Chem 1: 21-26.

11. Ngwa G (2010) Forced degradation studies as an integral part of HPLC stability indicating method development. Drug Deliv Technol 10: 56-59.

12. Blessy M, Patel RD, Prajapati PN, Agrawal YK (2014) Development of forced degradation and stability indicating studies of drugs- A review. J Pharm Anal 4: 159-165.

13. Bakshi M, Singh S (2002) Development of validated stability-indicating assay methods--critical review. J Pharm Biomed Anal 28: 1011-1040.

14. Singh S, Bakshi M (2000) Guidance on conduct of stress tests to determine inherent stability of drugs. Pharm Tech online 24: 1-14.

15. Brummer H (2011) How to approach a forced degradation study. Life Sci Technol Bull 31: 1-4.

13

A Validated Chiral Liquid Chromatographic Method for the Enantiomeric Separation of Lacosamide Drug Product and its Dosage Forms

Charagondla K *

Analytical Research & Development, InvaGen Pharmaceuticals Inc., 7 Oser ave, Hauppauge, New York, 11788, USA

Abstract

A new and simple, rapid, selective isocratic normal phase liquid chromatographic (NP-LC) method was developed for the chiral purity of Lacosamide ((R)-2-acetamido-N-benzyl-3- methoxypropionamide) and its undesired S-enantiomer ((S)-2-acetamido-N-benzyl-3-methoxypropionamide). Superior resolution between Lacosamide and its S-enantiomer was achieved on Cellulose tris (3,5-Dichlorophenylcarbamate) immobilisation of polysaccharide derivatives based Chiralpak IC (25 mm × 4.6 mm, 5 µm) column using n-hexane : ethanol (85:15, v/v) as a mobile phase at 27°C column oven temperature. The USP resolution between the enantiomers was found more than five. Mobile phase flow was fixed at a rate of 1.0 mL/min and elution was monitored at 210 nm. The test concentration was 1000 µg/ml, Limit of detection and quantitative of S-enantiomer is 0.17 µg/mL and 0.48 µg/mL respectively. The percentage RSD of the peak area of six replicate injections of S-enantiomer at LOQ concentration was 5.8. The percentage recoveries of S-enantiomer from Lacosamide (LAC) were ranged from 105%-107%. The test solution and mobile phase were observed to be stable up to 48 h on bench top. The method was found to be selective, precise, linear, accurate and also robust. This method was successfully validated according to the International Conference Harmonization (ICH) guidelines.

Keywords: Lacosamide; Enantiomers; Chiral liquid chromatography; Validation; Specificity

Introduction

Epilepsy is a major neurological disorder that affects all populations [1]. Epilepsy describes the types of recurrent seizures produced by paroxysmal excessive neuronal disorders in the brain [2-3]. The mainstay of treatment has been the long-term and consistent administration of anticonvulsant drugs [4-6]. Unfortunately, current medications are ineffective for approximately one-third of patients with epilepsy [7]. Many people continue to have seizures, while others experience disturbing side effects (e.g., drowsiness, dizziness, nausea, liver damage) [8]. The lead compound, (R)-lacosamide ((R)-2, (R)-N-benzyl 2-acetamido-3-methoxypropionamide, has recently gained regulatory approval for adjunctive therapy for partial-onset seizures in adults [9].

The biological activities of chiral substances often depend upon their stereochemistry, thus, showing significant enantioselective differences in pharmacokinetics and pharmacodynamics. There is a growing demand for the direct methods of enantiomeric separation of chiral drugs, as the US food and Drug Administration (USFDA) has issued an order to specify the enantiomeric purities of chiral drugs [10].

Enantiomeric separations have acquired importance in all stages of drug development and commercialization process. Therefore, the development of new methods for efficient chiral separations mainly based on HPLC, capillary electrophoresis (CE) or gas chromatography (GC) is more necessary [11,12]. The chromatographic separation of enantiomers using high-performance liquid chromatography (HPLC) with chiral stationary phases (CSPs) is one of the most useful and popular techniques for enantio-purity analysis in pharmaceutical preparations and biological fluids [13].

The thorough literature survey revealed that none of the most recognized pharmacopoeias or any journals including this drug for the determination chiral purity is not available on the enantiomeric separation of LAC drug substance (Figures 1A and 1B) and its penultimate stages by HPLC on the analytical method or any analytical tools. The chiral purity of drug is very important since only R-isomer

of it is active and S-isomer is inactive. The aim of this study was to develop a simple HPLC method capable to separate an enantiomer of the title compound. Our research aimed to evaluate the enantiomeric resolving capability of the immobilized-type Chiralpak IC CSP in the polar organic phase and normal-phase conditions. First, we selected a mobile phase capable of giving a baseline separation of LAC. The best analytical normal-phase condition was validated in terms of linearity, precision, accuracy, repeatability, ruggedness and robustness and limits of detection (LOD) and limit of quantification (LOQ) in order to quantify both enantiomers in active pharmaceutical ingredient and its drug dosage formulations.

This method can be able to separate all process impurities and degradants. It can also be able to separate its intermediate stages (penultimate stage) enantiomers. This method could be useful for reaction monitoring for synthesis of Lacosamide and its dosage forms. This method can be used for the routine regular analysis as well as the stability studies.

Experimental

Chemicals and reagents

N-Hexane was purchased from Merck (Worli, Mumbai, India), and Ethanol of AR grade (purity 99.9%) from Ching chaung Chemical, China. All the chemicals were used without further purification.

*Corresponding author: Charagondla Krishnaiah, Analytical Research & Development, InvaGen Pharmaceuticals Inc, 7 Oser ave, Hauppauge, New York, 11788, USA, E-mail: krishnaiah.charu@ gmail.com

Apparatus

The liquid chromatography systems Waters alliance 2695 separation module with 2998 diode array detector (DAD) and Agilent (Wilmington, Delaware, USA) 1100 system equipped with photo DAD and a variable wavelength detector (VWD), were used for method development and method validations. The output signal was monitored and processed by using Waters (Milford, MA, USA) Empower-2 software. Chiralpak-IC (250 mm × 4.6 mm, 5 μm), Immobilized cellulose tris 3,5 dichlorophenylcarbamate, coated on 5 μm silica-gel support, was purchased from Daicel Chemical Industries (Tokyo, Japan).

Chromatographic conditions

Enantioseparation was achieved on a Chiralpack-IC column (250 mm × 4.6 mm, 5 μm) by using the mobile phase consisting of n-hexane: ethanol (85:15, v/v), delivered at a flow rate of 1.0 mL/min. The column oven temperature was maintained at 27°C and the wavelength was monitored at 210 nm. The injection volume was 10 μL. 2-Propanol (IPA) was used as diluent.

Preparation of test solution and control solution

Stock solutions of (R)-LAC, (S)-LAC, (R)-LAC-1 and (S)-LAC-1 (1000 μg/mL) were prepared individually by dissolving the appropriate amount in IPA and filled to the mark with IPA, and further diluting to required concentration with IPA.

Commercially available tablets containing 5 mg of LAC (Vimpact) (n=10) were pulverized with a pestle in a porcelain mortar. The tablet powder was taken into a 100 mL of volumetric flask, made up to the mark with IPA and the suspension was sonicated for 10-12 min and filtered through 0.45 μm membrane filters. The IPA solution was injected to LC system.

Weight percentage of each enantiomer (R, S) present in the sample was calculated by area normalization method because the other isomer was having the same response to that of R-LAC.

Method validation

The method was validated (in terms of sensitivity, linearity, accuracy, precision, solution stability, mobile phase stability, and robustness) in accordance with International Conference Harmonization (ICH) guidelines.

Specificity/Selectivity: The specificity/selectivity of the developed LC method was determined in the presence of its related impurities (Figure 1C and 1D), LAC enantiomers and degradation products. Forced degradation studies were also performed on LAC bulk drug samples to provide an indication of the stability-indicating property and specificity of the proposed method. The stress conditions employed for degradation study included light (1.2 million lx h/m²/240 h), heat (90°C/10 days), humidity (40°C/75% RH/10 days), acid hydrolysis (0.1 M HCl/10 mL/3 h at 70°C), base hydrolysis (0.025 M NaOH/10 mL/90 min at RT), and peroxide degradation (6% H_2O_2/10 mL /9 h at 70°C). About 5%-15% degradation was observed in forced degradation studies. The typical chromatograms of control sample and stressed samples are represented in Figures 2A-2C. Peak purity of stressed sample of R-LAC was checked by using photo diode array detector (DAD/PDA). Purity angle was found less than purity threshold in all stress samples and no interference was found from blank demonstrating the analyte peak homogeneity.

LOD and LOQ: LOD and LOQ represent the concentration of the analyte that would yield a signal/noise ratio (S/N) of 3 and 10,

respectively. LOD and LOQ of (S)-LAC, (R)-LAC were determined by injecting a series of diluted solutions.

Precision and repeatability: Method precision was determined by measuring the repeatability (intra-day precision) and intermediate precision (inter-day precision) of retention times and peak areas for LAC enantiomers. The intra-day variability was performed by the

1A: LAC: (R)-2-acetamido-N-benzyl-3-methoxypropanamide.

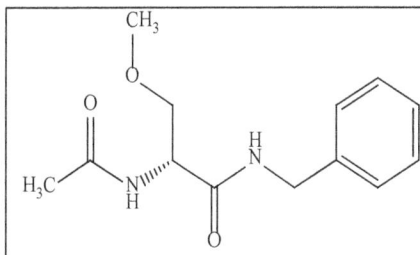

1B: LAC:(S)-2-acetamido-N-benzyl-3-methoxypropanamide.

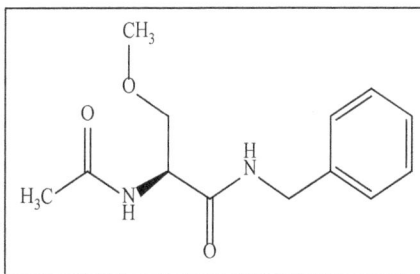

1C: (R)-Imp-1: (R)-2-acetamido-N-benzyl-3-hydroxypropanamide.

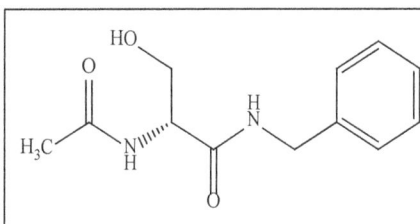

1D: (S)-Imp-1: (S)-2-acetamido-N-benzyl-3-hydroxypropanamide.

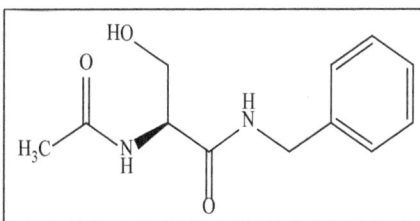

Figure 1: Lacosamide (LAC) and its impurities structures.

2A: Base degradation chromatogram.

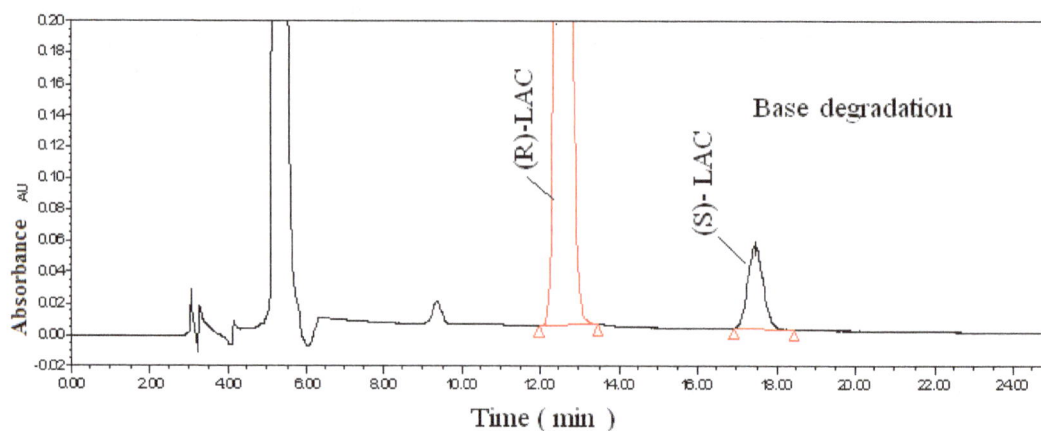

2B: Oxidative degradation chromatogram

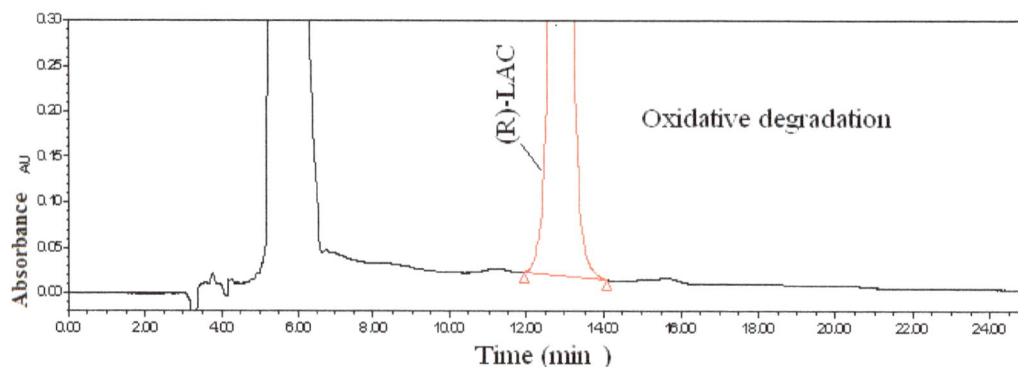

2C: Acid degradation chromatogram

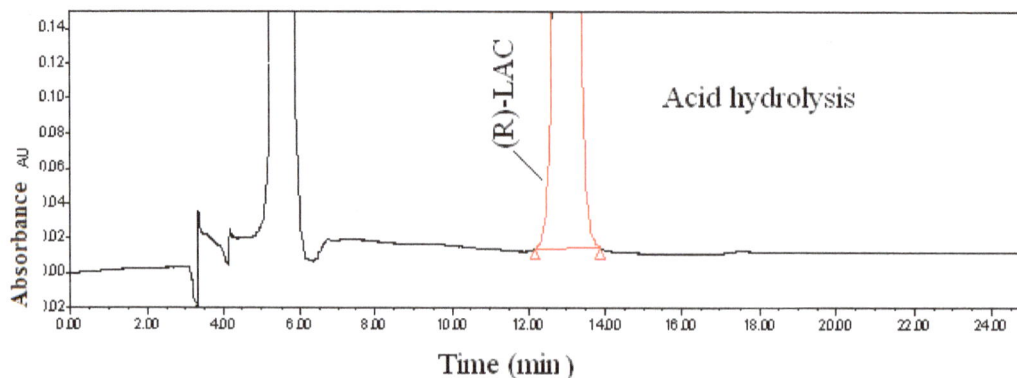

Figure 2: The degradation/specificity chromatograms of LAC.

same analyst over one day, while inter-day precision was carried out by another independent analyst over three days. In order to determine the repeatability of the method, replicate injections (n=6) of 1.5 μg/mL of LAC enantiomers were carried out. The intermediate precision was evaluated over three days by performing six consecutive injections each day. Precision was reported as % of relative standard deviation (%RSD).

Linearity: The linearity evaluation was performed with the spiked standard solutions of (S)-enantiomer at the concentrations ranging from six concentration (n=6) levels from LOQ to 150% (LOQ, 50, 75, 100, 125 and 150%). Three injections (n=3) of each solution were made under the chromatographic conditions described above, using an injection volume of 10 μL. The calibration curve was drawn by plotting the peak areas of (S)-LAC and (R)-LAC against its corresponding concentration. The %RSD and Y-intercept of the calibration curve (linear regression equation) were computed.

Accuracy: Standard addition and recovery experiments were conducted to determine accuracy of other enantiomer method for quantification of the impurities in bulk drug samples and dosage forms. The study was carried out in triplicate at 0.375, 0.75 and 1.125 μg/mL concentrations, whereas the analyte concentration was 1000 μg/mL. The percentage recoveries of (S) and (R)-LAC were calculated.

Solution stability and mobile phase stability: The solution stability of LAC and its enantiomers in this method was carried out by leaving spiked sample solution in tightly capped volumetric flask at room temperature (bench top) for 48 h. Contents of other enatiomer were determined for every 6 h interval. Mobile phase was also carried out for 48 h by injecting the freshly prepared sample solutions for every 6 h interval. Contents of other enatiomer were checked in the test solutions. Mobile phase prepared was kept constant during the study period.

Robustness: To determine the robustness of the developed method, the chromatographic conditions were deliberately altered and verified including the system suitability criteria. Establish the selectivity factor in all the robustness studies for enantiomer impurity and compare with regular experiment. The flow rate of mobile phase was 1.0 mL/min, to study the effect of flow rate on the peak USP tailing and USP resolution between enantiomers, flow rate was altered by 0.2 units, i.e, from 0.8 to 1.2 mL/min. The effect of composition of ethanol ratio on peak USP resolution and tailing of LAC were studied by varying ± 10% ethanol solvent. The effect of column temperature was studied at 27°C ± 5°C (22°C, 27°C and 32°C).

Results

Method development

The aim of this work was to separate the enantiomers for the accurate quantification of (S)-enantiomer. Racemic mixture solution of 1 mg/mL prepared in IPA was used in the method development. To develop a rugged and suitable LC method for the separation of LAC enantiomers, different mobile phases and stationary phases were employed. Initial development trials were carried out on the chiralpak AD-H column with the mobile phase combination of n-hexane and ethanol in the ratio of 90:10 (v/v). The peaks were retained very early and selectivity could be very poor. With further increasing the ratio of ethanol by ten percent volume, selectivity was remains same, but the peak shape was broad with tailing. After that introduced chiralcel OD-H and chiralpak-AS columns with different combinations of mobile phases but individual pair resolutions were observed with tailing but the selectivity was poor (Figure 3A and 3B), with merged (S)-LAC and (S)-Imp-1 peaks. Apply above mobile phase combination to phenominex Lux Cellulose-2 column and get individual separation for each enatiomers, but (R)-LAC and (R)–Imp-1 were merged Figures 3C and 3D. Finally introduce chiralpak-IC immobilized stationary phase initially with n-hexane and ethanol in the ratio of 50:50 (v/v), (R)-LAC and (S)-Imp-1 were not separated well (Figure 3E). Changing in mobile phase ratio with n-Hexane: Ethanol (7:3, v/v) (R)-LAC and (S)-Imp-1 completely merged (Figure 3F), trial with a mobile phase n-Hexane: Ethanol (9:1, v/v) combination achieved best enantiometric separation (Figure 3G). But retention time was high. Reducing the retention time by increasing ethanol 5% in the mobile phase combination, which is 85:15 (v/v) ratio of n-hexane and ethanol, eventually reduced retention time without compromising and affects on USP resolution (Rs > 3.0) with all pairs of enantiomers. In the optimized method, retention times of (R)-LAC, (S)-LAC and (R)-Imp-1 and (S)-Imp-1 its enantiomer were about 11.5 and 15.8 min; 9.5 and 12.1 respectively (Figure 3H).

However other verified mobile phase ratio combinations like 2-Propanol and Butanol, column oven temperature was also studied at room temperature 22°C and 40°C. In view of possible interference study, attempts were also made by adjusting the mobile phase flow rate to separate all process related impurities. But no significant impact was observed on the critical chromatographic parameters (resolution, USP tailing and retention time).

Influence of the mobile phase composition on enantioselectivity and resolution

Besides the already available Chiralpak IA, Chiralpak IB and Chiralpak IC CSPs, a new immobilised-type CSP for HPLC, the Chiralpak IC CSP, has been recently launched by Daicel Chemical Industries Ltd. The Chiralpak IC used cellulose tris (3,5- dichlorophenylcarbamate) immobilised onto silica 5 micron particles as a chiral selector. In the first part of our study, the resolving ability of the Chiralpak IC CSP towards LAC was investigated in the nonpolar organic mode using pure n-Hexane, n-Hexane and Ethanol with a ratio of 85:15 (v/v). The resolving power of the IC CSP was sufficiently high to achieve a baseline enantioseparation in each of the used conditions. The best resolution value was achieved by using the n-Hexane and Ethanol with a ratio of 85:15 (v/v) eluent with enantioselectivity factor (k) and resolution factor (Rs) values of 1.3 and 5.7, respectively (Figure 3I).

Chemoselective analysis

A critical aspect in the development of an analytical method to determine the enantiomeric purity of a chiral drug is the presence of impurities from the processes of synthesis, degradation or manufacturing either in the drug substance or in the finished product. The main organic impurities of LAC process in the manufacturing process are checked and found no interference from impurities. Stability studies, chemical development studies, routine batch analyses and quality control of the commercial product require a chemo- and enantioselective method capable of discriminating the enantiomeric API (Active Pharmaceutical Ingredient) from the related substances. Consequently, we turned our attention to examine the chemical selectivity of the Chiralpak IC using the n-Hexane- Ethanol (absolute 85:15, v/v) as a mobile phase. In these conditions, the LAC enantiomers were well separated and there was no overlapping with potential organic impurities. From Figure 3H, it can be seen that all the process and penultimate impurities were resolved with a k-value ranging from 1.1 to 1.3. The graphical approach for mobile phase composition verses USP resolutions (Figure 4).

Validation of the method

System suitability: The system suitability solution is prepared as mentioned in the preparation of the test solution and the control solution, to obtain a final concentration of LAC 1000 μg/mL spiked with 1.5 μg/mL of LAC enantiomer. The resolution is not less than 3.0 (Figure 3I).

LOD and LOQ: LOD and LOQ were established from S/N ratio methodology. LOD and LOQ were estimated to be 15 and 48 μg/mL for each enantiomer of (S)-LAC with good precision at LOQ. Precision at LOQ was established by calculating %RSD for replicate injections (n=6) of reference standard prepared at LOQ level. The determined LOD, LOQ limit of quantification and precision at LOQ values are reported in Table 1.

3A: Spiked chromatogram (Chiralcel-OD-H).

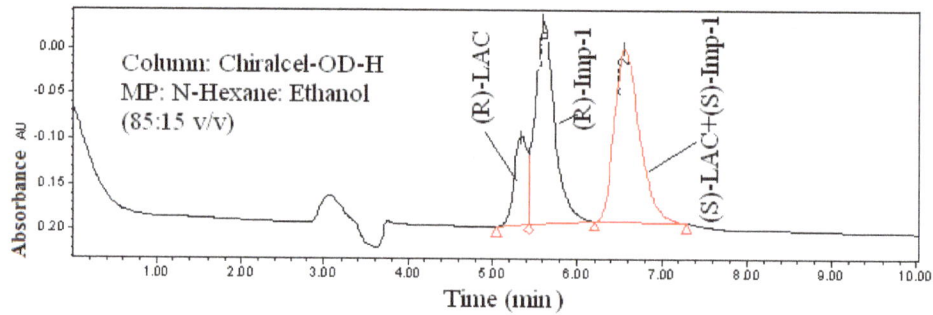

3B: Spiked chromatogram (Chiralpak-AS).

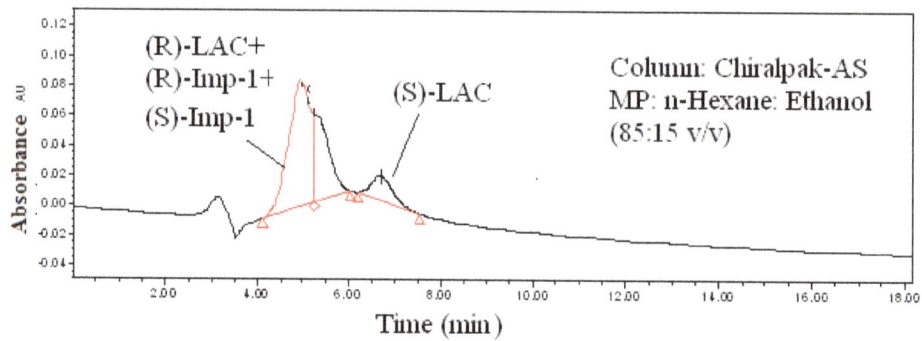

3C: (R), (S)- Imp-1 spiked chromatogram (Lux cellulose-2).

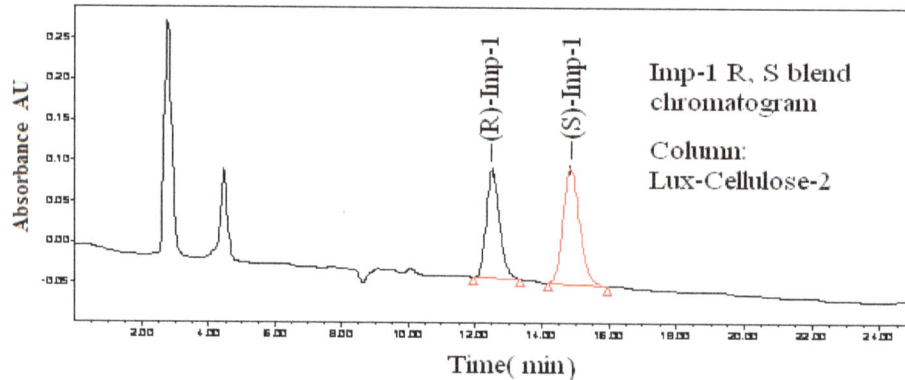

3D: (R), (S)- LAC spiked chromatogram (Lux cellulose-2).

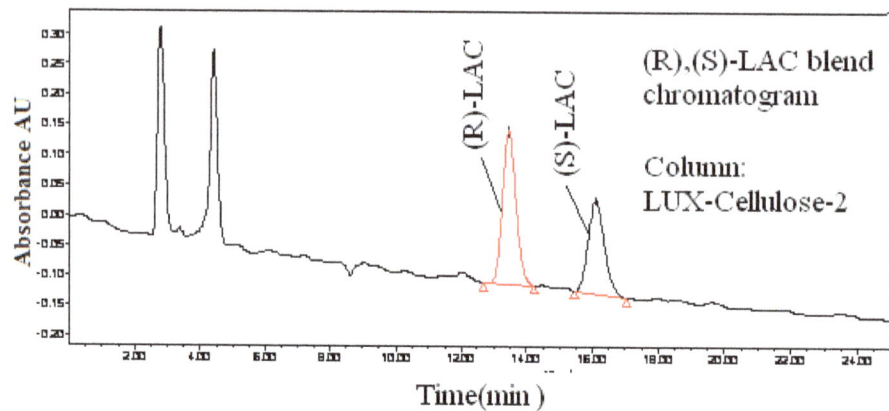

3E: Spiked chromatogram (Chiralpak-IC MP: n-Hexane: Ethanol (50:50 v/v).

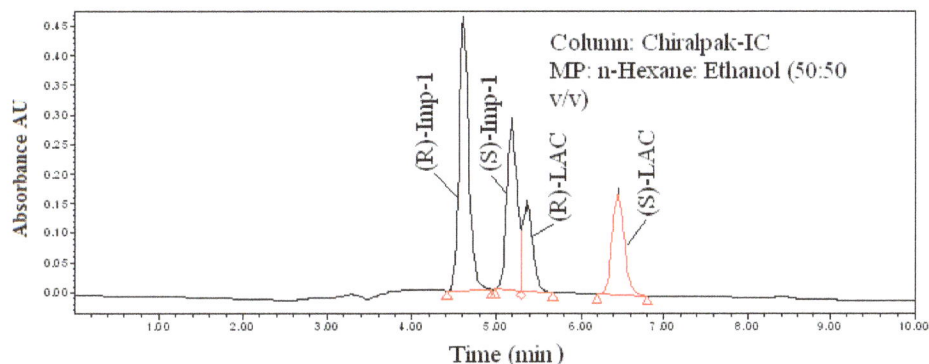

3F: Spiked chromatogram (Chiralpak-IC MP: n-Hexane: Ethanol (70:30 v/v).

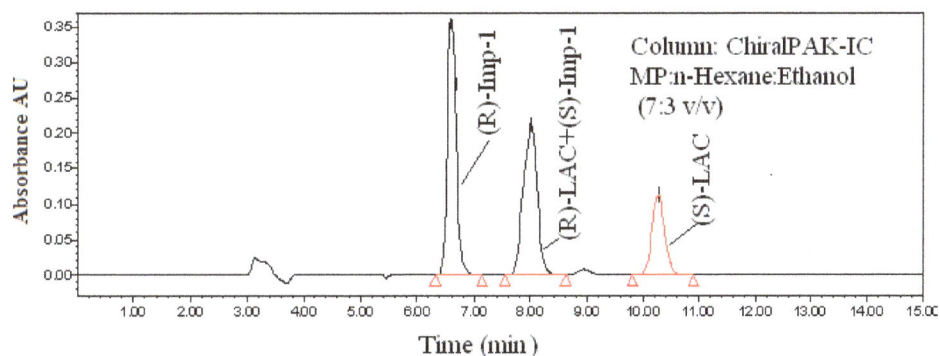

3G: Spiked chromatogram (Chiralpak-IC MP: n-Hexane: Ethanol (90:10 v/v).

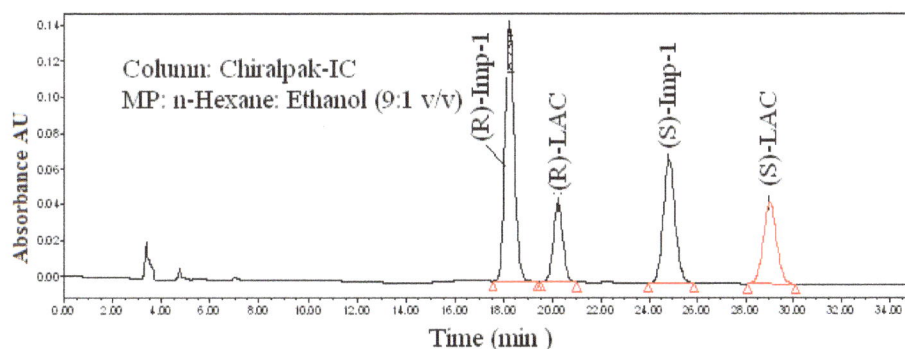

3H: Spiked chromatogram (Chiralpak-IC MP: n-Hexane: Ethanol (85:15 v/v).

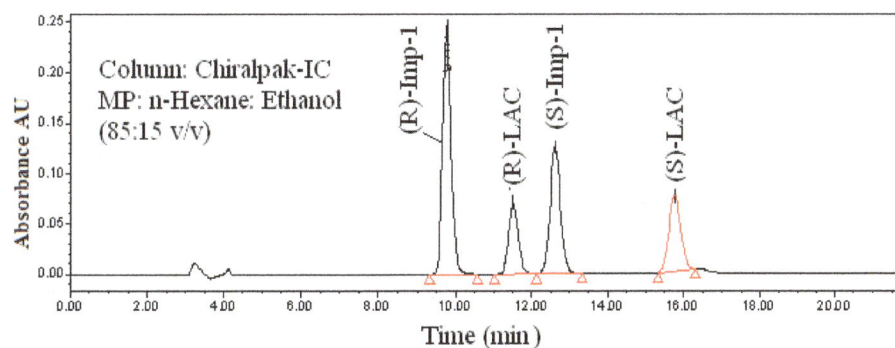

3I: System suitability chromatogram.

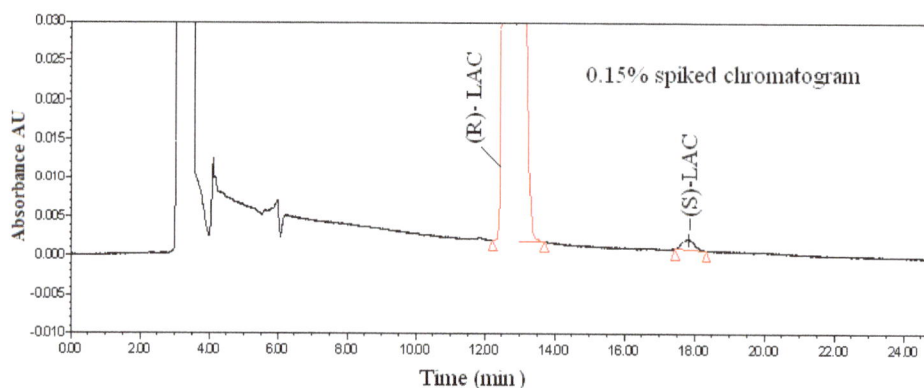

Figure 3: Blend Chromatogram of LAC and its impurities.

Mobile phase composition	(R)-Imp-1	(R)-LAC	(S)-Imp-1	(S)-LAC
n-Hexane: EtOH (1:1)	NA	2.4	0.6	4.7
n-Hexane: EtOH (7:3)	NA	3.7	0.0	5.1
n-Hexane: EtOH (85:15)	NA	4.1	2.2	5.7
n-Hexane: EtOH (9:1)	NA	5.1	3.5	6.5

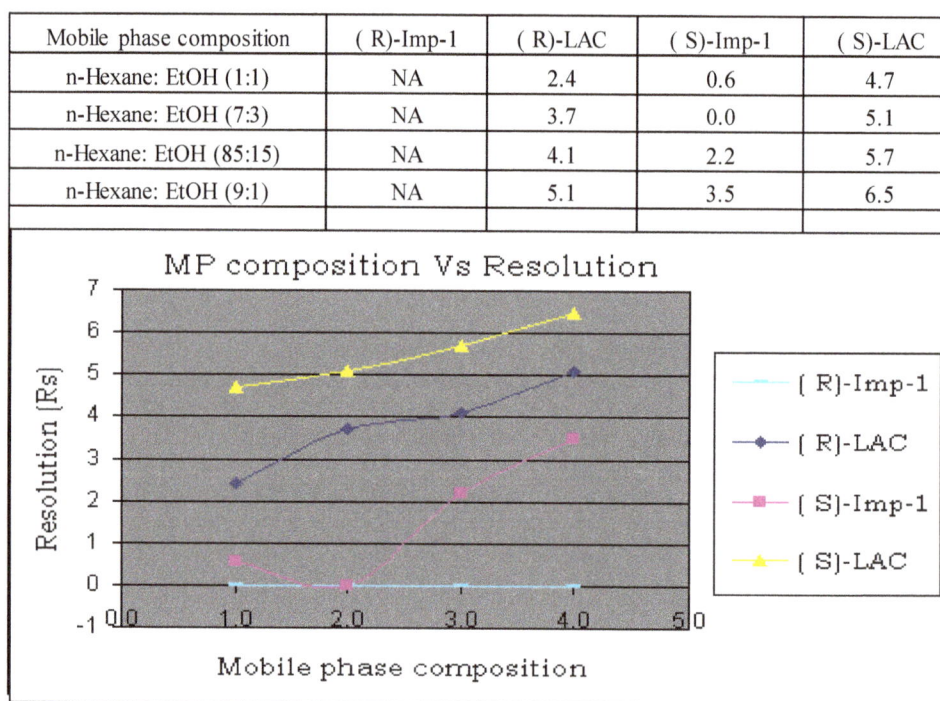

Figure 4: Graphical representation for mobile phase composition verses USP resolution.

Precision/Repeatability/Method: Method reproducibility was determined by measuring repeatability, precision, and intermediate precision (between-day precision). Repeatability of the method was determined by calculating %RSD for the enantiomer area from replicate injections (n=6) of the system suitability solution. The %w/w values of the enantiomer for six different preparations of the system suitability solution were calculated. Method precision was determined by calculating %RSD for these %w/w values Table 1.

Intermediate precision was determined by performing method precision by a different analyst on a different day. The results in Table 2 show that method reproducibility was good.

Linearity: The linearity of the HPLC method was evaluated by injecting standard concentrations of racemic samples with a concentration ranging from LOQ to 150%, i.e, 0.48 μg/mL and 2.25 μg/mL for single enantiomer. The peak area response of the both enantiomers was plotted versus the nominal concentration of the enantiomer. The linearity was evaluated by linear regression analysis, which was calculated by least squares regression method. The obtained calibration curve showed correlation coefficient greater than 0.9994 for (S)-LAC. The following regression equation was obtained $y=251717x - 73.289$ ($R^2=0.9988$) where y is the peak area of (S)-LAC and x is the concentration. For (R)-LAC the following regression equation was obtained $y=244532x + 398592$ ($R^2=0.9999$).

Accuracy: Accuracy for the determination of enantiomeric composition was determined by preparing three drug substance samples at 50%, 100% and 150% of the target concentration (0.075%,

0.15% and 0.225%). Apparent recovery was ranged from 105% to 107%. Overall percent recovery was 106 (RSD%-0.8) for the enantiomer (S)-LAC Table 3.

Solution stability: Stability of the sample in the diluent was evaluated by injecting the freshly prepared spiked test solution every 6 h up to 48 h, at room temperature bench top solution. Physical appearance, for example, color of the solution did not change; no extra peaks appeared; and the cumulative relative standard deviation for the peak area of LAC enantiomer was less than 2.0%. Peak purity obtained from the photodiode array detector for the LAC enantiomer peak, showed that the peak was pure (purity angle is less than purity threshold). The results indicated that the solutions were stable up to 48 h at room temperature bench top solution.

Mobile phase stability: Mobile phase was also carried out for 48 h by injecting the freshly prepared spiked sample solutions for every 6 h interval. Physical appearance, no turbidity/haziness were observed mobile phase during analysis, Mobile phase prepared was kept constant during the study period with no extra peaks appeared, and the cumulative relative standard deviation for the peak area of LAC enantiomer was less than 2.0%. Peak purity obtained from the photodiode array detector for the LAC enantiomer peak, showed that the peak was pure (purity angle was less than purity threshold) Table 4.

Robustness: The robustness of a method is the ability of the method to remain unaffected by small but deliberate changes in parameters such as flow rate, mobile phase composition, and column oven temperature. The effect of resolution for the system suitability solution was monitored in this study. The flow rate of the mobile phase is 1.0 mL/min and was varied to 0.8 mL/min and 1.2 mL/min. The effect of change in percent organic, i.e., ethanol, (9:1; 8:2 v/v) ratio was studied by varying its concentration to -10 and +10% absolute, i.e., n-hexane: ethanol (90%:10%, v/v); n-hexane: ethanol (80%:20%, v/v). The column temperature is ambient (oven temperature maintained at 27°C) and was varied to 32°C and 22°C. The effects of each of these parameters were studied while the rest of the parameters were kept constant, as mentioned in an above section. The resolution remained unaffected by these small changes as shown in Table 5, thus proving the method to be rugged.

Discussion

A simple and new analytical chiral method has been developed for Lacosamide intermediate stages, Lacosamide drug product and its dosage forms. This method can be able to separate all process impurities. This method has been validated as per ICH and proved for specific. This method could be useful for reaction monitoring for synthesis of Lacosamide and its dosage forms.

Conclusion

In conclusion, a simple, sensitive, chemo- and enantio selective LC method for the quantitative determination of the enantiomers of LAC was developed and validated using the immobilised-type

Situation	Station	Resolution (Rs)	Tailing factor (T)	Spiked precision (n=6), %RSD
Different column	Column-1	8.58	1.13	0.6
	Column-2	7.92	1.12	7.4
Different analyst	Analyst-1	8.58	1.13	0.6
	Analyst-2	7.96	1.12	5.0
Different day	Day-1	8.58	1.13	0.6
	Day-2	8.10	1.12	1.0

Table 2: Ruggedness (Intermediate precision).

Concentration (%)	Sample ID	Amount added(%m/m)	Amount found(%m/m)	Recovery (%)	Statistical analysis
50	1	10.25	10.80	105.4	Mean:106.4
	2	10.28	10.98	106.8	SD:0.81
	3	10.15	10.84	106.8	%RSD:0.8
100	1	10.19	10.88	106.8	Mean:106.4
	2	10.04	10.72	106.8	SD:0.81
	3	10.25	10.80	105.4	%RSD:0.8
150	1	10.11	10.76	106.4	Mean:105.7
	2	10.28	10.84	105.4	SD:0.52
	3	10.19	10.74	105.4	%RSD:0.5

Table 3: Recovery data of Enantiomer.

Station	Impurity	Initial	Day-1	%Variation (relative)	Day-2	%Variation (relative)
MS		0.154	0.148	0.006(+3.9%)	0.160	-0.006(-3.9%)
SS	S-Isomer	0.154	0.144	0.01(+6.5%)	0.155	-0.001(-0.6%)

Table 4: Mobile phase stability and Solution stability study data.

Condition	Station	USP Resolution (Rs)	USP Tailing factor (T)	Relative retention time (RRT)
Mobile phase composition change (%)	90	9.11	1.16	1.41
	100	8.58	1.13	1.39
	110	8.23	1.14	1.38
Column oven temperature(°C)	22	8.48	1.12	1.40
	27	8.58	1.13	1.39
	32	7.95	1.10	1.36
Column flow (mL/min)	0.8	9.27	1.14	1.39
	1.0	8.58	1.13	1.39
	1.2	7.92	1.11	1.38

Table 5: Robustness.

Chiralpak IC CSP in normal-phase conditions. The proposed method meets the requirements of the all guidelines and is a reliable way to determine the enantiomeric purity of LAC penultimate stages and LAC in bulk drugs and pharmaceutical dosage forms. This method can be used for the routine analysis as well as the stability studies to evaluate (R) →(S), (S) →(R) isomerisation in pharmaceutical quality control.

References

1. Hauser WA, Annegers JF, Kurland LT (1991) Prevalence of epilepsy in Rochester, Minnesota: 1940-1980.   Epilepsia 32: 429-445.

2. Evans JH (1962) Post-traumatic epilepsy. Neurology 12: 665-674.

3. Lindsay JM (1971) Genetics and epilepsy: a model from critical path analysis. Epilepsia 12: 47-54.

4. Rogawski MA, Porter RJ (1990) Antiepileptic drugs: pharmacological mechanisms and clinical efficacy with consideration of promising developmental stage compounds. Pharmacol Rev 42: 223-286.

5. Brodie MJ, Dichter MA (1996) Antiepileptic drugs. N Engl J Med 334: 168-175.

Parameter	Drug	Enantiomer
Calibration range (mg/mL)	LOQ-150%	LOQ-150%
Calibration points	6	6
Slope	244532	251717
Intercept	398592	-73.3
Residual sum of squares	0.9999	0.9988
Correlation coefficient	0.9999	0.9994

Table 1: Linearity of the developed chiral LC method.

6. Dichter MA, Brodie MJ (1996) New antiepileptic drugs.   N Engl J Med 334: 1583-1590.

7. McCorry D, Chadwick D, Marson A (2004) Current drug treatment of epilepsy in adults.   Lancet Neurol 3: 729-735.

8. Pellock JM, Willmore LJ (1991) A rational guide to routine blood monitoring in patients receiving antiepileptic drugs.   Neurology 41: 961-964.

9. Stöhr T, Kupferberg HJ, Stables JP, Choi D, Harris RH, et al. (2007) Lacosamide, a novel anti-convulsant drug, shows efficacy with a wide safety margin in rodent models for epilepsy.   Epilepsy Res 74: 147-154.

10. FDA Policy (1992) Statements for the Development of New Stereo isomeric Drugs U.S. Food & Drug Administration,;Rockville: MD.

11. Subrmanian G (2001) Chiral Separation Techniques, A Practical Approach Wiley-VCH, Germany.

12. Ahuja S (2000) Chiral Separations by Chromatography American Chemical Society, Washington, D.C. Org Proc Res Dev 5: 458-458.

13. USP Drug Index (2000) Micromedex, Inc.

14

A Simple and Modified Method Development of Vancomycin Using High Performance Liquid Chromatography

Meenakshi K Chauhan* and Nidhi Bhatt

NDDS Research Laboratory, Department of Pharmaceutics, Delhi Institute of Pharmaceutical Sciences and Research, New Delhi, India

Abstract

Vancomycin is well known as a prominent member of the glycopeptide class of antibiotics. In this paper, a rapid resolution method using high performance liquid chromatography is employed to identify the presence of Vancomycin. A robust method is established and validated for the simultaneous quantification of glycopeptides antibiotic like Vancomycin and applied for their pharmaceutical research. This method is simple, modified, user and environment friendly with new techniques and all sample preparations are performed in excellent manner. The method showed good sensitivity with 2 µg/mL limit of quantification (LOQ) and the calibration curve is linear in the range of 2-500 µg/mL concentration range. The within- run and between- run precision obtained is less than 2 and 4% respectively. The method is proposed to be applied in an experimental micro particulate carrier based oral study design. The results can proved to be specific, linear, accurate and sensitive and the available data can provide valuable information regarding the analysis and pharmacokinetics of Vancomycin after the oral administration.

Keywords: Vancomycin; Glycopeptide; High performance liquid chromatography; Antibiotics

Introduction

Vancomycin (VCM) is a glycopeptide antibiotic which is widely employed in the treatment of serious infections by Gram positive bacteria [1,2]. It was chosen for therapy of infections in patients allergic to β-lactam antibiotics. This antibiotic acts by preventing the peptidoglycan synthesis of bacterial cell wall [3]. Monitoring VCM in the biological fluids and in pharmaceutical products is of importance to prevent side effects in patients under treatment and to achieve optimum therapeutic concentrations [4]. Moreover, antibiotics including VCM have been found in the aquatic environment such as waste and polluted water resources. This compound can play a role in the maintenance or extension of antibiotic resistance bacteria, finally resulting in hazards to human health [5,6]. Therefore, screening VCM both in biological and environmental studies seems to be very important. Several different approaches in liquid chromatography have been developed to satisfy the demand for fast analysis without compromising separation efficiency and resolution. The main efforts have been focused on performing separations at high performance liquid chromatography [7].

The aim of our study is to develop a modified HPLC method for measuring total VCM concentrations with acceptable runtimes. The current method is easier to carry out the qualitative and quantitative analysis of vancomycin *in vivo*, by using new improved mobile phases in very short time. In addition, the clinical impact of developed method is planned to be investigated with the new oral formulation in our lab for validating the outcomes. This study might play a very significant role in certain therapeutic effects and warrant further study in the case of oral micro particulate based oral study design.

Experiment

Chemicals and materials

VCM was provided by Concord Biotech Limited. HPLC grade acetonitrile and Formic acid were obtained from RFCL limited. Water was purified by a Milli-Q system (Millipore, Billerica, MA, USA). All the used chemical reagents were of analytical grade and were used as received.

Instruments and chromatographic conditions

Chromatographic experiments were performed on Shimadzu UFLC system (SPD-M20A) equipped with a binary pump (model LC-20AD) with a diode array detector (Gro-Zimmern, Germany). The separation was carried out on Nucleosil 120 C_{18} 5 µm column at 35°C with a flow rate of 1 mL/min and Injection volume was 25 µL. Mobile phase was a mixture of Acetonitrile (A): Water, pH 2.0 (B) and Formic acid was used to adjust the pH of water. For data processing, a Class VP Data system (Shimadzu, Duisburg, Germany) was used. Binary pump was used for carrying two different solvents.

Preparation of standards, calibration standard and test samples

Standard stock solutions were prepared in ultra pure water. The stock solution was further diluted with water in formation of the following standard solutions, with the concentration of 2-20, 40-300 and 10-500 µg/mL, respectively. All the freshly prepared solutions were used for analysis. The chromatograms of different standard solutions are shown in Figure 1. The chromatograms were identified in linear relationship by their retention times and a response height with respect to concentration for the different calibration ranges which are put up in Figure 2.

---

*Corresponding author: Meenakshi K. Chauhan, NDDS Research Laboratory, Department of Pharmaceutics, Delhi Institute of Pharmaceutical Sciences and Research, Government of NCT of Delhi, Pushp Vihar, Sec-3, New Delhi-110 017, India, E-mail: meenakshindds@gmail.com

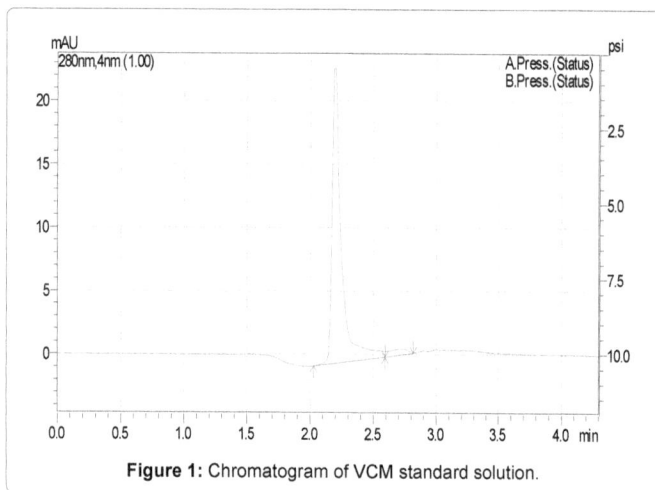

Figure 1: Chromatogram of VCM standard solution.

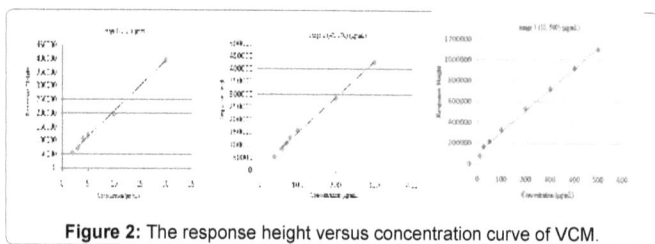

Figure 2: The response height versus concentration curve of VCM.

Results and Discussion

Optimization of chromatographic condition

To obtain reliable chromatographic results and appropriate ionization, several mobile phase systems (acetonitrile-0.2 M sodium sulfate buffer, acetonitrile-10 mM acetate buffer, pH 4.0, acetonitrile-water etc.) were tested and compared [8-19]. The results suggested that acetonitrile-acid aqueous solution was superior to the others. Meanwhile, formic acid was added into the mobile phase for pH adjustment and to improve the peak shape as well as to restrain the peak tailing. This finding was confirmed by the optimal solvent systems containing a mixture of 0.1% formic acid-acetonitrile (A) and water (B). For obtaining the effective separation, gradient elution technique was employed which guarantees high ionization, and minimize the ion suppression.

Optimization of sample preparation

Different methods of sample preparation were tested to select a coherent extraction method for VCM. Different volume ratios (50:50, 80:20, 70:30) of mobile phases were investigated. The results suggested that the volume ratio 50:50 was best among others.

Screening and analysis of optimized samples

The optimized samples were screened and analyzed. The results concluded that the retention time (2.5 min) for the drug was very less as compared to other previously reported methods in literature. The drug was very well separated and identified by their less retention time, with remarkable peak heights.

Method validation

The method was validated with regard to its specificity, linearity, accuracy and sensitivity (within and between days) [20-25].

Linearity and LOQ: Regression equations, linear ranges, correlation of coefficients are shown in Table 1. All calibrations consists excellent linearity with coefficients (r) higher than 0.995. As seen in Table 1 LOQ of three ranges were 2, 40 and 10 μg/mL respectively [26], with accuracy (recovery) between 74.5% and 111% this was absolute for quantification study. The correlation coefficient calculation and the regression analysis were performed without any type of mathematical transformation [27-31].

Extraction recovery: For checking the accuracy of the method, three standard samples with low intermediate and high concentrations of each range were analyzed. The concentrations were calculated from the corresponding calibration standard line (experimental concentrations) and were compared with the theoretical concentrations [32-38]. The extraction recovery was estimated according to Equation 1 and is shown in Table 2.

$$Recovery = ((C_{exp}/C_{teo}) \times 100) \qquad \dots 1$$

Where C_{exp}=Experimental concentration and

C_{teo}=Theoretical concentration

Assessment of precision, repeatability and reproducibility are listed in Tables 3 and 4, which were calculated on the same day and on 6 different days. Intra and inter- precision (relative standard deviation, RSD) of samples was less than 1.72% and 3.12%, respectively.

Conclusion

The chromatographic system was proved to be a rapid and efficient

Linear range (μg/mL)	Slope	Intercept	Correlation coefficient (r)
2-20	18612	22180	0.996
40-300	1403	8350	0.997
10-500	2036	10654	0.996

Table 1: Regression data and statistical analysis.

Measure conc. (μg/mL)	Accuracy (% Recovery)
2	74.5
10	95.5
20	101
40	75
150	105
300	102
10	111
250	92.4
500	83

Table 2: Recovery Data.

Spiked conc. (μg/mL)	Mean response ± SD	Intra-run (RSD%)
2	57164.4 ± 793.907	1.38
4	74595.83± 1288.45	1.72
10	221093.3 ± 3629.48	1.64
100	158255.2 ± 1668.99	1.05
200	296663 ± 2399.00	0.80
300	428491 ± 1244.61	0.29
400	887793.1 ± 7134.58	0.80
500	1122905 ± 8527.87	0.75

Table 3: Repeatability (Intra-run precision). SD: Standard deviation; RSD: Relative standard deviation.

Spiked conc. (µg/mL)	Mean response ± SD	Inter-run (RSD%)
2	57333.27 ± 1109.148	1.93
4	75966.8 ± 2371.276	3.12
10	221859.6 ± 5534.238	2.49
100	160598.5 ± 4399.284	2.73
200	293177 ± 2511.768	0.85
300	428532.9 ± 2820.084	0.65
400	892793.1 ± 7691.935	0.86
500	1119745 ± 8069.782	0.72

Table 4: Reproducibility (Inter-run precision). SD: Standard deviation; RSD: Relative standard deviation.

for research purpose of VCM formulation in systemic use. The method was established and validated for the simultaneous quantification of the drug and can be applied for their pharmacokinetic research. These works could provide more in- depth knowledge into the active components working *in vivo* with VCM and would be helpful for further investigation regarding the pharmacology and mechanism of VCM. This study would be helpful for explaining the metabolism of drug in rat and human plasma.

References

1. Nicolaou KC, Boddy CN, Bräse S, Winssinger N (1999) Chemistry, Biology, and Medicine of the Glycopeptide Antibiotics. Angew Chem Int Ed Engl 38: 2096-2152.

2. Nailor MD, Sobel JD (2011) Antibiotics for gram-positive bacterial infection: vancomycin, teicoplanin, quinupristin/dalfopristin, oxazolidinones, daptomycin, telavancin and ceftaroline. Med Clin North Am 95: 723-742.

3. Allen NE, Nicas T (2003) Mechanism of action of oritavancin and related glycopeptide antibiotics. FEMS Microbiol Rev 26: 511-532.

4. Zhang T, Watson DG, Azike C, Tettey JN, Stearns AT, et al. (2007) Determination of vancomycin in serum by liquid chromatography-high resolution full scan mass spectrometry. J Chromatogr B Analyt Technol Biomed Life Sci 857: 352-356.

5. Li B, Zhang T, Xu Z, Fang HH (2009) Rapid analysis of 21 antibiotics of multiple classes in municipal wastewater using ultra performance liquid chromatography-tandem mass spectrometry. Anal Chim Acta 645: 64-72.

6. Tamtam F, Mercier F, Le Bot B, Eurin J, Tuc Dinh Q, et al. (2008) Occurrence and fate of antibiotics in the Seine River in various hydrological conditions. Sci Total Environ 393: 84-95.

7. Belal F, el-Ashry SM, el-Kerdawy MM, el-Wasseef DR (2001) Voltametric determination of vancomycin in dosage forms through treatment with nitrous acid. Arzneimittelforschung 51: 763-768.

8. Jesús Valle MJ, López FG, Navarro AS (2008) Development and validation of an HPLC method for vancomycin and its application to a pharmacokinetic study. J Pharm Biomed Anal 48: 835-839.

9. Bijleveld Y, de Haan T, Toersche J, Jorjani S, van der Lee J, et al. (2014) A simple quantitative method analysing amikacin, gentamicin, and vancomycin levels in human newborn plasma using ion-pair liquid chromatography/tandem mass spectrometry and its applicability to a clinical study. J Chromatogr B Analyt Technol Biomed Life Sci 951-952: 110-118.

10. Júnior AR, Vila MMDC, Tubino M (2008) Green spectrophotometric method for the quantitative analysis of vancomycin in pharmaceuticals and comparison with HPLC. Anal Lett 41: 822-836.

11. El-Ashry SM, Belal F, El-Kerdawy MM, El Wasseef DR (2000) Spectrophotometric Determination of Some Phenolic Antibiotics in Dosage Forms. Microchim Acta 135: 191-196.

12. Vila MMDC, Salomão AA, Tubino M, (2008) Eclectic Chemistry. Eclet Quim 33: 67-72.

13. El-Didamony AM, Amin AS, Ghoneim AK, Telebany AM (2006) Indirect spectrophotometric determination of gentamicin and vancomycin antibiotics based on their oxidation by potassium permanganate. Cent Eur J chem 4: 708-722.

14. Sastry CSP, Rao TS, Rao PSNHR, Prasad UV (2002) Assay of Vancomycin and Dobutamine Using Sodium Metaperiodate. Microchim Acta 140: 109-118.

15. Pfaller MA, Krogstad DJ, Granich GG, Murray PR (1984) Laboratory evaluation of five assay methods for vancomycin: bioassay, high-pressure liquid chromatography, fluorescence polarization immunoassay, radioimmunoassay, and fluorescence immunoassay. J Clin Microbiol 20: 311-316.

16. Lam MT, Le Chris X (2002) Competitive immunoassay for vancomycin using capillary electrophoresis with laser-induced fluorescence detection. Analyst 127: 1633-1637.

17. Tetin SY, Swift KM, Matayoshi ED (2002) Measuring antibody affinity and performing immunoassay at the single molecule level. Anal Biochem 307: 84-91.

18. Furuta I, Kitahashi T, Kuroda T, Nishio H, Oka C, et al. (2000) Rapid serum vancomycin assay by high-performance liquid chromatography using a semipermeable surface packing material column. Clin Chim Acta 301: 31-39.

19. López KJ, Bertoluci DF, Vicente KM, Dell'Aquilla AM, Santos SR (2007) Simultaneous determination of cefepime, vancomycin and imipenem in human plasma of burn patients by high-performance liquid chromatography. J Chromatogr B Analyt Technol Biomed Life Sci 860: 241-245.

20. Hagihara M, Sutherland C, Nicolau DP (2013) Development of HPLC methods for the determination of vancomycin in human plasma, mouse serum and bronchoalveolar lavage fluid. J Chromatogr Sci 51: 201-207.

21. Backes DW, Aboleneen HI, Simpson JA (1998) Quantitation of vancomycin and its crystalline degradation product (CDP-1) in human serum by high performance liquid chromatography. J Pharm Biomed Anal 16: 1281-1287.

22. Farin D, Piva GA, Gozlan I, Kitzes-Cohen R (1998) A modified HPLC method for the determination of vancomycin in plasma and tissues and comparison to FPIA (TDX). J Pharm Biomed Anal 18: 367-372.

23. Saito M, Santa T, Tsunoda M, Hamamoto H, Usui N (2004) An automated analyzer for vancomycin in plasma samples by column-switching high-performance liquid chromatography with UV detection. Biomed Chromatogr 18: 735-738.

24. Ye G, Cai X, Wang B, Zhou Z, Yu X, et al. (2008) Simultaneous determination of vancomycin and ceftazidime in cerebrospinal fluid in craniotomy patients by high-performance liquid chromatography. J Pharm Biomed Anal 48: 860-865.

25. Tariq A, Siddiqui MR, Kumar J, Reddy D, Negi PS, et al. (2010) Development and validation of high performance liquid chromatographic method for the simultaneous determination of ceftriaxone and vancomycin in pharmaceutical formulations and biological samples. Sci Asia 36: 297-304.

26. Favetta P, Guitto J, Bleyzac N, Dufresne C, Bureau J (2001) New sensitive assay of vancomycin in human plasma using high-performance liquid chromatography and electrochemical detection. J Chromatogr B Biomed Sci Appl 751: 377-382.

27. Shibata N, Ishida M, Prasad YV, Gao W, Yoshikawa Y, et al. (2003) Highly sensitive quantification of vancomycin in plasma samples using liquid chromatography-tandem mass spectrometry and oral bioavailability in rats. J Chromatogr B Analyt Technol Biomed Life Sci 789: 211-218.

28. Cheng C, Liu S, Xiao D, Hollembaek J, Yao L, et al. (2010) LC-MS/MS method development and validation for the determination of polymyxins and vancomycin in rat plasma. J Chromatogr B 878: 2831-2838.

29. Seifrtová M, Nováková L, Lino C, Pena A, Solich P, et al. (2009) An overview of analytical methodologies for the determination of antibiotics in environmental waters. Anal Chim Acta 649: 158-179.

30. Abu-Shandi KH (2009) Determination of vancomycin in human plasma using high-performance liquid chromatography with fluorescence detection. Anal Bioanal Chem 395: 527-532.

31. Musenga A, Mandrioli R, Zecchi V, Luppi B, Fanali S, et al. (2006) Capillary electrophoretic analysis of the antibiotic vancomycin in innovative microparticles and in commercial formulations. J Pharm Biomed Anal 42: 32-38.

32. Bonnici PJ, Damen M, Waterval JC, Heck AJ (2001) Formation and efficacy of vancomycin group glycopeptide antibiotic stereoisomers studied by capillary electrophoresis and bioaffinity mass spectrometry. Anal Biochem 290: 292-301.

33. LeTourneau DL, Allen NE (1997) Use of capillary electrophoresis to measure dimerization of glycopeptide antibiotics. Anal Biochem 246: 62-66.

34. Heinisch S, Rocca JL (2009) Sense and nonsense of high-temperature liquid chromatography. J Chromatogr A 1216: 642-658.

35. Yea G (2008) Simultaneous determination of vancomycin and ceftazidime in cerebrospinal fluid in craniotomy patients by high-performance liquid chromatography. J Pharm Bio Analysis 48: 860-865.

36. Kitahashi T, Furuta I (2001) Determination of vancomycin in human serum by micellar electrokinetic capillary chromatography with direct sample injection. Clin Chim Acta 312: 221-225.

37. Hefnawy MM, Sultan MA, Al-Shehri MM (2007) HPLC separation technique for analysis of bufuralol enantiomers in plasma and pharmaceutical formulations using a vancomycin chiral stationary phase and UV detection. J Chromatogr B Analyt Technol Biomed Life Sci 856: 328-336.

38. Diana J, Visky D, Roets E, Hoogmartens J (2003) Development and validation of an improved method for the analysis of vancomycin by liquid chromatography selectivity of reversed-phase columns towards vancomycin components. J Chromatogr A 996: 115-131.

15

Development and Validation of High Performance Thin Layer Chromatography for Determination of Esomeprazole Magnesium in Human Plasma

Gosavi SM* and Tayade MA

Department of Pharmaceutical Chemistry, M. G. V.'s Pharmacy College, Panchavati, Mumbai, Agra Road, Nashik- 422003, Maharashtra, India

Abstract

A simple, sensitive, rapid and economic high performance thin layer chromatographic method has been developed for determination of Esomeprazole magnesium in human plasma by liquid-liquid extraction. The plasma sample was extracted using chloroform. A concentration range from 200-700 ng/spot of Esomeprazole magnesium was used for calibration curve. The percent recovery of Esomeprazole magnesium was found 101.61 percent. The mobile phase constitute of ethyl acetate: methanol: ammonia (32%) (9:1: 0.5 v/v). Densiometric analysis was carried out at wavelength 301 nm. The Rf value for Esomeprazole magnesium was found 0.54 ± 0.05. The stability of Esomeprazole magnesium in human plasma was confirmed during freeze thaw cycles at -30°C, on bench top during 24 h at room temperature and post preparative for 48 h. The proposed method was validated statistically by performing recovery study for determination of Esomeprazole magnesium in human plasma by liquid-liquid extraction.

Keywords: HPTLC; Esomeprazole magnesium; Human plasma; Liquid-liquid extraction

Introduction

Esomeprazole magnesium dihydrate1 (ESO), bis(5-methoxy-2-[(S)-[(4-methoxy-3,5dimethyl-2-pyridinyl) methyl] sulfinyl]-1-Hbenzimidazole-1-yl) magnesium dihydrate (Figure 1) a, is a compound that inhibits gastric acid secretion. ESO is cost-effective in the treatment of gastric oesophageal reflux diseases. ESO is the S-isomer of omeprazole, the first single optical isomer proton pump inhibitor, generally provides better acid control than current racemic proton pump inhibitors and has a favorable pharmacokinetic profile relative to omeprazole. Several methods have been employed for the estimation of ESO alone and combination with other drugs such as UV and RP-HPLC methods. Literature survey reveals that many analytical methods such as UV spectrophotometric [1-4], HPLC methods [5-10], LCMS [11], HPTLC [12-14] methods are reported for determination of Esomeprazole magnesium individually as well in combination. Specific and sensitive methods based on mass spectrometry methods were reported earlier. Earlier reports on HPLC based bioanalytical estimation of esomeprazole resulted in lesser sensitivity, and high noise in the base line indicating a need to develop a more efficient, sensitive, simple and rapid method in human plasma. To access the reproducibility and wide applicability of the developed method, it was validated as per FDA guidelines [15].

Materials and Methods

Instrumentation

HPTLC Camag with precoated silica gel Plate 60F254 (20 cm × 10 cm) 250 μm thicknesses (E. Merck, Darmstadt, Germany) was used as stationary phase. Sample application was done by using Camag 100 μl syringe and Camag Linomat V applicator. The sample was sprayed in the form of narrow bands of 8 mm length at a constant rate 2 μl/s. Linear ascending development was carried out in 20 cm × 10 cm twin trough glass chamber (Camag, Muttenz, Switzerland). The densitometric scanning was performed by using Camag TLC scanner III supported by win CATS software (V1.4.2.8121 Camag). Evaluation of chromatogram was done by using peak areas of drug.

Chemicals

Esomeprazole magnesium (A.S Bulk drugs, Hyderabad, India), was used as such by without checking their purity. The HPLC grade methanol and Analytical Reagent grade ethyl acetate, ammonia solution, chloroform was purchased from Modern sciences, Nashik, India. Human plasma used for research work was supplied by Arpan Blood Bank, Nashik, Maharashtra, India.

Chromatographic condition

Mobile phase was selected as mixture of Ethyl acetate: methanol: ammonia solution (32%) in the ratio of (9:1:0.5 v/v/v) for the development of plates. Time for chamber saturation was optimized to 20 min. The length of chromatographic development was 70 mm. The densitometric scanning was performed at 301 nm.

Preparation of stock solution and working standard solution

Stock solutions1 mg/ml each of Esomeprazole magnesium was prepared in methanol.

Preparation of plasma sample

To 1 ml of fresh human plasma in separate 15 ml of centrifuge tubes, spiked 1 ml of drug solution from 1000 mg/ml stock solution. Wait for C_{max} i.e up to T_{max} of drug (1 and ½ h) to reach equilibrium concentration. 5 ml extracting solvent chloroform was then added in

*Corresponding author: Gosavi SM, Department of Pharmaceutical Chemistry, M. G. V.'s Pharmacy College, Panchavati, Mumbai, Agra Road, Nashik- 422003, Maharashtra, India, E-mail: suvi.gosavi999@gmail.com

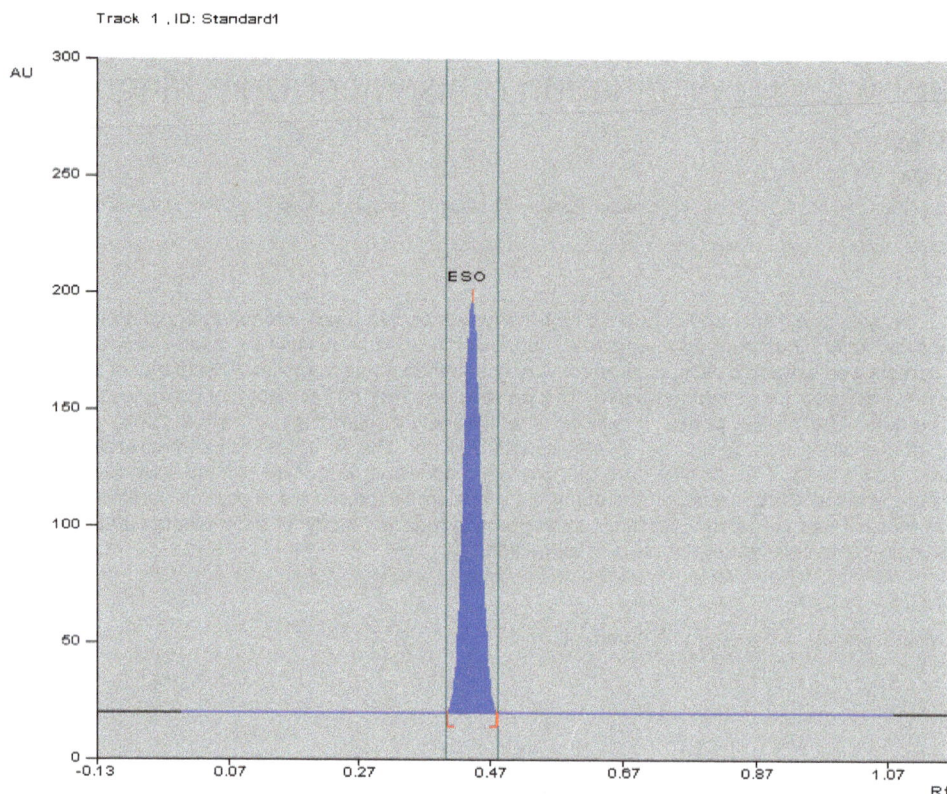

Figure 1: Typical densitogram of Esomeprazole magnesium in human plasma (300 ng/spot). Typical densitogram of Esomeprazole magnesium in human plasma (300 ng/spot).

centrifuge tube. Tube was cycomix for 5 min and then centrifuged at 3000 for 20 min. After centrifugation process, the supernatant liquid was collected in another tube and volume make up to 10 ml with methanol. The analysis was carried on HPTLC.

Method validation

Calibration plot: The calibration plot for the HPTLC method was constructed by analysis of six solutions containing different concentrations of Esomeprazole magnesium (200-700 ng/ml). The Esomeprazole magnesium can be determined at LLOQ 2 µl/ml. In the range 200-700 ng/ml the data were best fitted by a linear equation y=mx+b, the coefficient of determination (R2) was 0.9958.

Selectivity: For selectivity, analyses of blank sample of the appropriate biological matrix (plasma) were done. Each blank sample was tested for interference, and selectivity was ensured at the lower limit of quantification (LLOQ) i.e., 200 ng/ml [16,17].

Precision: Precision was measured using a minimum of three determinations per concentration (200, 400, 600 ng/spot). Intraday precision (Repeatability) was performed by taking three different concentrations (200, 00, 600 ng/spot) covering specified range in the triplicates and were analyzed three times within a day with same operator and with same equipment. Inter day precision was determined by analyzing three different concentrations (200, 400, 600 ng/spot) in triplicates on three different days within same laboratory conditions.

Accuracy: Accuracy was determined by replicate analysis of samples containing known amounts of the analyte (200, 400, 600 ng/spot). The study was determined by spiking known amount of standard

stock to the test solution prepared from tablet formulation at three different spiking level 80%, 100%, 120% of target concentration.

Recovery: Recovery experiments was performed by comparing the analytical result for extracted samples at three concentrations (200, 400, 600 ng/spot) with unextracted standards that represent 100% recovery.

Limit of detection (LOD) and limit of quantitation (LOQ): The parameters LOD and LOQ were determined using the signal-to-noise ratio by comparing results of the test of samples with known concentrations of analyte to blank samples. The analyte concentration that produced a signal-to noise ratio of 3:1 was accepted as the LOD. The LOQ was identified as the lowest plasma concentration of the standard curve that could be quantified with acceptable accuracy, precision and variability.

Stability

Bench top stability: A Stock solution of Esomeprazole magnesium was kept at room temperature for 24 hours.

Post preparative stability: A Stock solution of Esomeprazole magnesium was kept at room temperature for 48 days.

Freeze thaw stability: The stability of low and high quality control samples were determined after three freeze thaw cycles.

Results and Discussion

Extraction Procedure Optimization One of the most difficult task during the method development was to achieve a high and reproducible recovery from the solvent, which is used for extraction of the drug

(Table 1). Different solvents were tried for the extraction of from human plasma. Densitogram obtained shows the Rf at 0.52 and is shown in Figure 1. HPTLC densitogram of 200-700 ng/band was analyzed and observed 3-D view was shown in Figure 2.

Linearity

The linear plot was observed in the concentration range of 200-700 ng/spot. Results obtained are shown in Table 2 and calibration plot obtained was shown in Figure 2.

Precision

Intraday and interday precision of Esomeprazole magnesium in human plasma assures the repeatability of test results. The % RSD found was below 2. Results of intraday and interday precision were shown in Table 3 and Table 4, respectively.

Accuracy

Accuracy was studied by standard addition method and % recovery found was within acceptable limit. Results of recovery study are shown in Table 2 and Table 4 and statistical validation is shown in Table 5.

Recovery

The recovery of Esomeprazole magnesium for HPTLC recovery at the three concentrations 200, 400, 600 ng/spot were found to 87.82%, 83.84%and 95.27%, respectively (Table 6).

Selectivity

Analyses of blank sample of the appropriate biological matrix (plasma) were done. Each blank sample was tested for interference,

Sn no	Conc. in ng/spot	Area
1	200	3197.20
2	300	4317.58
3	400	5101.51
4	500	6038.19
5	600	7014.55
6	700	7795.89

Table 2: Data of calibration curve of Esomeprazole magnesium in human plasma.

Conc. in ng/spot	Mean	SD	% RSD	SE
200	3219.249	15.51388	0.48191	8.957203
400	5422.349	22.42588	0.413582	12.94797
600	7327.349	51.20786	0.698859	29.56574

Table 3: Data for intraday precision of Esomeprazole magnesium in human plasma by HPTLC method.

Conc. in ng/spot	Mean	SD	% RSD	SE
200	3280.24	51.75421	1.57762	29.88118
400	5654.73	23.75512	0.420093	13.71543
600	7675.251	95.36232	1.242465	55.05908

Table 4: Data for interday precision of Esomeprazole magnesium in human plasma by HPTLC method.

Level of addition	Tablet conc. (ng/band)	API conc. (ng/band)	Total conc. In ng/spot	% recovery
80%	300	240	540	95.47
	300	240	540	99.19
	300	240	540	100.2
100%	300	300	600	99.90
	300	300	600	101.2
	300	300	600	100.25
120%	300	360	660	99.80
	300	360	660	100.01
	300	360	660	98.59

Table 5: Data for recovery study of Esomeprazole magnesium in human plasma by HPTLC method.

Sn no	Concentration (ng)	% Recovery
1	200	87.82
2	400	83.84
3	600	95.27

Table 6: Result of recovery of Esomeprazole magnesium in human plasma.

Sn no	Solvents used	Recovery
1	Acetonitrile	No recovery
2	Ethyl acetate	No recovery
3	Dimethyl ether	No recovery
4	Chloroform 1 ml	20%
5	Chloroform 2 ml	40%
6	Chloroform 4 ml	70-80%
7	Chloroform 5 ml	90%

Table 1: Extraction procedure optimization.

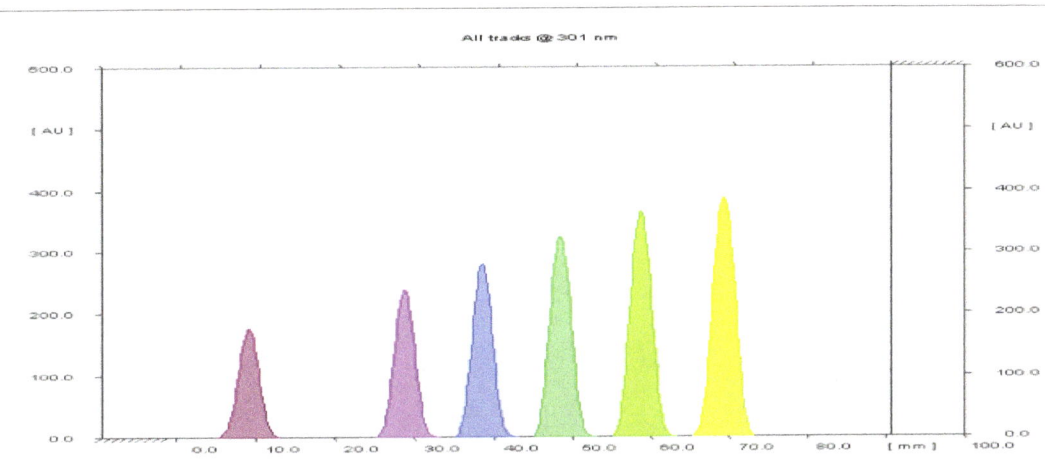

Figure 2: 3-D view of Esomeprazole magnesium 200-700 ng/band.

and selectivity was ensured at the lower limit quantification and chromatogram was shown in Figure 3.

Stability

Freeze thaw stability: The stability of low and high quality control samples were determined after three freeze thaw cycles (Table 7).

Short term stock stability: A Stock solution of Esomeprazole magnesium was kept at room temperature for 24 hours (Table 8).

Post preparative stability: A Stock solution of Esomeprazole magnesium was kept at room temperature for 48 days (Table 9).

Conclusion

The proposed HPTLC method for the estimation of Esomeprazole magnesium in human plasma is selective and sensitive. Sensitivity of the method is suitable for handling various plasma levels of the drug. The method is economical and faster than earlier published methods. In future, we can use this method for bioequivalence study.

Acknowledgement

The authors are thankful to the Management and Principal, Dr. Rajendra S. Bhambar of M G V's Pharmacy College, Nashik for providing necessary facilities for the research work. The authors are also thankful to Arpan Blood Bank, Nashik for providing human plasma, (A.S. Bulk Drugs, Hyderabad, India) for providing Esomeprazole Magnesium, as a gift sample for the research work.

Sn no	Conc. In ng/spot	Recovery (%) Thaw extract
1	200	94.23
2	400	100.69
3	600	100.2

Table 7: Result of freeze thaw stability of Esomeprazole in human plasma.

Sn no	Conc. In ng/spot	Recovery (%) Thaw extract
1	200	93.18
2	400	95.74
3	600	97.36

Table 8: Results of Bench top stability of ESO in human plasma.

Sn no	Conc. In ng/spot	Recovery (%) Thaw extract
1	200	99.173
2	400	100.5
3	600	98.40

Table 9: Results for post preparative stability of Esomeprazole in human plasma.

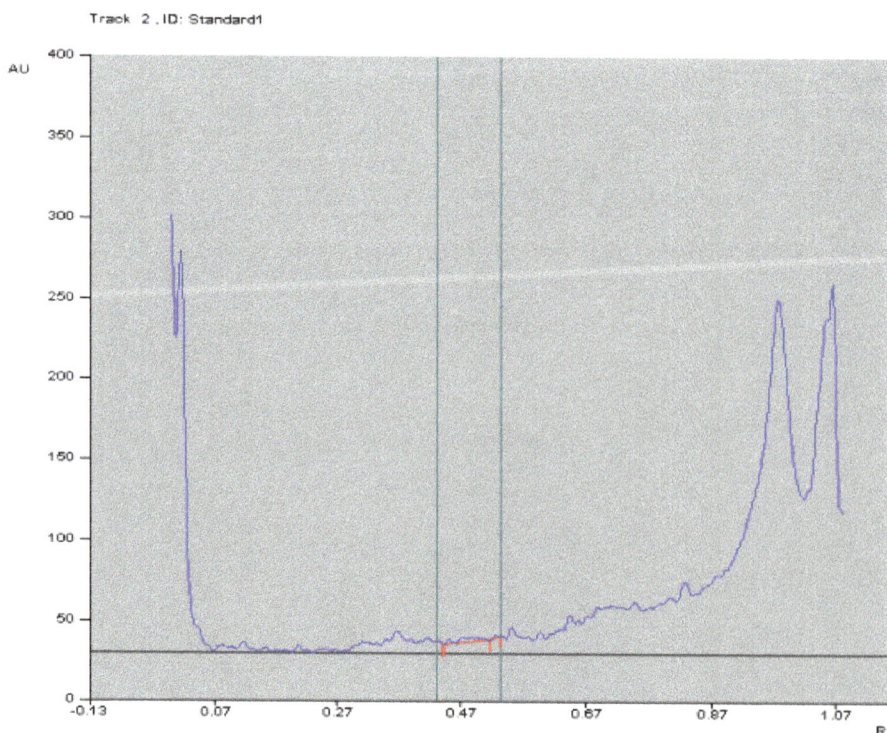

Figure 3: Chromatogram of Blank human plasma at 300 ng/spot.

References

1. Ashutosh KS, Manidipa D, Seshagiri R (2013) Stability indicating simultaneous estimation of assay method for esomeprazole and naproxen in bulk as well as in pharmaceutical formulation by using RP-HPLC. World J Pharm and Pharma Sci 4: 1897-1920.

2. Jain MS, Agrawal YS, Chavhan RB, Bari MM, Barhate SD (2012) UV Spectrophotometric Methods for simultaneous estimation of Levosulpiride and Esomeprazole in capsule dosage form. Asian J Pharm Ana 4: 106-109.

3. Putta RK, Somashekar S, Mallikarjuna GM, Shantakumar SM (2010) Physico-chemical characterization, UV spectrophotometric method development and validation studies of Esomeprazole Magnesium Trihydrate. J Chem Pharm Res 2: 484-490.

4. Lakshmana PS, Shirwaikar A, Annie S, Dineshkumar C, Joseph A, et al. (2008) Simultaneous estimation of Esomeprazole and Domperidon by UV spectrophotometric method. Indian J Pharm Sci 70: 128-131.

5. Singh S, Rai J, Inamullah, Nisha C, Hemendra G, et al. (2013) Stability Indicating Simultaneous Equation Method for Determination of Domperidone and (S)-Esomeprazole Magnesium in Capsule Dosage Form Using UV-Spectrophotometer. Br J Pharmaceut Res 3: 435-445.

6. Deepak KJ, Nitesh J, Rita C, Nilesh J (2011) The RP-HPLC method for simultaneous estimation of Esomeprazole and Naproxen in binary combination. Pharm Methods 2: 167-172.

7. Kumar SA (2016) Development and validation of a new bioanalytical method for simultaneous estimation of esomeprazole and naproxen in human plasma by using RP-HPLC. Br J Pharmaceut Res 4: 2312-2327.

8. Styadev TN, Sudhir KVR, Santosh T, Syam sundar B (2013) Development and validation of High performance liquid chromatographic technique for the determination of Esomeprazole in human plasma. Int J Res Pharma Chem pp: 552-559.

9. Palavai SR, Shakil S, Gururaj V, Badri V, Vure P, et al. (2011) Stability indicating simultaneous estimation of assay method for naproxen and esomeprazole in pharmaceutical formulations by RP-HPLC. Der Pharma Chemica 3: 553-564.

10. Santaji UN, Vangala RR, Dantu DR, Nagendrakumar M (2012) A validated stability indicating ultra performance liquid chromatographic method for determination of impurities in Esomeprazole magnesium gastro resistant tablets. J Pharm and Biom Anal 57: 109-114.

11. Mogili1 R (2011) Development and validation of Liquid Chromatography Tandem Mass Spectrometry for determination of Esomeprazole in Human plasma. Der Pharmacia Lettre 3: 138-145.

12. Sharma S, Sharma MC (2011) Densitometric method for the quantitative determination of Esomeprazole and Domperidon in dosage forms. Am Eur J Toxic Sci 3: 143-148.

13. Sharma S, Sharma MC (2011) Development and validation of densitometric method for quantitative determination of Esomeprazole and domperidone in dosage form. Am Eur J Toxic Sci 3: 143-148.

14. Shaikh NC, Bandgar SA, Kadam AM, Dhaneshwar SR (2012) Development and validation for simultaneous estimation of Esomeprazole and domperidone as bulk drug and in tablet dosage form by high performance thin layer chromatography. Int J Res Ayurveda Pharma 3: 421-424.

15. Validation of chromatographic methods (2001) (Center for drug evaluation and research CDER), pp: 2-26.

16. Lakshmana PS, Suriyaprakash TNK (2012) Extraction of Drug from the Biological Matrix: A Review. Biol Mat pp: 480-506.

17. Aryal B, Mamata SN, Kalpana D, Harish CN (2012) Pharmacokinetics of Esomeprazole in Oryctolagus Cuniculus Male Rabbit with Immobilized Stress. Int J Pharmaceut Biol Arch 3: 49-55.

16

A Simple Bioanalytical Method for the Quantification of Levetiracetam in Human Plasma and Saliva

Karas K[1], Kuczynska J[1], Sienkiewicz-Jarosz H[2], Bienkowski P[1] and Mierzejewski P[1]

[1]*Institute of Psychiatry and Neurology, Department of Pharmacology, Sobieskiego 9, 02-957, Warsaw, Poland*
[2]*Institute of Psychiatry and Neurology, Department of Neurology, Sobieskiego 9, 02-957, Warsaw, Poland*

Abstract

A novel, sensitive and selective ultra-high-performance liquid chromatography - coupled to electrospray ionization tandem mass spectrometry (HPLC-ESI-MS/MS) method was developed and validated for the quantification of levetiracetam (LEV), a broad-spectrum antiepileptic drug, in human plasma and saliva. A simple protein precipitation method with acetonitrile as precipitating solvent was used to extract LEV from human plasma and saliva. LEV and an internal standard (fluconazole, IS) were separated on a Kinetex analytical column (5 µm C18 100A, 100 × 2.1 mm) using an isocratic solvent system consisting of methanol, water and 100% formic acid at a ratio of 97:3:0.25 (v/v/v) and a flow rate of 0.2 ml/min over 2 min run time. Detection was performed using target ions of $[M+H]^+$ at m/z 171.0 for LEV and m/z 307.0 for IS in selective ion mode. The method was validated in the calibration range of 1.0-50.0 µg/mL for plasma and 0.5-30.0 µg/mL for saliva. Thus the present HPLC-MS/MS method for determination of LEV in human plasma and saliva, is highly sensitive, rapid with a short run-time of 2 min, can be suitable for high sample throughput. The developed method required minimal sample preparation and less plasma sample volume compared to earlier published LC-MS/MS methods. The validation parameters were found to be well within the acceptance limit.

Keywords: Levetiracetam; LC-MS/MS; Plasma; Saliva; Validation

Introduction

Levetiracetam, (S)-a-ethyl-2-oxo-1-pyrrolidine acetamide, is the (S)-enantiomer of the ethyl analogue of piracetam and shares its chemical structure with numerous nootropic drugs [1]. Levetiracetam (LEV) was developed as a new substance that acted on the central nervous system. LEV is an antiepileptic drug (AED) with a unique preclinical and pharmacological profile [2,3].

Levetiracetam has favorable pharmacokinetic parameters, including oral bioavailability of 100%, linear pharmacokinetics, minimal protein binding (10%), lack of hepatic metabolism, is not metabolized by CYP-dependent pathways, rapid achievement of steady-state concentrations, low potential for drug interactions, and a half-life of 6-8 hours [1,4]. It is primarily distributed in body water and has been shown to readily enter cerebrospinal fluid and the brain [1].

The saliva may be used as alternative matrix for drug monitoring concentration [2]. Saliva can be collected by non-invasive techniques and many samples can be obtained without exposing patients to discomfort, skin irritation and infection risk. This is of special importance for children and elderly patients. Also, when single or multiple serial samples are required from out-patients, the organization of collecting saliva samples is relatively simple, since they can be collected at home [5].

Several methods have been developed for quantification of LEV concentration in human plasma, e.g., gas chromatography (GC) with nitrogen-phosphorus detection or mass spectrometry [6], high-performance liquid chromatography (HPLC) with UV detection [7], and more recently liquid chromatography with tandem mass spectrometry (LC- MS/MS) [8] and ultra-performance liquid chromatography with tandem mass spectrometry (HPLC-MS/MS) [9]. A variety of analytical procedures have been developed for the determination of levetiracetam in saliva using gas chromatography (GC) with mass spectrometry [10] and high performance liquid chromatography-electrospray tandem mass spectrometry (HPLC-ESI-MS/MS) [11,12].

In this study a sensitive, simple and high-throughput LC-MS/MS method (optimized and validated according to EMEA/CHMP/EWP/192217/2009 guideline) was developed for the quantification of LEV concentration in small volumes of plasma and saliva. The method was used to monitor level of levetiracetam in patients of Departments of Neurology, Institute of Psychiatry and Neurology for purpose of clinical needs in epilepsy treatment. For extraction of levetiracetam from matrices a simple and direct protein precipitation with acetonitrile (ACN) with internal standard (IS) was used. Fluconazole as internal standard (IS) has been selected based on an analysis patient cards. Fluconazole was not used in these patients as a drug.

Materials and Methods

Chemicals and reagents

Levetiracetam (purity ≥ 98%) and fluconazole (purity ≥ 98%, the internal standard, IS) were purchased from Sigma-Aldrich (Steinheim, Germany). Methanol (LC-MS-grade), acetonitrile (LC-MS-grade), formic acid (LC-MS-grade) and water (LC-MS-grade) were also purchased from Sigma-Aldrich, (Steinheim, Germany). Pooled human plasma was obtained from Regional Blood Center (Warsaw, Poland). Blank human saliva samples were obtained from healthy individuals.

Instrumentation and chromatographic conditions

*Corresponding author: Katarzyna Karaś, Institute of Psychiatry and Neurology, Department of Pharmacology, Sobieskiego 9, Warsaw, 02-957, Poland
E-mail: kkaras@ipin.edu.pl

The LC-MS/MS system used for method development and validation was performed on the liquid chromatography system (Shimadzu, Duisburg, Germany) consisting of LCMS-8030 Triple Quadrupole Liquid Chromatography Mass Spectrometer (LC-MS/MS), LC-20AD binary gradient pump, SIL-20ACXR auto-sampler, DGU-20A3 degasser and a column oven CTO-20AC (Shimadzu Corporation, Kyoto, Japan). Data acquisition and processing were conducted using the LabSolution LC-MS 5.42 SP3 software package for quantification (Shimadzu).

Chromatographic conditions were identical for plasma and saliva samples. Chromatographic separation was achieved on a Kinetex analytical column, 5 μm C18 100A, 100 × 2,1 mm (Phenomenex, Torrance, CA, USA) with a corresponding guard column Security Guard Ultra UHPLC 2.1 ID (Phenomenex, Torrance, CA, USA). The temperature of column department was set to 40°C. The mobile phase was an isocratic solvent system consisting of methanol, water and 100% formic acid at a ratio of 97:3:0.25 (v/v/v) and a flow rate of 0.2 mL/min. The needle rinse solvent was methanol. The injection volume was 1 μL.

The mass spectrum machine was optimized to get the highest intensity for levetiracetam. Nitrogen was used as the nebulizing gas. Argon was used as the collision gas. Instrument settings were as follows: capillary voltage 4 kV, nebulizer gas flow N2 2.0 l/min, drying gas flow N2 10.0 l/min, DL temperature 200°C. Heat Block temperature 300°C and collision energy offset of -20.0 V for levetiracetam and at -20.0 for IS.

Sample analysis was performed in the multiple reaction monitoring mode (MRM) monitoring the transition of the m/z 171.2 precursor ion to the m/z 126.0 product ion for levetiracetam (171.0→126.0) and m/z 307.0 precursor ion to the m/z 220.0 product ion for IS (307.00→220.00). Quadrupoles Q1 and Q3 were set to unit resolution. The dwell time was 3 msec, pause time was 1 msec. The mass spectrometer was operated in a positive ion mode. The mass spectrometric signal was recorded after injection, and the total analytical run time was 2.0 min. Analyst software was used to control the system, data acquisition, integrate peak area and calculate the concentration of unknowns against a standard curve derived from calibrators analyzed within the same analytical run.

Preparation of stock and standard solutions

Standard stock solution and IS stock solutions of fluconazole were prepared from solid powders and dissolved into methanol. Standard stock solutions (levetiracetam 1.0 mg/mL) and IS stock solutions (fluconazole 1.0 mg/mL) were stored at ≤ 8.0°C. Standard curve and quality control (QC) samples were prepared by diluting stock solutions with blank human plasma or saliva. A calibration curve was constructed by plotting the peak area ratio of levetiracetam to IS against the nominal analyte concentration ratio using the following calibrator: 1.0, 5.0, 10.0, 20.0, 30.0, 40.0 and 50.0 μg/mL of levetiracetam in plasma and 0.5, 1.0, 2.0, 5.0, 10.0, 15.0, 20.0 and 30.0 μg/mL for saliva. Concentrations of 2.5, 15.0 and 45.0 μg/mL were chosen as the low QC level (LQC), mid QC level (MQC) and high QC level (HQC) values for plasma and 0.75, 7.5 and 25.0 μg/mL were set as the QC concentrations for saliva, respectively. IS stock solution was diluted to achieve a final concentration of 26.67 μg/mL for plasma and 13.33 μg/mL for saliva. Calibration and QC samples were freshly analyzed at the beginning of each batch of specimens and then used to determine concentrations of QC and unknown samples.

Sample preparation

Before analysis, the plasma and saliva samples were thawed to room temperature. Sample preparation was performed in a 1.5 mL polypropylene microcentrifuge tube. 20 μL matrix samples (plasma or saliva) were added to 20 μL of water and the mixture was shaken on a vibrax mixer for 30 s at 1500 rpm. Plasma or saliva aliquots were deproteinized by addition 200 μl of ACN spiked with IS, vortexed for 30 s and then centrifuged at 15,000 g for 15 min. The 50 μL clear organic supernatant was transferred to the 1.5 mL polypropylene microcentrifuge tube and was reconstituted with 250 μL mobile phase. The mixture was shaken on a vibrax mixer for 30 s at 1500 rpm. The solution was transferred to the autosampler vessel and 1.0 μL was injected into the LC-MS/MS system.

Data analysis

A LabSolutions LCMS 5.42 SP3 software was employed to obtain and analyze the chromatogram. Microsoft Office Excel 2007 was used to calculate intra- and inter-assay means, standard deviation (SD) and relative standard deviation (RSD), as well as coefficient of variation (CV), etc.

Method validation

Validation was based on the European Medicine Agency (EMEA) guideline - EMEA/CHMP/EWP/192217/2009. The validation was performed in order to evaluate the methods in terms of specificity, carry over, sensitivity, calibration curve (linearity of response), accuracy, precision, dilution integrity, matrix effect, recovery and stability.

Selectivity: Selectivity of the method was assessed by comparing chromatograms of blank plasma or saliva from each of six different sources was extracted without addition of any standards. Proteins were precipitated using acetonitrile without IS. The data of the chromatograms were processed and the integrated instrument of interfering compounds extracted from blank plasma or saliva response should not exceed 20% of the average integrated response of the LLOQ of LEV and 5% of the integrated response of IS.

Limit of detection (LOD) was calculated from the standard deviation (SD) of response and slope of the curve (S) using the equations: LOD=3.3 × (SD/S), according to ICH Q2 (R1) (Validation of analytical procedures: text and methodology).

Linearity: A total of six calibration lines, consisting of seven different concentrations, were prepared in blank human plasma or saliva and measured during six runs. Calibration curves were obtained by fitting the peak area ratios to a weighted (1/y) least squares regression model. Concentrations were evaluated on the basis of the corresponding calibration curve, and deviations from the theoretical concentrations were required to be within ± 20% for the LLOQ and within ± 15% for other calibration levels. Correlation coefficients (r) were required to be 0.99 or higher.

Inter-assay accuracy and precision: The intra-day and inter-day accuracy and precision were assessed by the replicate analysis (n=6) of QC samples (plasma or saliva) at three different concentrations on the same day and three consecutive days, respectively. Intra- and inter-day assay precisions were determined as % CV (coefficient of variance), and the %CV was required to not exceed 20% for the LLOQ and 15% for other concentrations. The intra-and inter-day assay accuracies were determined as the percent difference between the mean of observed concentrations and the theoretical concentration, and were required to be within ± 20% for the LLOQ and within ± 15% for other concentrations.

Recovery and matrix effect: For levetiracetam and IS, the matrix factor (MF) was measured in 6 different lots of plasma or saliva, by calculating the ratio of the peak area in the presence of matrix (measured by analyzing blank matrix spiked after extraction with levetiracetam), to the peak area in absence of matrix (pure solution for levetiracetam). The CV of theIS-normalized MF calculated from the 6 lots for each of plasma and saliva should not be greater than 15%. This determination should be done at a low and at a high QC level of concentration.

Stability: The stability of levetiracetam in matrix samples was assessed by analyzing replicates QC samples during the sample and storage procedure. For all stability studies, freshly prepared and stability testing QC samples were evaluated by using a freshly prepared standard curve for the measurement. The short-term stability was assessed after exposure of the matrix samples to room temperature for 24 h. The Freeze/thaw stability was determined after three freeze/thaw cycles. The concentrations obtained from all stability studies were compared to freshly prepared QC samples that were at room temperature for 4 h, and the percentage concentration deviation was calculated. In addition, long-term stability was assessed for QC samples stored at -14°C and at −60°C for 14 days. The analytes were considered stable in human plasma when the concentration difference was less than 15% between the freshly prepared samples and the stability testing samples.

Results and Discussion

Mass spectrometry

The mass spectrometer was operated in the positive ion MRM mode for both levetiracetam and fluconazole in the LC-MS/MS analysis. ESI spectra revealed higher signals at m/z 171.20 for levetiracetam. The full scan spectra was dominated by protonated molecules [M+H]$^+$. The product ion mass spectra of two protonated molecular ions are shown in Figure 1, in which the most abundant ions were observed at m/z 126.00 (levetiracetam). By monitoring the product ions, a highly sensitive assay for levetiracetam and fluconazole was developed. Additional tuning of ESI source and collision-induced dissociation parameters onto the transition m/z 171.20→126.00 (levetiracetam) further improved the sensitivity. Hence structurally related compound having similar chromatographic properties, mass spectrometric behavior and extraction characteristics was selected as IS. Therefore, fluconazole was chosen as IS for levetiracetam. In the positive ESI mode, fluconazole predominantly formed the protonated molecule [M+ H]$^+$ in full-scan spectra. To determine fluconazole using the MRM mode, full-scan and product ion spectra were investigated. Representative mass spectra of [M + H]$^+$ ions of fluconazole is shown in Figure 1. The major fragment ions at m/z 307.00→220.00 (fluconazole) were chosen in the MRM mode.

Liquid chromatography

The chromatographic conditions were investigated to optimize sensitivity, speed and peak shape. Different trails were carried out for effective separation of drug and internal standard without the interference of plasma by changing the composition of mobile phase, using different columns, altering the column temperature. Analyte and IS were free of interference from endogenous substances. The Kinetex C18 column (100 × 2.1 mm, 5 um) provided very good selectivity, sensitivity and peak shape for levetiracetam and IS as compared with other columns. The mobile phase consisting of methanol, water and 100% formic acid (97:3:0.25, v/v/v) with flow rate of 0.2 mL/min was found to be suitable during LC optimization. The retention time of leveiracetam and fluconazole was found to be 1.079 min and 1.072 min,

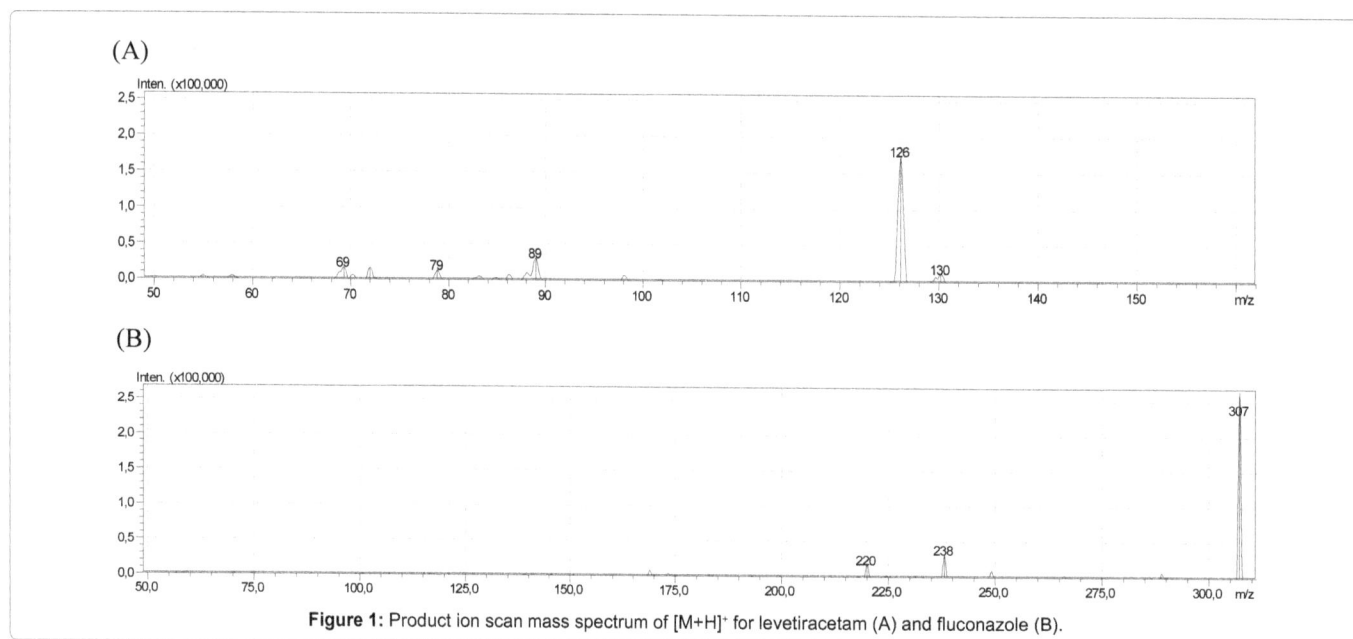

Figure 1: Product ion scan mass spectrum of [M+H]$^+$ for levetiracetam (A) and fluconazole (B).

Nominal concentration (µg/ml)	Plasma		Nominal concentration (µg/ml)	Saliva	
	Recovery (%)			Recovery (%)	
	LEV	IS		LEV	IS
2.5	87.87	87.23	0.75	92.83	98.81
15.0	79.87	82.78	7.5	94.37	101.15
45.0	81.84	84.66	25.0	94.50	99.86
n=6 for intra-day precision and accuracy; n=18 for inter-day precision and accuracy					

Table 1: Extraction recovery of levetiracetam in human plasma and saliva.

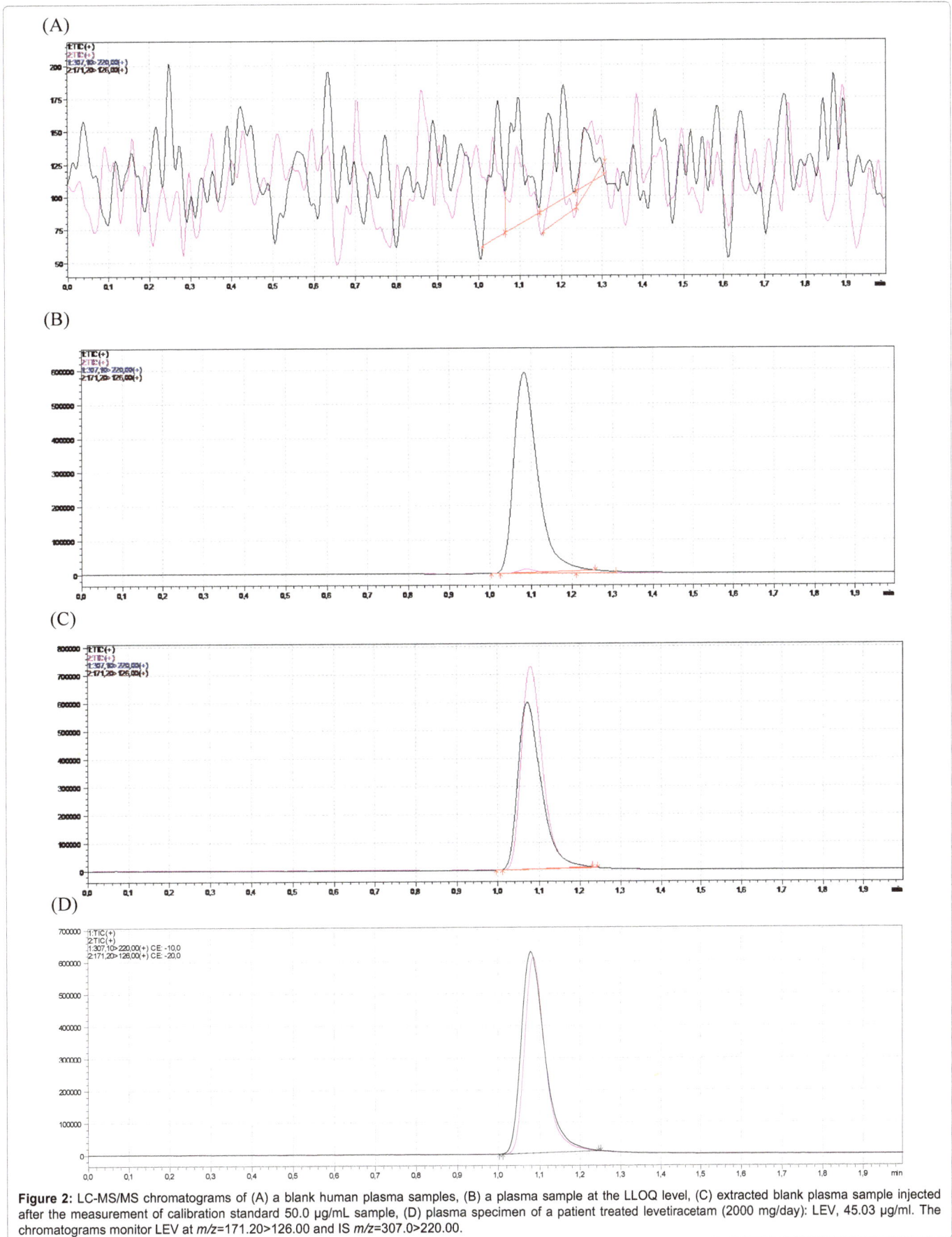

Figure 2: LC-MS/MS chromatograms of (A) a blank human plasma samples, (B) a plasma sample at the LLOQ level, (C) extracted blank plasma sample injected after the measurement of calibration standard 50.0 μg/mL sample, (D) plasma specimen of a patient treated levetiracetam (2000 mg/day): LEV, 45.03 μg/ml. The chromatograms monitor LEV at m/z=171.20>126.00 and IS m/z=307.0>220.00.

(A)

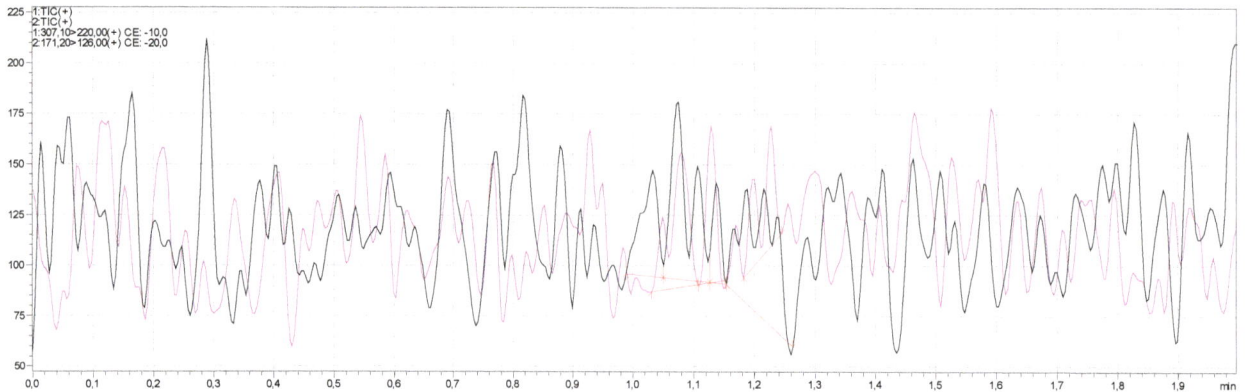

(B)

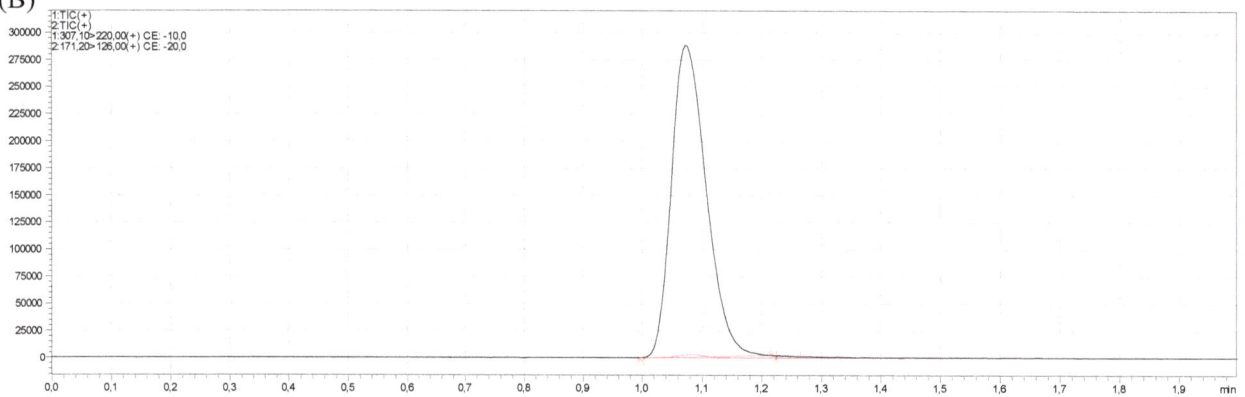

(C)

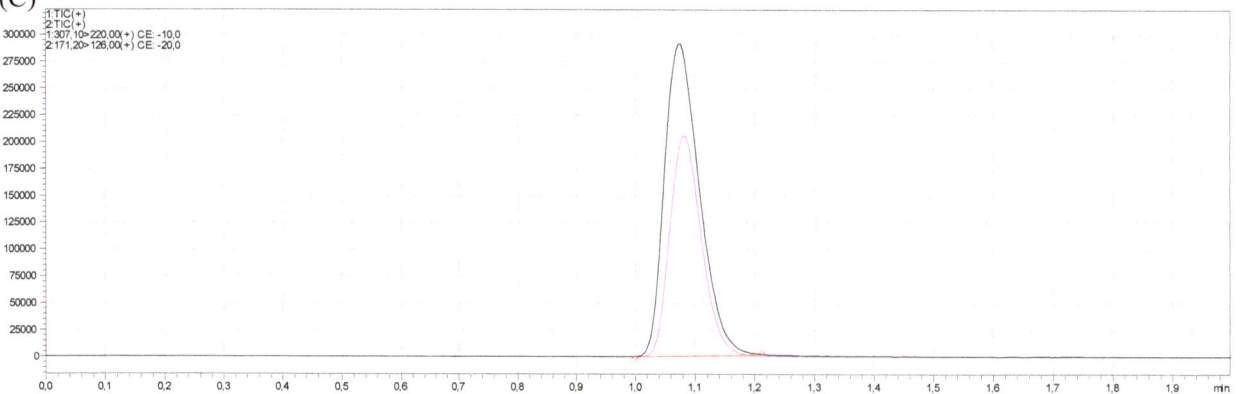

(D)

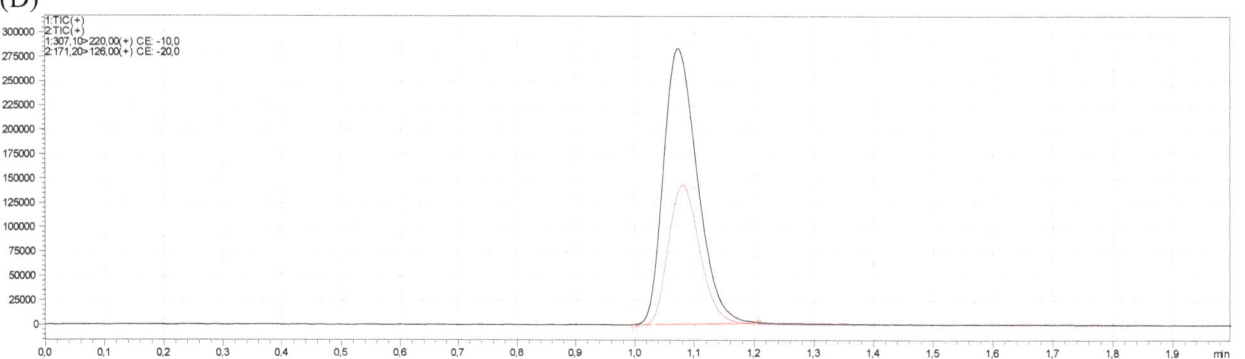

Figure 3: LC-MS/MS chromatograms of (A) a blank human saliva samples, (B) a saliva sample at the LLOQ level, (C) extracted blank saliva sample injected after the measurement of calibration standard 30.0 µg/mL sample, (D) saliva specimen of a patient treated levetiracetam (500 mg/day): LEV, 14.54 µg/ml. The chromatograms monitor LEV at m/z=171.20>126.00 and IS m/z=307.0>220.00.

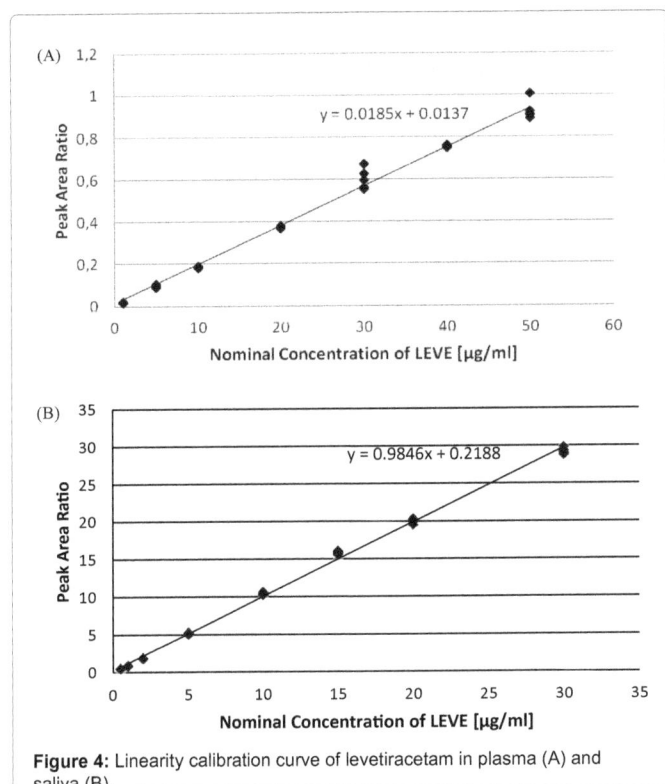

Figure 4: Linearity calibration curve of levetiracetam in plasma (A) and saliva (B).

respectively. The total run time was 2.0 min, which has advantages over other methods described in the literature [8,12].

Recovery

The results of the recovery tests for the three levels tested are shown in Table 1 for plasma and saliva. The extraction recoveries of LEV ranged from 79.87 to 87.87% for plasma and from 92.83 to 94.50% for saliva. The extraction recoveries of IS ranged from 82.78 to 87.23% for plasma and from 98.81 to 101.15% for saliva. The routine standard curve was calculated for each plasma and saliva sample. The extraction recoveries were consistent over the entire concentration range and comparable to the respective internal standard (Table 1).

Method validation

Selectivity: In the chromatograms from six lots of blank plasma samples or saliva samples, no interference peaks from endogenous substances were observed at the retention time of Levetiracetam (1.079 min) and IS (1.072 min) under the established chromatography condition, as shown in the representative chromatogram of a blank plasma sample in Figure 2 and a blank saliva sample in Figure 3. The LOD was determined by LOD=3 × (SD/S), where SD is the standard deviation of the response of the blank and S is the slope of calibration curve. The LOD values for levetiracetam were found to be 0.20 µg/ml for plasma and 0.11 µg/ml for saliva.

Matrix effect: The matrix factor of LEV (LQC) ranged from 0.974-1.146 for plasma and from 0.967-1.130 for saliva. The matrix factor of LEV (HQC) ranged from 0.952-0.990 for plasma and from 0.990-1.049 for saliva. These results were within the acceptance criteria of 0.80-1.20 indicating that no undetected co-eluting compounds that could influence the ionization of the analytes (Table 2).

Limit of quantification and linearity: The calibration curves of

LEV in both plasma and saliva were linear over the entire concentration ranges. After comparing the different weighting models, a regression equation with a weighting factor of $1/x^2$ of the drug to the IS concentration was found to produce the best fit for the concentration-detector response relationship. The mean correlation coefficient of the weighted calibration curves generated during the validation was ≥ 0.99. Calibration curve plot (Figure 4) is shown below. The calculated LLOQs were 1.0 µg/mL for LEV in plasma and 0.5 µg/mL in saliva; and thus below the concentration of the lowest calibration standard. The LLOQ in the plasma was <8.58% RSD and of 3.8% accuracy, respectively, and in the saliva was <2.57% RSD and of 3.4% accuracy, respectively. This assay was proved to be sufficiently sensitive for the pharmacokinetic analysis of levetiracetam in plasma and saliva. The chromatograms LLOQ of plasma (Figure 2) and saliva (Figure 3) are shown.

Carry-over and dilution integrity: No peaks in the retention time of LEV and IS were observed in the chromatograms of immediately injected blank plasma samples following the injection of ULOQ or HQC samples, which demonstrated that the carry-over effect could be negligible during the sample analysis. The upper concentration limits can be extended to 100.0 µg/mL for levetiracetam by ½ dilution with screened human blank plasma and saliva. The mean back calculated concentrations for ½ dilution samples were within 91.16-105.93% for plasma and 95.25-101.03% for saliva of their nominal concentrations. The %CV for ½ dilution samples were 4.97% for plasma and 3.48% for saliva and the nominal percentages for ½ dilution samples were 89.55% for plasma and 97.35% for saliva.

Accuracy and precision: Accuracy and precision data for intra-day and inter-day plasma and saliva samples for levetiracetam are summarized in Table 3 and Table 4. The intra-day accuracies varied from 103.8 and 107.4% for plasma and from 101.92% and 104.83% for saliva with a precision of 1.82-8.58% and of 1.01-2.89% for plasma and saliva respectively. Furthermore, the inter-day accuracies varied from 101.2 and 106.0% for plasma and from 100.68 and 104.82% for saliva, with a precision of 1.80-4.79% and of 1.90-3.66% for plasma and saliva, respectively. The assay values on both the occasions (intra-day and inter-day) were found to be within the accepted variable limits.

Stability: The stability of levetiracetam in matrix samples were evaluated under the conditions described in Table 5 and Table 6. The samples were stable under the storage conditions with acceptable accuracy and precision. The results of the tested samples were within the acceptance criteria.

Application of Method

S No	Plasma				Saliva			
	MF		IS MF		MF		IS MF	
	QC low	QC high	QC low	QC high	QC low	QC high	QC low	QC high
1.	0.974	0.957	1.016	0.973	1.004	0.990	0.993	0.982
2.	0.983	0.966	1.030	0.966	0.973	0.978	1.059	0.990
3.	1.146	0.952	1.003	0.961	0.967	1.049	1.016	1.011
4.	1.053	0.990	0.982	0.988	1.036	1.016	0.993	0.968
5.	1.079	0.955	1.013	0.961	1.130	1.004	1.020	1.036
6.	1.136	0.958	1.049	0.996	1.009	0.996	1.103	0.961
Mean	1.062	0.963	1.016	0.974	1.020	1.006	1.031	0.991
SD	0.073	0.014	0.023	0.015	0.060	0.025	0.043	0.028
CV%	6.91	1.46	2.25	1.51	5.84	2.47	4.16	2.83

Table 2: Matrix effects in human plasma and saliva represented as MF for levetiracetam and MF for IS.

Nominal concentration (µg/ml)	Intra-day			Inter-day		
	Determined concentration (µg/ml)	Precision (RSD, %)	Accuracy (RE, %)	Determined concentration (µg/ml)	Precision (RSD, %)	Accuracy (RE, %)
1.0	1.038	8.58	103.8	1.06	3.08	106.0
2.5	2.63	2.63	105.1	2.67	1.80	105.6
15.0	16.12	1.82	107.4	15.48	4.79	103.2
45.0	46.91	2.46	104.2	45.53	2.36	101.2
n=6 for intra-day precision and accuracy; n=18 for inter-day precision and accuracy						

Table 3: Precision and accuracy of levetiracetam in human plasma.

Nominal concentration (µg/ml)	Intra-day			Inter-day		
	Determined concentration (µg/ml)	Precision (RSD, %)	Accuracy (RE, %)	Determined concentration (µg/ml)	Precision (RSD, %)	Accuracy (RE, %)
0.5	0.52	2.57	103.42	0.51	3.66	101.85
0.75	0.77	2.89	103.05	0.79	2.36	104.82
7.5	7.86	1.66	104.83	7.62	1.90	101.57
25.0	25.48	1.01	101.92	25.17	2.65	100.68
n=6 for intra-day precision and accuracy; n=18 for inter-day precision and accuracy						

Table 4: Precision and accuracy of levetiracetam in human saliva.

Concentration (µg/ml)	Room temperature (4 h)		Auto-sampler (48 h)		Freeze-thaw		Long-term	
	Precision (RSD, %)	Accuracy (RE, %)	Precision (RSD, %)	Accuracy (RE, %)	Precision (RSD, %)	Accuracy (RE, %)	Precision (RSD, %)	Accuracy (RE, %)
2.5	1.45	102.15	2.78	102.02	5.26	100.7	1.61	105.1
15.0	1.69	101.40	1.93	101.28	4.96	97.3	5.00	96.0
45.0	2.14	101.02	4.32	99.66	2.34	100.9	2.64	99.2

Table 5: Summary of stability studiem of levetiracetam in human plasma.

Concentration (µg/ml)	Room temperature (4 h)		Auto-sampler (48 h)		Freeze-thaw		Long-term	
	Precision (RSD, %)	Accuracy (RE, %)	Precision (RSD, %)	Accuracy (RE, %)	Precision (RSD, %)	Accuracy (RE, %)	Precision (RSD, %)	Accuracy (RE, %)
0.75	2.08	102.98	1.28	103.45	2.40	102.50	1.07	102.77
7.5	1.90	101.57	2.17	100.50	2.11	101.90	1.67	100.95
25.0	1.26	100.35	1.69	100.77	2.00	102.13	1.35	101.72

Table 6: Summary of stability studiem of levetiracetam in human saliva.

Patient No.	Sex	Age (years)	Daily dose (mg/day)	LEV (mg/kg)	Plasma Level (mg/L)	Saliva Level (mg/L)	Other Antiepileptic Drugs
1	M	20	500	9.43	10.97	14.54	VPA
2	M	34	1000	11.63	11.98	13.93	VPA
3	F	28	2000	30.77	36.29	36.81	-
4	F	54	2000	40.82	45.03	44.94	VPA
VPA: Valproic acid							

Table 7: Subject Characteristics.

The validated method has been successfully used to analyse levetiracetam concentrations in four patients (Table 7). Patients gave their written informed consent to participate and the study procedures were approved by the local ethics committee. Blood samples were withdrawn in the morning, before the drug intake. Saliva samples were collected by the same physician within 5 to 10 minutes after blood sampling.

Conclusion

In summary, a selective and sensitive LC-MS/MS method for determination of levetiracetam in human plasma or saliva was successfully developed and validated. A simple extraction procedure and isocratic chromatography conditions provides an assay well suited for real-time analyses. The method provid a short analysis, superior sensitivity with the lower limit of quantitation as low as 1.0 µg/mL for levetiracetam in plasma and 0.5 µg/mL for drug in saliva. The assay requires only 20 µL plasma or saliva volume and is easy to perform with minimal sample preparation. The method adhered to the regulatory requirements for selectivity, sensitivity, linearity, precision, accuracy, recovery, matrix effect and stability. From the results of the validation parameters, we can conclude that the developer method can be useful for Bioequivalence studies and routine therapeutic drug monitoring with desired precision and accuracy.

Acknowledgements

This work was supported by the Institute of Psychiatry and Neurology, Warsaw, Poland (Grant No. 501-004-15-045).

References

1. Iwasaki T, Toki T, Nonoda Y, Ishii M (2015) The efficacy of levetiracetam for focal seizures and its blood levels in children. Brain Dev 37: 773-779.

2. Patsalos PN, Berry DJ (2013) Therapeutic drug monitoring of antiepileptic

drugs by use of saliva. Ther Drug Monit 35: 4-29.

3. Radtke RA (2001) Pharmacokinetics of levetiracetam. Epilepsia 42 Suppl 4: 24-27.

4. Patsalos PN (2004) Clinical pharmacokinetics of levetiracetam. Clin Pharmacokinet 43: 707-724.

5. Nunes LA, Mussavira S, Bindhu OS (2015) Clinical and diagnostic utility of saliva as a non-invasive diagnostic fluid: a systematic review. Biochem Med (Zagreb) 25: 177-192.

6. Greiner-Sosanko E, Giannoutsos S, Lower DR, Virji MA, Krasowski MD (2007) Drug monitoring: Simultaneous analysis of lamotrigine, oxcarbazepine, 10-hydroxycarbazepine, and zonisamide by HPLC-UV and a rapid GC method using a nitrogen-phosphorus detector for levetiracetam. J Chromatogr Sci 45: 616-622.

7. Contin M, Mohamed S, Albani F, Riva R, Baruzzi A (2008) Simple and validated HPLC-UV analysis of levetiracetam in deproteinized plasma of patients with epilepsy. J Chromatogr B Analyt Technol Biomed Life Sci 873: 129-132.

8. Matar KM (2008) Quantification of levetiracetam in human plasma by liquid chromatography-tandem mass spectrometry: application to therapeutic drug monitoring. J Pharm Biomed Anal 48: 822-828.

9. Blonk MI, van der Nagel BC, Smit LS, Mathot RA (2010) Quantification of levetiracetam in plasma of neonates by ultra performance liquid chromatography-tandem mass spectrometry. J Chromatogr B Analyt Technol Biomed Life Sci 878: 675-681.

10. Mecarelli O, Li Voti P, Pro S, Romolo FS, Rotolo M, et al. (2007) Saliva and serum levetiracetam concentrations in patients with epilepsy. Ther Drug Monit 29: 313-318.

11. Guo T, Oswald LM, Mendu DR, Soldin SJ (2007) Determination of levetiracetam in human plasma/serum/saliva by liquid chromatography-electrospray tandem mass spectrometry. Clin Chim Acta 375: 115-118.

12. Lins RL, Otoul C, De Smedt F, Coupez R, Stockis A (2007) Comparison of plasma and saliva concentrations of levetiracetam following administration orally as a tablet and as a solution in healthy adult volunteers. Int J Clin Pharmacol Ther 45: 47-54.

17

Development of a High Performance Liquid Chromatography Method for the Determination of Tedizolid in Human Plasma, Human Serum, Saline and Mouse Plasma

Santini DA[1], Sutherland CA[1], Nicolau DP[1,2]*

[1]Center for Anti-Infective Research and Development, Hartford Hospital, 80 Seymour Street, Hartford, CT 06102, USA
[2]Division of Infectious Diseases, Hartford Hospital, Hartford, CT, Hartford Hospital, 80 Seymour Street, Hartford, CT 06102, USA

Abstract

A simple and reliable high performance liquid chromatography method was developed and validated to analyze tedizolid in human plasma, human serum, saline and CD-1 mouse plasma. An ultra violet detector set at 251nm was used with a reverse phase column. The mobile phase consisted of sodium acetate, deionized water and acetonitrile at a flow rate of 1.0 ml/min. 4-nitroaniline was used as the internal standard. The standard curves were linear over a range of 0.2 to 5 µg/ml. Accuracy and precision for the validations were within the acceptable 10% limit. This method was successfully used to assay samples previously analyzed by liquid chromatography with tandem mass spectrometric detection.

Keywords: Tedizolid; HPLC; Validation

Abbreviations

UV: Ultra Violet; FDA: Food and Drug Administration; ABSSSI: Acute Bacterial Skin and Skin Structure Infection; HPLC: High Performance Liquid Chromatography; DMSO: Dimethyl Sulfoxide; K3EDTA: Tri-potassium Ethylene Di-amine Tetra Acetic Acid; QC: Quality control; LLQ: Lower Limit of Quantification; LC-MS/MS: Liquid Chromatography – Tandem Mass Spectrometry; SD: Standard Deviation

Introduction

Tedizolid phosphate (SIVEXTRO®, Cubist Pharmaceuticals) is a novel oxazolidinone antibiotic recently approved by the Food and Drug Administration (FDA) (Figure 1). Tedizolid phosphate is rapidly converted in vivo by phosphatases to the microbiologically active antibiotic tedizolid. Tedizolid phosphate is indicated for treatment of adults with acute bacterial skin and skin-structure infections caused by the following Gram-positive microorganisms, *Staphylococcus aureus* (including methicillin-resistant and methicillin-susceptible pathogens), *Streptococcus pyogenes*, *Streptococcus agalactiae*, *Streptococcus anginosus* group (including *Streptococcus anginosus*, *Streptococcus intermedius* and *Streptococcus constellatus*) and *Enterococcus faecalis* [1]. Tedizolid phosphate is available as both oral (i.e., tablet) and intravenous (administered over 1 hour) formulations both administered at the same dose of 200 mg once daily for the treatment of acute bacterial skin and skin structure infections (ABSSSI) [1]. The purpose of this present study was to develop a simple, reproducible, and selective high performance liquid chromatography (HPLC) method for the detection of tedizolid in human plasma, human serum, saline and mouse plasma that is consistent with FDA guidance [2].

Materials

Chemicals

Tedizolid standard powder was provided by Cubist Pharmaceuticals (Lexington, MA.). Commercially available 4-nitroaniline was purchased from Sigma (St. Louis, MO) and used as the internal standard. Sodium acetate, trichloroacetic acid, ethyl alcohol and dimethyl sulfoxide (DMSO) were also purchased from Sigma. Acetonitrile from Avantor Performance Materials, Inc. (Center Valley, PA) was used without

further purification. Deionized water was obtained from Millipore's Milli-Q analytical deionization system (Bedford, MA). 0.9% sodium chloride injection USP was purchased from Hospira, Inc., (Lake Forest, IL).

Instrumentation

An HPLC system consisting of a Waters 515 gradient pump (Waters Associates, Milford, MA) and WISP 717 plus autosampler (Waters) was equipped with a 5 µm SphereClone-80 ODS2 column (4.6 x 150 mm, Phemonenex, Torrance, CA) coupled to a Bondapak C_{18} Guard-pak precolumn (Waters). The autosampler was cooled to 10°C. The column was maintained at room temperature. A programmable Hitachi L 2400 ultra violet (UV) detector (Model 526; ESA Inc., Chelmsford, MA) set at 251nm was used to detect the analytes. The run time for each human plasma and serum sample was 28 minutes. The run time for each saline and mouse plasma sample was 20 minutes. The EZChrom Elite chromatography data system (Scientific Software Inc., Pleasanton, CA) was used to quantify the peak heights.

Mobile phase preparation

The mobile phase consisted of a mixture of 0.0192 M sodium acetate buffer and 23% acetonitrile. The flow rate was 1.0 ml/min. All chromatographic procedures were performed at room temperature.

Standard solutions and control

Tedizolid standard solutions were made in a volumetric flask using DMSO to dissolve and dilute the stock standard. The internal

***Corresponding author:** David P Nicolau, FCCP, FIDSA, Center for Anti-Infective Research and Development, Hartford Hospital, 80 Seymour Street, Hartford, CT 06102, E-mail: David.Nicolau@hhchealth.org

standard, 4-nitroaniline 6 μg/ml, was prepared in ethanol according to the manufacturer's recommendations. Drug free adult human plasma with tri-potassium ethylene di-amine tetra acetic acid (K3EDTA) and serum from 6 different individuals as well as the CD-I non-immunized mouse plasma (K3EDTA) were purchased from Bioreclamation Inc. (Hicksville, NY). Separate standard curves were made for tedizolid in pooled human plasma, mouse plasma and saline. Tedizolid was spiked into human plasma and saline to make 8 standard solutions (0.2, 0.4, 0.6, 0.8, 1.5, 2, 3, and 5 μg/ml) and three quality controls (QC) (0.3, 1, and 4 μg/ml) for each validation. Tedizolid was spiked into mouse plasma to make 6 standard solutions (0.2, 0.4, 0.8, 2, 3, and 5 μg/ml) and two quality controls (0.3 and 4 μg/ml). Additional quality controls (0.3, 1, and 4 μg/ml) were made in human serum for a cross matrix validation against the human plasma standard curve. Aliquots of the standards and internal standard were stored at -80°C until analysis.

Method Validation

200 μl of sample and 50 μl of internal standard was pipetted into a polyethylene tube. 150 μl of a 7.5% tri chloro actetic acid was then carefully added down the side of the tube to deproteinize the sample. The sample was vortexed for 30 seconds and then centrifuged for 10 minutes at 3600rpm. 200μl of the sample was then placed into a WISP vial (Waters) for injection.

Assay validation

The peak height ratio of tedizolid and the internal standard was plotted to generate the calibration curve. The linear regression equation was generated by applying the weighted (1/concentration) least square regression analyses. Linear regression was used to calculate the concentrations of the quality controls and unknown samples. Linearity of the standard curve was assessed with the correlation coefficient by plotting the peak height ratio of tedizolid versus the theoretical concentrations.

A human plasma and saline calibration curve consisted of a blank sample (matrix sample processed without internal standard), a zero sample (matrix sample processed with internal standard), and eight non-zero samples covering the expected concentration range of the unknown samples, including the lower limit of quantification (LLQ) for the human plasma and saline validations. The precision and accuracy were evaluated with the three quality controls with low, middle, and high concentrations. Precision was determined by taking the standard deviation divided by the average. Accuracy was determined by the relative error from the theoretical concentrations taking the calculated concentration divided by the theoretical concentration and subtracting from 100%. The LLQ for the assay was evaluated on five samples with 0.2 μg/ml of tedizolid for each validation.

Human plasma and saline quality control samples were run in triplicate to determine tedizolid stability at room temperature, freeze and thaw, in stock solution, and stability in the autosampler. Room temperature studies were done at 23°C. Each quality control sample was thawed and kept at room temperature for 6 hours which is the maximum length of time it takes to process the samples in a run. Freeze and thaw stability was assessed by completely thawing the quality controls at room temperature and then refreezing them at -80°C for 24 hours. This freeze and thaw cycle was repeated two more times and analyzed after the third freeze and thaw cycle. The stability of the drug stock solution and the internal standard were evaluated at room temperature for 6 hours and compared to that of freshly prepared solutions. The autosampler stability of a sample after extraction was verified to ensure that all samples stored in the cooled auto

sampler during a run do not experience degradation in the time required to complete the assay. This stability assessment was performed on three aliquots of each quality control concentration. To determine long term stability, aliquots of each of the quality control concentrations were stored at -80°C. The concentrations of the long term stability samples were compared to the mean of back-calculated values for the appropriate standards from the first day.

Recovery experiments were performed in triplicate for each of the human plasma and saline quality control concentrations. Values of percent recoveries of tedizolid were calculated by comparing the peak height ratio of tedizolid and the internal standard in plasma to that of non-extracted pH adjusted saline solutions for human plasma. For the saline assay values of percent recoveries of tedizolid were calculated by comparing the peak height ratio of tedizolid and the internal standard in saline to that of water.

Since tedizolid is likely to be determined in other matrices, a cross matrix validation was performed in human serum against the human plasma standard curve. A full validation of tedizolid was conducted in human plasma and saline. A partial validation was conducted in mouse plasma.

To assess the absolute recovery, precision, and accuracy of the current analytical methodology, we re-assayed human plasma samples from a previously conducted pharmacokinetic study. These samples were originally assayed using a validated liquid chromatography – mass spectrometry and liquid chromatography with tandem mass spectrometry detection (LC-MS/MS) method and stored at -80°C [3].

Results

Chromatography

Human plasma from six individuals was tested for interference using this assay. Figure 2 represents a typical chromatogram with blank human plasma. The chromatogram shows no interfering peaks with tedizolid or the internal standard. The retention time of the internal standard and tedizolid were 8.8 and 15.1 minutes for the plasma assay. Figure 3 represents a typical chromatogram with blank saline and shows no interfering peaks with tedizolid or the internal standard. A slight shift in retention time was seen on the saline standard curve. The internal standard and tedizolid were 9.4 and 16.3 minutes respectively. The retention time of the internal standard and tedizolid were 8.1 and 14.2 minutes for the mouse plasma standard curve. Figure 4 represents a chromatogram with blank mouse plasma.

Linearity, precision and accuracy

Linearity was demonstrated with the correlation coefficient (r) for each calibration curve of ≥0.999 for plasma (n=11) and saline (n=6).

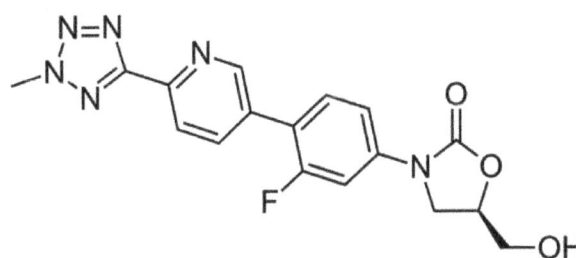

Figure 1: Chemical structure of Tedizolid.

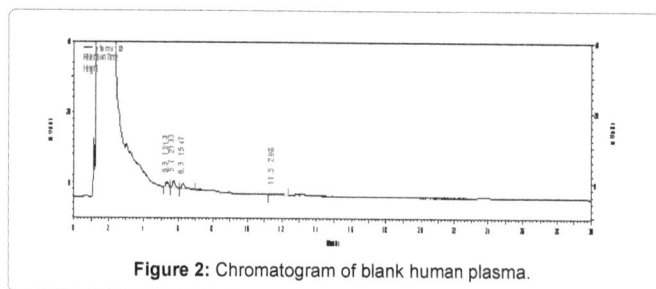

Figure 2: Chromatogram of blank human plasma.

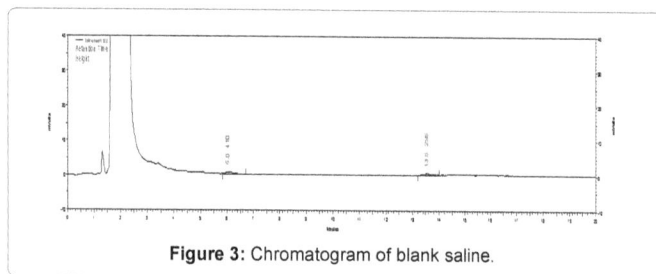

Figure 3: Chromatogram of blank saline.

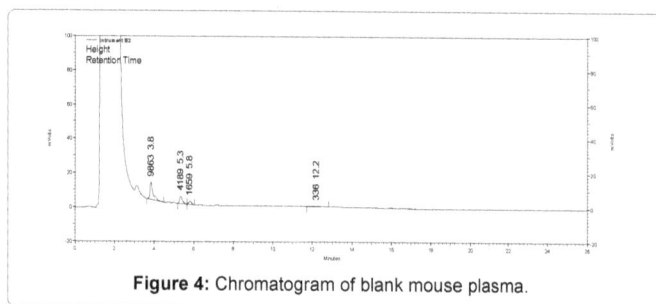

Figure 4: Chromatogram of blank mouse plasma.

The plasma slope was 0. 4350 ± 0.013 (mean ± standard deviation (SD)) and the intercept was 0.0041 ± 0.005 (mean ± SD). The saline slope was 0.4904 ± 0.066 (mean ± SD) and the intercept was 0.0017 ± 0.004 (mean ± SD). Linearity for the mouse plasma (n=3) was demonstrated with the correlation coefficient (r) for each calibration curve of ≥0.999. The mouse plasma slope was 0.4796 ± 0.033 (mean ± SD) and the intercept was 0.0078 ± 0.006 (mean ± SD).

The summary data for the inter- and intra-day precision and accuracy in tedizolid-spiked human plasma are shown in Table 1. Summary data for the saline inter- and intra-day precision and accuracy are shown in Table 2. Summary data for the mouse plasma standard curve inter- and intra-day precision and accuracy are shown in Table 3.

Lower limit of quantification and recovery

The LLQ of 0.2 μg/ml for tedizolid was chosen as the concentration for the lowest standard sample (Figure 5). The precision and accuracy of LLQ (n=5) for tedizolid in human plasma was 2.25% and 3.75%, respectively while the precision and accuracy of the LLQ (n=5) for tedizolid in saline (Figure 6) was 0.50% and -8.44%.

Recovery experiments were performed in triplicate by comparing the analytical results for extracted plasma samples at the quality control concentrations with unextracted controls in a 0.1M phosphate buffer of pH 8 that represents 100% recovery. The pH of 8 was chosen to closely resemble that of the plasma and serum used. The recovery of the 0.3, 1 and 4 μg/ml samples was 81.09% 0.045%, 77.22% ± 0.039%, and 82.66% ± 0.004%, respectively. Similarly recovery experiments

were performed in triplicate for extracted saline samples at the quality control concentrations with unextracted controls in water that represents 100% recovery. The recovery of the 0.3, 1 and 4 μg/ml samples was 104% ± 0.0283, 102.5% ± 0.042, and 101.8% ± 0.017.

Stability

Room temperature stability was determined in triplicate for each of the three quality control concentrations. Each sample was thawed at room temperature (23°C) and kept at this temperature for 6 hours. Tedizolid plasma quality controls were stable at room temperature with <9% degradation. Tedizolid saline quality controls were stable at room temperature with < 10 % degradation. The 4-nitroaniline standard solution of 6 μg/ml was stable at room temperature for 6 hours with <3% degradation. Tedizolid plasma and saline quality controls were stable for three freeze thaw cycles with <10% degradation. Stock solution stability of the drug and the internal standard were evaluated on each of the three aliquots of quality control concentration at room temperature for 6 hours with <9% degradation. The autosampler stability of a sample after extraction is verified to ensure that all samples stored in the cooled autosampler during a run do not experience degradation in the time required to complete the assay. The stability was performed on three aliquots of each quality control concentration. The plasma and saline quality control samples after extraction in the 10°C autosamplerr were stable for 24 hours with <8% degradation. Tedizolid plasma quality controls were stable at -80°C for 28 days with <9% degradation while the tedizolid in saline quality controls were stable at -80°C for 11 days with <9% degradation

Cross matrix validation

The summary data for the human serum inter- and intra-day precision and accuracy for tedizolid are shown in Table 4.

Comparison of HPLC and mass spectrometry

A total of 9 samples were assayed with concentrations ranging from 0.36 to 2.97 μg/ml. Average recovery of these samples was 91.01% ± 0.052 as shown in Table 5. Precision and accuracy were 5.71% and -9.16% respectively [3].

Conclusion

An efficient and reliable HPLC method has been developed to assess tedizolid concentrations in human plasma, human serum, saline, and mouse plasma. Precision and accuracy of the QC samples were within the acceptable limits of 10% as defined by FDA guidelines [2]. Moreover, this methodology was successfully used to replicate the results obtained previously using a LC-MS/MS detection method. While LC-MS/MS detection is more sensitive, this current HPLC method is more cost effective and has the advantage of less variability.

	Tedizolid concentration (μg/ml)		
	Low (0.3)	Medium (1)	High (4)
Inter-run (n=11)			
Mean	0.31	1.01	4.20
SD	0.007	0.028	0.133
Precision	2.16%	2.72%	3.16%
Accuracy	4.88%	1.02%	4.92%
Intra-run (n=10)			
Mean	0.32	1.06	4.31
SD	0.009	0.026	0.062
Precision	2.75%	2.48%	1.43%
Accuracy	6.40%	5.65%	7.86%

Table 1: Precision and accuracy of Tedizolid in human plasma.

	Tedizolid concentration (µg/ml)		
	Low (0.3)	Medium (1)	High (4)
Inter-run (n=6)			
Mean	0.31	1.04	4.18
SD	0.010	0.010	0.072
Precision	3.30%	1.01%	1.72%
Accuracy	4.44%	3.50%	4.45%
Intra-run (n=10)			
Mean	0.30	1.04	4.24
SD	0.013	0.019	0.104
Precision	4.35%	1.86%	2.5%
Accuracy	0.46%	4.45%	6.03%

Table 2: Precision and accuracy Tedizolid in saline.

	Tedizolid concentration (µg/ml)	
	Low (0.3)	High (4)
Inter-run (n=3)		
Mean	0.30	4.12
SD	0.006	0.080
Precision	1.95%	1.95%
Accuracy	-1.11%	2.92%
Intra-run (n=10)		
Mean	0.29	4.11
SD	0.012	0.058
Precision	4.30%	1.4%
Accuracy	-3.86%	2.73%

Table 3: Precision and accuracy of Tedizolid in mouse plasma.

	Tedizolid concentration (µg/ml)		
	Low (0.3)	Medium (1)	High (4)
Inter-run (n=4)			
Mean	0.32	1.01	4.11
SD	0.008	0.039	0.066
Precision	2.55%	3.92%	1.61%
Accuracy	6.67%	0.75%	2.62%
Intra-run (n=10)			
Mean	0.31	0.96	4.06
SD	0.010	0.027	0.120
Precision	3.09%	2.79%	2.95%
Accuracy	4.05%	-3.54%	1.51%

Table 4: Precision and accuracy of Tedizolid in human serum.

Subject	Sample Time	Re-assayed Concentration	Original Concentration	%Accuracy	%Recovery
104	0h	0.36	0.37	-3.25%	96.23%
104	1h	2.97	3.25	-8.50%	91.50%
104	8h	1.25	1.46	-14.32%	85.68%
101	0h	0.48	0.48	-0.32%	100.52%
101	1h	1.88	2.10	-10.29%	89.71%
101	8h	1.23	1.39	-11.41%	88.59%
118	0h	0.45	0.48	-5.49%	94.12%
118	1h	2.42	2.88	-15.81%	84.19%
118	8h	1.37	1.55	-11.45%	88.55%

Table 5: Comparison of tedizolid concentrations (µg/ml) in HPLC and mass spectrometry.

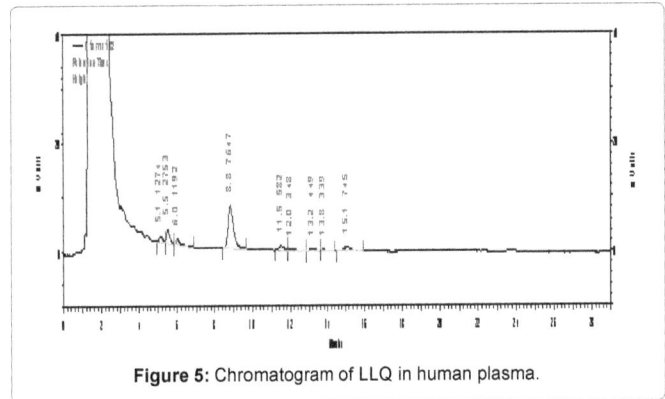

Figure 5: Chromatogram of LLQ in human plasma.

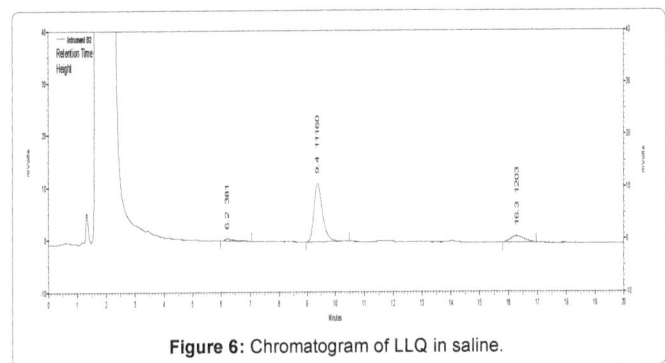

Figure 6: Chromatogram of LLQ in saline.

Acknowledgement

This study was funded by Cubist Pharmaceuticals, Lexington, MA. We would like to thank Cubist Pharmaceuticals for providing analytical standard and supporting the assay development. David Nicolau is on the speaker bureau for Cubist and has received grants from Cubist. The other authors have nothing to disclose.

References

1. SIVEXTRO® (tedizolid phosphate) prescribing information (2014)Cubist pharmaceuticals.

2. Guidance for Industry Bioanalytical Method Validation (2001) U.S. Department of Health and Human Services,Food and Drug Administration, Center for Drug Evaluation and Research (CDER), Center for Veterinary Medicine (CVM).

3. Housman S, Pope J, Russomanno J, Salerno E, Shore E, et al. (2012) Pulmonary disposition of tedizolid following administration of once-daily oral 200 mg tedizolid phosphate in healthy adult volunteers. Antimicrob Agents Chemother 56: 2627- 2634.

18

Purification of A1PI from Human Plasma-An Improved Process to Achieve Therapeutic Grade Purity

Nuvula Ashok Kumar*, **Korla Lakshmana Rao**, **Zinia Chakraborthy** , **Archana Giri** and **Komath Uma Devi**
*Centre for Biotechnology, JNTUH, Hyderabad, India

Abstract

Alpha-1-proteinase inhibitor (A1PI) also known as α-anti trypsin (AAT) is a member of serine protease inhibitors super family present in human plasma. A1PI is the only known inhibitor of elastase but also inhibits trypsin and chymotrypsin. A1PI is a molecule with lot of therapeutic importance and is widely used in the treatment of emphysema or chronic obstructive pulmonary disease (COPD). There are many methods detailed in scientific articles and patents to purify A1PI but the current research describes a simple three step chromatography procedure to fractionate A1PI directly from human plasma. The current purification process starts from human plasma unlike most of the existing methods where Cohn fraction IV-1 paste is the starting material. This method excludes the use of more methods like Cohn ethanol fractionation, precipitation or affinity capture. The A1PI purified by this scheme is characterized and found to be a single chain glycoprotein with a molecular weight of 52 KDa with increased purity. As the process is devoid of precipitation and affinity steps, this scheme is easily scalable, economical to adapt for manufacturing A1PI with good yields of 0.6-0.7 gm of A1PI/L of plasma.

Keywords: Alpha-1-proteinase inhibitor; DEAE sepharose; Fractogel TMAE; Capto adhere

Abbreviations: A1PI: Alpha-1-Proteinase Inhibitor; PEG: Poly Ethylene Glycol; DEAE: Diethyl Amino Ethyl; TMAE: Tri Methyl Amino Ethyl; BAPA: Benzoyl-DL-Arginine-p-Nitroanilide.

Introduction

Human Alpha-1-proteinase inhibitor (A1PI) is a well characterized multifunctional protease inhibitor, also known as α-antitrypsin (AAT). It is a glycoprotein with molecular weight of about 52000 Daltons and is a single polypeptide chain to which several oligosaccharide units are covalently bound. It belongs to the serine protease inhibitor (serpin) super family, which in addition to A1PI also includes α1-antichymotrypsin, antithrombin, plasminogen activator inhibitor, C1 esterase inhibitor [1,2]. A1PI is comprised of 394 amino acid residues, including one cysteine, 2 tryptophans and 9 methionine residues [3,4]. Three N-linked glycans attached to asparagine residues 46, 83, and 247 represent ~12% of A1PI by molecular weight [5,6,7]. It synthesized in the liver and is present in the serum at level between 150 to 350 mg/dl (30-80 μM) when assayed with plasma standard. The major physiological role of A1PI is to inhibit neutrophil elastase (NE) in the lungs. A1PI is an inhibitor of proteases such as trypsin, chymotrypsin, pancreatic elastase, skin collagenase, rennin, urokinase and protease of polymorphonuclear lymphocytes. Alpha-1-proteinase inhibitor deficiency is an autosomal co-dominant, hereditary disorder characterized by low serum and lung levels of alpha1-PI. Severe forms of the deficiency are frequently associated with progressive, moderate to severe emphysema that may become apparent in adults at any age regardless of smoking history, potentially resulting in a lower life expectancy. In the absence of A1PI, neutrophil elastase, released by lung macrophages and neutrophils, damages protein components of the alveolar wall leading to elastin breakdown and the loss of lung elasticity and lung tissue. Excessive elastolytic activity underlies progressive emphysema in patients with low plasma A1PI levels (<11 μM). A1PI augmentation therapy specifically aims to restore the protease-antiprotease balance in lung tissue of such patients and to attenuate the progression of pulmonary emphysema. Based on the A1PI serum concentration, a common classification to define A1PI deficiency includes the four major categories: (1) normal (with A1PI serum levels not lower than 20 μM); (2) deficient (with A1PI concentrations in serum lower than 20 μM); (3) dysfunctional (with normal A1PI level, but lost or lower inhibitory activity); and (4) null (with A1PI serum concentrations below the detectable level) [8]. Currently, A1PI therapeutic preparations are licensed exclusively for one indication, i.e. chronic augmentation and maintenance therapy in individuals with emphysema due to congenital A1PI deficiency but as evident from the available literature, due to the multiple biological activities of A1PI, it has been associated with other lung diseases (cystic fibrosis) and many non-pulmonary diseases like vasculitis, fibromyalgia, asthma, pancreatitis, rheumatoid arthritis, atherosclerosis [8]. Currently, there are four US Food and Drug Administration (US FDA) approved preparations of A1PI obtained from Cohn fraction IV-1 paste, which differ in the purification methods, concentrations, and dosage forms. Purification methods involve Cohn fractionation, polyethylene glycol (PEG) precipitation or affinity chromatography or by using multiple chromatography steps. Saklatvala and Wood et al. adopted a five step procedure to isolate and purify A1PI which includes precipitation with $(NH4)_2SO_4$, DEAE cellulose and affinity chromatography on concanavalin A sepharose [9]. Most of the inventions in the patents detail the purification methods of A1PI where the starting material is Cohn fraction IV-1 paste and precipitation agents like PEG and $ZnCl_2$ are added to precipitate impurities and A1PI respectively [10,11,12]. The precipitation methods suffer from low yields and less specific activity besides having an additional step included in the process in the form of filtration or centrifugation to separate the precipitate from the supernatant. The methods in which affinity chromatography are used are not suitable for therapeutic protein manufacture due to

*Corresponding author: Ashok Kumar N, Centre for Biotechnology, JNTUH, Hyderabad, Telangana, India, E-mail: ashnuvvula@gmail.com

the strong possibility of ligand leaching. The ligand concanavalin A is listed as being potentially mutagenic and teratogenic [13]. Cohn's ethanol fractionation is the method used by most of the manufacturers. Using ethanol in the process is known to lead to C-terminal lys truncation of A1PI. A linear increase of lys removal with increasing ethanol concentration was reported indicating the usage of ethanol in the process is not desirable to have an intact end product without lysine truncation [14]. The process described herein for human plasma derived A1PI eliminates the need for PEG, ZnCl$_2$, ammonium sulphate precipitation, Cohn's ethanol fractionation or affinity chromatography to purify this protein in high yields. A more efficient process helps to ensure optimal use of a scarce and valuable commodity like human plasma to produce the end product of therapeutic grade.

Materials and Methods

General equipment's in the laboratory

XK16/20 chromatography columns, AKTA Purifier chromatography system, DEAE sepharose resin, capto adhere resin from GE Healthcare, Uppsala, Sweden and Fractogel TMAE resin from EMD-Millipore.

All equipments were from standard manufacturers as given below:

UV Spectrophotometer (GE Healthcare)

pH and Conductivity meter (Thermo Scientific)

Ultra filtration unit-Stirred cell (Amicon, Millipore)

Mini Pellicon TFF system (Millipore)

Electrophoresis unit and power supply (Tarsons)

HPLC system (SHIMADZU-LC2010CHT)

ELISA plate Reader-MULTISKAN EX (Thermo scientific).

Reagents

Substrate: Nα-Benzoyl-DL-arginine-4-nitroanilide hydrochloride (BAPA- Sigma/B4875-100 mg)

Trypsin from bovine pancreas (Sigma/T1426-100 mg)

DMSO (Dimethyl sulphoxide) (Sigma/D8418-50 ml)

Primary antibody: Anti human alpha-1-antitrypsin produced in rabbit (A0409-Sigma Aldrich)

Secondary antibody: Anti rabbit IgG ALP conjugated, produced in goat (A3812-Sigma Aldrich)

Coomassie brilliant blue R-250 (11255300251730-Merck)

Chemicals

a) Sodium dihydrogen phosphate monohydrate GR, Di sodium hydrogen phosphate anhydrous GR, Ammonium sulphate GR, Glycine GR, Tris buffer, Sodium acetate trihydrate, Sodium dodecyl sulphate.

A1PI reference standard

α1-Antitrypsin was procured from Sigma-Aldrich. A human plasma-derived protein which is 52 KDa under reducing conditions on SDS-PAGE (Catalog No A9024, Lot No: 018K7546). This was used as a reference standard in this study.

Human plasma

Recovered plasma was procured from licensed and audited blood banks through cold chain shipment. The serology test reports for each of the donor units were verified as negative for HIV 1 and 2 antibodies, HBS Ag, HCV antibody, VDRL and malaria parasite. Plasma bags were also tested negative by nucleic acid testing (NAT for HIV 1 and 2, HBV and HCV). These tested units were taken for the process.

Methodology adopted to purify A1PI

All steps were carried out at room temperature (25°C -27°C). Prior to the purification process, the plasma bags were removed from the freezer (-40°C) and left at room temperature i.e. 25°C for thawing for about 2 hr and then kept for further thawing in water bath set at 25°C -30°C for 30 min. After thawing, the bags were cut open under laminar air flow and the plasma from the bags was pooled in a collection vessel. The pooled plasma was filtered through 40 μm, polypropylene filter (Domnick hunter) followed by a second 15 μm, polypropylene filter (Domnick hunter) to remove the clots or any particulates from plasma. The filtrate was loaded on to chromatography columns to purify A1PI. Filtration was carried out using a peristaltic pump connected to the filters through silicone tubing. Given below was the sequence adopted to purify A1PI to homogeneity (Figure 1).

Capture step of A1PI:

Various chromatography resins were screened to check the efficiency to bind A1PI from human plasma. To capture A1PI from plasma, cation exchange resins like CM sepharose, SP sepharose, eshmunoS; anion exchange resins-DEAE sepharose, eshmuno Q. Hydrophobic interaction chromatography–butyl sepharose, phenyl sepharose and mixed mode resins like capto MMC were tried. Among all the resins, anion exchangers have the ability to bind A1PI and in the group of anion exchangers, particularly DEAE sepharose resin has the highest ability to bind A1PI at pH 5.2 and 2 mS/cm conductivity. So, the column packed with DEAE sepharose resin was optimized as the step to capture A1PI.

The column was packed by flow-packing method at a constant flow rate. Packing buffer was mixed with the resin to form 50-70% slurry. The slurry was poured into the column in a single continuous motion against the column wall to minimize the introduction of air bubbles. After pouring the resin, the adaptor was mounted on the top of the column and connected to a pump. The bottom outlet of the column

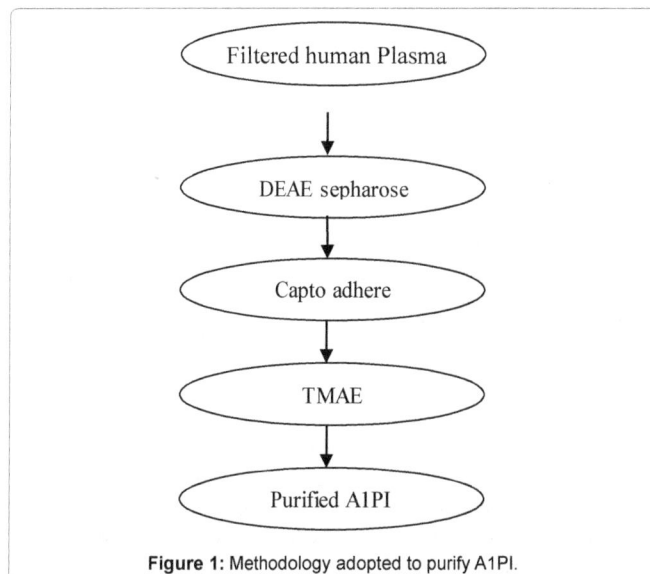

Figure 1: Methodology adopted to purify A1PI.

was opened and the pump was set at flow rate 30% greater than that used during operation. Mobile phase was passed through the column in order to compress the resin. After the bed had compressed, the pump was paused and the adaptor was lowered onto the compressed bed, to complete the packing of the column. The efficiency of the packed column was monitored by determining HETP and asymmetry factor. 1% acetone solution was prepared and applied to the column. The column was washed with water at a flow rate of 30 cm/hr and the washing was continued till the UV peak at 280 nm reached the baseline. HETP and asymmetry values were calculated for this peak to determine the efficiency of the packed column

The filtered plasma without any particulate matter was loaded on to DEAE- sepharose resin packed in XK 16/20 column to a height of 10 cm and this column was equilibrated thoroughly before loading the sample with 50 mM sodium acetate pH 5.2 buffers which is having a conductivity of 2 mS/cm. The column was run at 90 cm/hr-120 cm/hr for loading, washing and elution. The column was washed with the same equilibration buffer to remove the other unbound plasma proteins; 50 mM sodium acetate pH 4.5 buffer was passed through the column to remove other major proteins like albumin from the column. The fraction containing the A1PI protein was eluted with the 50 mM sodium acetate pH 8.0 buffer containing 120 mM NaCl. The column was washed with 0.5M NaCl to remove other tightly bound proteins and 0.5M NaOH was used for regeneration and sanitization.

Other proteins collected in the flow through, wash fraction and regeneration step were not individually characterized in the current research but SDS-PAGE and Western blot analysis indicated that these could be certain major proteins in plasma like albumin, IgG, fibrinogen and others. SDS-PAGE analysis and Western blotting were performed for the column fractions to track A1PI. Western blot analysis confirmed the presence of A1PI in elution peak and SDS-PAGE analysis indicated an A1PI protein band with electrophoretic mobility matching the reference standard

Intermediate purification step of A1PI

A1PI after the initial capture step was loaded on to a second chromatography column for further purification. This step was mainly to separate other co-eluting plasma proteins from A1PI. Q-sepharose, SP sepharose, capto adhere (GEHC), capto MMC phenyl sepharose (GEHC), macro prep methyl (Bio-Rad), cellufine phenyl (JNC) were the different resins tried out for the intermediate purification step of A1PI. Among all the resins, capto adhere which is a strong anion exchanger with multimodal functionality resin was found to give good yield, purity and had better removal capacity for impurities at higher flow rates when compared to other resins which was confirmed by SDS-PAGE analysis. Column was packed with capto adhere resin in XK 16/20 column to a height of 10 cm by flow-packing method as detailed in the above section. 25 mM sodium phosphate buffer containing 20 mM NaCl, pH 7.0 was used for equilibrating the column. DEAE eluate was loaded on to capto adhere column after the conductivity was brought down to 5 mS/cm by diluting with 2 mM sodium phosphate, pH 7.0 buffer. It was observed that A1PI bound to the column and was eluted with 25 mM sodium phosphate buffer containing 150 mM NaCl, pH 7.0. Before elution, the column was washed with 25 mM sodium phosphate buffer containing 70 mM NaCl, pH 7.0 to remove certain impurities whereas the other plasma proteins bound to the column were eluted from the column in 0.5M NaCl wash and regeneration steps. Other conditions tried out were loading conductivity at 2 ms/cm, 10 ms/cm and at pH 8.0 and pH 6.0; elution buffer conductivity of 25 mS/cm, 35 mS/cm but under these conditions A1PI yield and purity

was less than the finally chosen method. So, finally loading at pH 7.0, 5 ms/cm and eluting with buffer of pH 7.0 was found to be optimal condition to get purified A1PI in the Elution fraction and to separate it from other plasma proteins. All the fractions from this column were analyzed on SDS-PAGE to check for purity and composition.

Polishing step of A1PI

After the intermediate purification step, A1PI was found to have few more extra bands which were carried over from plasma. In order to purify this further, A1PI protein collected in elution fraction in above step was taken and diluted to 2 mS/cm with 2 mM Tris buffer pH 8.5. The capto adhere column eluate was loaded on to various anion exchange resins to remove the other few impurities and among all the resins, TMAE resin was able to selectively bind A1PI protein. The TMAE resin was packed in XK 16/20 column to a height of 10 cm and equilibrated with 50 mM Tris buffer PH 8.5, conductivity of 2 mS/cm, The above diluted and pH adjusted sample was loaded onto TMAE column at a flow rate of 90 cm/hr. After loading, the column was washed with equilibration buffer to remove any unbound proteins and more impurities were removed from the column by washing the column with 50 mM Tris buffer pH 8.5, containing 75 mM NaCl. Bound A1PI is eluted from the column by changing the pH and ionic strength of the elution buffer and to bring the pH shift, the column was prewashed with 50 mM Tris buffer pH 7.0, conductivity of 2 mS/cm. After the column has attained pH 7.0, elution buffer i.e. 50 mM Tris buffer containing 120 mM NaCl, pH 7.0 was passed through the column to elute A1PI. The impurities which bound to column were observed in the wash fraction and in the cleaning and regeneration peaks. A1PI which was eluted by the buffer of higher ionic strength was found to be free of impurities and homogenous on SDS-PAGE.

Characterization of the purified A1PI

The purified A1PI preparation was analyzed on SDS-PAGE under reducing and non-reducing conditions and on a size exclusion HPLC (SE-HPLC) column. The activity of the purified sample was measured by an enzyme assay method. All the analytical studies were carried out alongside a reference standard (obtained commercially from Sigma).

SDS-PAGE analysis: This was performed to analyze the column fractions and the final purified sample in comparison with a reference standard.

SDS-PAGE methodology: 10% SDS-PAGE gel was made where resolving gel was prepared by adding 3.1 ml of 30% Bis-Acrylamide solution, 1.9 ml of 1.5 M Tris buffer pH 8.8 to 2.5 ml of water. To this solution 37.5 μl of 10% APS and 10 μl TEMED were added and poured into the assembled glass plates and allowed to polymerise. Stacking gel was prepared by adding 0.4 ml of 30% Bis-Acrylamide solution, 0.75 ml of 0.5M Tris buffer pH 6.8 to 1.8 ml of water. To this solution 20 μl of 10% APS and 5 μl TEMED were added and poured on top of the resolving gel. Immediately a clean comb was inserted into the stacking gel carefully avoiding trapping of air bubbles. The gel was allowed to polymerize. After the gel was polymerized the comb was removed carefully. Meanwhile, 40 μl of the samples to be loaded on the gel were aliquoted into micro centrifuge tubes and to that 10 μl of sample buffer (containing Tris, SDS, bromophenol blue, β-mercapto ethanol and glycerol) was added and kept in boiling water bath for 5 min. After boiling, the samples were removed and centrifuged at 10,000 rpm for 2 min. Once the 10% gel was ready, the samples were loaded into the wells slowly using micropipette. The casting unit was filled with tank buffer (containing Tris, glycine and SDS) and electrophoresis was carried out at 300 V for 45 min till the tracking dye (bromophenol blue) had run out of the gel. Then the gel was carefully removed and stained with Coomassie brilliant blue R-250 dye prepared in methanol and acetic acid. Once the protein bands were developed, the gel was destained using solution containing methanol and acetic acid. Then the gel was scanned and documented.

Western blot methodology: The protein bands were separated on 10% SDS-PAGE as detailed above and the gel was trans-blotted onto a nitrocellulose (NC) membrane. Transfer was carried out for 2 hrs at 100 V in a buffer containing Tris–HCl, glycine and methanol. After transfer, the NC membrane was incubated in 50 ml TBST (Tris buffered saline with Tween-20) containing 5% skimmed milk powder for 60 min. This step was to block additional protein binding sites. Following the incubation, the NC membrane was washed with TBST twice for 10-15 min. After washing, the membrane was incubated with 25 mL anti human alpha-1-antitrypsin antibody produced in rabbit (Polyclonal IgG, 0.5 mg/mL stock diluted 1:10,000 in TBST buffer containing 0.25% Bovine serum albumin) for 1 hr at room temperature with gentle shaking. The membrane was washed thrice for 10 min in 50 ml TBST and then incubated the membrane in 25 mL Anti rabbit IgG ALP Conjugated antibody (1:30,000 dilution with TBST) at room temperature for 1 hr with gentle shaking. After 1 hr, the blot was washed with TBST and developed with 3,3′,5,5′N-tetramethylbenzene substrate which is a chromogenic substrate added to develop the specific protein bands which reacted with the antibodies.

SE-HPLC: SE-HPLC method was used to characterize the purified molecule. The analysis was carried on a Shodex (vendor) Protein KW-803 Shodex column using sodium phosphate pH 7.0 buffer.

Specific activity: The alpha1-antitrypsin (AAT) activity of the purified A1PI was measured using an enzyme assay method which is based on the inhibition of trypsin activity by AAT and estimation of the residual trypsin activity using a synthetic substrate benzoyl-DL-arginine-P-nitroanilide (BAPA). The level of the product (p-Nitroaniline) formed was estimated by measuring the absorbance at 410 nm. The absorbance was read at 410 nm after the reaction was terminated using MULTISKAN EX reader (Thermo scientific) 0.15 ml of trypsin was mixed with 0.006 ml of test sample or standard and 0.144 ml of 50 mM Tris buffer, pH 8.2 with 20mM $CaCl_2$ and this mixture was incubated for 5 minutes at 37°C. Then 3.5 ml of benzoyl-DL-arginine-P-nitroanilide (BAPA) was added to the above solution and incubated for 10mins at 37°C. The reaction was stopped by addition of 0.5 ml of 30% acetic acid and the OD was measured at 410 nm against a suitable blank [15,16].

Excess Trypsin+ AAT →Trypsin-AAT complex+ Residual Trypsin

Residual Trypsin+BAPA→Nα-Benzoyl-DL-arginine+P-Nitro aniline

Results and Discussion

DEAE sepharose step to capture A1PI

DEAE-sepharose resin was packed in XK 16/20 column to a height of 10 cm and this column was equilibrated with 50 mM sodium acetate pH 5.2 buffers with a conductivity of 2 mS/cm. The column was run at 90 cm/hr-120 cm/hr for loading, washing and elution. The column was washed with the same equilibration buffer to remove the other unbound plasma proteins and then passed wash-2 buffer i.e. 50 mM sodium acetate pH 4.5 buffer to remove other bound proteins like albumin and clotting factors. A1PI which was bound to the column was eluted with the 50 mM sodium acetate pH 8.0 buffer containing 120 mM NaCl, conductivity of 18 mS/cm (DEAE Chromatogram

Figure 2a). The column was washed with buffer containing 0.5M NaCl to remove the tightly bound proteins and 0.5M NaOH for regeneration and sanitization. SDS-PAGE analysis and western blotting was performed for the column fractions to track for A1PI (Figures 2b and 2c). Western blot analysis indicated the presence of A1PI band in the eluate fraction and its mobility is matching to the reference standard.

Capto adhere step

The DEAE- sepharose eluate with other impurities was loaded on to capto adhere multi modal strong anion exchange column after the conductivity and pH were adjusted to 5 ms/cm and 7.0 respectively. The column was equilibrated with 25 mM sodium phosphate buffer containing 20 mM NaCl, pH 7.0 and Conductivity of 5 mS/cm. The sample was loaded on to the equilibrated column at 90 cm/hr flow rate and under the applied conditions A1PI bound to the column and some impurities observed in Flow through fraction. The column was washed with wash-2 buffer of higher conductivity to remove other proteins (Figure 3a). A1PI bound to the column was eluted with buffer at 20 mS/cm conductivity. Eluate sample was analyzed on SDS-PAGE. The A1PI band was enriched in intensity but it was also found to have a prominent low molecular weight impurity and few high molecular weight impurities (Figure 3b). In order to remove these impurities, the eluate sample is further loaded on to TMAE column.

TMAE step to purify A1PI to homogeneity

The TMAE resin was packed in XK 16/20 column to a height of 10 cm and was equilibrated with 50 mM Tris pH 8.5 buffers which are having a conductivity of 2 mS/cm. The eluate from capto adhere column after adjusting its conductivity to 2 mS/cm was loaded onto TMAE column at a flow rate of 90 cm/hr and after loading the column was washed with equilibration buffer and wash-2 buffer to remove any unbound proteins and to separate the other bounded impurities from the A1PI. After passing wash-2 buffer, pre elution buffer was passed through the column to shift the pH of column to 7.0. A1PI bound to the column was eluted by changing the pH and ionic strength of the buffer and elution was carried out with 50 mM Tris pH 7.0 buffer containing 120 mM NaCl, conductivity of 14 mS/cm. The strongly bound impurities were removed by washing the column with 0.5M NaCl and 0.5M NaOH solutions (Figure 4a). All the fractions from the column were analyzed on SDS-PAGE and the analysis indicated that the low molecular weight impurity bound strongly to the column under the applied conditions and was observed in 0.5M NaCl wash. The sample from the elution peak showed a single band on SDS-PAGE (Figure 4b). The eluate from TMAE column is concentrated and subsequently exchanged into 20 mM sodium phosphate buffer pH 6.8, containing 0.1M sodium chloride. This buffer exchange is carried out using 10

Dilution Folds	Content of total protein in reaction(mg)-test sample	OD at 410 nm	Active A1PI (mg)	Active A1PI per mg of total A1PI	% of active A1PI
Neat	0.0047	0.902	0.0045	0.957	95.7

Table 1a: The data indicates that 0.957 mg of active A1PI is present in 1 mg of total protein.

Dilution folds	Content of Total Protein in Reaction(mg)-Test sample	Active A1PI (mg)	Content of Trypsin Inhibited (mg)	A1PI inhibiting 1mg of trypsin (mg)
Neat	0.0047	0.0045	0.0026	1.8

Table 1b: The data indicates that 0.957 mg of active A1PI is present in 1 mg of total protein.

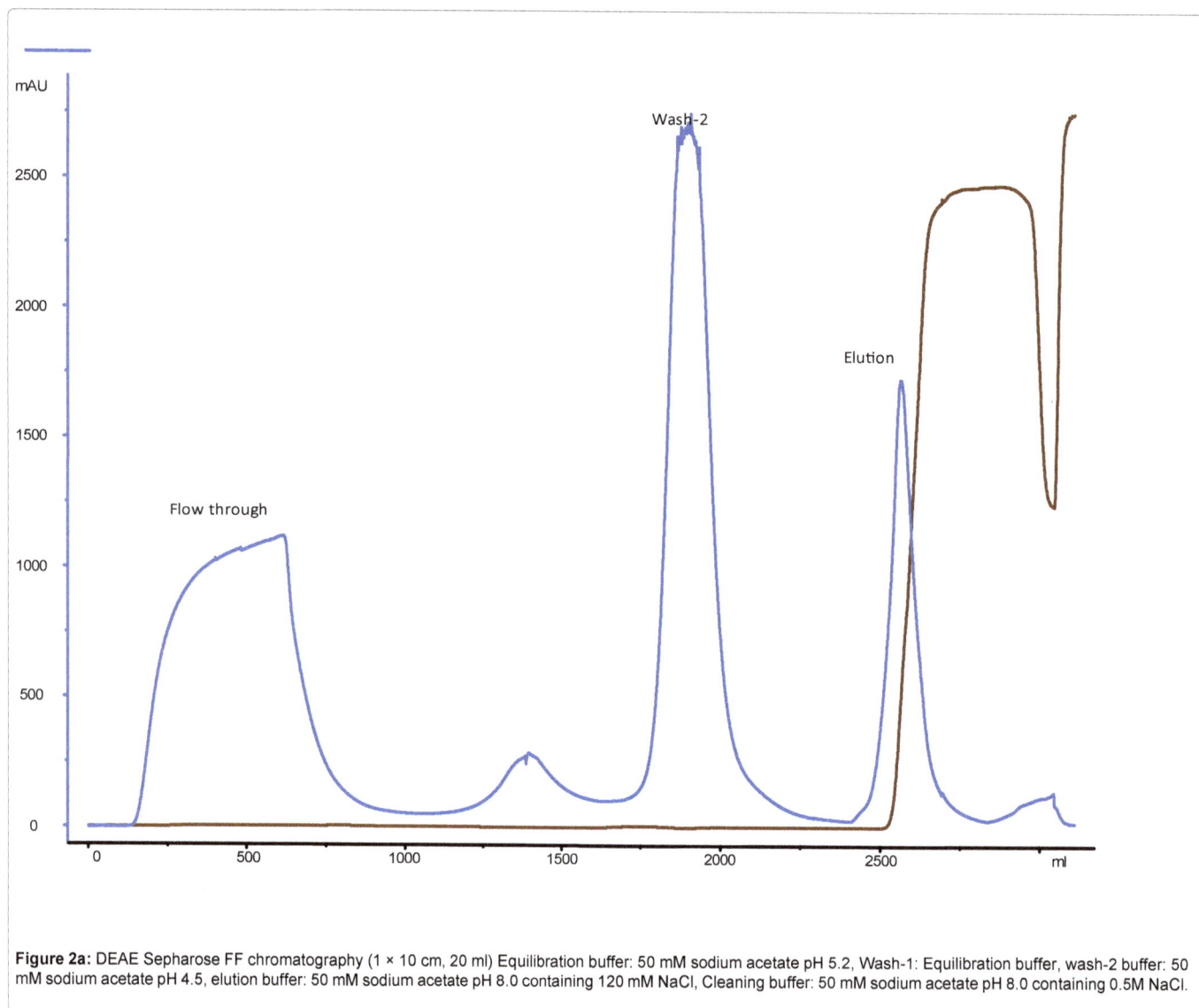

Figure 2a: DEAE Sepharose FF chromatography (1 × 10 cm, 20 ml) Equilibration buffer: 50 mM sodium acetate pH 5.2, Wash-1: Equilibration buffer, wash-2 buffer: 50 mM sodium acetate pH 4.5, elution buffer: 50 mM sodium acetate pH 8.0 containing 120 mM NaCl, Cleaning buffer: 50 mM sodium acetate pH 8.0 containing 0.5M NaCl.

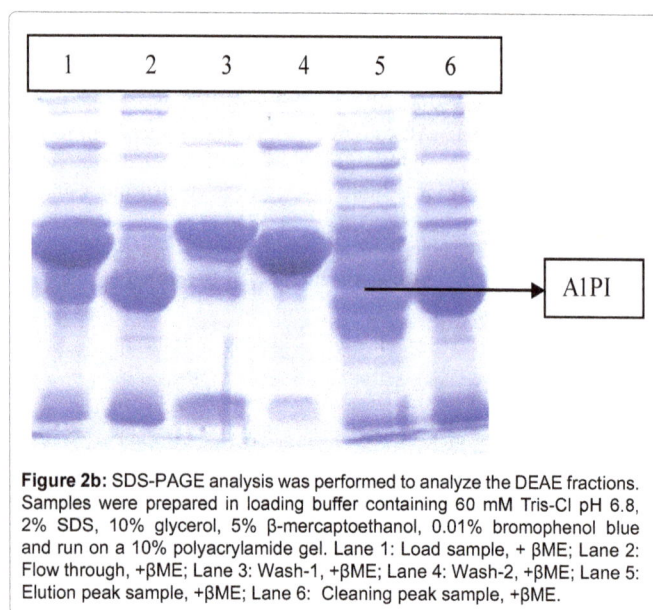

Figure 2b: SDS-PAGE analysis was performed to analyze the DEAE fractions. Samples were prepared in loading buffer containing 60 mM Tris-Cl pH 6.8, 2% SDS, 10% glycerol, 5% β-mercaptoethanol, 0.01% bromophenol blue and run on a 10% polyacrylamide gel. Lane 1: Load sample, + βME; Lane 2: Flow through, +βME; Lane 3: Wash-1, +βME; Lane 4: Wash-2, +βME; Lane 5: Elution peak sample, +βME; Lane 6: Cleaning peak sample, +βME.

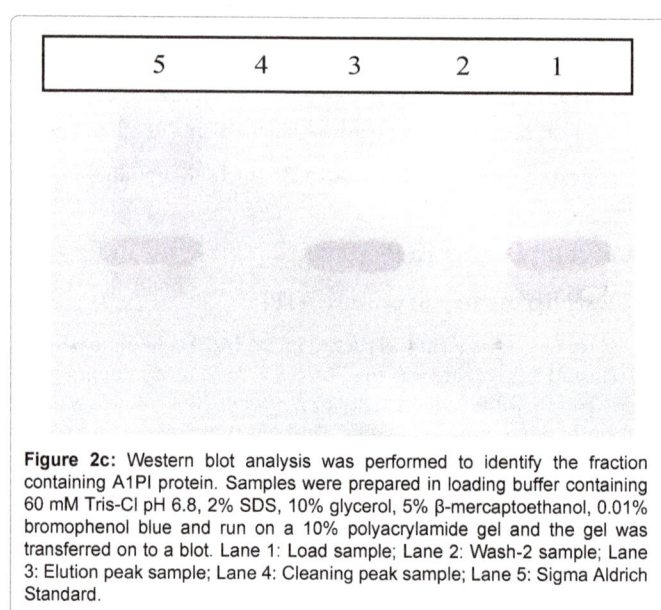

Figure 2c: Western blot analysis was performed to identify the fraction containing A1PI protein. Samples were prepared in loading buffer containing 60 mM Tris-Cl pH 6.8, 2% SDS, 10% glycerol, 5% β-mercaptoethanol, 0.01% bromophenol blue and run on a 10% polyacrylamide gel and the gel was transferred on to a blot. Lane 1: Load sample; Lane 2: Wash-2 sample; Lane 3: Elution peak sample; Lane 4: Cleaning peak sample; Lane 5: Sigma Aldrich Standard.

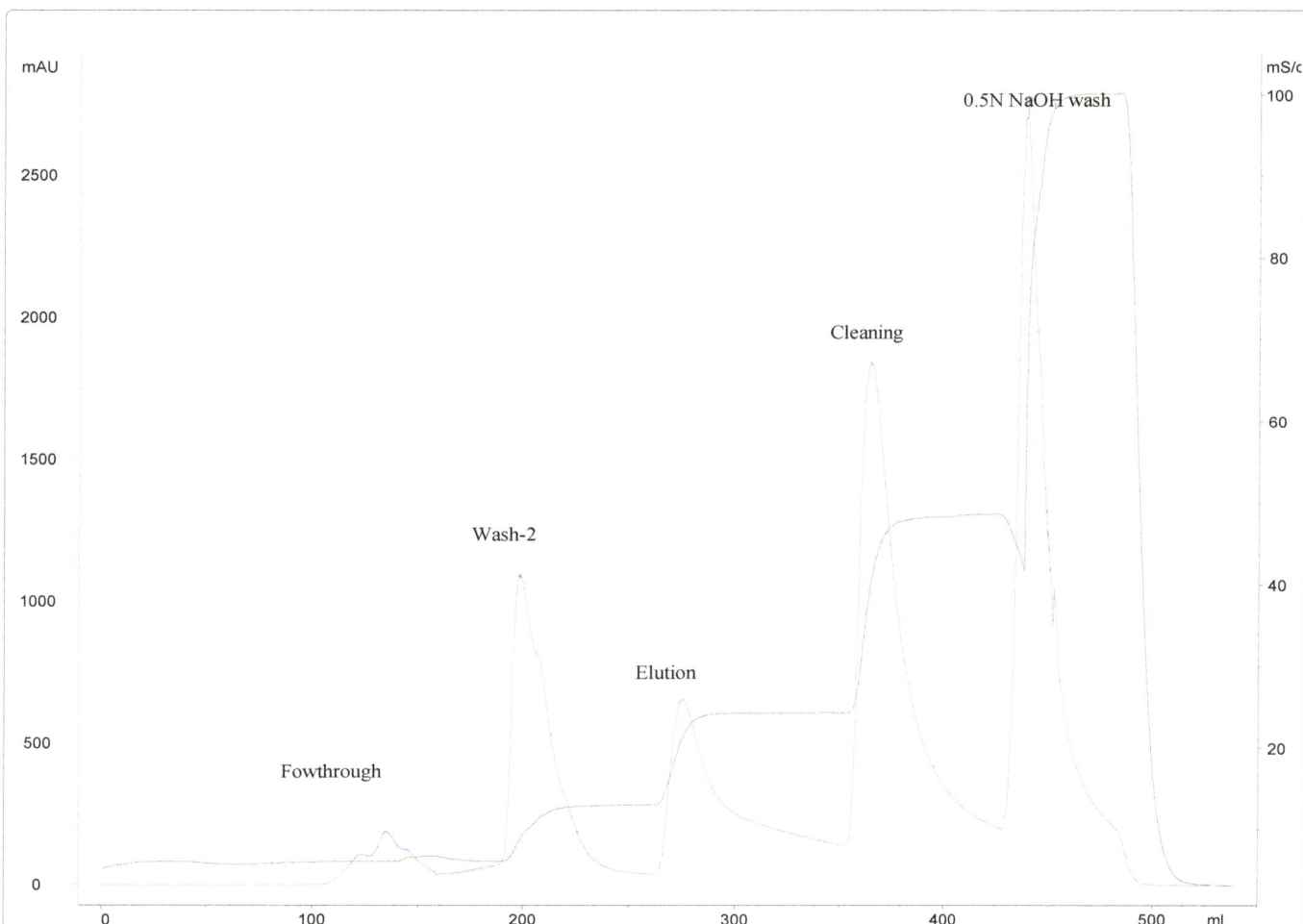

Figure 3a: Capto adhere multi modal chromatography for the DEAE elution fraction on a 20 ml column. A1PI bound to the column and was observed in the elution fraction.

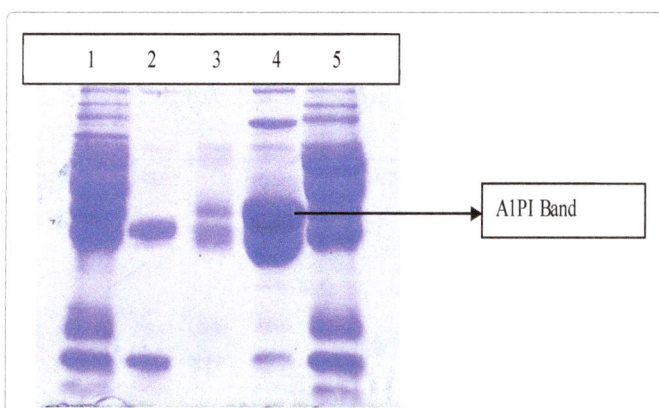

Figure 3b: SDS-PAGE analysis was performed to analyze the capto adhere column fractions. Samples were prepared in loading buffer containing 60 mM Tris-Cl pH 6.8, 2% SDS, 10% glycerol, 5% β-mercaptoethanol, 0.01% bromophenol blue and run on a 10% polyacrylamide gel. Lane 1: Load sample, + βME; Lane 2: Flow through, +βME; Lane 3: Wash-2, +βME; Lane 4: Lane 4: Elution peak sample, +βME; Lane 5: Cleaning peak sample, +βME.

KDa TFF cassettes on a mini Pellicon TFF system. The observed yield is 0.6-0.7 g of A1PI per liter of human plasma by adapting the developed process and the purified molecule is characterized by the methods described in the following section.

SDS-PAGE analysis of the purified A1PI

The eluate after TMAE step is analyzed on SDS- polyacrylamide gel electrophoresis (Figure 5) and is found be more than 95% pure. On comparison with reference standard its molecular weight is around 52 kDa under both reducing conditions and non-reducing conditions and the mobility matched that of the reference standard.

SE-HPLC analysis of the purified A1PI

The identity and purity check by SE-HPLC analysis in comparison to standard indicated that the purified A1PI protein is close to 93% pure (Figure 6a and 6b). The analysis was carried out on Protein KW-803 Shodex column (8.0 mm I.D × 300 mm length) with Particle size of 5 μm. The retention time of the purified A1PI peak (9.313 min) closely matched the retention time of the reference standard (9.293 min) and also showed a better purity profile (93%) with a major single peak and a minor peak at 7.771 min, when compared to the reference standard (75% purity) with at least three other contaminants on SE-HPLC chromatogram. Sodium phosphate buffer containing NaCl was the buffer used for running the column at 1 ml/min flow rate. A pilot study was carried out by ZLB Behring LLC to compare the purity, functionality and isoform composition of the available market products (Aralast, Prolastin and Zemaira). SEC-HPLC comparison indicated that Zemaira was 95.98%, Prolastin 79.00% and Aralast 63.55% monomeric [17]. In-house purified A1PI was found to have 93% monomeric form.

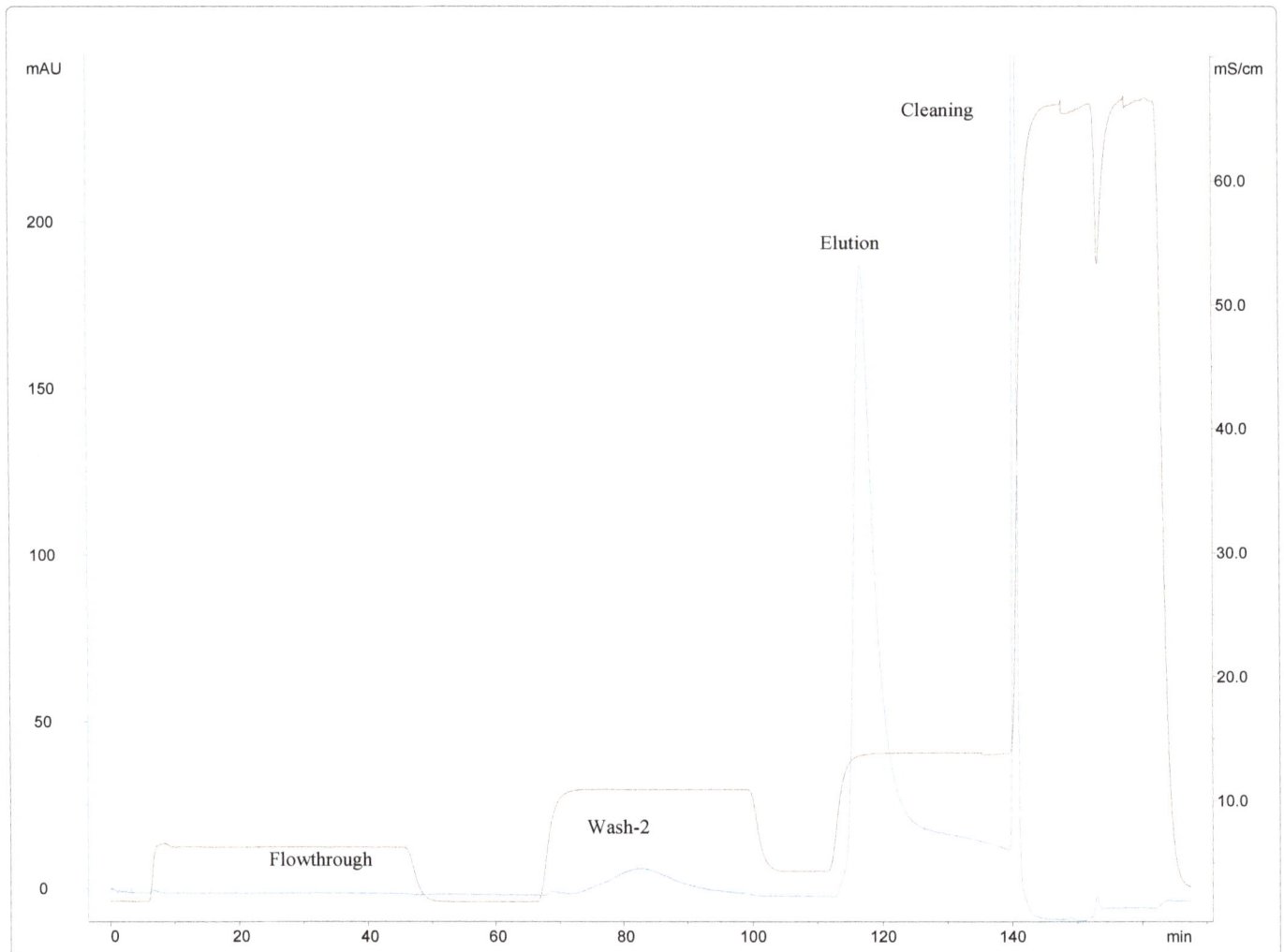

Figure 4a: TMAE chromatography for the capto adhere eluate on a (1 × 10 cm, 20 ml) column. Equilibration: 50 mM Tris buffer PH 8.5, conductivity 2 mS/cm. Wash-2: 50 mM Tris buffer PH 8.5, containing 75 mM NaCl.10 mS/cm. Wash-3: 50 mM Tris buffer PH 7.0, conductivity 2 mS/cm. Elution: 50 mM Tris buffer containing 120 mM NaCl, pH 7.0 and conductivity 14 mS/cm.

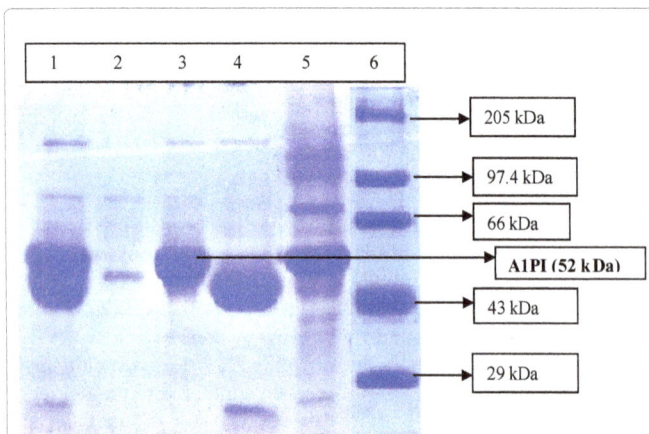

Figure 4b: SDS-PAGE analysis was performed to analyze the fractions from TMAE column. Samples were prepared in loading buffer containing 60 mM Tris -Cl pH 6.8, 2% SDS, 10% glycerol, 5% β-mercaptoethanol, 0.01% bromophenol blue and run on a 10% polyacrylamide gel. Lane 1: Load sample, +βME; Lane 2: Wash-2 Peak, +βME; Lane 3: Elution peak sample; Lane 4: 0.5M NaCl wash peak, +βME; Lane 5: 0.5N NaOH wash peak, +βME; Lane 6: Molecular weight marker, 29-205 KDa.

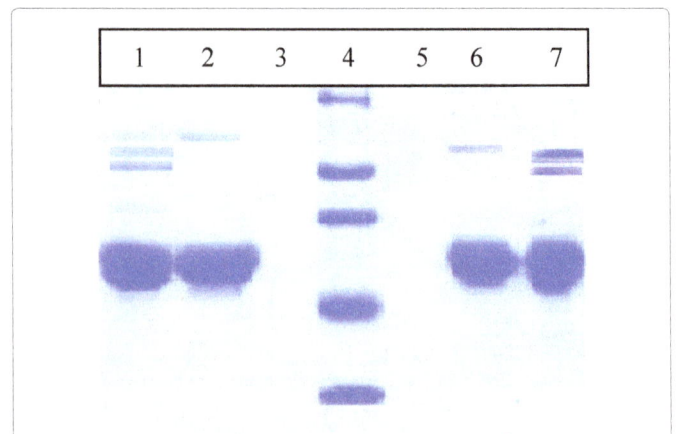

Figure 5: SDS-PAGE analysis was performed to check the purity of A1PI protein in reference to standard (Sigma). Samples were prepared in loading buffer containing 60 mM Tris-Cl pH 6.8, 2% SDS, 10% glycerol, 5% β-mercaptoethanol, 0.01% bromophenol blue and run on a 10% polyacrylamide gel. Lane 1: Reference standard 15 µg, -βME; Lane 2: Purified sample 15 µg, -βME; Lane 3: Empty well; Lane 4: Molecular weight marker, 29-205 KDa; Lane 5: Empty well; Lane 6: Purified sample 15 µg, +βME; Lane 7: Reference standard 15 µg, +βME.

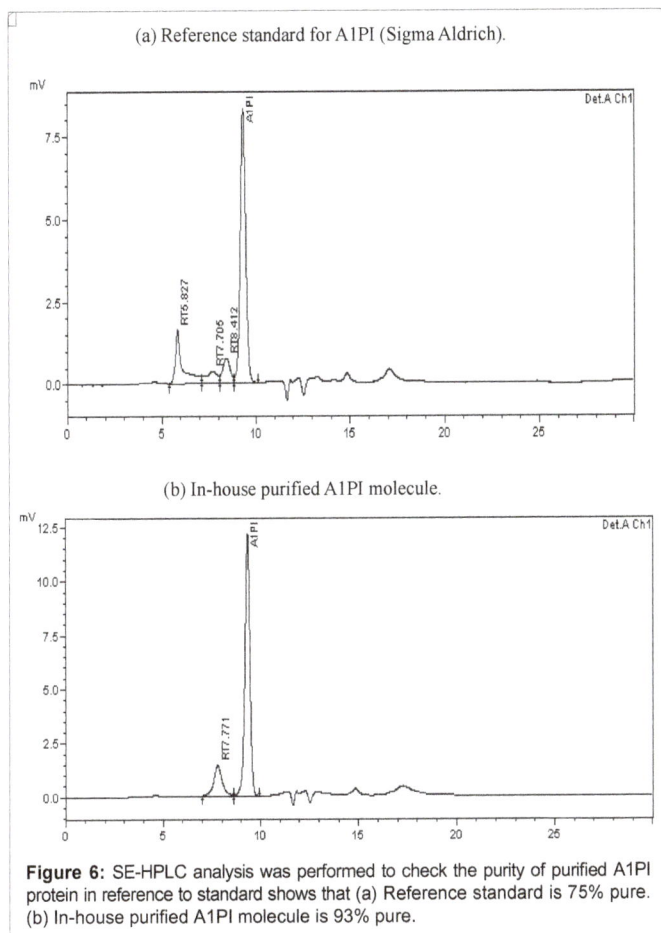

Figure 6: SE-HPLC analysis was performed to check the purity of purified A1PI protein in reference to standard shows that (a) Reference standard is 75% pure. (b) In-house purified A1PI molecule is 93% pure.

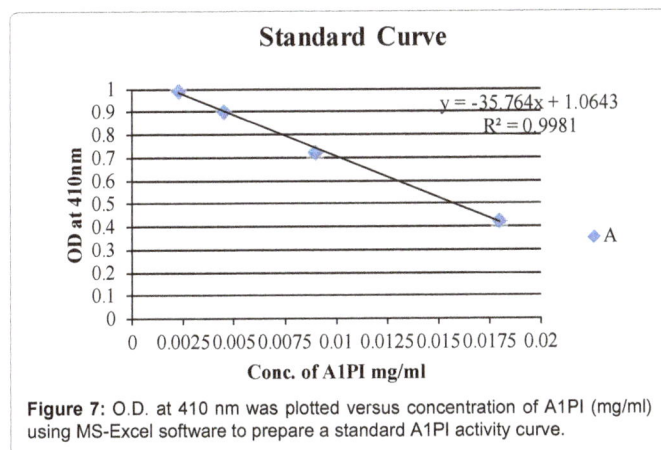

Figure 7: O.D. at 410 nm was plotted versus concentration of A1PI (mg/ml) using MS-Excel software to prepare a standard A1PI activity curve.

Dilution folds	Content of ref. std. in reaction mixture (mg)	Content of trypsin inhibited (mg)	OD at 410 nm
1.7	0.018	0.0105	0.427
3.4	0.009	0.0053	0.727
6.8	0.0045	0.0026	0.904
13.6	0.00225	0.0013	0.992

Table 2: A standard curve was plotted at different concentrations of A1PI as given below:

Specific activity

The specific activity (mg of A1PI/mg of protein) of purified A1PI is determined with reference to a standard A1PI activity curve. This assay was carried out along with the reference standard from Sigma and considered the below information for the assay - alpha-1-antitrypsin from human plasma (Sigma) was used and as per the label claim 1.7 mg of alpha 1 proteinase inhibitor inhibits 1.0 mg of trypsin (Figure 7 and Table 2). The purified and concentrated TMAE eluate was tested in this assay for its ability to inhibit trypsin and to calculate the percent of active protein in the purified fraction (Table 1a and 1b). Pilot study carried out by ZLB Behring LLC to compare the specific activity of the available market products indicated the specific activity (activity/mg protein) values were 0.85, 0.65 and 0.56 for Zemaira, Aralast and Prolastin respectively [17]. So, the in-house purified A1PI has specific activity comparable to Zemaira.

Conclusion

Since 1987, several A1PI products derived from pooled human plasma have been approved and are currently available to slow down the progression of emphysematous conditions in A1PI deficient patients. In addition, due to its multiple physiological activities, A1PI has been identified for its involvement in several other rare diseases, the treatment of which may possibly benefit from A1PI based therapies. From this standpoint, the product quality of the purified preparations for therapeutic use is very important [8]. There are many reports of purification of A1PI by using precipitation methods with PEG and $ZnCl_2$ [18-22]. In the above experiments, an attempt was made to look for a purification scheme that is both economical and scalable for the purification of A1PI, without compromising on yields, purity and its other therapeutic characteristics. After an initial capture on DEAE resin, two different new generation resins were chosen for their ability to selectively bind and purify A1PI from among the other plasma proteins. Figures 3a and 3b indicate how the use of capto adhere resin was able to purify A1PI to a significant extent with the SDS-PAGE (Figure 3b) analysis revealing only a few protein impurities. As a polishing step for purity, the A1PI sample after Capto adhere step was loaded on Fractogel TMAE resin, which was successful in increasing the purity to a significant extent as shown in lane 3 of Figure 4b. The yield of the final purified A1PI by this process was found to be in the range 0.6-0.7 g/litre of human plasma. This increased recovery and purity helped us nail this as an ideal and improved process scheme for therapeutic manufacturing, subject to the other biochemical characteristics being within the required specifications. Results of the characterization studies on purified A1PI indicated the comparability of the protein with an existing reference standard with respect to its purity on SDS-PAGE, SE-HPLC analysis and specific activity (Figures 4-7). The specific activity results show that the functionality of the protein has not been compromised by this purification scheme. A scheme that uses only chromatography steps is easily scalable unlike the earlier reported procedures of precipitation and affinity capture. The increased yields and purity show that further studies can be done to adapt the method for industrial manufacturing of this very important therapeutic protein from human plasma.

References

1. Stein PE, Carrell RW (1995) What do dysfunctional serpins tell us about molecular mobility and disease? Nat Struct Biol 2: 96-113.

2. Silverman GA, Bird PI, Carrell RW, et al. (2001) The serpins are an expanding superfamily of structurally similar but functionally diverse proteins: evolution, mechanism of inhibition, novel functions, and a revised nomenclature. J Biol Chem 276: 33293-33296.

3. Carp H, Miller F, Hoidal JR, Janoff A (1982) Potential mechanism of emphysema: alpha 1-proteinase inhibitor recovered from lungs of cigarette smokers contains oxidized methionine and has decreased elastase inhibitory capacity. Proc Natl Acad Sci 79: 2041-2045.

4. Johnson D, Travis J (1979) The oxidative inactivation of human alpha-1-proteinase inhibitor. Further evidence for methionine at the reactive center. J Biol Chem 254: 4022-4026.

5. Mega T, Lujan E, Yoshida A (1980a) Studies on the oligosaccharade chains of humanα1-protease inhibitor: I. Isolation of glycopeptides. J Biol Chem 255: 4053-4056.

6. Carrell RW, Jeppson JO, Vaughan L, Brennan SO, Owen MC, et al. (1981) Human α1-antitrypsin: carbohydrate attachment and sequence homology. FEBS Lett 135: 301-303.

7. Carrell RW, Jeppson JO, Laurell CB, Brennan SO, Owen MC (1982) Structure and variation of human α1-antitrypsin. Nature 298: 329-334.

8. Elena K (2011) Center for Biologics Evaluation and Research USA, Recent Advances in the Research and Development of Alpha-1 Proteinase Inhibitor for Therapeutic Use, pp. 83-104.

9. Bagdasarian A (1981) Isolation of alpha 1-protease inhibitor from human normal and malignant ovarian tissue. J clin invest 67: 281-291.

10. Peter M (2010) Method for the purification of Alpha-1 proteinase inhibitor. US Patent 7807435 B2.

11. Wytold RL (1997) Purification of A1PI using novel chromatographic separation conditions. US Patent 5610285.

12. Duk SH (1997) Process for separating A1PI from Cohn fraction IV1 and IV4 paste. US Patent 5616693.

13. Joseph B, Neil G, John C (2012) Production of Plasma Proteins for Therapeutic Use, pp: 512.

14. Peter M (2007) Ethanol dependence of A1PI C-terminal Lys truncation by basic carboxypeptidases. US Patent 2007/0274976 A1.

15. Arjumand SW (1989) Alpha-1-Antitrypsin In Saudi Population. Journal of Islamic Academy of Sciences 2: 85-88.

16. Peter M (2007) Method for the purification of A1PI. US Patent 2007/0037270 A1.

17. Cowden DI, Fisher GE, Weeks RL (2005) A pilot study comparing the purity, functionality and isoform composition of alpha-1-proteinase inhibitor (human) products. Curr Med Res Opin 21: 877-883.

18. Scott MK (2010) Method for purification of alpha-1-antitrypsin. US Patent 7777006 B2.

19. Nathan B (2011) Methods for the purification of Alpha 1-antitrypsin and Apolipoprotein A-1. US Patent 2011/0087008 A1.

20. Tom (2001) Process for separating α1-proteinase inhibitor from Cohn fraction IV1 and IV4 paste. US Patent 6284874 B1.

21. Duk SH (1999) Process for increasing the yield of a protein which has been subjected to viral inactivation. US Patent 5981715.

22. Michael HC (1987) Method of preparing A1PI, US patent 4697003.

19

HPLC Method for Quantification of Halofuginone in Human Ureter: *Ex-Vivo* Application

Sassi A[1], Hassairi A[1]*, Kallel M[2], Jaidane M[1] and Saguem S[1]

[1]Metabolic laboratory of Biophysics and Applied Occupational and Environmental Toxicology, Faculty of Medicine, university of Sousse, Tunisia
[2]Chemical, pharmacological and pharmaceutical drug development, Faculty of Pharmacy, Monastir, Tunisia

Abstract

A new high-performance liquid chromatography method was developed to study the diffusion of halofuginone in the wall of the ureter. The human ureter extracts were prepared by trypsin digestion of the tissues followed by liquid–liquid extraction using the isopropanol after precipitating the proteins. The method used a reversed-phase C18 column with a mobile phase delivered to the analytical column according to a gradient program starting at a composition of ammonium acetate (pH 4.7; 10 mM)-acetonitrile-triethylamine (70:30:0.2, v/v/v) and linear changes to 90% of acetonitrile at 11 min. Liquid-liquid extraction proved to be selective for the HFG and provided a high recovery rate of 97.7%. The HPLC method was successfully validated by applying the novel validation protocol using the accuracy profile based on a new concept, that of the total error. The protocol V4, with five levels of concentration and 105 trials, was selected according to the algorithm designed by the SFSTP 2003 committee. Acceptance limits were set up at ± 20%, while the risk was settled at 5%. The method was found accurate over a concentration range of 0.2–10 µg/ml. The limit of detection for HFG was 0.06061 µg/ml. In order to demonstrate the applicability of this method in *ex vivo* application, the quantification of HFG in the wall was applied to study its distribution from a gel (0.03% w/w of HFG) in the human ureter.

Keywords: Halofuginone; HPLC-UV; Wall ureter

Introduction

Halofuginone (HFG) is a coccidiostat used in veterinary medicine (Figure 1). It is a halogenated derivative of febrifugine, a natural quinazolinone alkaloid [1]. Furthermore, it inhibits the expression of the collagen type 1 gene and is becoming one of the most intersting novel antifibrotic [2-6].

The efficacy of halofuginone in preventing the occurrence and recurrence of urethra narrowing has been proven in animal studies using HFG orally or locally [7,8]. Indeed, HFG is a promising compound in the treatment of the human urethra structures when used as an intraurethral gel to attain a local effect.

Before starting the clinical phase 1 trials in humans, *in vitro* studies, using an experimental model simulating the different layers of the human urethra, are needed to determine the qualitative and quantitative composition of the gel. Indeed, the gel formula used is one that would allow rapid diffusion of HFG in the wall without passage into the bloodstream. The human ureter has the same composition as the urethra and may be obtained following renal transplantation. Thus, it is used in *ex vivo* experiments to simulate the different layers of the urethra.

In this context, the quantitative determination of HFG in the ureter wall is essential to study its distribution from different gels in the human ureter *ex vivo*.

Several methods are available for determining HFG in feeds and tissues [9-16]. In fact, in broiler production, HFG is used during almost all the breeding periods except for pre-slaughter withdrawal. There is a considerable risk of HFG residues in edible tissues. The concern for consumers' safety was the reason for the establishment of maximum residue limits for most of the coccidiostats in tissues [17,18]. Therefore, the monitoring of residues of coccidiostats is performed in all European countries [17,18]. As far as we know, this is the first study to date aiming at determining HFG while using a new method of extraction from the wall of a human ureter, developed and validated using high-performance liquid chromatography (HPLC). In order to

demonstrate the applicability of this method in *ex vivo* application, the quantification of HFG in the wall was applied to study its distribution from a gel (0.03% w/w of HFG) in the human ureter in *ex vivo*

Experimental Part

Materials

HFG hydrobromide standard was obtained from Discovery Fine Chemicals, Unit 4A, Old Forge Roadster, Ferndown Industrial Estate, Wimborn, Dorset, United Kingdom. The internal standard (IS) imipramine hydrochloride was purchased from Sigma Ultra (England). HPLC grade acetonitrile (ACN) and propan-2-ol was purchased from Lab Scan. Ammonium acetate, sodium carbonate, and trypsin were purchased from Sigma Aldrich. Acetic acid was purchased from Merck

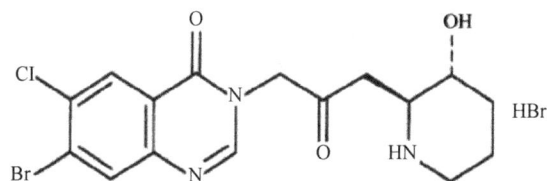

Figure 1: Structure of halofuginone

---

*Corresponding author: Dr. Saâd SAGUEM, Metabolic laboratory of Biophysics and Applied Occupational and Environmental Toxicology, Faculty of Medicine, Sousse, Tunisia, E-mail: saadsaguem_yahoo.fr

and triethylamine from Fluka. Ammonium sulfate was obtained from Riedel-de Haên. Purified water was obtained by the distillation apparatus Millipore (Arium 611DI/611UV) in our laboratory. Acetate buffer (10 mM, pH=4.3) was prepared by dissolving 19.27 g of ammonium acetate and 25 ml of acetic acid in distilled water, and by diluting them to 1 L of water. Sodium carbonate solution (10% w/v) was prepared by dissolving 100 g of sodium carbonate in distilled water and diluting it to 1 L. The pH was adjusted to 8 - 8.5 with acetic acid. All other chemicals used were of an analytical grade.

Liquid chromatography instrumentations

Chromatographic analysis was performed using Agilent 1200 HPLC equipped with G1311A quaternary pump, G1322A degasser , G1315D diode array detector, thermostated column compartment, manual injectors (syringe volume =5 µl). Separation was performed at 30°C using Lichrospher® C18 analytical column with 5 µm particle size, 4 mm internal diameter, and 250 mm in length, and using Lichrospher® 100 RP18 guard column (4 x 4 mm) with 5 µm particle size.

HPLC conditions

HFG and IS were eluted in gradient mode. The mobile phase consists of 10 mM ammonium acetate buffer (pH adjusted to 4.3 with acetic acid), triethylamine, and acetonitrile. It is delivered to the analytical column according to a gradient program, starting at a composition of ammonium acetate (pH 4.7; 10 mM)-ACN-triethylamine (70:30:0.2, v/v/v) and linear changes to ammonium acetate (pH 4.7; 10 mM)-ACN-triethylamine (10:90:0.2, v/v/v) at 11 min. The flow rate was 1 ml/min. A 5µl injection volume was used for each analytical run. The detector was set at 243 nm.

Preparation of standards

Stock solutions of HFG and IS were prepared at a concentration of 0.5 and 1 mg/ml, respectively, in ammonium acetate buffer (pH 4.3; 25 mM). These were stored at 4°C. The stock standard solution was stable for 3 months [19].

A series of standard HFG solutions were prepared by the appropriate dilution with mobile phase to obtain concentrations across a range of 0.2-10 µg/ml in human ureter extracts.

A working internal standard solution was freshly prepared every day from stocks at 0.1 mg/ml in distilled water.

Sample preparation

Fresh, surgically excised samples of human ureter were obtained directly after kidney transplant surgery. All the specimens were transferred to our laboratory within 1 h after being placed in a transport fluid (Eurocollins`) to provide 48 hours conservation. Ethical approval for the use of the ureter was provided by the Research Ethics Committee of the Hospital. After trimming away excess connective and adipose tissue, all specimens were immersed in physiologic water and distilled water, respectively, for 5 seconds. Human ureters were taken up into a Whatman filter paper to dry. Samples were cut into circular discs by means of a pair of scissors and a pair of tweezers (exposed areas 50 mm²).

Extraction

The extraction of HFG from the human ureter was described for the first time in our study. Put each circular disc of human ureters into a 2-ml Eppendorf tube. One milliliter of sodium carbonate solution (10% w/v) was added to each sample. Manual grinding was carried out using scissors. We added 100 µl of 0.1 mg/ml IS and HFG to the homogenate. Enzymatic hydrolysis was performed by 1 mg of trypsin for digesting the proteins. The homogenate was incubated for 3 h in a bath at a temperature of 37°C. The mixture was transferred into 15-ml tubes and 600 mg of ammonium sulfate was added to precipitate the proteins. It was mixed for 1 minute on a vortex. Four milliliters of propan-2-ol were added. The mixture was homogenized for 20 minutes by agitation. Then, samples were centrifuged for 15 min at 4000 rpm. An aliquot of 2 ml of supernatant was removed and evaporated to dryness at 55°C under reduced pressure, using a rotary evaporator. The residue was dissolved in 100 µl of the mobile phase. Five microliters were injected into the HPLC system.

To determine the extraction yield, samples containing HFG and IS drugs with internal standard were injected into the HPLC system, with and without extraction. The direct injection samples contained an equivalent amount of drugs as spiked in the extracted samples. Extraction efficiency is given as: *extraction efficiency % = (peak area of extracted sample/peak area of un-extracted sample)*100.*

Validation of analytical method

The validation was carried out in accordance with the performance criteria described in SFSTP 2003 using human ureter tissue [20]. This validation is based on the use of the accuracy profile based on a new concept, that of the total error [20,21]. This approach has been used in a wide range of methods such as liquid chromatography (LC-UV, LC–MS), spectrophotometry, and ELISA [22]. The accuracy profile of the analytical procedure is based on the expectation tolerance interval and the concept of total error (bias + standard deviation). It allows not only bringing together the objectives of the procedure with those of the validation but also visually grasping the capacity of the procedure to fulfill its objectives and to control the risk associated with its use in routine [20,21,23-25].

There are five experimental validation protocols [26]. Protocol V4 was selected according to the algorithm designed by the committee SFSTP 2003 [20,26]. The following criteria were considered: a prior knowledge of the assay procedure of HFG by HPLC; the matrix effect is possible and unknown; and calibration is not done at a single level of concentration.

The V4 protocol includes:

- Calibration standards at five concentration levels (0,2, 1, 2.5, 5 and 10 µg/ml) with and without matrix with two independent replicates for each standard.

- Standards validation, five concentration levels (0,2, 1, 2.5, 5 and 10 µg/ml) with the matrix with three independent replicates for each standard.

All these standards were prepared in 3 days with a total of 105 trials.

To test the selectivity/specificity of the method, representative blank samples (six human ureters from different patients) were analyzed and checked for interferences (peaks) at the retention times of the HFG and IS. Peak identification was performed by spectral information provided by a diode array detector.

During the validation step, acceptance limits were enlarged in accordance with a risk decrease. The first ones were set up at ± 20%, a value adopted by the American Association of Pharmaceutical Scientists guidelines when the validation results are expressed in terms of total error, while the risk was settled at 5% [27].

Results and Discussion

Development and optimization of liquid-liquid extraction of HFG

The wall of a human ureter is a complex biological matrix. No previous study has described the extraction of HFG from this matrix. A new method of HFG extraction was established after performing different screening runs based on the previously reported liquid-liquid extraction method for HFG extraction [9,14,28]. After a homogenization step and enzymatic hydrolysis performed by trypsin, a chemical deproteinization with ammonium sulfate was used. Indeed, the salt plays a dual role; not only does it allow a differential precipitation of the proteins but it also permits separation of the aqueous phase and organic phase by increasing the ionic strength of the solution. In fact, the extraction solvent selected during optimization is isopropanol, a solvent which is miscible with water. However, the presence of ammonium sulfate ensures the trapping of water molecules to a separate organic phase.

Ethyl acetate is the most commonly used solvent in the literature [9,14,28]. We tested different extraction solvents ethyl acetate, dichloromethane and isopropanol . However, isopropanol gave the best extraction efficiency of HFG. Indeed, it is a slightly more polar solvent than ethyl acetate with an additional protic effect. Isopropanol is a hydrogen bond donor, thus it can form hydrogen bonds with two oxygen acceptors of HFG. This may explain why isopropanol was the solvent of choice to extract the HFG. However, the disadvantage of a polar solvent lies in it's poor selectivity toward polar compounds. But, the specificity of the method has been proven and there is no interference with endogenous peaks at 243 nm. Increasing the volume of extraction and using of a slow mechanical stirring by inversion increases the extraction yield of HFG and IS. The extraction was carried out using a carbonate alkaline solution. At this pH, HFG and IS have an overall neutral charge and are extractable by organic solvents. The extraction yield is 97.7% and 90.0% for HFG and IS, respectively. This method is simple and has a high extraction yield.

Optimization of the chromatographic method

The assay of HFG by HPLC was commonly used [9,11-15]. The absorption maxima of HFG are at 243, 285, 315, and 325 nm (Figure 2). In the literature, most authors have used a wavelength of 243, which allows a maximum absorption of the molecule and is in agreement with the absorbance of the IS, which has an absorption maximum at 250 nm [9,14,16].

HFG and IS have two cyclic structures with a basic nitrogen. These are moderately polar molecules that have the same chromatographic behavior (Figure 3). We then opted for reverse-phase chromatography for this molecule. We have chosen, as a stationary phase, a column based on octadecylsilane bonded silica C18 (25 cm × 4 mm, 5 μm). The choice of the composition of the mobile phase was directed by the literature. Indeed, many authors use a mixture of acetate buffer and acetonitrile [9,11-14,16,29]. The tests were carried out on solutions of 20μg/ml HFG in a solution of ammonium acetate 0.25 M. Preliminary tests were performed by varying several parameters: the elution system of the mobile phase, the proportion of organic solvent, the concentration of the buffer, and the pH of mobile phase.

By varying the proportions of the organic solvent, we concluded that the HFG is better eluted with the aqueous phase. Indeed, with a high proportion of acetonitrile, HFG peaks are wider and the retention times are slightly higher, whereas IS is less polar than HFG and presents

less restraint with high proportions of acetonitrile. In order to have an acceptable analysis time, we experimented with a gradient elution, which allowed us to have a retention time for 4.9 min and 9.2 min of HFG and IS, respectively. Indeed, starting with a high proportion of acetate buffer, the HFG, more polar than the internal standard, is eluted first; and increasing the proportion of acetonitrile decreases the retention time of IS.

The two molecules analyzed are basic compounds. Streaks were observed in preliminary trials. Indeed, they are attributed to interactions with residual silanol remains. Two factors have enabled us to refine the peaks: increasing the concentration of the acetate buffer and adding triethylamine, which is a modifier whose role is to saturate the remaining sites. A pH of 4.3 allows a sharing phenomenon between the ionized form and the ideal molecular shape for optimal retention of the two molecules.

The best chromatogram was obtained with a mobile phase composition starting at a composition of ammonium acetate (pH 4.7; 10 mM)-ACN-triethylamine (70:30:0.2, v/v/v) and linear changes to ammonium acetate (pH 4.7; 10 mM)-ACN-triethylamine (10:90:0.2, v/v/v) at 11 min. In these conditions, HFG and IS have retention times of 4.9 and 9.2 min, respectively (Figure 4).

Validation of analytical method

Selectivity: A diode array test of purity was applied to verify the purity of HFG and IS. Referring to the chromatogram of blank extracts, no interference was observed from these extracts at the retention times of the peak corresponding to HFG and IS.

Response function: The response function of an analytical method is, within the range selected, the existing relationship between the response (signal) and the amount (quantity) of the analyte in the sample system [26]. The validation results for the response function

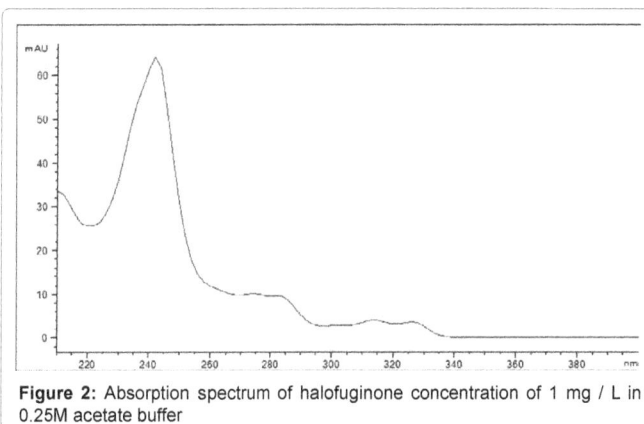

Figure 2: Absorption spectrum of halofuginone concentration of 1 mg / L in 0.25M acetate buffer

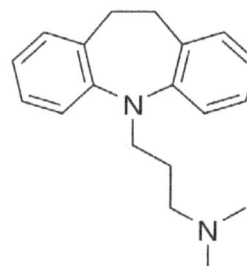

Figure 3: Structure of imipramine

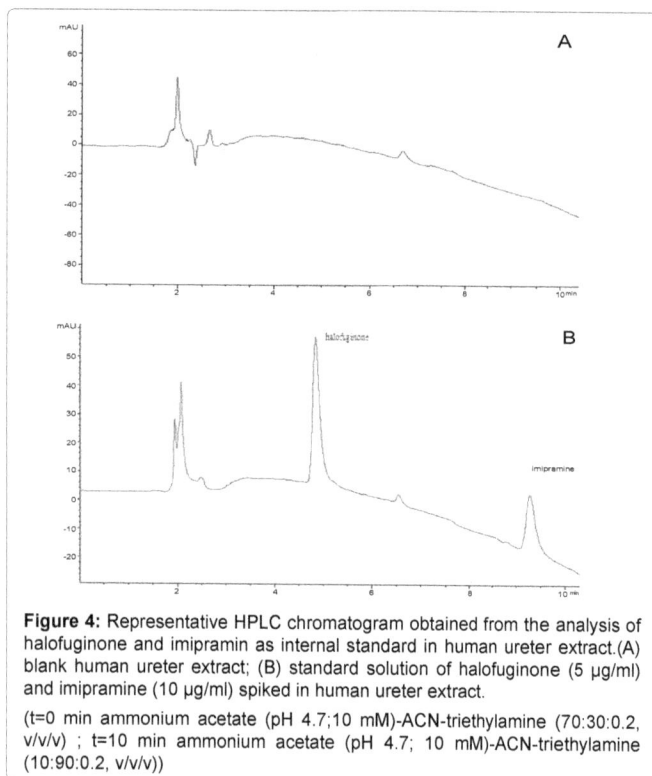

Figure 4: Representative HPLC chromatogram obtained from the analysis of halofuginone and imipramin as internal standard in human ureter extract.(A) blank human ureter extract; (B) standard solution of halofuginone (5 µg/ml) and imipramine (10 µg/ml) spiked in human ureter extract.

(t=0 min ammonium acetate (pH 4.7;10 mM)-ACN-triethylamine (70:30:0.2, v/v/v) ; t=10 min ammonium acetate (pH 4.7; 10 mM)-ACN-triethylamine (10:90:0.2, v/v/v))

in the present study are presented below in Table 1. A weighted (1/X) quadratic regression (with matrix) model was used for the determination of HFG in the human ureter.

Trueness and precision: Trueness refers to the closeness of agreement between a conventionally accepted value and a mean experimental one [26,30]. Precision is the closeness of agreement among measurements from multiple sampling of a homogenous sample under recommended conditions. Trueness and precision are not considered here as decisional parameters but their determination allows assessing the quality of the analytical method as they correspond respectively to random errors and systematic ones [20]. As can be seen in Table 1, the overall relative bias does not exceed 6.4% and the relative standard deviation for repeatability and intermediate precision are below 6.8%.

Accuracy profile: Accuracy takes into account the total error, which is the sum of systematic and random errors related to the test result [26]. The upper and lower β expectation tolerance limits expressed in µg are presented in Table 1 as a function of the introduced amounts. As can be seen from these results, the proposed method was accurate, because the different tolerance limits did not exceed the acceptance limits of total error for all amount levels tested, including the lowest one. The tolerance interval was within the 20% acceptance limit at all amounts. This validation suggests that the method offers sufficient guarantees, providing results ± 20% of the true value in at least 95% of cases (Figure 5).

Linearity: The linearity of an analytical method is its ability within a definite range to obtain results directly proportional to the quantities of the analyte in the sample [26].Therefore, a linear model was fitted in a calculated amount of the validation standards for all series as a function of the introduced amounts. The regression equation is presented in Table 1. A linear regression model (Figure 6) is fitted on the back-calculated amounts as a function of the introduced amounts

in order to obtain the following equation:

$$Y = 0.08998 + 0.9973\,X$$

where Y = back-calculated amounts (µg) and X = introduced amount (µg). The coefficient of determination (r^2) is equal to 0.9979. The residual sum of squares (RSS) is equal to 1.154.

In order to prove method linearity, the absolute β expectation tolerance interval was applied. The linearity of the HFG HPLC method was also demonstrated since the β-expectation tolerance limits were included in the absolute acceptance limits for the whole concentration range investigated as shown in Figure 6.

Detection and quantification limits: The limit of detection (LOD) is the smallest quantity of the targeted substance that can be detected, but cannot be accurately quantified in the sample. The LOD was estimated using the mean intercept of the calibration model and the residual variance of the regression [20,22,25]. By applying this computation method, the LOD was found to be equal to 0.06061 µg/ml.

The lower limit of quantification (LOQ) is the smallest quantity of the targeted substance in the sample that can be assayed under experimental conditions with well-defined accuracy (22). *LOQ is given as: LOQ = LOD * 3. With this method, the lower LOQ was equal to 0.2 µg/ml.*"

Ex-vivo model application

The purpose of this application was to evaluate a gel formulation of HFG, an inhibitor of collagen synthesis, for the treatment and the prevention of occurrence and recurrence of urethra narrowing. The effectiveness of the therapy, following the intraurethral injection, is based on the availability of a pharmacologically active element, able to reach its biological target, and penetrate the wall. HFG should penetrate into the deep structures and should remain there at an effective concentration. The kinetics of diffusion in this case is important as the patient can only refrain from urinating, and in turn eliminating the gel, for two hours maximum. HPLC assay was applied to kinetic study to ensure adequate release kinetics which guarantees the therapeutic effect.

The male urethra is composed of several layers: epithelium transitional, chorion, muscular and adventitia. These four layers are also found in the human ureter, which has a very similar composition. The male urethra is surrounded by the anterior part of the corpus spongiosum, a vascular structure. To simulate in *ex vivo* the male urethra, wall absorption has been studied on samples of fresh human ureter mounted on modified Franz cells (exposed areas 50 mm²). The fresh, surgically excised samples of human ureter used for the tests were obtained following kidney transplant surgery kept in a fluid conservation (Eurocollins˙).

Four cells were performed for each ureter to quantify HFG in the wall after 30, 60, 90, and 120 minutes and were immediately tested. The ureter was cut and clamped between the receptor and donor compartments. The receptor compartment was filled with 2.5 ml of diffusion medium (phosphate buffer pH 7.4) through sampling port taking care to remove all the air bubbles. The contents were stirred by small magnetic bead to keep them well mixed.

Three hundred milligrams of the gel were filled in the upper compartment (on the internal surface of ureter). The cells, with stirring, were stabilized in a water bath at 37°C and closed with Parafilm˙ to avoid formula evaporation. The experiment was repeated five times

Response function (p=3; n=2)		Weighted (1/X) Quadratic Regression (within matrix)		
		Calibration range (m=5): 0.2-10 µg/ml		
		series 1	series 2	series 3
	Slope	0.4921	0.4729	0.4972
	Intercept	-0.05278	-0.04351	-0.05301
	r^2	0.9939	0.9974	0.9958

Trueness   (p=3; n=2)

	Absolute bias (µg/ml)	Relative bias (%)	Recovery (%)
0.1 µg/ml	0.004295	2.148	102.1
1 µg/ml	0.05569	5.569	105.6
2.5 µg/ml	0.07220	2.888	102.9
5 µg/ml	0.3170	6.339	106.3
10 µg/ml	-0.04940	-0.4940	99.51

Precision (p=3; n=2)

	Repeatability (RSD%)	Intermediate precision (RSD%)	
0.1 µg/ml	6.822	6.822	
1 µg/ml	3.179	3.421	
2.5 µg/ml	1.892	1.892	
5 µg/ml	0.8241	1.447	
10 µg/ml	1.865	2.281	

Accuracy (p=3; n=2)

	Beta-expectation tolerance limits (µg/ml)	Relative Beta-expectation tolerance limits (%)	
0.1 µg/ml	[0.1709 , 0.2377]	[-14.54 , 18.84]	
1 µg/ml	[0.9690 , 1.142]	[-3.101 , 14.24]	
2.5 µg/ml	[2.456 , 2.688]	[-1.742 , 7.517]	
5 µg/ml	[5.066 , 5.568]	[1.325 , 11.35]	
10 µg/ml	[9.325 , 10.58]	[-6.751 , 5.763]	

Linearity   (p=3; n=2)

	Range (µg/ml)	0.2-10	
	Slope	0.9973	
	Intercept	0.08998	
	r^2	0.9979	
LOD (µg/ml)		0.06061	
LOQ µg/ml)		0.2000	

P: number of series of analysis; m: number of amount levels; n: number of replicates; RSD: Relative Standard Deviation.

Table 1: Validation results

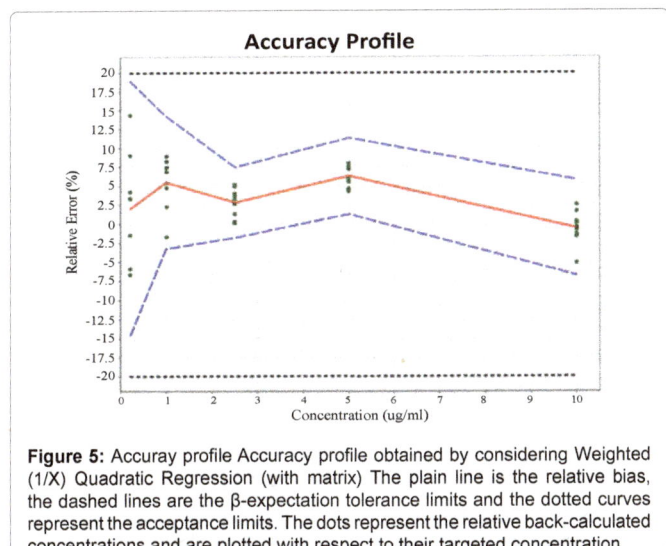

Figure 5: Accuray profile Accuracy profile obtained by considering Weighted (1/X) Quadratic Regression (with matrix) The plain line is the relative bias, the dashed lines are the β-expectation tolerance limits and the dotted curves represent the acceptance limits. The dots represent the relative back-calculated concentrations and are plotted with respect to their targeted concentration.

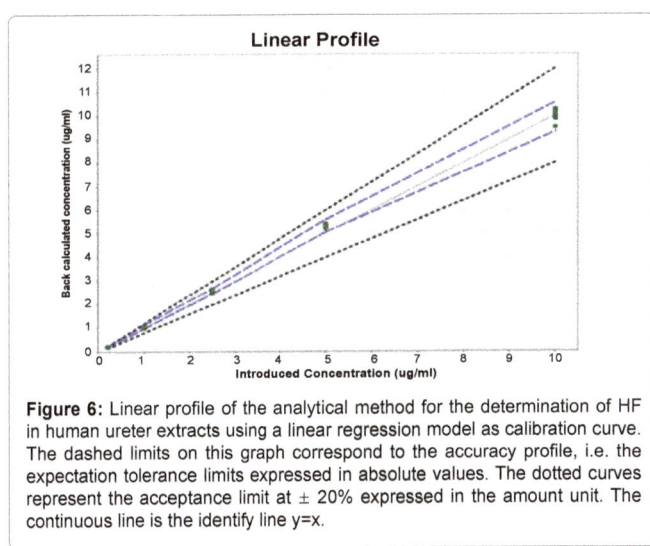

Figure 6: Linear profile of the analytical method for the determination of HF in human ureter extracts using a linear regression model as calibration curve. The dashed limits on this graph correspond to the accuracy profile, i.e. the expectation tolerance limits expressed in absolute values. The dotted curves represent the acceptance limit at ± 20% expressed in the amount unit. The continuous line is the identify line y=x.

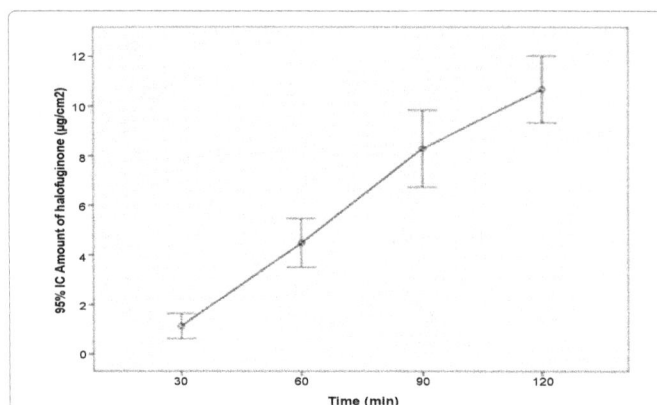

Figure 7: Permeation profile of halofuginone (through wall ureter from a gel at a concentration of 0.03% w/w of halofuginone). Applied dose: 300mg of gel. The cumulative amount of halofuginone (μg) that penetrated wall (per cm2 of wall). Each data point represents the averaged value of samples from wall replicates (n=5). The vertical error bars represent the 95% confidence intervals.

with five different ureters.

Wall diffusion was determined by measuring the amount of drug residing in the ureter layer. After the application, wall samples were washed off with water and dried with absorbent paper. HFG content in the various samples was measured by HPLC. Gel formulation was prepared by dissolving HFG at concentration of 0.03% w/w in water with carboxymethylcellulose 0.06% w/vol.

The HFG content of the wall was expressed as mean (μg/cm²) ± standard error of mean. Amount of HFG measured in the wall (μg/cm²) was plotted as a function of time.

As demonstrated in Figure 7, the results of the curve shows that the amount diffused in the wall increases during the two hours after application of the gel to reach an amount of 10.6 μg/cm² at 120 min. Various parameters can influence the diffusion of a molecule in the wall. This method could, in more extended studies, examine different gel formulations to keep the one whose diffusion kinetics ensures optimal therapeutic effect during the two hours of application.

Conclusion

The determination of HFG is essential to study its diffusion from different gels in the human ureter *in vitro* experiments. In this work, we have developed and validated a method for the determination of HFG from the wall of a human ureter by HPLC. To the best of our knowledge, this method is the first assay for the quantitative determination of HFG in the human ureter. Liquid-liquid extraction proved to be selective for the halofuginone and provided a high recovery rate of 97.7%. A rapid and reliable RP-HPLC assay has been developed and validated in accordance with the performance criteria described in SFSTP 2003. The assay provides a linear response across a range of concentrations 0.2-10 μg/ml. The assay is fast, accurate, and precise for the quantification of HFG in human ureter tissue extracts. The method of determination of HFG developed in this study allowed us to check the diffusion of HFG from a gel 0.03% w/w in the wall of the ureter.

References

1. Behr KP, Lüders H, Plate C (1986) Tolerance of halofuginone (Stenorol) by geese (Anser anser f. dom.), flight ducks (Cairina moschata f. dom.) and Peking ducks (Anas platyrhynchos f. dom.). Dtsch Tierarztl Wochenschr 93: 4-8.

2. Pines M, Nagler A (1998) Halofuginone: a novel antifibrotic therapy. Gen Pharmacol 30: 445-450.

3. Nagler A, Genina O, Lavelin I, Ohana M, Pines M (1999) Halofuginone, an inhibitor of collagen type I synthesis, prevents postoperative adhesion formation in the rat uterine horn model. Am J Obstet Gynecol 180: 558-563.

4. Nagler A, Rivkind AI, Raphael J, Levi-Schaffer F, Genina O, et al. (1998) Halofuginone--an inhibitor of collagen type I synthesis--prevents postoperative formation of abdominal adhesions. Ann Surg 227: 575-582.

5. Pines M, Knopov V, Genina O, Lavelin I, Nagler A (1997) Halofuginone, a specific inhibitor of collagen type I synthesis, prevents dimethylnitrosamine-induced liver cirrhosis. J Hepatol 27: 391-398.

6. Halevy O, Nagler A, Levi-Schaffer F, Genina O, Pines M (1996) Inhibition of collagen type I synthesis by skin fibroblasts of graft versus host disease and scleroderma patients: effect of halofuginone. Biochem Pharmacol 52: 1057-1063.

7. Jaidane M, Ali-El-Dein B, Ounaies A, Hafez AT, Mohsen T, et al. (2003) The use of halofuginone in limiting urethral stricture formation and recurrence: an experimental study in rabbits. J Urol 170: 2049-2052.

8. Krane LS, Gorbachinsky I, Sirintrapun J, Yoo JJ, Atala A, et al. (2011) Halofuginone-coated urethral catheters prevent periurethral spongiofibrosis in a rat model of urethral injury. J Endourol 25: 107-112.

9. Anderson A, Goodall E, Bliss GW, Woodhouse RN (1981) Analysis of the anti-coccidial drug, halofuginone, in chicken tissue and chicken feed using high-performance liquid chromatography. J Chromatogr 212: 347-355.

10. Anderson A, Christopher DH, Woodhouse RN (1979) Analysis of the anti-coccidial drug, halofuginone, in chicken feed using gas-liquid chromatography and high-performance liquid chromatography. J Chromatogr 168: 471-480.

11. Tillier C, Cagniant E, Devaux P (1988) Determination of halofuginone in poultry feeds by high-performance liquid chromatography. J Chromatogr 441: 406-416.

12. Holland DC, Munns RK, Roybal JE, Hurlbut JA, Long AR (1995) Liquid chromatographic determination of the anticoccidial drug halofuginone hydrobromide in eggs. J AOAC Int 78: 37-40.

13. No authors listed (1983) Determination of halofuginone hydrobromide in medicated animal feeds. Analyst 108: 1252-1256.

14. No authors listed (1984) Collaborative study of a method for the determination of residues of halofuginone in chicken tissue. Analytical Methods Committee, Royal Society of Chemistry. Analyst 109: 171-174.

15. Yamamoto Y, Kondo F (2001) Determination of halofuginone and amprolium in chicken muscle and egg by liquid chromatography. J AOAC Int 84: 43-46.

16. Kinabo LD, McKellar QA, Murray M (1989) Determination of halofuginone in bovine plasma by competing-ion high performance liquid chromatography after solid phase extraction. Biomed Chromatogr 3: 136-138.

17. Stolker AA, Brinkman UA (2005) Analytical strategies for residue analysis of veterinary drugs and growth-promoting agents in food-producing animals--a review. J Chromatogr A 1067: 15-53.

18. Aerts MM, Hogenboom AC, Brinkman UA (1995) Analytical strategies for the screening of veterinary drugs and their residues in edible products. J Chromatogr B Biomed Appl 667: 1-40.

19. Halofuginone: determinative method (1991) Analytical Chemistry Laboratory Guidebook: Residue Chemistry. USDA, FSIS, Science and Technology Program, Washington DC, USA.

20. Hubert P, Nguyen-Huu J, Boulanger B, et al. (2003) Validation of quantitative analytical procedures, steps Harmonisation. STP Pharma Pratiques 13.

21. Hubert P, Nguyen-Huu JJ, Boulanger B, Chapuzet E, Chiap P, et al. (2004) Harmonization of strategies for the validation of quantitative analytical procedures. A SFSTP proposal-Part I. J Pharm Biomed Anal 36: 579-586.

22. Hubert P, Nguyen-Huu JJ, Boulanger B, Chapuzet E, Cohen N, et al. (2007) Harmonization of strategies for the validation of quantitative analytical procedures. A SFSTP proposal-part III. J Pharm Biomed Anal 45: 82-96.

23. Feinberg M (2007) Validation of analytical methods based on accuracy profiles. J Chromatogr A 1158: 174-183.

24. Chandran S, Singh RS (2007) Comparison of various international guidelines for analytical method validation. Pharmazie 62: 4-14.

25. Hoffman D, Kringle R (2007) A total error approach for the validation of quantitative analytical methods. Pharm Res 24: 1157-1164.

26. Hubert P, Nguyen-Huu JJ, Boulanger B, Chapuzet E, Chiap P, et al. (2007) Harmonization of strategies for the validation of quantitative analytical procedures. A SFSTP proposal-part II. J Pharm Biomed Anal 45: 70-81.

27. Guidance for Industry: Bioanalytical Method Validation (2013) US Department of Health and Human Services, Food and Drug Administration, Center for Drug Evaluation and Research (CDER), Center for Veterinary Medecine (CVM), Rockville.

28. Dubreil-Cheneau, E, Bessiral M, Roudaut B, Verdon E, Sanders P (2009) Validation of a multi-residue liquid chromatography-tandem mass spectrometry confirmatory method for 10 anticoccidials in eggs according to Commission Decision 2002/657/EC. J Chromatogr A1216: 8149-8157.

29. Ding S, Hou Y, Wu N, Shen J (2005) Residue analysis for halofuginone in sturgeon muscle by immunoaffinity cleanup and liquid chromatography. J AOAC Int 88: 1644-1648.

30. Hubert P, Nguyen-Huu JJ, Boulanger B, Chapuzet E, Cohen N, et al. (2008) Harmonization of strategies for the validation of quantitative analytical procedures: a SFSTP proposal part IV. Examples of application. J Pharm Biomed Anal 48: 760-771.

20

Investigation of Cells Migration Effects in Microfluidic Chips

Yu-Sheng Lin*, Wenting Liu and Chunfei Hu

Division of Nanobionic Research, Suzhou Institute of Nano-Tech and Nano-Bionics, Chinese Academy of Sciences, Suzhou, Jiangsu, China

Abstract

Microfluidic chips offer the unique opportunity to establish novelty *in vitro* cells models where the *in vivo* cells microenvironment could be precisely reconstituted. Although they have significances in the fields of cell biology, the real applications are extremely limited by ambient materials for cells culture, i.e., cells morphology and migration are greatly influenced by substrate material. In this study, we investigated the cells migration effects in four kinds of microfluidic chips with the same geometry. They are PDMS mold structures bonded on culture dish, glass slide, and PDMS substrates, respectively, and another PMMA mold structure bonded on PMMA substrate. For convenient description, we denoted these four chips as PDMS-DISH, PDMS-GLASS, PDMS-PDMS, and PMMA-PMMA, respectively. We compared and summarized the relationship of cells migration effects on different substrate. The cells are initially introduced into the culture area. The experiment results indicate that cells spreading time, spreading area and cells migration on these chips has obvious diversities. To further investigate the cells migration in these chips, a new model is prepared using trypsin/EDTA solution and cell culture medium. It shows a good repeatability. Most of cells could be formed a good morphology and monolayer growth in these microfluidic chips. The cells migrated furthest is in the PDMS-DISH chip after monitored 24 hours. The migration rates are 20.30 µm/h, 18.63 µm/h, 15.00 µm/h, and 10.75 µm/h in PDMS-DISH, PDMS-GLASS, PDMS-PDMS, and PMMA-PMMA, respectively. This study turns open up opportunities for new biochips in prospective applications of wound healing and anti-scarring expected in drug screening and the related fields.

Keywords: Microfluidic chip; Cell migration; Cell spreading; Laminar flow

Introduction

The cells dynamic behaviors in culture has been the subject of research over several decades. Cells culture and cells migration *in vitro* is the basis of modern biology. Cells culture has made substantial contribution to understanding of many phenomena, such as intracellular enzyme activities, and cells interactions [1,2]. Cells migration is defined as the movement of individual cell, cell sheet and cluster from one location to another [3-6]. It is critical to a variety of different pathologic and physiologic processes across many areas of biology including wound healing, cancer metastasis, inflammation, cells growth and differentiation [7-9]. The conventional *in vitro* cells culture are used culture dishes or microliter plates and the typical cells migration assays are used Boyden chamber and physical scraping [10,11]. Although these methods are widely used, there are several limitations in this conventional methods [12,13]. For example, a traditional cells culture *in vitro*, there is neither cellular structure nor extracellular matrix for the cells to attach to, neither interaction nor communication with other cells. Cells lines and those in organs are different a lot in behavior, which growth rate, morphology, and intracellular metabolic activity will change correspondingly [14]. Furthermore, the membrane of Boyden chamber is not transparent. Therefore, it is difficult to perform the microscopy. In addition, the method is limited to tracing migration of individual cell. The cells are at the wound edges that can be damaged by physical scraping, which are important to cells migration assays [12,13]. In order to overcome these drawbacks, a solid and reliable approach for cells migration assay is necessary.

Recently, microfluidic systems have significant implications for cell biology and cell-based assay. They can provide precise cells pattern and controllable reagents distribution in a reproducible platform, which is not easily achieved by standard culture dish. In addition, they enable conventional assays to be performed using an automated and high-throughput approach [15]. In the past few years, microfluidic chips have been applied for use in cell culture and wound healing assay [1,16-19]. Unlike conventional *in vitro* cell culture methods, microfluidics can provide small and complex structures mimicking *in vivo* environment. Recent research has shown that microfluidic cell culture systems convey more reliable results due to their ability to grow cells as biological systems, and they could outperform those from conventional cell cultures and assay systems [20]. A microfluidic wound healing assay was reported in the literature [21-23]. In an attempt to reconstruct a wound edge, a confluent cell sheet inside a microchannel was partially digested using multiple laminar flows with and without trypsin/EDTA. By using these operations, the interface resembled the actual wound edge was fabricated by physical scraping, without damaging to the cells on the leading edge [12].

In the study, we used the classical Y-shape microfluidic chip to investigate the cells normal physiological functions, such as migration, growth, reproduction and wound healing. We found that activities of cells are related to chemical composition and physical properties of base material of culture. In the process of cells growth, cells spreading is the first step in the interaction of cells and extracellular matrix (extracellular matrix, ECM). Cells adhesion and spreading on substrates are not only associated with the chemical properties of the substrate, but also related to surface physical properties of substrates, such as stiffness, hydrophobicity and geometric structure. We investigated the cells migration effects in four types of microfluidic chips. For this purpose, Human lung fibroblasts cell (HLF) was utilized, they were

*Corresponding author: Yu-Sheng Lin, Division of Nanobionic Research, Suzhou Institute of Nano-Tech and Nano-Bionics, Chinese Academy of Sciences, Suzhou, Jiangsu, 215123, China, E-mail: yslin2016@sinano.ac.cn

inoculated into the culture district. Laminar flow in the microchannel is used to injure cell layer to form a wound. Two of these flows contained protease trypsin, whereas one of them contained culture medium. Cells that were exposed to the trypsin-containing flow detached from the surface, whereas cells that were exposed to normal medium remained attached. Cell spreading, monolayer growth and cell migration in the microchannel were monitored.

Preparation and Fabrication of Microfluidic Chips

Four classical Y-shape microfluidic chips with the same feature sizes were used in this study. They are PDMS mold structures bonded on culture dish, glass slide, and PDMS substrates, respectively, and PMMA layer bonded on PMMA substrate. For convenient description in this study, they are denoted as PDMS-DISH, PDMS-GLASS, PDMS-PDMS☒and PMMA-PMMA, respectively. The feature sizes of four chips are the main channel with 900 μm × 100 μm × 5 mm (width × height × length), three inlet channels with 300 μm × 100 μm × 5 mm (width × height × length), respectively. The fabrication process of PDMS mold structures were prepared by a silicon mold with micrometer-sized SU-8 exposed to chlorotrimethylsilane vapor for 3 minutes and then released after carrying out the baking process. Secondly, PDMS prepolymer and curing agent (8:1 w/w, RTV615A, RTV615B) were mixed at an 8:1 ratio and then poured onto the mold. Thirdly, the samples were degassed under vacuum for 30 minutes and then baked for 45 minutes at 85°C. Fourthly, the cured PDMS layers with the desired structures were peeled and punched using a metal pin at the terminals of the inlet and outlet channels as shown in Figure 1. Finally, the PDMS layers were bonded on a culture dish, glass slide, and PDMS substrates, respectively. The fourth microfluidic chip is used PMMA material. This chip consists of two pieces of PMMA bonded by an adhesive film (model: ARcareMH-90445). Three parallel streams were observed by means of a fluorescence microscope equipped with a CCD digital camera (model: SS-300, TUSUN Co. Ltd.). Cells culture and migration were investigated by a phase contrast microscope.

HLF cells were cultured in Dulbecco's modified Eagle's medium (DMEM) supplemented with 10% fetal bovine serum (FBS), 100 units/mL penicillin, and 100 μg/mL streptomycin at 37°C and 5% CO_2. When the cells reached confluence, they were dissociated from culture dishes with trypsin/EDTA. The cell suspension was then centrifuged at a rotational speed of 1,000 rpm for 5 min, after which it was re-suspended in DMEM containing FBS. The microfluidic channels were pretreated with DMEM containing FBS before cell seeding. The cell suspension ($1 × 10^7$ cell/mL) were carefully pipetted into four chips with different substrate material using a 3 μL micropipette. Due to pressure difference from micropipettes, cells flowed from the inlets to the outlets. To facilitate cell adhesion on the microchannel of chip, the microfluidic chip was kept stationary for 10 min at 37°C. Next, fresh medium using a 2 μL micropipette was then introduced. After cell adhesion on the inlet, the medium was replaced per every 6 hour. The growth and proliferation of cells was monitored at 0, 4, 8, 12, 16, 20, and 24 hours using an inverted fluorescence microscope.

Cells were ready for migration experiments when 80%-90% confluence was established in four Y-shape microfluidic chips. To prepare a wound model of monolayer cells, each of three inlets of the microfluidic channel was washed with PBS without divalent cations and trypsin/EDTA solution was introduced from two inlets, while DMEM containing 10% FBS was introduced from the middle simultaneously. The outlet was connected to its own 5 mL syringe in a syringe pump using polyethylene tubing, which were pulled through the microfluidic channel at a total rate of 20 μL/min for 15 min. After this, the pump was

stopped, and the channels were flushed with DMEM containing 10% FBS, and stored in the incubator.

Experimental Results and Discussion

The schematic of the computing model for the cells average migration distance in four microfluidic chips is shown in Figure 2. The cells average migration distance can be expressed by $D=d-A/L$, where d is the distance between baseline and the edge of microchannel, L is the length of microchannel, and A is the no cell zone. Cells propagation on chip generally includes the following steps (1) The adhesion between cell and substrate initializes. (2) Cells protruded around. (3) New adhesion sites form between pseudopodia tips and substrate. (4) Cell spreading driving force and resistance achieve a balance, and cells no longer spread out forward. The initial contact of cell and the substrate is mainly a passive process, and only depend on physical properties of cell and the substrate, such as the viscoelasticity of cell, cell membrane tension and substrate stiffness.

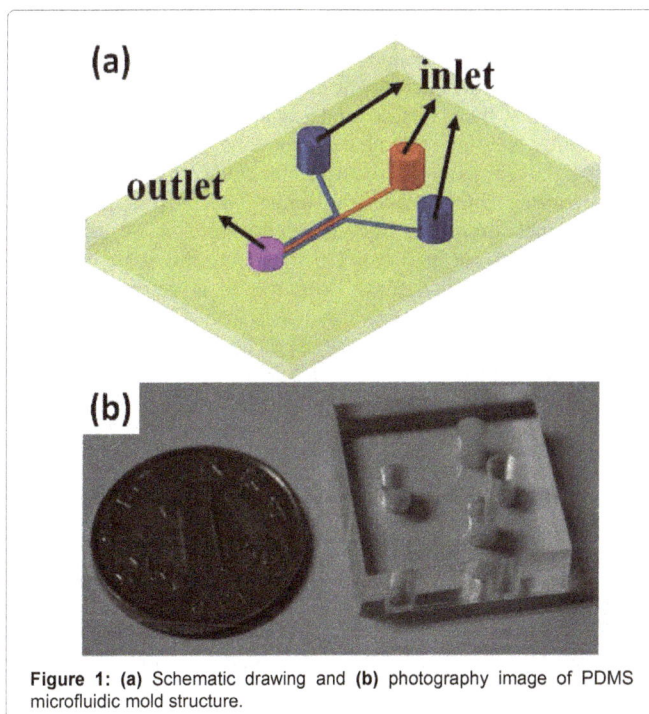

Figure 1: (a) Schematic drawing and **(b)** photography image of PDMS microfluidic mold structure.

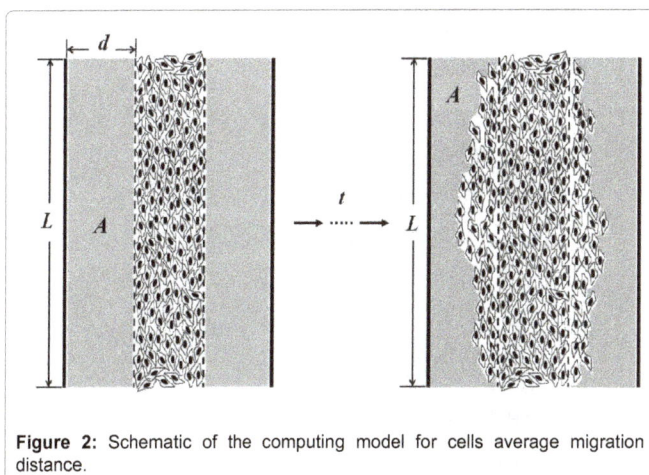

Figure 2: Schematic of the computing model for cells average migration distance.

During the migration assay, the cells were monitored and photographed at 0, 4, 8, 12, 16, 20, and 24 hours under a phase contrast microscope with a 80 times multiplied objective. The optical images of fixed positions in the wounds were taken with a CCD digital camera that was mounted on the microscope. Each experiment was repeated at least three times. Subsequently, the wound area in each image was determined by outlining the wound and measuring the area using Image-Pro Plus 6.0 software. From the wound area, the average wound width could be obtained by dividing the area by the length of the analyzed region. Figure 3 shows cell spreading and monolayer growth in four individual microfluidic chips at different culture time. Cell image analysis indicated that most of cells could be formed a good morphology and monolayer growth status by culture time. Therefore, these four substrate materials have biological safety and are suitable for cell growth. However, cell spreading time and spreading area on four microfluidic chips with different substrate material is distinct. Cell spreading speed in the increasing order is PDMS-DISH, PDMS-GLASS, PMMA-PMMA and PDMS-PDMS as shown in Figure 4a.

In order to observe material effect on cell growth and obtain monolayer for cell migration, bare material and cell seeding are very important. At the beginning of the test, a syringe was used to force air bubbles out of the microfluidic chip. When cells were introduced into a channel without surface modification, the phenomenon of cell aggregation occur for intercellular adhesion. Therefore, before the seeding of the cells in the microchannel for culture, the microchannels were pretreated with DMEM containing FBS. It is 3 µL of cells were introduced into each inlet at a density of 1×10^7 cell/mL. A micropipette was put at the outlet to aspirate the waste and fluid making them flowing out quickly. With the aid of pressure difference, cells flowed from the inlets to the outlets with a fast speed. When the liquid flow was stopped in the microchannel and the cells were uniformly distributed on the substrate surface, aspirating was stopped and the chip was put into the incubator to further incubate. Dynamic

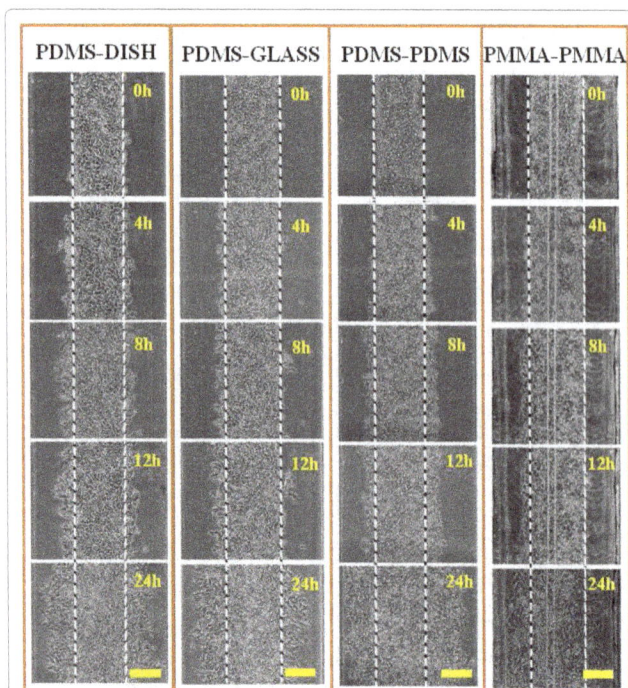

Figure 4: (a) The relationship of cells average migration distance and cultured time. (b) Cells migration rates in four microfluidic chips (n=3).

Figure 3: Optical images of human lung fibroblasts (HLF) cells migration in the four chips taken at 0, 4, 8, 12, and 24 hours, respectively. The white dashed lines are the baselines for the migration assay (Scale bar is 200 µm).

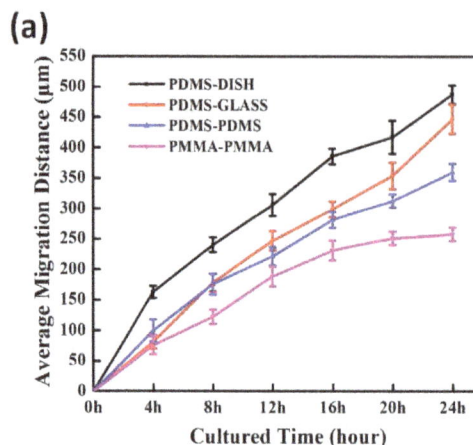

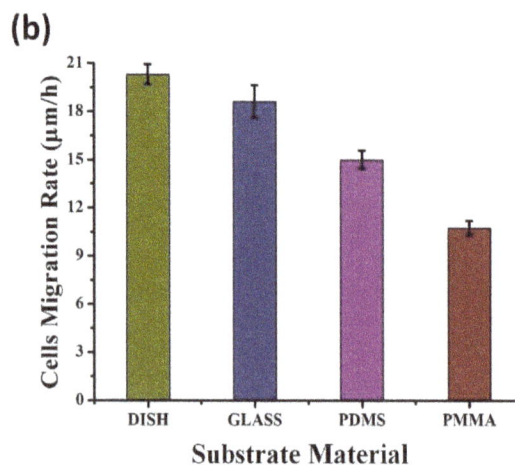

culture was offered by supplying fresh culture medium every 6 hour to provide enough nutrition for cells to grow and reduce cell injury. Cell morphology and migration movement is influenced to a great degree by substrate mechanical properties, such as permeability or hardness. The wound had been introduced in the monolayer. The migration rate was quantified by calculating migration distance of images at fixed position at fixed intervals. To quantify cell migration in this experiment, the initial wound edge was defined as the baseline. The cell migration distances were calculated from the baseline and the cell area move. The number of cells on the both sides of the baseline was also included. The photo images of cell migration at different times were taken at 0, 4, 8, 12, 16, 20, and 24 hours analyzed using Image-Pro Plus 6.0 software. Figure 4a indicates that four materials have distinct effects on cell migration. After cultured for 24 hours, cells migrated furtherest in the PDMS-DISH chip. The migration rates are 20.30 µm/h, 18.63 µm/h, 15.00 µm/h, and 10.75 µm/h in PDMS-DISH, PDMS-GLASS, PDMS-PDMS, and PMMA-PMMA, respectively, as summarized in Figure 4b. These differences may be caused by different materials with different hydrophilicity and hardness stiffness. The culture dish material is made of high quality crystal polystyrene, covalent binding of hydrogel layer on surface, also has good hardness and hydrophilicity. Glass material has better hardness and hydrophilicity than PDMS and PMMA materials. These results provide supportive evidence that cell migration distance associates with substrate material properties, such as hardness and hydrophilicity of culture material.

Conclusion

In this study, we investigated cells migration effects in four types of microfluidic chips. In cell biology, the real applications are extremely limited by ambient materials for cells culture, i.e., cells morphology and migration are greatly influenced by substrate material. A new wound model is prepared using trypsin/EDTA solution and culture medium, showing good repeatability. Most of cells could be formed a good morphology and monolayer growth in these microfluidic chips. For cells migration, cells spreading time and spreading area on these chips with different substrate material are distinct. The furtherest cell migration rate is 20.30 μm/h in the PDMS-DISH chip. This assay paves a way to deepen our understanding of biological behavior in cells culture and aid in developing novel material for biochip to achieve different aims. For example, the migration of cancer cells would help to understand cancer invasion, mechanism of metastasis and anti-cancer drugs treatment and so on.

References

1. Zhang B, Kim MC, Thorsen T, Wang Z (2009) A self-contained microfluidic cell culture system. Biomed Microdevices 11: 1233-1237.

2. Abeille F, Mittler F, Obeid P, Huet M, Kermarrec F, et al. (2014) Continuous microcarrier-based cell culture in a benchtop microfluidic bioreactor. Lab Chip 14: 3510-3518.

3. Friedl P, Brocker EB (2000) The biology of cell locomotion within three-dimensional extracellular matrix. Cellular and Molecular Life Sciences 57: 41-64.

4. Friedl P, Wolf K (2003) Tumour-cell invasion and migration: Diversity and escape mechanisms. Nature Reviews Cancer 3: 362-374.

5. Lauffenburger DA, Horwitz AF (1996) Cell migration: A physically integrated molecular process. Cell 84: 359-369.

6. Ridley AJ, Schwartz AM, Burridge K, Firtel AR, Ginsberg HM, et al. (2003) Cell migration: Integrating signals from front to back. Science 302: 1704-1709.

7. Horwitz R, Webb D (2003) Cell migration. Current Biology 13: R756-R759.

8. Eccles SA, Box C, Court W (2005) Cell migration/invasion assays and their application in cancer drug discovery. Biotechnology Annual Review 11: 391-421.

9. Luster AD, Alon R, von Andrian UH (2005) Immune cell migration in inflammation: present and future therapeutic targets. Nature Immunology 6: 1182-1190.

10. Liang CC, Park AY, Guan JL (2007) In vitro scratch assay: a convenient and inexpensive method for analysis of cell migration in vitro. Nat Protoc 2: 329-333.

11. Boyden S (1962) The chemotactic effect of mixtures of antibody and antigen on polymorph nuclear leucocytes. The Journal of Experimental Medicine 115: 453.

12. Nie FQ (2007) On-chip cell migration assay using microfluidic channels. Biomaterials 28: 4017-4022.

13. Wei Y, Chen F, Zhang T, Chen D, Jia X, et al. (2015) A Tubing-Free Microfluidic Wound Healing Assay Enabling the Quantification of Vascular Smooth Muscle Cell Migration. Scientific Reports 5: 14049.

14. Kuraitis D, Giordano C, Ruel M, Musaro A, Suuronen EJ (2012) Exploiting extracellular matrix-stem cell interactions: A review of natural materials for therapeutic muscle regeneration. Biomaterials 33: 428-443.

15. Hulkower KI, Herber RL (2011) Cell migration and invasion assays as tools for drug discovery. Pharmaceutics 3: 107-124.

16. El-Ali J, Sorger PK, Jensen KF (2006) Cells on chips. Nature 442: 403-411.

17. Van der Meer AD, Vermeul K, Poot AA, Feijen J, Vermes I (2010) A microfluidic wound-healing assay for quantifying endothelial cell migration. Am J Physiol Heart Circ Physiol 298: H719-725.

18. Murrell M, Kamm R, Matsudaira P (2011) Tension, free space, and cell damage in a microfluidic wound healing assay. PLoS One 6: e24283.

19. Xi YL (2012) Wound Healing Assay Using Microfluidic Channels. Chemical Research in Chinese Universities 28: 472-476.

20. Yeon JH , Park JK (2007) Microfluidic cell culture systems for cellular analysis. Biochip Journal 1: 17-27.

21. Zhao Q, Wang S, Xie Y, Zheng W, Wang Z, et al. (2012) A Rapid Screening Method for Wound Dressing by Cell-on-a-Chip Device. Advanced Healthcare Materials 1: 560-566.

22. Zhang M, Li H, Ma H, Qin J (2013) A simple microfluidic strategy for cell migration assay in an in vitro wound-healing model. Wound Repair and Regeneration 21: 897-903.

23. Chen X, Gumbiner BM (2006) Crosstalk between different adhesion molecules. Current Opinion in Cell Biology 18: 572-578.

21

New Validated Stability Indicating RP-HPLC Method for Simultaneous Estimation of Metformin and Alogliptin in Human Plasma

Ashutosh KS[1*], Manidipa D[2], Seshagiri RJVLN[3] and Gowri SD[4]

[1]Department of Pharmaceutical Analysis and Quality Assurance, AKRG College of Pharmacy, Nallajerla, West Godavari, Andhra Pradesh, India
[2]Department of Pharmaceutics, AKRG College of Pharmacy, Nallajerla, West Godavari, Andhra Pradesh, India
[3]Srinivasarao College of Pharmacy, Pothinamallayyapalem, Madhurawada, Visakhapatnam, Andhra Pradesh, India
[4]College of Pharmaceutical Sciences, Andhra University, Visakhapatnam, Andhra Pradesh, India

Abstract

A validated new stability indicating RP-HPLC method for the quantitative determination of metformin and alogliptin in human plasma was developed as per US-FDA guidelines. The drug was spiked in the plasma and extracted with mobile phase by precipitation method. The extracted analyte was injected into X-Terra C18 (4.6 × 150 mm, 3.5 µm, Make: ACE) or equivalent, maintained at 25°C temperature and effluent was monitored at 235 nm. The mobile phase was consisted of sodium dihydrogen ortho phosphate [pH 4.0]:acetonitrile [HPLC Grade] (70:30 v/v). The flow rate was maintained at 1.0 mL/min. The calibration curve for metformin and alogliptin was linear from 300.0 to 700.0 µg/mL (r^2=0.997) and 7.5 to 17.5 µg/mL (r^2=0.998) respectively. The inter-day and intra-day precision was found to be within limits. The Lower limit of quantification (LLOQ) for metformin and alogliptin were 5.936 and 1.983 µg/mL respectively. The average % recovery for metformin and alogliptin were 100.17 and 99.40-99.55% respectively and reproducibility was found to be satisfactory. This RP-HPLC method is suitable for determining the concentration of metformin and alogliptin in human plasma and it can applied for routine analysis for determination of the metformin and alogliptin from dosage form during pharmacokinetic study.

Keywords: Metformin; Alogliptin; RP-HPLC; ICH-guidelines; Validation; Human Plasma; US-FDA guidelines

Introduction

Diabetes is a chronic condition associated with abnormally high levels of sugar (glucose) in the blood. The production of insulin by the pancreas lowers blood glucose level. Due to the absence or insufficient production of insulin causes diabetes. There are two types of diabetes which are referred to as type 1 and type 2 [1]. Formerly names for these conditions were insulin-dependent and non-insulin-dependent diabetes, or juvenile onset and adult onset diabetes. The symptoms of diabetes include increased urine output, thirst, hunger, and fatigue. Diabetes is diagnosed by blood sugar (glucose) testing. The number of individuals affected by diabetes is continuing to increase worldwide; the need for effective management assumes ever greater urgency [2]. Newer classes of medications, particularly those which work via the incretin pathway, achieve glucose lowering and minimizing risks associated with more traditional therapies. Ideally, combination therapies should be well tolerated, convenient to take, have few contraindications, have a low risk of hypoglycemia and weight gain, and be reasonably effective over both the short and long term such as the combination of metformin (MF) and the dipeptidyl peptidase-4 (DPP-4) inhibitor alogliptin (ALG). The chemical structure of metformin and alogliptin were represented in Figures 1 and 2 respectively. Alogliptin [3-5] is a selective, orally-bioavailable inhibitor of enzymatic activity of dipeptidyl peptidase-4 (DPP-4). DPP-4 inhibitors represent a new therapeutic approach to the treatment of type 2 diabetes that functions to stimulate glucose dependent insulin release and reduce glucagons levels. This is done through inhibition of the inactivation of in cretins, particularly glucagon-like peptide-1 (GLP-1) and gastric inhibitory polypeptide (GIP). Alogliptin inhibits dipeptidyl peptidase 4 (DPP-4), which normally degrades the incretin glucose-dependent insulin tropic polypeptide (GIP) and glucagon like peptide 1 (GLP-1), thereby improving glycemic control [3-5]. Metformin hydrochloride (MTF) [6-9] ($C_4H_{11}N_5$.HCl) is 1:1 dimethylbiguanidine monohydrochloride is an anti-diabetic drug from the biguanide class of oral Hypoglycaemic agents, given orally in the treatment of non-insulin-dependent diabetes mellitus. The main action of Metformin HCl is in increasing glucose transport across the cell membrane in skeletal muscle. Several analytical methods based on UV [9,10], Spectroflourimetry [11,12], Reverse Phase-HPLC [13-23], LC-MS/MS [24-27], HPTLC [28] were reported for the determination of metformin either in alone or in combination with other drugs. Although literature survey reveals that no methods were reported for metformin (MTF) and alogliptin (ALG) in combination form.

Materials and Methods

Chemicals and reagents

The reference sample of metformin and alogliptin were supplied by M/s Pharma Train, Hyderabad. HPLC grade water (prepared by using 0.45 Millipore Milli-Q) was procured from Standard Reagents, Hyderabad. HPLC grade acetonitrile was purchased from Merck, Mumbai. The chemicals used for preparation of buffer include sodium dihydrogen ortho phosphate (Finar Chemicals, Ahmedabad), and orthophosphoric acid (Standard Reagents, Hyderabad). 0.45 µm membrane filters (Advanced Micro Devices Pvt. Ltd., Chandigarh, India) were used for filtration of various solvents and solutions intended for injection into the column [29-32].

---

*Corresponding author: Ashutosh Kumar S, Department of Pharmaceutical Analysis and Quality Assurance, AKRG College of Pharmacy, Nallajerla, West Godavari, Andhra Pradesh-534 112, India
E-mail: ashu.mpharm2007@gmail.com

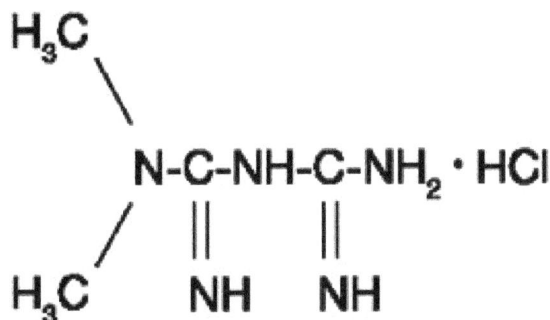

Figure 1: Chemical structure of Metformin HCl.

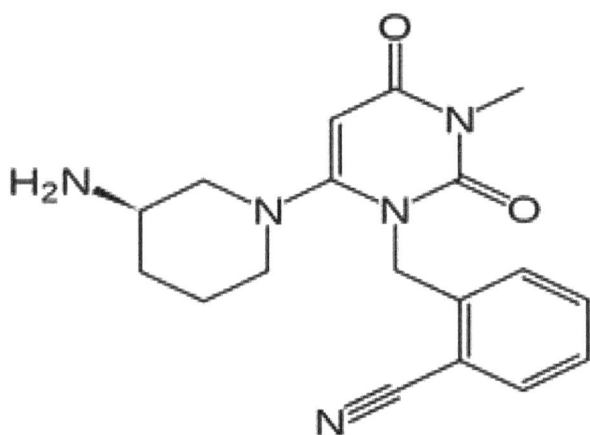

Figure 2: Chemical structure of Alogliptin.

Instrumentation

A Waters Alliance 2695 separation module equipped with a 2487 UV detector was employed throughout this study. Column that was employed in the method was X-Terra C_{18} (4.6 × 150 mm, 3.5 μm, Make: ACE). The samples were injected with an automatic injector. The 20 μL volume of sample was injected. The input and output operations of the chromatographic system were monitored by Waters Empower software. The flow rate selected was 1.0 mL per min. The detection was done at 235 nm. The temperature and run time was monitored at 25°C and 8.0 min respectively. The ultra violet spectra of the drugs used for the investigation were taken on a Lab India UV 3000 spectrophotometer for finding out their λ_{max} values. Solubility of the compounds was enhanced by sonication on an ultra sonicator. (Model: Power Sonic 510, Hwashin Technology). All the weighings in the experiments were done with an Afcoset electronic balance. The HermLe microlitre centrifuge Z100 (model no 292 P01) was used for the centrifugation process and Remi equipments (model no- CM101DX) Cyclomixer was used.

Glassware

All the volumetric glassware used in the study was of Grade A quality Borosil.

Preparation of sodium phosphate buffer (pH 4.0)

The buffer solution was prepared by dissolving 2.5 g of sodium dihydrogen ortho phosphate in 900 mL of HPLC grade water in a 1000

mL clean and dry flask. The mixture was stirred well until complete dissolution of the salt. The volume was made upto the mark with water. The pH was adjusted to 4.0 with 1% ortho phosphoric acid.

Preparation of mobile phase

The mobile phase was prepared by mixing 700 mL phosphate buffer (pH 4.0) and 300 mL of acetonitrile in a 1000 mL clean and dry flask. The resultant mobile phase was filtered through a 0.45 μm membrane filter (Advanced Micro Devices Pvt. Ltd., Chandigarh, India) under vacuum. The resultant mobile phase was degassed in an ultra sonicator for 5 min.

Preparation of diluent

The diluent was prepared by mixing phosphate buffer (pH 4.0) and acetonitrile (HPLC grade) in the ratio of 70:30 v/v. This solution was used to dilute the drug solutions in the study.

Preparation of standard solution of metformin and alogliptin

10 mg metformin was weighed accurately and transferred into a 10 mL clean and dry volumetric flask. Initially, the drug was mixed with 7 mL of diluent. The solution was sonicated for 15 min for complete dissolution of the drug. The final volume was made up to the mark with the same solvent. Similarly, about 10 mg alogliptin was weighed accurately and transferred into a 100 mL clean and dry volumetric flask. Initially, the drug was mixed with 70 mL of diluent. The solution was sonicated for 15 min for complete dissolution of the drug. The final volume was made up to the mark with the same solvent. From the above prepared stock solutions 5.0 mL of metformin and 1.25 mL of alogliptin were pipetted out into a 10 mL clean and dry volumetric flask and it was diluted up to the mark with diluent. This mixed stock solution contains 500.0 μg/mL of metformin and 12.5 μg/mL of alogliptin.

Spiking of metformin and alogliptin into plasma and their extraction from plasma (By precipitation method)

From the above prepared mixed stock solution (500.0 μg/mL of metformin and 12.5 μg/mL of alogliptin), 0.5 mL was pipetted out and spiked into 0.5 mL of plasma in a polypropylene tube (Torson's). Then the tube was cyclo mixed for 5 min. Then 1.0 mL of acetonitrile was added to the tube and centrifuged for 20 min at 3000 rpm. Further the supernatant liquids were collected in another Eppendorf tube and 20 μL supernatant was injected into the analytical column.

Validation Development

Selectivity

An aqueous mixture of metformin and alogliptin (500.0 μg/mL of metformin and 12.5 μg/mL of alogliptin) was prepared and injected into the column and the retention times were checked and any interference at the retention times was checked by comparing the response in the blank. No interference was observed at the retention times for metformin and alogliptin extracted from plasma [33-38]. The method was found to be precise and specific. A typical chromatogram of metformin and alogliptin in plasma is shown in Figure 3.

Sensitivity

To determine the sensitivity in terms of LLOQ, 'Lower Limit of Quantification' where the response of LLOQ must be at least five times greater than the response of interference in blank matrix at the retention time of the analyte(s). The LLOQ obtained by the proposed method were 5.936 and 1.983 μg/mL for metformin and alogliptin respectively.

Precision

To check the intra and inter-day variations of the method, solutions containing 500.0 μg/mL of metformin and 12.5 μg/mL of alogliptin were subjected to the proposed HPLC method of analysis and results obtained were noted. The precision of the proposed method i.e., the intra and inter-day variations in the peak areas of the drugs solutions in plasma were calculated in terms of percent relative standard deviation (RSD) and the results are represented in Tables 1, 2, 3 and 4. A statistical evaluation revealed that the percentage relative standard deviation of the drugs at linearity level for 6 injections was less than 2.0. Typical chromatogram of metformin and alogliptin in plasma for intra and inter-day precision are shown in Figures 4 and 5.

Accuracy

To determine the accuracy of the proposed method, recovery studies were carried out by analyzing (8.0, 10.0, 12.0 mg of metformin and alogliptin) of pure drugs. The solutions were suitably diluted at linearity level (500.0 μg/mL of metformin and 12.5 μg/mL of alogliptin). Then each dilution was injected thrice (n=3). The percent recoveries of the drugs were determined. The results are shown in Tables 5 and 6.

Linearity

In order to find out the linearity range of the proposed HPLC method in plasma, curves were constructed by plotting peak areas obtained for the analyte against their concentrations. A good linear relationship (r^2=0.997) was observed between the concentrations of metformin and alogliptin and their corresponding peak areas. The relevant regression equations were y=1192.x-38173 for metformin (r^2=0.997) and y=2802x-1113 for alogliptin (r^2=0.998) (where y is the peak area and x is the concentration of metformin and alogliptin (μg/mL)). The slope, intercept and the correlation coefficient of the plots are shown in Tables 7 and 8. The linearity ranges for metformin and alogliptin and their corresponding graphs are shown in Figures 6 and 7.

Injection	Retention Time	Area
Injection-1	2.734	553001
Injection-2	2.731	559476
Injection-3	2.738	553994
Injection-4	2.738	552953
Injection-5	2.731	553012
Injection-6	2.732	553032
Average	2.734	554245
Standard Deviation	0.003	2593.64
%RSD	0.12	0.47

Table 1: Intra-day precision of the proposed method for Metformin in plasma.

Injection	Retention Time	Area
Injection-1	4.451	34316
Injection-2	4.452	34431
Injection-3	4.454	34085
Injection-4	4.459	34350
Injection-5	4.452	34814
Injection-6	4.458	34360
Average	4.454	34392.7
Standard Deviation	0.003	237.7
%RSD	0.076	0.7

Table 2: Intra-day precision of the proposed method for Alogliptin in plasma.

Days	Retention Time	Area
Day-1*	2.734	563116
Day -2*	2.736	563076
Day -3*	2.735	561049
Average	2.735	562414
Standard Deviation	0.001	1182
%RSD	0.04	0.21

Table 3: Inter-day precision of the proposed method for Metformin (on three consecutive days n=6) in plasma.

Days	Retention Time	Area
Day-1*	4.451	33218
Day-2*	4.456	33876
Day-3*	4.451	33790
Average	4.452	33628
Standard Deviation	0.003	358
%RSD	0.06	1.1

Table 4: Inter-day precision of the proposed method for Alogliptin (on three consecutive days n=6) in plasma. *Average of Six injections.

Conc. Level	% Recovery	Avg. % Recovery	Amount Recovered (mg)	SD	% RSD
80%	100.05	100.17	8.0	0.032	0.4
	99.88		7.99		
	100.59		8.05		
100%	100.05	100.17	10.01	0.036	0.36
	99.88		9.99		
	100.59		10.06		
120%	100.05	100.17	12.01	0.042	0.35
	99.88		11.99		
	100.59		12.07		

Table 5: Accuracy data of the proposed method for Metformin in plasma.

Conc. Level	% Recovery	Avg. % Recovery	Amount Recovered (mg)	SD	% RSD
80%	98.48	99.55	7.88	0.119	1.50
	101.23		8.1		
	98.94		7.91		
100%	98.48	99.55	9.85	0.146	1.46
	101.23		10.12		
	98.94		9.89		
120%	98.48	99.40	11.82	0.178	1.49
	101.23		12.15		
	98.48		11.87		

Table 6: Accuracy data of the proposed method for Alogliptin in plasma.

Concentration (μg/mL)	Area	Statistical Analysis
300	308855	Slope=1192
400	445785	Intercept= -38173
500	564758	C. C=0.997
600	684752	
700	785468	

Table 7: Linearity range of Metformin in plasma.

Concentration (μg/mL)	Area	Statistical Analysis
7.5	19414	Slope=2802
10	27219	Intercept= -1113
12.5	34464	C. C=0.998
15	40829	
17.5	47634	

Table 8: Linearity range of Alogliptin in plasma.

S. No.	Standard Sample	Freeze and Thaw Stability Sample	Short Term Stability Sample	Long Term Stability Sample
1.	201585	185470	195862	187452
2.	204758	189562	194258	187569
3.	206984	188475	196541	187414
Mean	204442	187835.7	195554	187478
SD	2713	2120	1172	81
%RSD	1.33	1.13	0.60	0.04
Assay		91.88	95.65	91.70

Table 9: The Stability data for Metformin in plasma.

S. No.	Standard Sample	Freeze and Thaw Stability Sample	Short Term Stability Sample	Long Term Stability Sample
1.	13985	12014	12956	12014
2.	13586	12036	12947	12036
3.	13632	12159	12933	12159
Mean	13734	12069.67	12945	12070
SD	218	78	12	78
%RSD	1.59	0.65	0.09	0.65
Assay		87.88	94.26	87.88

Table 10: The Stability result for Alogliptin in plasma.

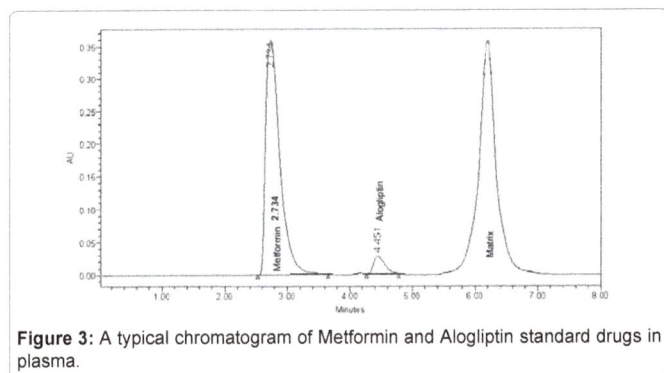

Figure 3: A typical chromatogram of Metformin and Alogliptin standard drugs in plasma.

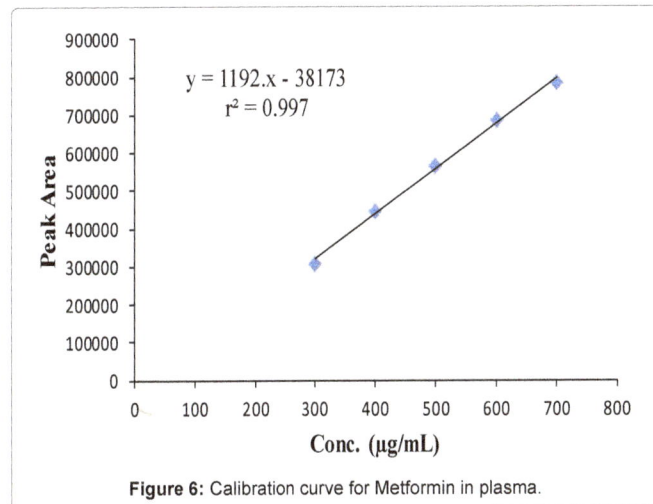

Figure 4: Typical chromatogram of metformin and alogliptin in plasma for intra-day precision.

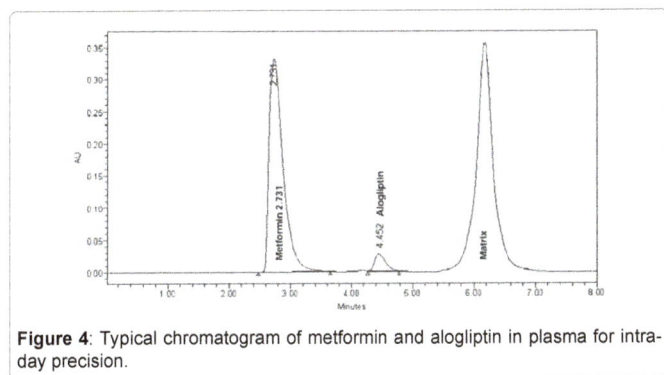

Figure 5: Typical chromatogram of metformin and alogliptin in plasma for inter-day precision.

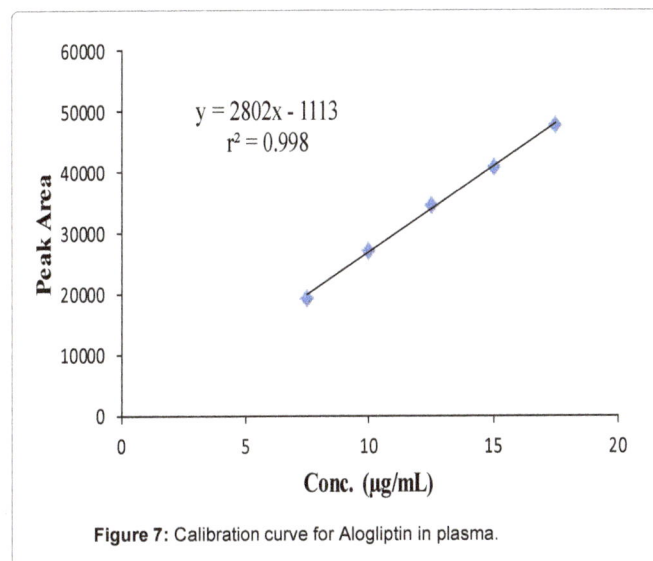

$$y = 1192.x - 38173$$
$$r^2 = 0.997$$

Figure 6: Calibration curve for Metformin in plasma.

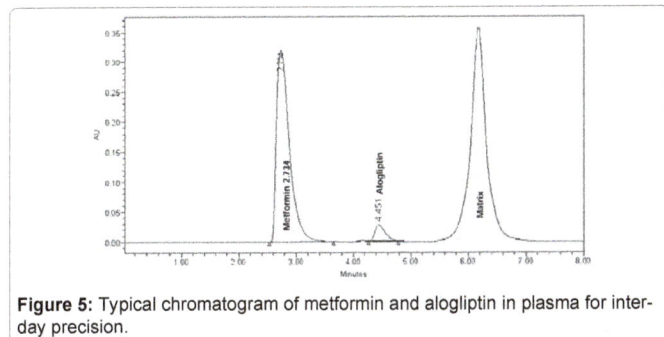

$$y = 2802x - 1113$$
$$r^2 = 0.998$$

Figure 7: Calibration curve for Alogliptin in plasma.

Stability

All stability determinations used a set of samples prepared from a freshly made stock solution of the analyte in the appropriate analyte-free, interference-free biological matrix [39]. The stock solutions of the analyte for stability evaluation were prepared in an appropriate solvent at known concentrations. To test the stability of the drug extract, it was subjected to

α) Freeze and thaw stability at -20°C ± 2°C,

β) Short term stability for period of 24 hours stored at room temperature,

χ) Long term stability for period of 15 days stored at 4°C.

Similar to the preparation of the standard preparation, the above samples were spiked into the plasma and extracted and collected in vial and injected into HPLC system. All the stability samples compared against the standard stock solution assessed for stability. The results are presented in Tables 9 and 10 (the figures in the table are in peak area units). Typical chromatograms for standard samples, freeze and thaw stability samples, short term stability samples and long term stability samples were represented in Figures 8, 9, 10 and 11.

Results and Discussion

To optimize the mobile phase, various proportions of sodium phosphate buffer (pH 4.0) with acetonitrile (HPLC Grade) were tested. The use of sodium phosphate buffer (pH 4.0) and acetonitrile (HPLC Grade) in the ratio of 70:30 v/v resulted in peak with good shapes and resolution. A flow rate of 1.0 mL /min was found to be optimum in the 0.4-1.5 mL/min range resulting in short retention time, baseline stability and minimum noise.

The LLOQ obtained for metformin and alogliptin by the proposed method in plasma was 5.936 and 1.983 µg/mL respectively. The retention times obtained for metformin and alogliptin in plasma were 2.734 and 4.451 respectively. Quantitative linearity of drugs in plasma was obeyed in the concentration range of 300-700 µg/mL for metformin and 7.5-17.5 µg/mL for alogliptin respectively. The relevant regression equations were y=1192.x-38173 for metformin (r²=0.997) and y=2802x-1113 for alogliptin (r²=0.998) (where y is the peak area and x is the concentration of metformin and alogliptin (µg/mL)). The intra-day and inter-day drugs variations in plasma by the proposed method showed percentage relative standard deviation were less than 2%, indicating that the method is precise. The corresponding mean recoveries of the drugs in plasma were 98.40-100.17%. This reveals that the method is quite accurate. The percentage relative standard deviation obtained for the drugs spiked in plasma for stability studies were less than 2%.

Conclusion

The proposed HPLC method was found to be simple, precise, accurate and sensitive for the simultaneous determination of metformin and alogliptin. The method was validated as per ICH guidelines and all the parameters met within the acceptance criteria. Applicability of this method for simultaneous estimation of metformin and alogliptin in plasma was confirmed.

Acknowledgements

The authors are thankful to M/s Pharma Train, Hyderabad for providing reference samples of metformin and alogliptin and processed plasma.

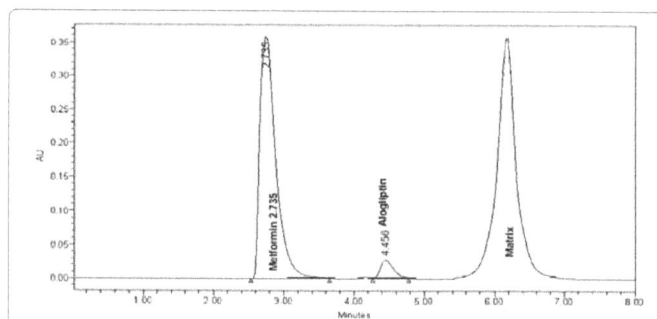

Figure 8: Typical chromatogram for standard samples in plasma for stability studies.

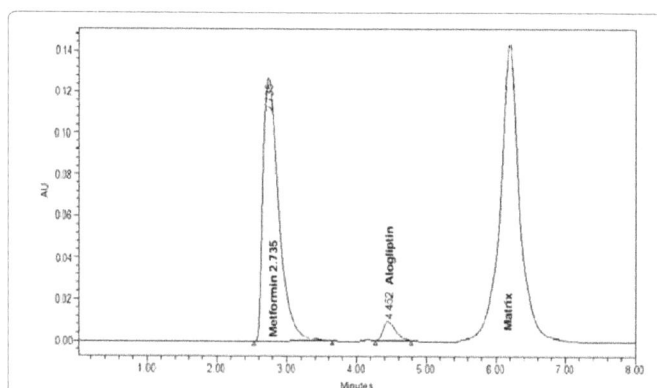

Figure 9: Typical chromatogram for freeze thaw samples in plasma for stability studies.

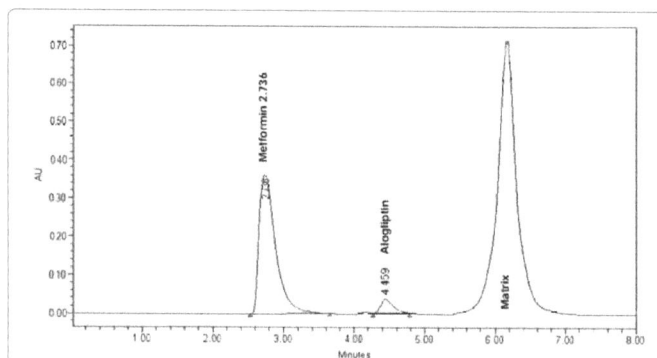

Figure 10: Typical chromatogram for short term samples in plasma for stability studies.

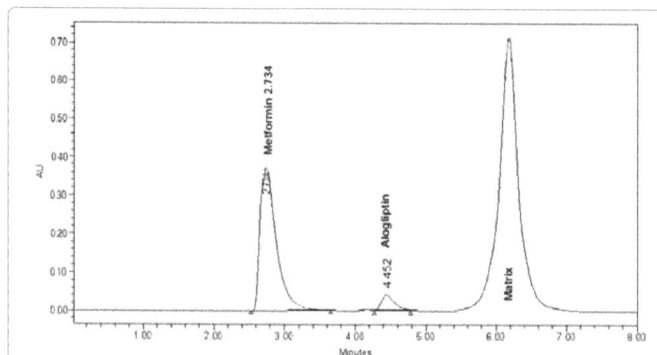

Figure 11: Typical chromatogram for long term samples in plasma for stability studies.

References

1. Medicinenet (2015) Diabetes (Type 1 and Type 2). www.medicinenet.com/diabetes_mellitus/article.htm. Accessed on: August 2015.

2. Medicalnewstoday (2015) Diabetes: Symptoms, Causes and Treatments. http://www.medicalnewstoday.com/info/diabetes. Accessed on: August 2015.

3. Wikipedia (2015) Alogliptin. http://en.wikipedia.org/wiki/alogliptin. Accessed on: August 2015.

4. Drugbank (2015) http://www.drugbank.ca/drugs. Accessed on: August 2015.

5. Druginformation (2015) http://www.druginformation.com/. Accessed on: August 2015.

6. Rxlist (2015) http://www.rxlist.com/nesina-drug/indications-dosage. Accessed on: August 2015.

7. Wikipedia (2015) Metformin. http://en.wikipedia.org/wiki/metformin. Accessed on: August 2015.

8. Rxlist (2015) http://www.rxlist.com/fortamet-drug.htm. Accessed on: August 2015.

9. Khan G, Agrawal YP, Sabarwal N, Jain A, Gupta A K (2011) Simultaneous Estimation of Metformin and Sitagliptin In tablet dosage form. Asian J Biochem Pharma Res 1: 352-358.

10. Patil SS, Bonde C G (2009) Development and Validation of analytical method for simultaneous estimation of Glibenclamide and Metformin HCl in Bulk and Tablets using UV visible spectroscopy. Int JChem Tech Res 1: 905-909.

11. Ramzia El -Bagary, Ehab EF, Bassam AM (2011) Spectroflourometric and Spectrophotometric methods for the determination of Sitagliptin in binary mixture with Metformin and ternary mixture with Metformin and Sitagliptin Alkaline Degradation Product. Int J Biomed Sci 7: 62-69.

12. Hassasaad SM, Mahmoud WH, Elmosallamy MA, Othman AH (1999) Determination of Metformin in pharmaceutical preparations using potentiometry, spectrofluorimetry and UV–visible spectrophotometry. Anal Chimic 378: 299-311.

13. Ravi PP, Sastry BS, Rajendra PY, Appala RN (2011) Simultaneous Estimation of Metformin HCl and Sitagliptin Phosphate in tablet dosage forms by RP-HPLC. Res J Pharm Tech 4: 646-649.

14. Shyamala M, Mohideen S, Satyanarayana T, Narasimha R, Suresh K (2011) Validated RP-HPLC for simultaneous estimation of Sitagliptin phosphate and Metformin in hydrochloride in tablet dosage form. American J Pharm Tech Res 1: 93-101.

15. Freddy HH, Dharmendra VL (2010) Simultaneous estimation of Metformin hydrochloride, rosiglitazone and pioglitazone hydrochloride in the tablets dosage form. Int J App Bio Pharm Tec 1: 1000-1005.

16. Aburuz S, Millership J, McElnay J (2005) The development and validation of liquid chromatography method for the simultaneous determination of metformin and glipizide, gliclazide, glibenclamide or glimperide in plasma. J Chromatogr B Analyt Technol Biomed Life Sci 817: 277-286.

17. Al-Rimawi F (2009) Development and validation of an analytical method for metformin hydrochloride and its related compound (1-cyanoguanidine) in tablet formulations by HPLC-UV. Talanta 79: 1368-1371.

18. Lakshmi KS, Rajesh T, Sharma S, Lakshmi S (2009) Development and Validation of Liquid Chromatographic and UV Derivative Spectrophotometric Methods for the Determination of Metformin, Pioglitazone and Glimepiride in Pharmaceutical Formulations. Der Pharma Chemica 1: 238-246.

19. Jain D, Jain S, Jain D, Amin M (2008) Simultaneous estimation of metformin hydrochloride, pioglitazone hydrochloride, and glimepiride by RP-HPLC in tablet formulation. J Chromatogr Sci 46: 501-504.

20. Lakshmi KS, Rajesh T, Sharma S (2009) Simultaneous determination of Metformin and pioglitazone by reversed phase HPLC in pharmaceutical dosage forms. Int J Pharm Pharm Sci 1: 162-166.

21. Alexandar S, Diwedi R, Chandrasekar M (2010) A RP-HPLC method for simultaneous estimation of Metformin and pioglitazone in pharmaceutical formulation. Res J Pharm Bio Chem Sci 1: 858-866.

22. Florentin T, Monica A (2007) Specificity of an analytical HPLC assay method of Metformin hydrochloride. Revue Roumainede Chimie 52: 603-609.

23. Pawar S, Meshram G, Jadhav R, Bansal Y (2010) Simultaneous determination of Glimepiride and Metformin hydrochloride impurities in sustained release pharmaceutical drug product by HPLC. Der Pharma Chemica 2: 157-168.

24. Zeng W, Xu Y, Constanzer M, Woolf EJ (2010) Determination of sitaglitinin human plasma using protein precipitation and tandem mass spectrometry. J Chromatogr B Analyt Technol Biomed Life Sci 878: 1817-1823.

25. Zeng W, Musson DG, Fisher AL, Chen L, Schwartz MS, et al. (2008) Determination of sitagliptin in human urine and hemodialysate using turbulent flow online extraction and tandem mass spectrometry. J Pharm Biomed Anal 46: 534-542.

26. Georgita C, Albu F, David V, Medvedovici A (2007) Simultaneous assay of metformin and glibenclamide in human plasma based on extraction-less sample preparation procedure and LC/(APCI)MS. J Chromatogr B Analyt Technol Biomed Life Sci 854: 211-218.

27. Nirogi R, Kandikere V, Mudigonda K, Komarneni P, Aleti R (2008) Sensitive liquid chromatography tandem mass spectrometry method for the quantification of Sitagliptin, a DPP-4 inhibitor, in human plasma using liquid–liquid extraction. Biomed Chromatogr 22: 214-222.

28. Havele S, Dhaneshwar S (2010) Estimation of Metformin in Bulk Drug and in Formulation by HPTLC. J Nanomedic Nanotechnolo 1: 102.

29. Ashutosh KS, Manidipa D, Seshagiri RJVLN (2013) Simultaneous Estimation of Metformin Hydrochloride and Sitagliptin Phosphate Monohydrate in bulk as well as in Pharmaceutical formulation by RP -HPLC. AM J Pharm Tech Res 3: 556-575.

30. Ashutosh KS, Manidipa D, Seshagiri RJVLN (2013) Development of stability indicating RP-HPLC method for simultaneous estimation of Metformin Hydrochloride and Sitagliptin Phosphate Monohydrate in bulk as well as in Pharmaceutical Formulation. Der Pharmacia Sinica 4: 47-61.

31. Satya SG, Ashutosh KS, Saravanan J, Manidipa D, Greeshma V, et al. (2013) A new RP-HPLC method development for simultaneous estimation of metformin and alogliptin in bulk as well as in pharmaceutical formulation by using PDA detector. World Journal of Pharmacy and Pharmaceutical Sciences 2: 6720-6743.

32. Satya SG, Ashutosh KS, Saravanan J, Manidipa D, Greeshma V, et al. (2013) A new stability indicating RP-HPLC method development for simultaneous estimation of metformin and alogliptin in bulk as well as in pharmaceutical formulation by using PDA detector. Indo American Journal of Pharmaceutical Research 3: 9222- 9241.

33. Validation of analytical procedure: Methodology Q2B, (1996) ICH Harmonized Tripartite Guidelines 1-8.

34. International Conference on Harmonization of Technical Requirements for Registration of Pharmaceuticals for Human Use ICH Harmonized tripartite guideline Validation of analytical procedures: Text and Methodology Q2 (R1) 6 November 1996.

35. Ravichandran V, Shalini S, Sundram KM, Harish R (2010) Validation of analytical methods – strategies & importance. Int J Pharmacy and Pharm Sci 2: 18-22.

36. Tangri P, Rawat PS, Jakhmola V (2012) Validation: A Critical Parameter for Quality Control of Pharmaceuticals. Journal of Drug Delivery & Therapeutics 2: 34-40.

37. ICH, Validation of Analytical Procedure, Text and Methodology Q2 (R1) (2005) International conference on Harmonization, IFPMA, Geneva, Switzerland.

38. ICH harmonized tripartite guideline. Impurities in New Drug products Q3B (R2) current step 4 versions dated 2 June 2006.

39. International Conference on Harmonization, ICH Q1 A(R2); Stability Testing of New Drug Substances and Products 2003.

22

Efficient, Baseline Separation of Pyrethrins by Centrifugal Partition Chromatography

Alan Wong* and Jan A Glinski

Planta Analytica LLC, New Milford, Connecticut, USA

Abstract

Pyrethrins are natural pesticides present in the oil extracted from the flowers of *Chrysanthemum cinerariaefolium*. They are potent ion-channel neurotoxins which have higher selective toxicity for insects compared to mammals. These phytochemicals have an exceptionally safe environmental profile and are an attractive natural alternative to organophosphate insecticides currently used in agriculture. The oil extract contains six pyrethrins compounds, which are structurally related esters. Isolation of reference standards for pyrethrins and their derivatives in multi-gram quantities is a prerequisite for studying toxicity and soil degradation for these natural pyrethrins. Chromatography is difficult because of the high hydrophobicity and close structural similarity. The isolated pyrethrin compounds also have poor stability because they are photosensitive. To facilitate these studies, we have developed a two-step purification process, in which the two groups were first separated by normal phase liquid chromatography on silica gel. The separation of pyrethrin, cinerin, and jasmolin within each group was achieved by centrifugal partition chromatography (CPC). We explored and successfully optimized solvent systems that were subsequently applied to achieve multi-gram scale separation. The pyrethrins I group compounds were separated using a solvent system containing heptane-methanol-acetonitrile (6:1:2, v/v) in ascending mode. The pyrethrins II group compounds were separated using a solvent system containing heptane-terbutylmethyl ether-acetonitrile-water (8:1:5:1.5, v/v) also in ascending mode. In both cases, gram quantities of individual pyrethrins with purity exceeding 99% were produced.

Keywords: Pyrethrin; Jasmolin; Cinerin; Centrifugal partition chromatography; Countercurrent chromatography; Natural pesticides; Chrysanthemum

Highlights

Centrifugal partition chromatography was applied for the separation of six natural pesticidal pyrethrins. Solvent systems were optimized for the separation of pyrethrins I group and pyrethrins II group. Pyrethrin I, cinerin I, and jasmolin I were separated in gram quantity per single CPC run. Similarly, pyrethrin II, cinerin II, and jasmolin II were purified nearly base-line.

Introduction

Pyrethrins are natural pesticides obtained from the extraction of *Chrysanthemum cinerariaefolium* flowers, which are cultivated largely in Kenya. Pyrethrin derivatives have been developed for agriculture and household pesticide products. In recent years, the use of natural non-derivatized pyrethrins for agriculture is increasing in popularity. Pyrethrin compounds have a good environmental safety profile stemming from the fact that these esters are photosensitive and have a short half-life following application to crops. The basic issues related to production, chemistry, and toxicology were reviewed by Maciver [1], while issues related to extraction, refining and analysis were reviewed by Carlson [2]. The Pyrethrum oil contains six pyrethrins belonging to two groups. Pyrethrin I (**1**), cinerin I (**2**), and jasmolin I (**3**) are esters of *trans*-chrysanthemic acid (**7**) and constitute group I. The pyrethrins II group are esters of carbomethoxychrysanthemic acid (pyrethric acid) (**8**) and comprise of pyrethrin II (**4**), cinerin II (**5**), and jasmolin II (**6**). The pyrethrins, cinerins, and jasmolins differs by the alkyl chain present in the alcohol portion, which is called pyrethrolone (**9**), cinerolone (**10**), and jasmolone (**11**), respectively (Figure 1).

Pyrethrins are potent ion-channel neurotoxins with higher selective toxicity for insects compared to mammals, and compared to other insecticides used in agriculture they have less toxicity towards humans. To facilitate the studies of toxicity and the environmental fate of pyrethrins, we needed to purify gram quantities of these esters. The

unmodified phytochemicals are also used for the semi-synthesis of their environmental degradation products. The gram quantity separation was achieved by developing appropriate solvent systems for centrifugal partition chromatography (CPC), which gave baseline separation of the natural esters in high purity. CPC, also called liquid-liquid partitioning, belongs to the family of countercurrent chromatography techniques – the separation takes place in a series of small mixing chambers. The mobile liquid phase interacts with an immiscible stationary liquid phase during passage, carrying the analyte through the system based on the partition coefficient. In the CPC instrument, the stationary phase is immobilized by centrifugal force and the mobile phase diffuses through [3]. Retention time is determined by the partition coefficients of the individual components in the solvent system [4]. One of the greatest advantages of countercurrent chromatography is an absence of chemical degradation often seen with solid phase chromatography, allowing full recovery of the constituents of the mixtures [3].

A comprehensive environmental study required pure pyrethrins for use as reference standards and, additionally, as the starting materials to produce a variety of derivatives originating during environmental degradation. To accomplish this goal, we estimated the required quantity for pyrethrins to be in the range of 10 - 20 g, and 2 - 3 g for each of the cinerins and jasmolins. These were significant and unprecedented quantities to isolate. In early approaches, these esters were obtained by a semi-synthetic methodology, reacting acid chlorides with pyrethrolone (**9**), cinerolone (**10**), and jasmolone (**11**)

---

*Corresponding author: Alan Wong, Planta Analytica LLC, New Milford, Connecticut, USA, E-mail: awong39@uic.edu

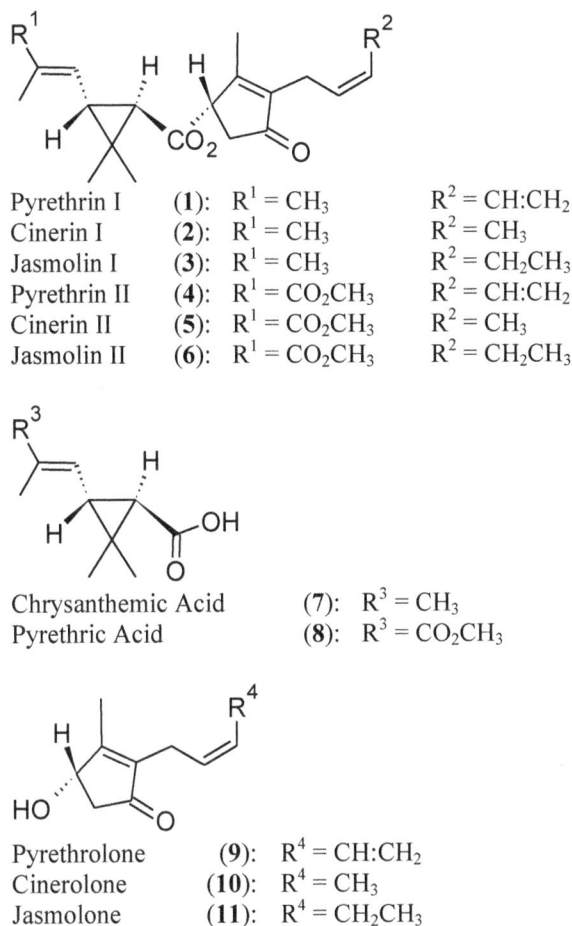

Pyrethrin I   (**1**): $R^1 = CH_3$ $R^2 = CH{:}CH_2$
Cinerin I   (**2**): $R^1 = CH_3$ $R^2 = CH_3$
Jasmolin I   (**3**): $R^1 = CH_3$ $R^2 = CH_2CH_3$
Pyrethrin II   (**4**): $R^1 = CO_2CH_3$ $R^2 = CH{:}CH_2$
Cinerin II   (**5**): $R^1 = CO_2CH_3$ $R^2 = CH_3$
Jasmolin II   (**6**): $R^1 = CO_2CH_3$ $R^2 = CH_2CH_3$

Chrysanthemic Acid   (**7**): $R^3 = CH_3$
Pyrethric Acid   (**8**): $R^3 = CO_2CH_3$

Pyrethrolone   (**9**): $R^4 = CH{:}CH_2$
Cinerolone   (**10**): $R^4 = CH_3$
Jasmolone   (**11**): $R^4 = CH_2CH_3$

Figure 1: Structures of pyrethrins, with their constitutive acids and alcohols.

[5]. Until recently, purification of reference standards of pyrethrins directly from the extract have been described only a few times [6]. Rickett successfully used liquid-liquid partitioning on diatomaceous earth carrying nitromethane as the stationary phase, while hexane and carbon tetrachloride acted as the mobile phase, separating 270 mg of total pyrethrins in a 17-hour experiment [7]. In 2001, separation of pyrethrins was achieved by preparative HPLC using silica gel column [6]. The method was 100 min long, and after multiple injections, the collected amount varied between 11.4 mg for jasmolin II (**6**) (smallest component) and 93.4 mg for pyrethrin I (**1**) (largest component). The isolated pyrethrins were used for determining spectral properties and standardization of the pyrethrin oils. Scaling up any of the prior approaches to yield multi-gram quantity would be a daunting effort with a low chance of success. In addition, reference standards of these esters were often adulterated with stabilizing agents due to poor stability – the stabilizing agents will interfere with environmental degradation studies.

Pyrethrins are generally considered to be of strong hydrophobic character. The higher hydrophobicity of pyrethrins I group comes from the chrysanthemic acid (**7**) moiety, in comparison to the pyrethric acid (**8**) moiety of their group II counterparts. This difference in

hydrophobicity is sufficient to separate the two ester groups by medium pressure chromatography on silica gel. Silica gel also induces partial separation within the group constituents, but under preparative conditions it is impractical to achieve a full separation. Analytical monitoring of the fractionation can be done by NP HPLC on Cyano columns or by RP HPLC on ODS columns. A review of chromatographic methods for pyrethrins was published by Gullick [8] and Chen [9].

Materials and Methods

Partition coefficients were measured by dissolving 5 mg of pyrethrin mixture in 2 mL of a bi-phasic solvent system. After equilibrating, even volumes (0.5 mL) of each phase were separated and evaporated to dryness under the stream of nitrogen. The dry residues were dissolved in 1.00 ml of MeOH and analyzed by high performance liquid chromatography coupled to a diode array detector (HPLC-DAD). A ratio of the peak areas from the stationary phase to the mobile phase constituted a partition coefficient for each compound.

Thin layer chromatography (TLC) analyses used silica gel plates developed in hexane-acetone (3:1), followed by spraying with cerium sulfate dissolved in 10% sulfuric acid and heated with a heat gun until the spots developed.

Pyrethrum oil sample (MGK Pyrocide 50%, code 224-14) was obtained from the McLaughlin Gormley King Co., Minneapolis, MN, USA. Medium pressure chromatography with silica gel (Premium Rf, 60 Å, 40 - 75 μm, Sorbent Technologies, Atlanta, GA) was performed on a glass column 4" ID × 47" (Spectrum Chromatography, Houston, TX), with a pump solvent delivery (model # 1212100 from Thomson Instrument Company). The HPLC analyzes were carried out on Agilent HPLC model 1100 (quaternary pump G1314, autosampler 1313A, DAD G1315A, Shimadzu degasser DGU-14). Normal phase HPLC analysis was done on a 4.6 × 100 mm Cyano column (YMC CN12S031046WTA), 3 μm, 120 Å, in a step gradient. Solvent A was hexane and solvent B was 10% THF in hexane. The flow rate was 2.0 mL/min: 0.0 min [1% B], 5.0 min [5.5% B], 5.1 min [12.0% B], 10.6 min [25% B], 13.0 min [25% B]. Reverse phase HPLC analysis was done on a XTerra MS C18, 5 μm, 2.1 × 100 mm; solvent A was 5% MeCN and 95% water with 0.1% TFA, solvent B was MeCN. The flow rate was 0.4 mL/min: 0.0 min [45% B], 12.0 min [75% B]. Centrifugal partition chromatography experiment uses the Kromaton FCPC-A (Rousselet Robatel, Annonay, France) with a 1 L rotor. Eluent was collected into test tubes of 16 or 18 mm × 20 cm using a fraction collector (Varian ProStar 701). The liquid phases were pumped into the FCPC instrument using a pump (Waters HPLC pump model 590). Solvents were purchased from Pharmco-AAPER, Brookfield, CT.

Results

In the initial fractionation, 300 g of Pyrethrum oil concentrate was dissolved in 2.7 L of hexane and pumped into a glass medium pressure column filled with 4 kg of silica gel (40 - 75 μm) in hexane. The solvent was pumped at the rate of 60 mL/min. Initially, the column was fed with 6 L of hexane; for every 2 L after, acetone content in hexane was increased by 0.33% v/v. The back pressure remained below 30 psi. Fractions were collected in 2 L increments. The pyrethrins I group, eluted in the fractions 11 - 15, had a total mass of 117 g. The pyrethrins II group, in the fractions 19 - 25 has a total mass of 78 g. In both cases the total mass comprises pyrethrins as well as other constituents. The earlier fractions in each sequence were enriched into jasmolin and cinerin and the later fractions contained almost exclusively pyrethrins.

Experiments on finding an appropriate solvent system for CPC separation of pyrethrin I group led to the solvent system SS-02, containing Hept-MeOH-MeCN (6:1:2, v/v) (Table 1). SS-02 offered good spread of the three esters (Figure 2A) within a run time of 1.5 hours at flow rate of 33 mL/min. All non-pyrethrin impurities had partition coefficients different from 1 - 3 and eluted either with the front of the mobile phase or were retained beyond the band of pyrethrin I. Further experimentation led to a discovery of an alternative solvent system, SS-03, containing Hept-MTBE-MeCN (8:1:5, v/v), in which the partition coefficients for 1 and 3 are near identical as in SS-02, but dramatically different for 2, resulting in its early elution ahead of 1 (Figure 2C).

In a representative CPC process, first, the rotor was filled at 200 rpm with stationary phase (lower layer). With the valve configured to ascending run and the rotational speed increased to 1500 rpm, a solution of 7 - 8 g of a mixture of pyrethrins I group dissolved in a mixture of 20 mL of upper and 50 mL lower phase was injected into the rotor. Upon completion of the injection, the mobile phase was introduced to the system at a flow rate of 33 mL/min and the eluent was collected into 16 mm × 20 cm test tubes at 20 mL per tube. Only minimal loss of the stationary phase occurred during the run. The positions of eluted bands were determined by normal phase (NP) HPLC analysis on a YMC Cyano column. Retention times were as follows: 3 [4.42 min], 2 [4.63 min], and 1 [4.89 min], 6 [7.17 min], 5 [7.30 min], and 4 [7.48 min]. In reverse phase (RP) HPLC the retention times were: 5 [7.24 min], 4 [7.48 min], 6 [8.13 min], 2 [9.89 min], 1 [10.05 min], and 3 [10.65 min]. After analysis, the test tubes with similar content were combined, evaporated to dryness on a rotary evaporator and transferred to a freezer for storage. Each collected fraction from the silica gel column contained varying ratios of pyrethrins. The results illustrated in Figure 2A were of a fraction containing representative proportions of the three components found in pyrethrin I group, of which we obtained 3 [517 mg], 2 [726 mg], and 1 [3393 mg].

The CPC purification of pyrethrins II group was performed on these fractions from the silica gel fractionation that had a slightly higher content ratio of jasmolin and cinerin to pyrethrin (fractions 19 - 22). The selected solvent system SS-12, containing Hept-MTBE-MeCN-H_2O (8:1:5:1.5, v/v) was used in ascending mode. The run time was approximately 2 hours with a flow rate of 33 mL/min, and fractions were collected in 25 mL increments using 18 mm × 20 cm test tubes. A representative spread of the eluted bands of the pyrethrins II group is shown in Figure 2B, from which we obtained 6 [507 mg], 5 [967 mg], and 4 [2316 mg].

Discussion

Countercurrent chromatography is one of the gentlest of all chromatographic approaches capable of exploiting miniscule structural differences [4], and it provides separations on a scale much larger than preparative HPLC. The separations are carried out using solvent systems comprising several solvents, which form two phases after mixing. In ascending mode, the upper layer is used as mobile phase, while in descending, the lower layer is the mobile phase. In our experiment design, ascending mode was preferred since the eluted organic phase is easier to evaporate than the stationary phase. The mobile phase elutes components of a mixture according to the increasing partition coefficient ($K=[A]_{stationary}/[A]_{mobile}$). Considering close structural similarity of pyrethrins, it was necessary to find a solvent system that will maximally differentiate the components (increase the selectivity) without unnecessarily lengthening the run time. Tables 1-3 provide the partition coefficients (K) for each of the esters in selected solvent

Solvent System	Hept	MTBE	MeOH	MeCN	H_2O	Cpd	K	$\alpha=K_n/K_1$
SS-01	1.5	--	1	--	0.03	1	0.43	1.00
						2	0.38	0.88
						3	0.30	0.70
SS-02	6	--	1	2	--	1	2.63	1.00
						2	2.08	0.79
						3	1.61	0.61
SS-03	8	1	--	5	--	1	2.64	1.00
						2	0.68	0.26
						3	1.64	0.62

Table 1: Partition coefficients (K) for the pyrethrins I group in three selected solvent systems. K_n is the partition coefficient of a given pyrethrin

systems (SS), and their selectivity ($\alpha=K_n/K_1$ or K_n/K_4) to pyrethrin I (**1**) or pyrethrin II (**4**). The solvent systems with the largest differences in the normalized partition coefficients offer the best practical separations. Desired partition coefficients do not guarantee expected CPC separation, and only an actual run will verify if hydrodynamic properties of such a solvent system will perform.

Purification of pyrethrins I group by CPC

Large-scale chromatographic fractionation of hydrophobic substances faces limitations caused by a smaller selection of suitable solvents as well as usable chromatographic media. This applies to the liquid-liquid partitioning of the pyrethrins groups by imposing limitations regarding what solvents could be utilized to create the necessary bi-phasic solvent system. Hexane and water are found in the majority of the solvent systems used in countercurrent chromatography. We found, however, that hexane was not protecting the phases against homogenization in the presence of a sample load higher than 5%. Water, on the other hand, adds strong polar character to the lower phase and effectively excluding the pyrethrins and significantly alter the partition coefficient (K) values. To ameliorate these effects, we substituted hexane with heptane, which prevented homogenization and the MeCN was selected as the main component of the lower phase in place of water (Table 1). A solvent system based only on heptane and MeCN was very retentive and required addition of solvents that would increase polarity

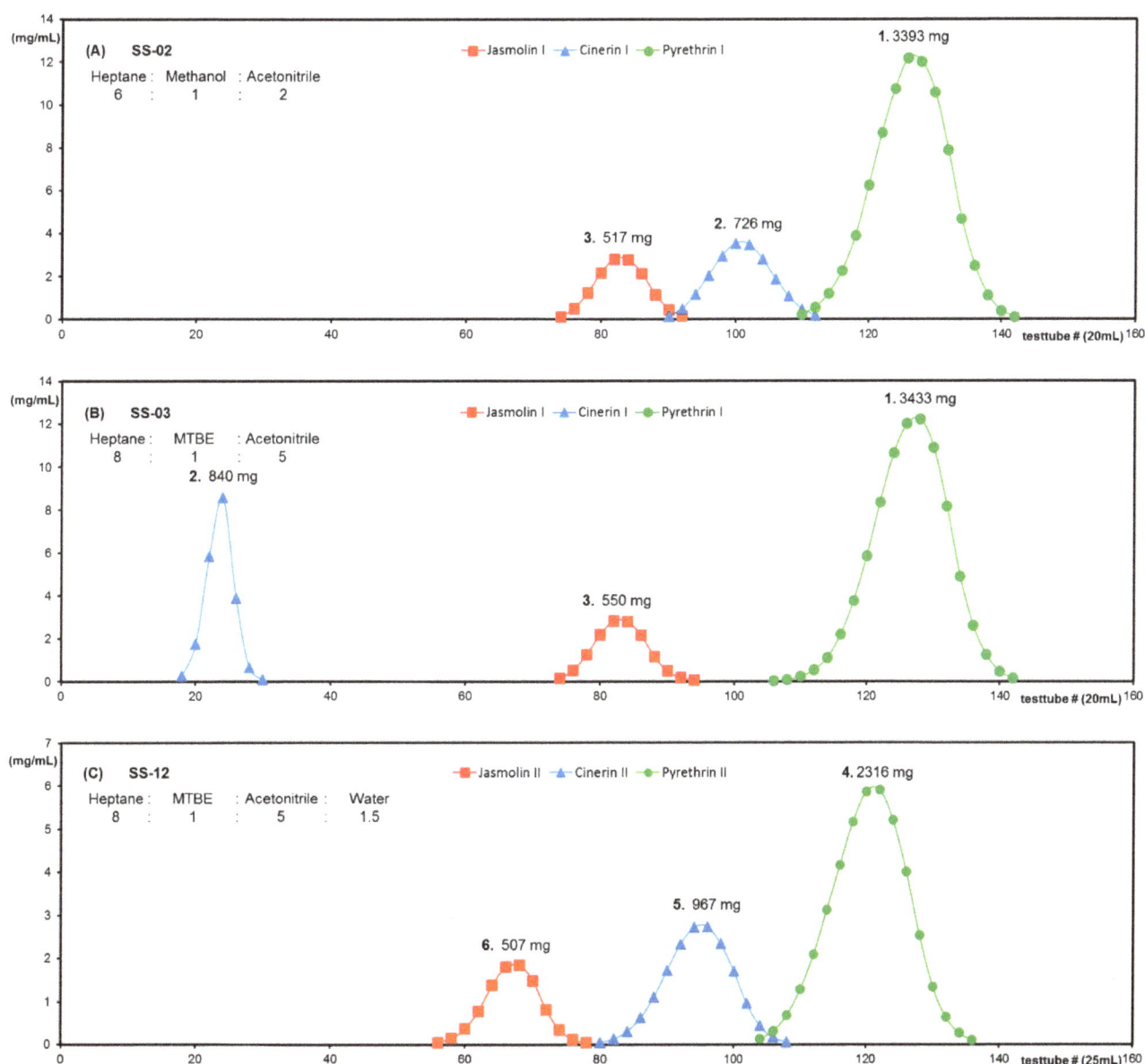

Figure 2: CPC chromatogram of pyrethrins I group in (A) SS-02, and (B) SS-03, and of pyrethrins II group in (C) SS-12. Concentrations of pyrethrin components in the collected fractions (20 mL for group I and 25 mL for group II) are measured using HPLC-DAD quantitation. Pyrethrin I (1), cinerin I (2), jasmolin I (3), pyrethrin II (4), cinerin II (5), and jasmolin II (6).

Solvent System	Hept	MTBE	MeOH	MeCN	H_2O	Cpd	K	$\alpha = K_n/K_4$
SS-04	5	--	2.5	0.5	--	4	6.25	1.00
						5	5.26	0.84
						6	4.76	0.76
SS-05	5	--	1.5	1.5	--	4	12.53	1.00
						5	9.53	0.76
						6	7.16	0.57
SS-06	5	--	1	2	--	4	8.82	1.00
						5	6.63	0.75
						6	4.85	0.55
SS-07	6	1	2	1	--	4	3.41	1.00
						5	3.10	0.91
						6	2.98	0.87
SS-08	6	0.5	2	1	0.5	4	2.94	1.00
						5	2.44	0.83
						6	1.79	0.61
SS-09	8	1	--	5	--	4	13.42	1.00
						5	9.87	0.74
						6	7.33	0.55
SS-10	8	2	--	4	--	4	7.84	1.00
						5	6.18	0.79
						6	4.90	0.63
SS-11	7	2	--	5	--	4	9.36	1.00
						5	4.17	0.76
						6	5.58	0.59
SS-12	8	1	--	5	1.5	4	4.54	1.00
						5	3.53	0.78
						6	2.26	0.50

Table 2: Partition coefficients (K) measured for the pyrethrins II group in selected solvent systems.

Solvent System	Hept	MTBE	MeOH	MeCN	H_2O	Cpd	K	$\alpha = K_n/K_{1,4,9}$
SS-02	6	--	1	2	--	1	2.63	1.00
						2	2.08	0.79
						3	1.61	0.61
						4	7.69	1.00
						5	6.67	0.87
						6	6.06	0.79
						9	136.0	1.00
						10	114.0	0.84
						11	79.11	0.58
SS-12	8	1	--	5	1.5	1	0.60	1.00
						2	0.14	0.24
						3	0.29	0.48
						4	4.54	1.00
						5	3.53	0.78
						6	2.26	0.50
						9	56.69	1.00
						10	48.40	0.85
						11	36.27	0.64

Table 3: Comparison of partition coefficients for the pyrethrins I group and pyrethrins II group and their respective alcohol moieties.

of the upper phase. This was accomplished by adding MeOH (SS-02). Experimentation with the pyrethrins I group was brief because SS-02 was simple in design and provided satisfactory CPC separation. It possessed the best combination of practical benefits: moderate run time and phase stability tolerating a sample load of 8 g (larger sample loads were not tested). This allowed isolation of up to 4 g of individual compounds per load, eluting in the order of jasmolin>cinerin>pyrethrin, with most of them at >99% purity (Figure 2A).

An alternative solvent system was found which incorporated MTBE (SS-03) instead of MeOH. The system also had good hydrodynamic properties, selectivity, and stability in the presence of a sample load. While the values of K for jasmolin I (**3**) and pyrethrin I (**1**) were similar to the values in SS-02 ($K_3=1.61\rightarrow1.64$ and $K_1=2.63\rightarrow2.64$), the elution of cinerin I (**2**) was significantly earlier ($K_2=2.08\rightarrow.68$) (Figure 2B). The change in elution order was unexpected and we cannot offer a good hypothesis to explain this phenomenon. A disadvantage of SS-03 resulted from co-elution of cinerin I (**2**) with non-pyrethrin components in the early fractions. The different selectivity of the two solvent systems can be exploited if they are used in succession for achieving superior purity with increased sample load.

Purification of pyrethrins II group by CPC

During the development stage of a suitable solvent system (SS) for pyrethrins II group, we prepared and tested 33 SS (only the data from selected SS are shown in Table 2, measuring K for pyrethrin II (**4**), cinerin II (**5**), and jasmolin II (**6**) in order to assess their effectiveness for CPC separation. In addition, hydrodynamic behavior was evaluated by observation. The first stage in developing a solvent system involves searching for a combination of individual solvents that offer useful selectivity ($\Delta\alpha > 0.2$) for separation. Next, we fine-tuned K to an appropriate range by adjusting ratios of solvents; this is to maintain satisfactory separation while having moderate experiment time ($2<K<3$). The last step is to determine if the SS has suitable hydrodynamic properties for CPC use. In SS-04 to SS-06, the systems were generally excessively retentive (Table 2). Attempts to lessen retentivity by addition of EtOAc or CH_2Cl_2 failed due to loss of selectivity ($\Delta\alpha < 0.2$). In addition, the hydrodynamic properties of these solvents were poor, resulting in excessive loss of the stationary phase during CPC experimentation; with sample loads higher than 5%, phase homogenization was observed. By addition of MTBE to adjust K to an acceptable range, SS-07 had improved hydrodynamic properties and phase stability. However, the selectivity was poor, which would result in significant band overlapping. With additional fine-tuning (SS-08), the selectivity was improved, but baseline separation of **5** and **4** was not possible. The removal of MeOH from the SS resulted in significant improvements to selectivity (SS-09). Decrease of retentivity was achieved by multiple changes, involving an increase in the MTBE-MeCN ratio (SS-10), increase in the MTBE-Hept ratio (SS-11), or addition of water (SS-12). By adjusting the ratio of the Hept-MTBE-MeCN, K decreased, but with the tradeoff of decreased selectivity. With the addition of water, the partition coefficients can be brought into the practical range, while at the same time maintaining excellent selectivity. SS-12 provided the optimal efficiency with the ratio of Hept-MTBE-MeCN-H_2O at 8:1:5:1.5, v/v (Table 2).

Contributions of each moieties to K

To understand the contributions of the acid and alcohol moieties to the partition coefficients, we measured K of the esters **1-6** as well as their constitutional alcohols **9, 10, 11** in SS-02 and SS-12 (Table 3). K for the three alcohols are very high, as their polarity is higher than those

of the esters. The more hydrophobic pyrethrins I group have lower partition coefficients than their group II counterparts. This reflects the dominant influence of the acid moiety with the hydrophobicity of these compounds. Because of this, chromatographic separation of the pyrethrin I and II groups is relatively easy on some chromatographic processes, including silica gel. Separation of the individual esters within each group is significantly more challenging, because of the subtle structural differences between them, laying with the alkyl chains of the constitutional alcohol moieties. K measured in SS-02, for the alcohols **9**, **10**, and **11**, as well as for both sets of the esters **1-6** show the same trend, eluting in the order of jasmolin>cinerin>pyrethrin. Thus, the structural differences in the alkyl chains are amplified in this particular solvent system (Table 3).

When the K values of the same nine compounds were measured in SS-12 (MTBE-based SS), we noticed an anomalous response only within the pyrethrins I group, expressed by a rearranged elution order of **2>1>3**. This anomalous effect of **2** was also observed in SS-03 (Table 1) This change in elution order is only observed with pyrethrin I group in MTBE-based SS; this effect was not observed with either the pyrethrin II group or the constitutional alcohols.

Conclusion

CPC is an effective method in separating structurally related compounds. The partition coefficients are dependent on the composition of the solvent system and its interactions with the analytes. Through extensive experimentation in manipulating partition coefficients, we can tailor solvent systems for specific compounds of interest. In this experiment, CPC coupled with optimized solvent systems allowed for baseline separation each natural pyrethrin esters; the minute difference is in the length of the alkyl chain or the degree of unsaturation. Using this method, gram quantities in high purity can be quickly obtained without the use of stabilizing agents, which would affect subsequent environmental studies.

By using a systematic approach to developing countercurrent solvent systems, it is possible to tailor an experiment to meet specific needs. The liquid-liquid systems developed for the lipophilic esters of pyrethrins can also be applied to the separation of other hydrophobic chemicals. A repository of solvent systems (with each providing unique selectivity) is necessary for countercurrent chromatography to be an effective and scalable alternative to solving challenging isolation problems.

References

1. Maciver DR (1995) Constituents of Pyrethrum Extracts. In: Casida JE, Quistad GB (eds.). Pyrethrum Flowers. Production, Chemistry, Toxicology, and Uses. Oxford: Oxford University Press, pp: 108-122.

2. Carlson D (1995) Pyrethrumm Extraction, Refining, and Analysis. In: Casida JE, Quistad GB (eds.). Pyrethrum Flowers, Production, Chemistry, Toxicology, and Uses. New York: Oxford University Press, pp: 97-107.

3. Foucault AP (1994) Theory of Centrifugal Partition Chromatography. Centrifugal Partition Chromatography. Chromatographic Science Series 68: 25-50.

4. Berthod A, Chang CD, Armstrong DW (1994) Operating the Centrifugal Partition Chromatograph. In: Foucault AP (ed). Centrifugal Partition Chromatography. Chromatographic Science Series 68: 1-24.

5. LaForge FB, Barthel WF (1947) Constituents of pyrethrum flowers; the partial synthesis of pyrethrins and cinerins and their relative toxicities. J Org Chem 12: 199-202.

6. Essig K, Zhao ZJ (2001) PMethod development and validation of a high-performance liquid chromatographic method for pyrethrum extract. J Chromatogr Sci 39: 473-480.

7. Rickett FE (1972) Preparative scale separation of pyrethrins by liquid-liquid partition chromatography. J Chromatogr 66: 356-360.

8. Gullick DR, Mott KB, Bartlett MG (2016) Chromatographic methods for the bioanalysis of pyrethroid pesticides. Biomed Chromatogr 30: 772-789.

9. Chen ZM, Wang YH (1996) Chromatographic methods for the determination of pyrethrin and pyrethroid residues in the crops, foods and environmental samples. J Chromatogr A 754: 367-395.

23

Quantification of Newer Anti-Cancer Drug Clofarabine in their Bulk and Pharmaceutical Dosage Form

Brijesh D Patel, Usmangani K Chhalotiya* and Dhruv B Patel

Department of Pharmaceutical Chemistry and Analysis, Indukaka Ipcowala College of Pharmacy, New Vallabh Vidyanagar, Anand, Gujarat, India

Abstract

Quantitative determination of Clofarabine in their Pharmaceutical Dosage form was carried out by Liquid chromatographic method. The aim of the present work was to develop a RP-HPLC method for Clofarabine (CFB). The column used was Luna C-18 column (250 mm × 4.6 mm id, 5 m particle size) using Water: Methanol (60:40 v/v) as a mobile phase at a flow rate of 1.0 ml/min. Quantification was achieved with UV detection at 266 nm over the concentration range 0.05-20 µg/ml and % recovery was found to be in the range of 99.58-100.95 for CFB by the RP-HPLC method. The proposed method was validated with respect to linearity, accuracy, precision, selective, sensitive and robustness. The method was fruitfully applied for the estimation of clofarabine in injection dosage forms.

Keywords: Clofarabine; RP-HPLC; Validation

Introduction

Chemically Clofarabine (CFB) is (2R,3R,4S,5R)-5-(6-amino-2-chloropurin-9-yl)-4-fluoro-2- (hydroxymethyl) oxolan-3-ol [1] with an empirical formula of $C_{10}H_{11}C_{l}FN_{5}O_{3}$, and having molecular weight of 303.67 g/mol (Figure 1) [1]. Intracellularly, clofarabine is metabolized to active 5'-monophosphate metabolite by deoxycytidine kinase and 5'-triphosphate metabolite by mono- and di-phospho-kinases. This metabolite inhibits DNA synthesis through an inhibitory action on ribonucleotide reductase, and by terminating DNA chain elongation and inhibiting repair through competitive inhibition of DNA polymerases which leads to the depletion of the intracellular deoxy-nucleotide triphosphate pool and the self - potentiating of clofarabine triphosphate incorporation into DNA, thereby intensifying the effectiveness of DNA synthesis inhibition. In preclinical models, clofarabine has demonstrated the ability to inhibit1 DNA repair by incorporation into the DNA chain during the repair process. Clofarabine 5'- triphosphate is also disrupts the integrity of mitochondrial membrane, leading to the release of the pro-apoptotic mitochondrial proteins, cytochrome C and apoptosis-inducing factor, leading to programmed cell death [2].

A literature survey revealed that no official methods for estimation of CFB, but there are few reported methods for the estimation of CFB, ultra performance convergence chromatographic stability indicating method for determination of CFB in injections [3].

Clofarabine triphosphate is used in leukemia cell by isocratic HPLC [4]. Estimation of clofarabine triphosphate concentration in mononuclear cells has been done by LC-MS/MS [5]. Simultaneous determination of Fludarabine and Clofarabine in human plasma was done by LC-MS/MS [6].

Literature survey revealed that no HPLC method is reported for estimation of clofarabine in their pharmaceutical dosage form. As per literature review, all the reported methods for estimation of clofarabine were carried out biological sample. The intention of present work was to develop a new, precise, accurate, selective and sensitive method RP-HPLC for estimation of Clofarabine. The proposed method was validated according to ICH guidelines [7,8] and its updated international convention ICH guideline on analytical method validation [9].

Experimental

Apparatus

A Series 200 HPLC system (PerkinElmer, Shelton, CT) equipped with a Series 200 UV detector, Series 200 quaternary gradient pump, Series 200 column oven, manual injector rheodyne valve) with 20 µL fixed loop, Total-Chrom navigator software (Version 6.1.1.0.0:K20), and Luna C18 column (250 mm × 4.6 mm id, 5 µm particle size) was used.

Reagents and materials

Pure samples and marketed sample: Analytically pure powder and formulation of Clofarabine was procured as gratis samples from one of the reputed Pharmaceuticals Limited, India. The purity of reference standard is in the range of 99.60%-101%.

Chemicals and reagents: HPLC grade water, Methanol was purchased from E. Merck (Mumbai, India).

Chromatographic conditions

At ambient temperature, The Luna C18 column was used. Water: Methanol (60:40, v/v) is used as mobile phase and the flow rate were maintained at 1 ml/min. The mobile phase was kept in sonicator for 15 min and was passed through nylon 0.45 µm-47 mm membranes filter, degassed before use. The elution was monitored with UV detector at 266 nm and 20 µL was the injection volume.

HPLC methods were depends upon the nature of the sample (ionic or ionizable or neutral molecule), its molecular weight and solubility. In the direction to optimize the chromatographic conditions the effect of chromatographic variables such as mobile phase, pH, flow rate and solvent ratio were studied. The resulting chromatograms

*Corresponding author: Usmangani K Chhalotiya, Department of Pharmaceutical Chemistry and Analysis, Indukaka Ipcowala College of Pharmacy, Beyond GIDC Phase IV, Vithal Udyognagar, New Vallabh Vidyanagar-388 121, Anand, Gujarat, India, E-mail: usmangani84@gmail.com

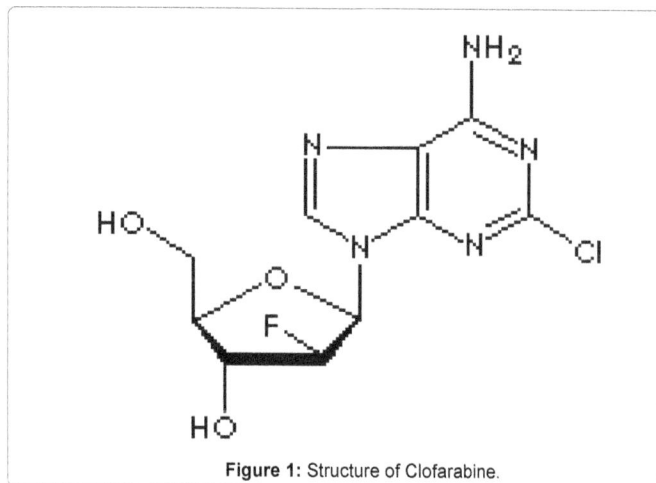

Figure 1: Structure of Clofarabine.

were recorded and the chromatographic parameters such as capacity factor, asymmetric factor, and resolution and column efficiency were calculated. The condition that gave the satisfactory resolution, symmetry and capacity factor was selected for estimation of active drugs.

Preparation of CFB standard stock solutions (100 µg/ml)

Weighed accurately, 25 mg of CFB transferred to a 25 ml volumetric flask containing few ml (10.0 ml) of methanol, swirled to dissolve and diluted up to the mark with methanol to obtain a standard solution of 1000 µg/ml. Pipette out (1.0 ml) of standard solution and transfer it into 10 ml volumetric flask with make up the volume up to the mark with methanol to obtain a Working standard stock solution of 100 µg/ml for the RP- HPLC method.

Preparation of sample solutions

From the above stock solution, appropriate aliquots of CFB working standard solution was transfer into different 10 ml volumetric flasks. The volume was make up to the mark with mobile phase to obtain final concentrations of 0.05, 0.1, 0.5, 1, 5, 10 and 20 µg/ml of CFB, respectively. The solutions were injected using a 20 µL fixed loop system and chromatograms were recorded. Calibration curves were constructed by plotting peak area versus concentrations of the drug and regression equations was computed for CFB.

Method validation

Linearity and range: Calibration curves were constructed by plotting peak areas versus concentrations of CFB, and the regression equations were calculated. The calibration curves were plotted over the concentration range 0.05-20 µg/ml. Aliquots (20 µL) of each solution were injected under the proposed operating chromatographic conditions as described above.

Accuracy (Recovery): The accuracy of the proposed method was determined by calculating % recovery of CFB by the standard addition method. Known amounts of standard solutions of CFB (50, 100 and 150%) were added to prequantified sample solutions of Injection. The amounts of CFB were quantified by applying these values to the regression equation of the calibration curve.

Method precision (repeatability): The precision of the instruments was checked by repeatedly injecting (n=6) solutions of CFB (10 µg/ml) and measure peak area by using proposed RP-HPLC method.

Intermediate precision: By intraday and inter day precision, we can evaluate precision. The intra-day and inter-day precision study of CFB was carried out by estimating the corresponding responses three times on the same day and on three different days for three different concentrations of CFB. The results are reported in terms of percentage relative standard deviation (%RSD).

Robustness: Robustness of the method was studied by observing the stability of the sample solution at 25 ± 2°C for 24 h, change in flow rate at 1 ± 0.1 ml, temperature of working area ± 5°C, and change in mobile phase ratio. Low value of relative standard deviation was indicating that the method was robust.

LOD and LOQ: The detection limit is defined as the lowest concentration of an analyte that can reliably be differentiated from background levels. Limit of quantification is the lowest amount of analyte that can be quantitatively determined with suitable precision and accuracy. LOD and LOQ were calculated using following equation as per ICH guidelines.

LOD=3.3 × σ/S, LOQ=10 × σ/S

Where σ is the standard deviation of y-intercepts of regression lines and S is the average slope of the calibration curves.

Specificity: Specificity is the ability to assess unequivocally the analyte in the presence of components, which may be expected to be present. The specificity was estimated by spiking commonly used excipients (starch, talc and magnesium stearate) into a pre weighed quantity of drug. The chromatogram was taken by appropriate dilutions and the quantities of drugs were determined.

System suitability: A system suitability test was an integral part of the method development to verify that the system is adequate for the quantification of CFB to be performed. The suitability of the chromatographic system was demonstrated by comparing the obtained parameter values with the acceptance criteria of the U.S. Food and Drug Administration, Center for Drug Evaluation and Research guidance document (U.S. Food and Drug Administration). A system suitability test of the chromatography system was performed before each validation run. Six replicate injections of a system suitability/calibration standard and one injection of a check standard were made. Area, retention time (RT), tailing factor, asymmetry factor, and theoretical plates for the six suitability injections were determined.

Stability of standard and sample solutions: Stability of standard and sample solution of CFB was evaluated at room temperature for 48 hr at 2 hr time interval measure the peak area of standard and sample solution. The relative standard deviation was found below 2.0%. It showed that both standard and sample solution were stable up to 48 hr at room temperature.

Determination of CFB in injection: The injection has the strength of 1 mg/ml solution. From marketed injection, 1.0 ml of aliquot was pipette out and transferred into 10 ml volumetric flask and then volume was adjusted upto the mark with mobile phase to obtain concentration of 100 µg/ml CFB. From above solution, 1.0 aliquot ml was transferred to another 10 ml volumetric flask and volume was adjusted up to the mark to obtain 10 µg/ml CFB. The solution was sonicated for 10 min. Solution were injected as per the above chromatographic condition and the peak areas were measure and quantification was carried by keeping these values to straight line equation of calibration curve.

Results and Discussion

Optimization of the chromatographic conditions

HPLC method: Optimization of mobile phase was performed based on resolution of the drug, asymmetric factor and theoretical plates obtained for CFB. Water: Methanol (60:40, v/v) was selected as a mobile phase at a flow rate of 1.0 ml/min was found to be satisfactory and gave well resolved and symmetric peak for CFB. The retention time for CFB was 5.39 min. For the selection of detection wavelength, the spectrum of 10 μg/ml CFB solution revealed that, at 266 nm the drug possesses significant absorbance. So considering above fact, 266 nm was selected as a detection wavelength for estimation of CFB using HPLC. Complete resolution of the peaks with clear baseline separation was obtained (Figure 2). The system suitability test parameters are shown in Table 1. Analytically pure CFB was identified by melting point study and infrared spectroscopy study.

Validation of the proposed methods: The developed method was validated, as described below, for various parameters like linearity and range, accuracy, precision, ruggedness, system suitability, specificity, LOQ, and LOD.

Linearity and range: Linearity of the method was evaluated at seven concentration levels by diluting the standard stock solution to give solutions in the range of 0.05-20 μg/ml. The calibration curve for CFB was prepared by plotting area v/s concentration. Calibration data for CFB was shown in Table 2. The linearity plot of CFB was found to be linear with the linear equation y=60725x+26878 and correlation coefficient 0.998 for CFB. Linearity was observed in the expected concentration range, demonstrating suitability of the method for analysis (Figure 3). This indicates that the method is linear in the specified range for the analysis of CFB in dosage form.

Accuracy: The recovery experiments were carried out by the standard addition method. The method was found to be accurate with % recovery 99.58%-100.95% the recoveries were obtained by the proposed RP-HPLC method for estimation CFB are shown in Table 3.

Precision: Precision was calculated as repeatability and intraday and interday precision for estimation of CFB. The method was found to be precise with %RSD 0.59-1.37 for intraday (n=3) and %RSD 0.87-1.77 for interday (n=3) for CFB. The low value of %RSD (i.e., NMT 2%) has observed for the three level results (Table 4) hence it concluded that the method is precise for the analysis of CFB.

Specificity (Placebo interference): There is no interference of mobile phase, solvent and placebo with the analyte peak and also the peak purity of analyte peak which indicate that the method is specific for the analysis of CFB in their dosage form.

Robustness: The method was found to be robust, as small but deliberate changes in the method parameters have no detrimental effect on the method performance as shown in Table 5. The low value of relative standard deviation was indicating that the method was robust.

Standard and sample solution stability: Standard and Sample solution stability was evaluated at room temperature for 48 hr. The relative standard deviation was found below 2.0%. It showed that both standard and sample solutions were stable up to 48 hr at room temperature.

LOD and LOQ: These data show that the method is sensitive for

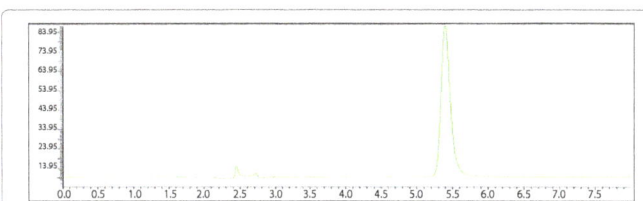

Figure 2: Chromatogram of CFB (RT 5.39 min) on C_{18} Luna column using Water: Methanol (60:40, v/v) as the mobile phase.

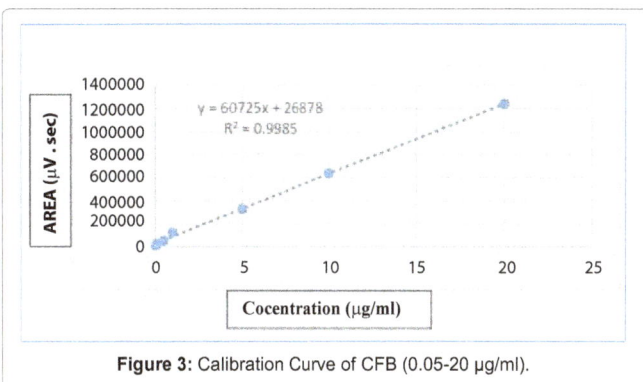

Figure 3: Calibration Curve of CFB (0.05-20 μg/ml).

Parameter	Clofarabine
Retention times (R$_T$)	5.39 Min
HPLC Plate Count	8292
Tailing factor	1.14
Capacity factor	1.20
Base width (sec)	14.22

Table 1: System suitability test parameters for CFB the proposed HPLC method.

Parameter	Clofarabine
Linearity (μg/ml)	0.05-20
Correlation co-efficient (r)	0.998
Slope of Regression (S)	60725
Intercept of Regression	26878
Standard deviation of slope	1080.58
Standard deviation of intercept	2471.16

Table 2: Regression analysis of CFB by proposed HPLC method.

the determination of CFB. The LOD and LOQ were carried out by visual method 0.01 and 0.05 μg/ml, respectively.

Analysis of a formulation: The proposed method was applied for the determination of CFB in injection of Clolar. The percentage amount of drugs was found to be 99.59% of the label claim for the formulation. The results of the assay indicate that the method is selective for the assay of CFB without interference from excipients used in the injection (Table 6).

Conclusions

The proposed validated RP-HPLC analytical method has been developed for the quantification of CFB in bulk and injectable dosage form. Validation of method undertaken according to the ICH guidelines revealed that the method is selective. The proposed method is sensitive, accurate, precise, and specific. The method is suitable for the routine analysis of CFB in injection and other pharmaceutical dosage form.

Amount of Sample (µg/ml)	Sets	Amount drug of spiked (µg/ml)	Area (n=3)	Average amount recovered (µg/ml)	% Recovery	Mean % Recovery	% RSD[c]
5	1	0	329568.26	4.97	99.69	99.58	1.13
	2	0	332514.87		100.66		
	3	0	325649.82		98.40		
5	1	2.5	481698.54	7.48	99.79	99.71	0.64
	2	2.5	479365.96		99.03		
	3	2.5	483245.71		100.30		
5	1	5	632569.41	10.04	99.48	100.95	1.65
	2	5	635964.09		100.60		
	3	5	642568.55		101.78		
5	1	7.5	791563.67	12.51	101.85	100.29	1.47
	2	7.5	786354.22		100.13		
	3	7.5	782589.38		98.89		

[c]RSD=Relative standard deviation

Table 3: Data derived from accuracy of Clofarabine by proposed HPLC method.

Parameters	Clofarabine
LOD (µg/ml)[d]	0.01
LOQ (µg/ml)[e] (n=5)	0.05
Accuracy, %	99.58-100.95
Repeatability, (% RSD, n=6)	1.27
Precision (% RSD) Interday (n=3)	0.87-0.1.77
Intraday (n=3)	0.59-1.37

[d]LOD=Limit of detection
[e]LOQ=Limit of quantitation

Table 4: Summary of validation parameters.

Parameters	Normal condition	Change in condition	Change in % RSD of Peak area
Flow Rate	1.0 ml/min	0.8 ml/min	1.45
		1.2 ml/min	1.30
Detection wavelength	266	264	1.36
		268	1.80
Mobile phase ratio	60:40	58:42	1.21
		62:38	1.41

Table 5: Robustness study of Clofarabine by proposed HPLC method.

Injection	Concentration (µg/ml)	Amount recovered (µg/ml)	Amount of Drug Found ± SD[a] (n=3), %
A	10	9.95	99.59 ± 0.88

[a]SD=Standard deviation

Table 6: Assay Results of marketed formulation.

Acknowledgements

The authors are thankful to reputed pharmaceuticals Ltd., India for providing gift sample of Clofarabine. The authors are very thankful to Indukaka Ipcowala College of Pharmacy, New Vallabh Vidyanagar, Anand, Gujarat, India for providing necessary facilities to carry out research work.

References

1. Chemical and Physical properties of Clofarabine (2004) Chem Spider.

2. De Gennero LJ, Raetz E (2014) Acute Lymphoblastic Leukemia.

3. Venkata NR, Jalandhar D, Gnanadey G, Bandari R, Manoi P (2013) Development of supercritical Fluid (Carbon dioxide) based ultra-performance convergence chromatographic stability indicating assay method for determination of clofarabine in injection. Analytical methods 5: 7008-7013.

4. Takahia Y, Yamauchi T, Rie N, Takanori U (2011) Determination of Clofarabine triphosphate concentration in leukemia cell using sensitive, isocratic High Performance Liquid Chromatography. Anticancer research 31: 2863-2867.

5. Xiowei TU, Youming LU, Xioyan CH, Dafang ZH, Yifan ZN (2014) A sensitive LC-MS/MS method for quantifying Clofarabine triphosphate concentration mononuclear cells. Journal of chromatography B 1964: 202-207.

6. Huang L, Lizak P, Dvorak C, Long-Boyle J (2014) Simultaneous determination of Fludarabine and Clofarabine in human plasma by LC-MS/MS. Journal of Chromatography B Analyt Technol Biomed Life Sci 960: 194-199.

7. Validation of Analytical Procedures: Text and Methodology (1996) International Conference on Harmonization (ICH), Geneva, Switzerland.

8. Guideline on Analytical Method Validation (2002) International Conference on Harmonization. International Convention on Quality for the Pharmaceutical Industry, Toronto, Canada.

9. ICH Guidelines (2005) Validation of Analytical Procedures: Text and Methodology. Geneva, Switzerland.

24

Method Validation for Stability Indicating Method of Related Substance in Active Pharmaceutical Ingredients Dabigatran Etexilate Mesylate by Reverse Phase Chromatography

Dare M[1], Jain R[2*] and Pandey A[2]

Department of chemistry, Dr. Hari Singh Gour University Sagar India, Department of Analytical development API Micro Labs Limited-API Banglore 560001, India

Abstract

To develop the stability indicating, highly accurate precise and linear method for related substance of the Dabigatran through the reverse phase high performance liquid chromatography and it is validated as per the current ICH guideline. The optimize method uses a reverse phase column, Poroshell 120 EC -18 (150 mm × 4.6 mm, 2.7µ), Mobile Phase of hexane-1 Sulfonic acid sodium salt monohydrate (6.5 ± 0.05) and Methanol through gradient flow rate of 0.6ml/min. Keeping the column at temperature 30°C and using the sample amount 10 µL at 5°C and detected all the impurities at 230 nm by the UV detector. In the developed method, elution of Dabigatran was at 26.9 min and all the eluted Impurities were well separated and met the system suitability criteria. The precision is exemplified by relative standard deviation of 0.40%. Method percentage mean recoveries of all the impurities are within the range (90.0% to 115.0%) as per the protocol. Resolution results show that the method is robust. Linearity coefficient for all Impurities is more than 0.999. The LOD obtained is 0.01% and LOQ is 0.03%.

A new highly accurate, precise and stability indicating method was developed for related substance of Dabigatran with all the impurities well separate from the degradation product of Dabigatran. It is ready to use for the routine analysis of related substance of Dabigatran in Pharmaceutical industry.

Keywords: Dabigatran etexilate mesylate; HPLC; Method development; Method validations; ICH

Abbreviations: DBA: Dabigatran Etexilate Mesylate; ICH: International committee of Harmonization; HPLC: High Performance Liquid Chromatography; Impurity-A: 3-(2-((4-Amidinophenylamino) methyl)-1-methyl-N-(pyridin-2-yl)-1H-benzo[d]imidazole-5-carboxamido)propanoic acid; DAB-2B: Ethyl 3-(2-((4-amidinophenylamino)methyl)-1-methyl-N-(pyridin-2-yl)-1H-benzo[d]imidazole-5-carboxamido)propanoate; Impurity-C : Ethyl 3-[[[2-[[[4-[[[(ethylloxy)carbonyl]amino]iminomethyl]phenyl] amino]methyl]-1-methyl-1H-benzimidazol-5-yl]carbonyl](pyridin-2-yl)amino] propanoate; Impurity-D: 3-(2-(((4-(N-((hexyloxy) carbonyl)carbamimidoyl)phenyl)amino)methyl)-1-methyl-N-(pyridin-2-yl)-1H-benzo[d]imidazole-5-carboxamido)propanoic acid; Impurity-E: Hexyl ((4-(((5-((3amino-3-oxopropyl) (pyridin-2-yl)carbamoyl)-1-methyl-1H-benzo[d]idazol-2-yl) methyl)amino)phenyl)(imino)methyl)carbamate; Impurity-G: Ethyl 2-(((4-(N-((hexyloxy)carbonyl)carbamimidoyl)phenyl) amino)methyl)-1-methyl-1H-benzo[d]imidazole-5-carboxylate; Impurity-H: 2-(((4-(N-((hexyloxy)carbonyl)carbamimidoyl) phenyl)amino)methyl)-1-methyl-1H-benzo[d]imidazole-5-carboxylic acid; Impurity-J: Ethyl 3-(2-(((4-(((hexyloxy) carbonyl)carbamoyl)phenyl)amino)methyl)-1-methyl-N-(pyridin-2-yl)-1H-benzo[d]imidazole-5-carboxamido)propanoate; Impurity-K: Ethyl 3-[[[2-[[[4-[[[(pentyloxy)carbonyl]amino]iminomethyl]phenyl] amino]methyl]-1-methyl-1H-benzimidazol-5-yl]carbonyl](pyridin-2-yl)amino]propanoate

Introduction

Dabigatran is a newly available oral direct thrombin inhibitor approved for anticoagulation therapy to prevent strokes in patients with non-valvular arterial fibrillation. It is being studied for various clinical indications and in some cases it offers an alternative to Warfarin as the preferred orally administered anticoagulant ("blood thinner") because it does not require frequent blood tests

for international normalized ratio (INR) monitoring while offering similar results in terms of efficacy. Dabigatran Etexilate chemically Ethyl 3-(1-{2-[(({4[Amino({[(Hexyloxy)Carbonyl] Imino})Methyl] Phenyl}Amino)Methyl]-1-Methyl-1H-1,3-Benzodiazol-5-Yl}-N-(Pyridin-2 Yl)Formamido)Propanoate (Figure 1).

Dabigatran (then compound BIBR 953) was discovered from a panel of chemicals with similar structure to benzamidine-basedthrombin inhibitor α-NAPAP (N-alpha-(2-naphthylsulfonylglycyl)-4-amidinophenylalanine piperidide), since 1980s this panel of chemicals had been known as powerful inhibitors of various serine proteases, specifically of thrombin, but also of trypsin [1-3].

The most commonly reported side effect of dabigatran is GI upset. When compared to people anticoagulated with warfarin, patients taking dabigatran had fewer life-threatening bleeds, fewer minor and major bleeds, including intracranial bleeds, but the rate of GI bleeding was significantly higher. Dabigatran capsules contain tartaric acid, which lowers the gastric pH and is required for adequate absorption. The lower pH has previously been associated with dyspepsia; some hypothesize that this plays a role in the increased risk of gastrointestinal

*Corresponding author: Jain R, Department of chemistry, Dr. Hari Singh Gour University Sagar India, Department of Analytical development API Micro Labs Limited-API Banglore 560001, India, E-mail: roli.92@rediffmail.com

bleeding [4-10].

Dabigatran has about nine reported impurities chemicals. The chemical structure and the origin of impurities reported under Table 1 [11-13]. Any impurity is defined as a substance coexisting with the original drug, such as starting material or intermediates or that is formed, due to any side reactions. Analysis of impurities in drug substance plays a key role in the novel journey of any drug substance in the pharmaceutical industry. Several analytical method have been reported in literature for the Assay determination of Dabigatran through different analytical methods but no literature reveals the accurate and precise method for the determination of related substances by the RP-HPLC [14-16].

Reverse Phase liquid chromatography has been proven as a versatile, sensitive, reproducible and highly precise method for its ability to separate the complex mixture of Drug Substances with impurities and its easy handling [17,18]. All these advantages of RP-HPLC make this the first choice of modern chemists. So in this scenario a reproducible and accurate method of analysis with proper documented validation give huge support to Pharmaceutical industry.

Material and Methods

A Sample of Dabigatran etexilate mesylate is provided by Micro labs Bangalore with purity of more than 99.0%. All the used Impurities are as per the reference standard. All the solvent and reagent those are used during the analysis are highly pure and HPLC graded. Instrument HPLC with UV/PDA Detector Shimadzu LC-2010CHT, HPLC with UV Detector Waters Alliance Analytical Balance, Micro Balance, PH meter were used during the analysis. All instruments were calibrated prior to Method development and validation.

Buffer preparation

1.88 g hexane-1 sulfonic acid sodium salt monohydrate was dissolved in 1000 mL of water. Add 1 mL of triethylamine to it and adjust the pH to 6.5 ± 0.05 with diluted orthophosphoric acid. Filter through 0.45 μ filter paper.

Time	0.01	15	20	35	45	46	55
% of Mobile phase A	60	30	20	15	15	60	60
% of Mobile phase B	40	70	80	85	85	40	40

Mobile Phase preparation

Mobile phase A prepared as degas Mixture of 20% of methanol and 80% Buffer and Mobile phase B prepared as a degas Mixture 80% methanol and 20% Buffer and both mobile phases use as per the below gradient for obtaining better separation . Poroshell 120 EC -18 (150 mm × 4.6 mm, 2.7μ) column is used with flow rate 0.6 ml/min. Keep the column at 30°C and using the sample amount 10 μL at 5°C and detected all the impurities at 230 nm by the UV detector on the HPLC.

System suitability solution

Accurately weighed about 1 mg each of DAB reference/working standard and DAB Impurity-J reference standard and transferred into a 100 mL volumetric flask, added about 25 mL of diluent, dissolve the sample and make the volume up to 100 ml with diluent. Shake well to mix.

Sample solution

Accurately weigh about 20 mg of sample and transfer into a 100 mL volumetric flask. Add about 25 mL of diluent to it, dissolve and dilute to volume with diluent. Shake well to mix.

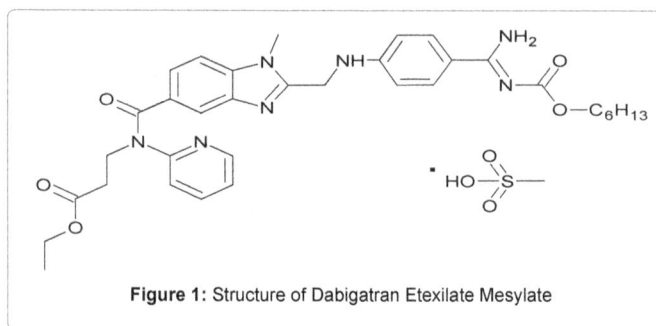

Figure 1: Structure of Dabigatran Etexilate Mesylate

Procedure

The system was Equilibrate until stable baseline was achieved. Inject 2 injection of Blank (Diluent), single injection of system suitability solution and single injection of sample solution. Record the chromatogram and report the result by area normalization method. Eliminate the peaks due to blank. Measure the system suitability criteria. The resolution between DAB and DAB Impurity-J observed in the chromatogram of system suitability solution should be not less than 2.5. Above method was finalized after lots of trial and this method would be validated on different parameters as per the ICH guideline.

Validation procedure and result are described under the result and discussion section.

Method Validation

Solution stability

Prepared blank, (as diluents) system suitability solution as per the method and spiked all the impurities which are mentioned in Table 1 as per the specification label. Sample solution was preparing freshly and injected. This study was repeated at the interval of 6 hours till 24 hours to measure the solution stability at 5°C temperature.

Specificity

Specificity is the ability to assess unequivocally of analytic in the presence of components which may be expected to be present. In this study Specificity has been performed as interference study and forced degradation study using HPLC system with PDA detector. For the interference study of DAB drug substance Prepared and injected blank, system suitability solution and spiked all the Impurities as per specification label. Checked the system suitability criteria and recorded the observations.

Force degradation

It is a measurement of specificity and it's also help to proper selection of stability-indicating analytical procedures. A literature revel the Stress studies should be done in different pH solutions, in the presence of oxygen and light, and at elevated temperatures and humidity levels to determine the stability of the drug substance. These stress studies are conducted on a single batch. This information will in turn help improve the formulation manufacturing process and determine the storage conditions. As no specific set of conditions is applicable to all drug products and drug substances and the regulatory guidance does not specify about the conditions to be used, this study requires the experimenter to use common sense [13]. In our study, stress conditions were performed in presence of Acid, Alkali, Oxygen and light (Photolytic), temperature/humidity and thermal conditions.

S.No	Impurity Code	Structure	Chemical Name	Remarks
1	DAB Impurity -A		3-(2-((4-Amidinophenylamino)methyl)-1-methyl-N-(pyridin-2-yl)-1H-benzo[d]imidazole-5-carboxamido)propanoic acid	Process related , and **Metabolite**
2	DAB Impurity -B		Ethyl 3-(2-((4-amidinophenylamino)methyl)-1-methyl-N-(pyridin-2-yl)-1H-benzo[d]imidazole-5-carboxamido)propanoate	Process related, **Degradation** (oxidative) and **Metabolite**
3	DAB Impurity -C		Ethyl 3-[[[2-[[[4-[[[(ethylloxy)carbonyl]amino]iminomethyl]phenyl]amino]methyl]-1-methyl-1H-benzimidazol-5-yl]carbonyl](pyridin-2-yl)amino]propanoate	Process related
4	DAB Impurity -D		3-(2-(((4-(N-((hexyloxy)carbonyl)carbamimidoyl)phenyl)amino)methyl)-1-methyl-N-(pyridin-2-yl)-1H-benzo[d]imidazole-5-carboxamido)propanoic acid	Process related , **degradation** (acid,base) and **Metabolite**
5	DAB Impurity -E		Hexyl ((4-(((5-((3amino-3oxopropyl)(pyridin-2yl)carbamoyl)-1methyl-1H-benzo[d]idazol2yl)methyl)amino)henyl)(imino)methyl)carbamate	Process related
6	DAB Impurity -G		Ethyl 2-(((4-(N-((hexyloxy)carbonyl)carbamimidoyl)phenyl)amino)methyl)-1-methyl-1H-benzo[d]imidazole-5-carboxylate	Process related

7	DAB Impurity -H		2-(((4-(N-((hexyloxy)carbonyl) carbamimidoyl)phenyl)amino)methyl)-1-methyl-1H-benzo[d]imidazole-5-carboxylic acid	Process related
8	DAB Impurity -J		Ethyl 3-(2-(((4-(((hexyloxy)carbonyl) carbamoyl)phenyl)amino)methyl)-1-methyl-N-(pyridin-2-yl)-1H-benzo[d] imidazole-5-carboxamido)propanoate	Process related and Degradation (Temp.& humidity)
9	DAB Impurity -K		Ethyl 3-[[[2-[[[4-[[[(pentyloxy)carbonyl] amino]iminomethyl]phenyl]amino] methyl]-1-methyl-1H-benzimidazol-5-yl]carbonyl](pyridin-2-yl)amino] propanoate	Process related

Table 1: Origin and pathway of DBA impurities

Limit of detection

The detection limit of an individual analytical procedure is the lowest amount of analyte in a sample which can be detected but not necessarily quantitated as an exact value. The detection limit is determined by the analysis of samples with known concentrations of analyte and by establishing the minimum level at which the analyte can be reliably detected. The detection limit (DL) may be expressed as [19-21]:

$$DL = \frac{3.3\ \sigma}{S}$$

Where σ = the standard deviation of the response

S = the slope of the calibration curve

The slope S may be estimated from the calibration curve of the analyte. The estimate of σ may be carried out in a variety of ways.

Limit of quantization

The quantitation limit of an individual analytical procedure is the lowest amount of analyte in a sample which can be quantitatively determined with suitable precision and accuracy. LOQ determined by several methods [22,23].

The detection limit (DL) may be expressed as:

$$QL = \frac{10\ \sigma}{S}$$

Where σ = the standard deviation of the response

S = the slope of the calibration curve

The slope S may be estimated from the calibration curve of the analyte. The estimate of σ may be carried out in a variety of ways.

In presented study author Determined the limit of quantification (LOQ) of DAB (un-known impurity) by the analysis of samples with known concentrations of analyte by establishing the concentration which will give a concentration of about 3.0 times of LOD with an RSD of 10% for peak area of DAB peak. The % RSD of 6 replicate injection of spiked sample is less than 4%.

Linearity

A linear relationship should be evaluated across the range (see section 3) of the analytical procedure. It may be demonstrated directly on the drug substance (by dilution of a standard stock solution) and/or separate weighing of synthetic mixtures of the drug product components, using the proposed procedure. The latter aspect can be studied during investigation of the range. As per the ICH guideline Linearity established for all the Impurities and for DAB form different concentration levels that are start form LOQ to 50%, 80%, 100%, 120%, 150% of Impurities as per their shelf specification level [24,25].

Accuracy

The accuracy of an analytical procedure expresses the closeness of agreement between the value which is accepted either as a conventional true value or an accepted reference value and the value found. In the present study author check the accuracy for all the Impurities as well as DAB on different specification level that are LOQ, 50%, 100% and

150%. All the preparation of test solution were prepared in triplicate (n=3).

Precision

It was performed by carrying out the analysis of the six homogenous solutions of same test sample and content of impurities was calculated. The determinations were carried out one after the other under conditions as similar as possible.

Intermediate precision

The intermediate precision of the method was checked by determining precision on a different instrument, analysis being performed in on a different day. The % content of all DAB impurities and total impurities was found to be below quantification level even when it is performed on a different instrument. The method is said to be precise with respect to the criteria of the intermediate precision.

Robustness

According to ICH, the robustness of an analytical procedure is a measure of its capacity to remain unaffected by small but deliberate variations in method parameters and provides an indication of its reliability during normal use. To establish the robustness of method here, make the changes in HPLC method as per the ICH guidelines. These changes are listed below. Compare the resolution of all Impurities in each condition and data are listed under Table 10

Condition (1): Altered the mobile phase flow by ± 0.1 mL/ minute

a. Altered the mobile phase flow to 0.5 mL instead of 0.6 mL.

b. Altered the mobile phase flow to 0.7 mL instead of 0.6 mL.

Condition (2): Altered the column oven temperature by ± 2°C

a. Altered the column oven temperature to 28°C instead of 30°C.

b. Alter the column oven temperature to 32°C instead of 30°C.

Condition (3): Altered the column with different lot number.

Condition (4): Altered the pH of the buffer solution by ± 0.2

c. Altered the pH of the buffer solution to 6.30 instead of 6.50.

d. Altered the pH of the buffer solution to 6.70 instead of 6.50

Result and Discussion

The present investigation optimum result was obtained under the prescribed chromatographic condition. Retention time of DAB was obtained around at 26.97 min that is well separated form eluting peak of DAB Impurity J. Obtained resolution between DAB and DAB impurity J is 10.02 that are more than required system suitability criteria under the procedure (Figure 2).

The Solution stability results indicate that system suitability solution, sample solution is stable at 5°C up to 18 hours from the time of preparation. All obtained result summarize in Table 2. Specificity result indicate no interference observed at the retention times of known impurities due to blank. Observed purity threshold is more than purity angle for all known impurities and DAB peak in the spiked sample solution All the Result of peak purity shown under (Table 2, Figure 3). So it is concluded the method is specific and selective.

The selection of stress study for DAB is given in (Tables 3 and

3.1). The acceptation criteria is, it should meet the system suitability requirement as per the method as well as percent degradation should be between 5 to 30% and Peak purity of main compound and known impurity peaks should pass. The peak purities of all known impurity peaks and DAB peak in each degradation condition are meeting the criteria. It indicates that there is no co-elution at the retention times of any of the known impurities and DAB peak. Dabigatran Etexilate Mesylate drug substance shows significant degradation in basic, oxidative conditions and slight degradation shown in acid degradation.

In the study of limit of detection injected spiked solution of impurities, we observed the chromatograms and conclude the System suitability criteria of DAB peak is pass. The result of LOD is given in Table 4.

In presented study author Determined the limit of quantification (LOQ) of DAB (un-known impurity) by the analysis of samples with known concentrations of analyte by establishing the concentration which will give a concentration of about 3.0 times of LOD with a RSD of 10% for peak area of DAB peak. The %RSD of 6 replicate injection of spiked sample is less than 4% , all the results of LOD is tabulated in Table 4. The results of LOQ (LOQ for all known impurities and DAB peak is 0.03%) determination indicate that the method is capable of quantitatively determining analytic in a sample with suitable precision and accuracy.

The correlation coefficient obtained was greater than 0.999 for all the impurities and for DAB. This shows that an excellent correlation existed between the peak area and the concentration of impurity A, impurity B, impurity C, impurity D, impurity E, Impurity G, Impurity H, Impurity J, Impurity K and DAB. Calculate the relative response factor form the linearity curve that is shown in Table 5.

Linearity result shows in Table 6 that indicate Correlation co-efficient is not be less than 0.99 for all known impurities and Dabigatran Etexilate Mesylate peak .The % y-intercept as obtained from the linearity data in range of ± 3 for all impurities as well as for the Dabigatran Etexilate Mesylate . So results indicate that the method has ability to generate responses which are directly proportional to the concentration of analytes present in the sample (Figure 4). The Accuracy result is shown in Table 7. All the results indicate that the method has expressed the closeness of agreement between observed value and true value. Hence, Accuracy has been established across the specified range (LOQ to 150%) of analytical method.

The relative standard deviation was calculated during the study of precision and all the obtained observations given under the Table 8 that concludes that method is precise.

Under the intermediate precision calculate the % content of all DAB impurities and total impurities was found to be below quantification level even when it is performed on a different instrument. The method is said to be precise with respect to the criteria of the intermediate precision. Obtained result shown in Table 9.

The results of robustness study indicate that the method is capable of demonstrating its capacity to remain unaffected by small variations in method parameters and provided an indication of its reliability during normal usage. All the result of robustness study summarized under the Table 10.

Conclusion

The analytical method used for determination of related substances in Dabigatran Etexilate Mesylate complies with the acceptance criteria

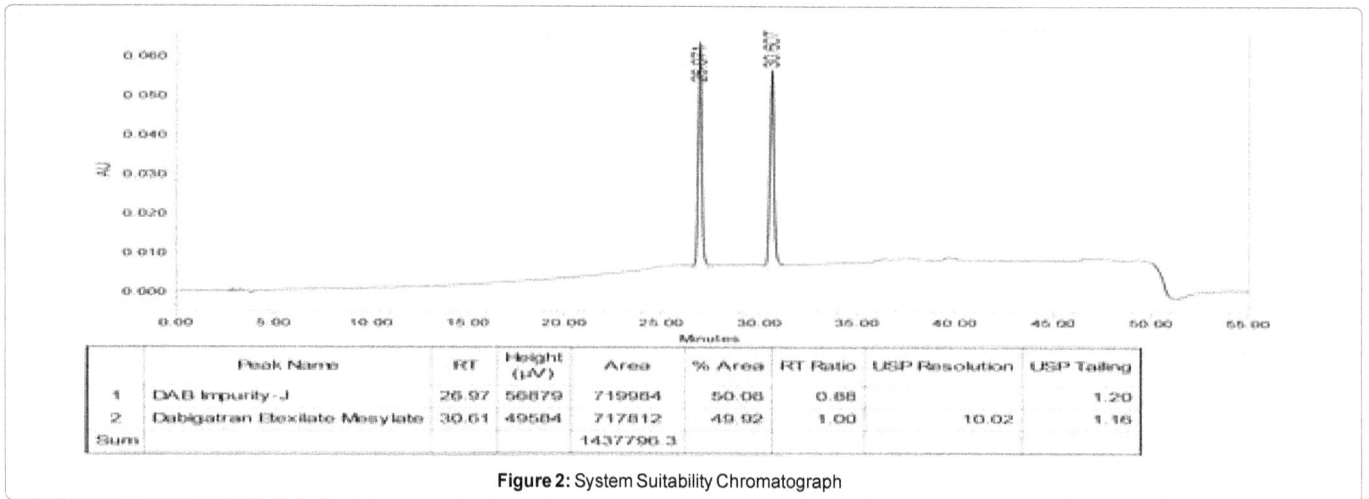

Figure 2: System Suitability Chromatograph

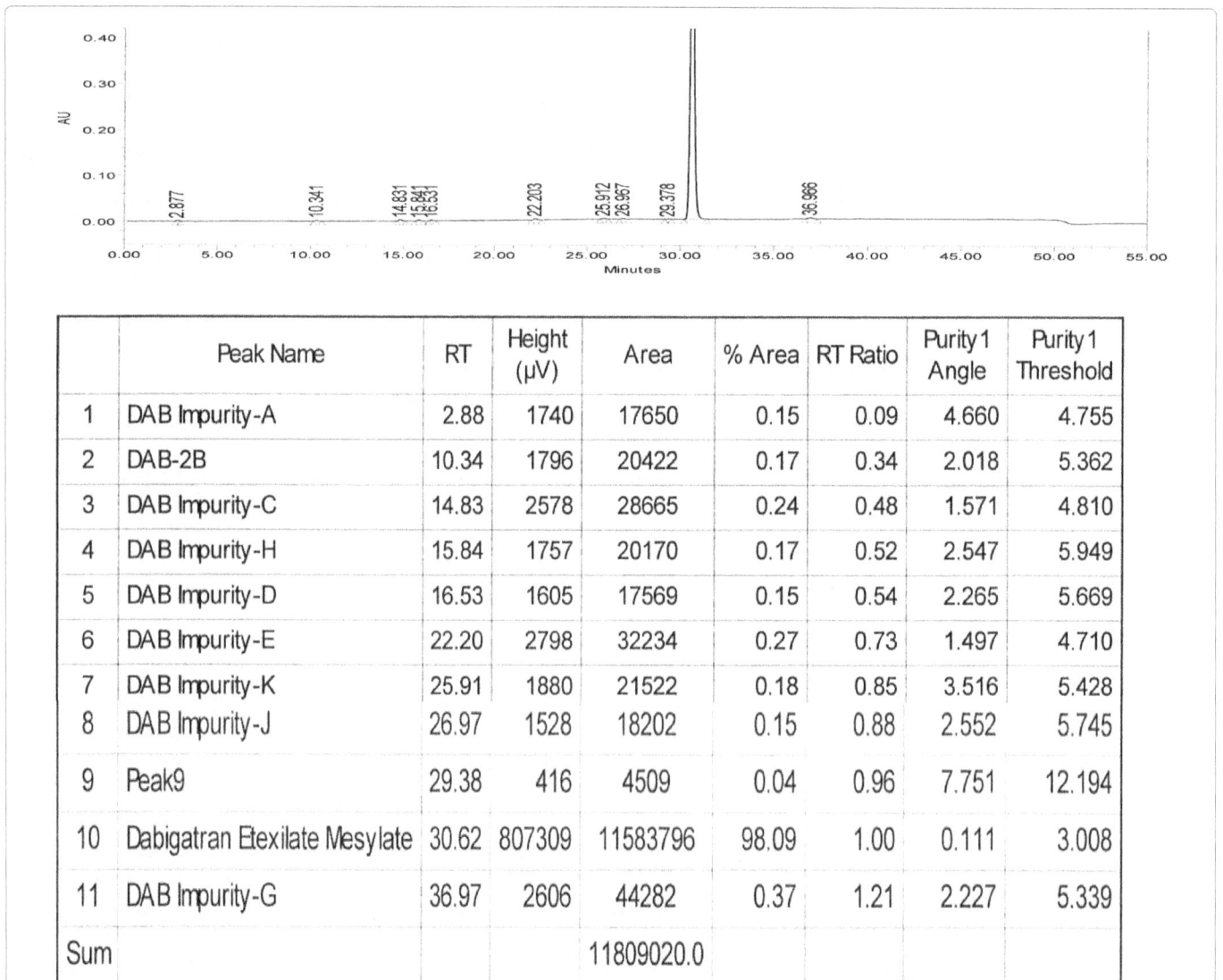

	Peak Name	RT	Height (µV)	Area	% Area	RT Ratio	Purity 1 Angle	Purity 1 Threshold
1	DAB Impurity-A	2.88	1740	17650	0.15	0.09	4.660	4.755
2	DAB-2B	10.34	1796	20422	0.17	0.34	2.018	5.362
3	DAB Impurity-C	14.83	2578	28665	0.24	0.48	1.571	4.810
4	DAB Impurity-H	15.84	1757	20170	0.17	0.52	2.547	5.949
5	DAB Impurity-D	16.53	1605	17569	0.15	0.54	2.265	5.669
6	DAB Impurity-E	22.20	2798	32234	0.27	0.73	1.497	4.710
7	DAB Impurity-K	25.91	1880	21522	0.18	0.85	3.516	5.428
8	DAB Impurity-J	26.97	1528	18202	0.15	0.88	2.552	5.745
9	Peak9	29.38	416	4509	0.04	0.96	7.751	12.194
10	Dabigatran Etexilate Mesylate	30.62	807309	11583796	98.09	1.00	0.111	3.008
11	DAB Impurity-G	36.97	2606	44282	0.37	1.21	2.227	5.339
Sum				11809020.0				

Figure 3: Chromatograph shown the Specificity as spiked all the impurities in sample solution

Linearity curve for DAB

$$y = 145701x - 298.16$$
$$R^2 = 0.9832$$

Figure 4: Linearity Curve of DAB

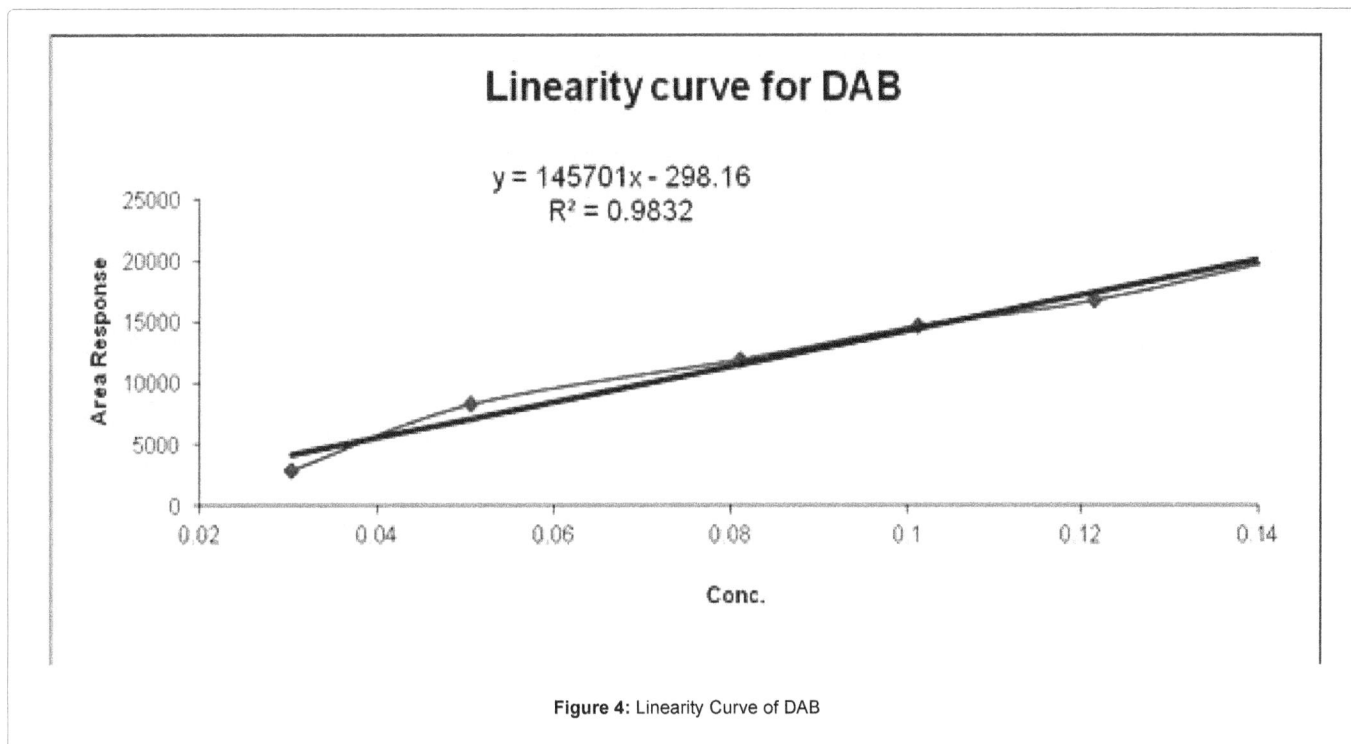

Name of component	Initial	After 6 hours	After 12 hours	After 18 hours	After 24 hours
DAB Impurity-A	-	-0.72	-1.24	-0.68	4.42
DAB-2B	-	0.95	-1.70	-0.45	-0.90
DAB Impurity-C	-	0.00	0.98	-0.24	-1.14
DAB Impurity-H	-	-0.51	-0.53	-1.05	-0.41
DAB Impurity-D	-	0.44	-1.36	0.66	-0.27
DAB Impurity-E	-	-0.91	-0.43	-0.97	-1.30
DAB Impurity-K	-	0.18	-0.60	-0.66	-0.01
DAB Impurity-J	-	-2.34	-7.19	-8.60	-15.89
DAB Impurity-G	-	1.75	1.88	-1.24	-2.61
DAB	-	0.01	0.01	0.03	0.03

Table 2: Summarize the result study of Solution stability of impurities in sample solution.

S. No.	Name of impurity		Conditions for degradation		
		Control sample (%)	Acid degradation (1N HCl / at RT / 1Hour)	Base degradation (1N NaOH / at RT/ 2 hours)	Oxidative degradation (3%v/v H_2O_2 / at RT / 12 Hours)
1	DAB Impurity-A	ND	ND	0.31	
2	DAB-2B	ND	0.01	0.01	1.93
3	DAB Impurity-C	0.08	0.08	0.06	0.08
4	DAB Impurity-H	ND	0.04	0.04	0.25
5	DAB Impurity-D	ND	4.55	9.37	0.04
6	DAB Impurity-E	0.09	0.11	0.09	0.77
7	DAB Impurity-K	0.01	0.01	0.01	0.01
8	DAB Impurity-J	ND	0.03	0.01	0.27
9	DAB Impurity-G	0.04	0.05	0.03	0.71
10	Total unknown impurities	0.08	0.12	0.14	1.30
11	Total impurities	0.30	4.99	10.07	5.36
12	% of Degradation	NA	4.70	9.80	5.07

Table 3: Result of Force Degradation study in Dabigatran etexilate mesylate

S. No.	Name of impurities	Photolytic degradation (%) (1.2Million lux hours & 200 WH/m2)	Temperature/ Humidity degradation (%) (60°C ± 2 and 80%RH ± 5% for 48 hours)	Thermal degradation (%) (105°C±2 for 24 hours)
1	DAB Impurity-A	ND	ND	ND
2	DAB-2B	ND	0.04	0.21
3	DAB Impurity-C	0.09	0.08	0.10
4	DAB Impurity-H	ND	ND	ND
5	DAB Impurity-D	0.02	ND	ND
6	DAB Impurity-E	ND	0.09	0.11
7	DAB Impurity-K	0.01	0.01	ND
8	DAB Impurity-J	0.02	1.32	0.02
9	DAB Impurity-G	0.06	0.04	0.71
10	Total unknown impurities	0.39	0.12	0.10
11	Total impurities	0.59	1.71	1.24
12	% of Degradation	0.29	1.41	0.94

Table 3.1: Result of Force Degradation study in Dabigatran etexilate mesylate

Name of impurity	LOD (%w/w)	Area of injection-1	Area of injection-2	Area of injection-3
DAB Impurity-A	0.01	2069	1930	1906
DAB-2B	0.01	1380	1359	1270
DAB Impurity-C	0.01	1362	1188	1067
DAB Impurity-H	0.01	1531	1460	1513
DAB Impurity-D	0.01	1208	1129	1147
DAB Impurity-E	0.01	1100	1336	1305
DAB Impurity-K	0.01	1190	1124	1169
DAB Impurity-J	0.01	1938	1766	1948
Dabigatran Etexilate Mesylate	0.01	2914	2286	3114
DAB Impurity-G	0.01	1656	1273	1334

Table 4: Result of the study of Limit of detection

Name of impurity	LOQ (%w/w)	Area of inj. -1	Area of inj. -2	Area of inj. -3	Area of inj. -4	Area of inj. -5	Area of inj. -6	% RSD
DAB Impurity-A	0.03	5432	5435	5448	5420	5541	5547	1.05
DAB-2B	0.03	3759	3720	3670	3626	3653	3633	1.43
DAB Impurity-C	0.03	3760	3750	3716	3843	3750	3656	1.64
DAB Impurity-H	0.03	4608	4635	4669	4451	4563	4514	1.77
DAB Impurity-D	0.03	3388	3488	3349	3327	3373	3428	1.71
DAB Impurity-E	0.03	4002	3873	3767	3896	3821	3893	2.04
DAB Impurity-K	0.03	3953	3892	3824	3949	4069	4035	2.28
DAB Impurity-J	0.03	3403	3358	3504	3331	3443	3478	1.99
Dabigatran Etexilate Mesylate	0.03	2910	2883	2805	3031	2907	3032	3.03
DAB Impurity-G	0.03	6703	6792	6804	6789	6989	6899	1.46

Table 5: Result of the study of Limit of Quantization

Name	Limit of detection (% w/w)	Limit of quantitation (%w/w)	Retention time (minutes)	Relative retention time	Relative response factor	Co-relation coefficient (R²)
DAB Impurity-A	0.01	0.03	2.88	0.09	1.30	1.0000
DAB-2B	0.01	0.03	10.34	0.34	0.91	0.9962
DAB Impurity-C	0.01	0.03	14.83	0.48	0.93	1.0000
DAB Impurity-H	0.01	0.03	15.84	0.52	1.42	1.0000
DAB Impurity-D	0.01	0.03	16.53	0.54	0.93	0.9999
DAB Impurity-E	0.01	0.03	22.20	0.73	0.95	1.0000
DAB Impurity-K	0.01	0.03	25.91	0.85	0.96	0.9979
DAB Impurity-J	0.01	0.03	26.97	0.88	0.86	0.9958
Dabigatran Etexilate Mesylate	0.01	0.03	30.62	1.00	1.0	0.9832
DAB Impurity-G	0.01	0.03	36.97	1.21	1.70	0.9895

Table 6: Result of Linearity and relative response factor for all impurities

Sno.	Name of Impurity/ Analyte	Accuracy @ % LOD Level	Accuracy @ 50% Level	Accuracy @100% Level	Accuracy @150% Level
1.	DAB Impurity-A	109.52	115.23	91.85	115.55
2.	DAB Impurity-B	114.94	110.96	113.56	111.62
3.	DAB Impurity-C	103.57	110.33	111.50	114.55
4.	DAB Impurity-D	109.88	110.78	115.93	110.05
5.	DAB Impurity-E	95.44	113.51	113.15	111.59
6.	DAB Impurity-H	101.45	112.87	113.57	115.88
7.	DAB Impurity-K	113.10	109.52	114.09	116.19
8.	DAB Impurity-J	113.79	115.74	116.43	112.29
9.	DAB Impurity-G	110.34	112.33	115.63	115.75

Table 7: Mean (n=3) Result of Accuracy @ different level with respect to specification level

Sno.	Name of Impurity/ Analyte	% Mean (n=6) of each Impurity	%RSD (n=6) of total Impurity
1.	DAB Impurity-A	0.128	
2.	DAB Impurity-B	0.164	
3.	DAB Impurity-C	0.225	
4.	DAB Impurity-D	0.128	
5.	DAB Impurity-E	0.265	1.618
6.	DAB Impurity-H	0.132	
7.	DAB Impurity-K	0.160	n=6 Preparation
8.	DAB Impurity-J	0.166	
9.	DAB Impurity-G	0.194	

Table 8: Mean (n=6) Result of precision of six different preparations

Preparation	Total impurities (%)	
	Analyst-1	Analyst-2
Preparation-01	1.625	1.664
Preparation-02	1.611	1.668
Preparation-03	1.614	1.666
Preparation-04	1.612	1.664
Preparation-05	1.622	1.664
Preparation-06	1.625	1.664
Mean	1.618	1.665
% RSD	0.41	0.10
Cumulative mean	1.642	
Cumulative %RSD	0.40	
Acceptance criteria for %RSD	Not more than 10.0%	

Table 9: Result of Intermediate precision

S. No.	Name of component	Resolution					
		Idea condition	Flow decrease	Flow increase	Temp. decrease	Temp. increase	Column change
1	DAB Impurity-A	-	-	-	-	-	-
2	DAB-2B	38.83	36.91	37.87	39.15	37.40	34.00
3	DAB Impurity-C	21.34	20.13	21.80	21.47	21.97	21.54
4	DAB Impurity-H	4.15	4.04	3.31	4.58	3.91	3.83
5	DAB Impurity-D	3.13	2.73	3.41	2.97	3.11	3.21
6	DAB Impurity-E	26.79	24.38	26.67	26.48	25.89	27.29
7	DAB Impurity-K	16.62	15.24	16.97	15.93	16.00	16.92
8	DAB Impurity-J	4.59	4.23	4.41	4.48	4.37	4.66
9	Dabigatran Etexilate Mesylate	4.43	4.15	4.24	4.46	4.18	4.59
10	DAB Impurity-G	18.76	17.22	17.87	19.32	18.17	19.35
S. No.	Name of component	Resolution					
		Idea condition	Flow decrease	Flow increase	Temp. decrease	Temp. increase	Column change
1	DAB Impurity-A	-	-	-	-	-	-
2	DAB-2B	38.83	36.91	37.87	39.15	37.40	34.00

3	DAB Impurity-C	21.34	20.13	21.80	21.47	21.97	21.54
4	DAB Impurity-H	4.15	4.04	3.31	4.58	3.91	3.83
5	DAB Impurity-D	3.13	2.73	3.41	2.97	3.11	3.21
6	DAB Impurity-E	26.79	24.38	26.67	26.48	25.89	27.29
7	DAB Impurity-K	16.62	15.24	16.97	15.93	16.00	16.92
8	DAB Impurity-J	4.59	4.23	4.41	4.48	4.37	4.66
9	Dabigatran Etexilate Mesylate	4.43	4.15	4.24	4.46	4.18	4.59
10	DAB Impurity-G	18.76	17.22	17.87	19.32	18.17	19.35

Table 10: Robustness result in each condition

of the analytical parameters such as System suitability, Specificity, Precision, Linearity, Accuracy, Range, Intermediate precision (Ruggedness), Robustness, Solution Stability and Mobile Phase Stability. As the carryover of Dabigatran Extexilate Mesylate peak has been observed and it is not impacting the analytical results, so Dabigatran Extexilate Mesylate peak area in the blank chromatogram has to be disregarded. Thus the method stands validated and can be used for routine analysis to determine the related substances of Dabigatran Etexilate Mesylate.

Acknowledgement

I am Thankful to the management of Micro Labs for providing the opportunity to all the facility and a part of this project.

References

1. Balaji N, Sivaraman VR, Neeraja P (2012) GC quantification of residual hexylmethane sulfonate in dabigatran etexilate mesylate. J Apply Che 2: 48-51.

2. www.drugbank.ca/drugs/DB06695

3. Prajapati A , Kumar S, Sen AK, Zanwar A, Seth AK (2014) Spectrophotometric method for estimation of dabigatran etexilate in bulk and its pharmaceutical dosage form . Pharma science monitor 5: 31.

4. Patel B, Ram P, Khatri T, Ram V, Dave P (2014) Study of Anticoagulant Dabigatran by Analytical Instrumentation. International Letters of Chemistry, Physics and Astronomy 11: 233-242.

5. https://www.pradaxa.com

6. Bernardi RM, Froehlich PE, Bergold AM (2013) Development and Validation of A Stability-Indicating Liquid Chromatography Method for the Determination of Dabigatran Etexilate in Capsules. J AOAC Int. 96: 37-41.

7. Shelke PG, Chandewar AV (2014) Validated stability-indicating high performance liquid chromatographic assay method for the determination of Dabigatran Etexilate Mesylate. Research Journal of Pharmaceutical, Biological and Chemical Sciences 5: 1637-1644.

8. Zhe-Yi Hu, Robert B. Parker, Vanessa L. Herring, and S. Casey Laizure, (2013) Conventional liquid chromatography/ triple quadruple mass spectroscopy based metabalite identification and semi quantitative estimation approach in the investigation of in-vitro dabigatran etexilate metabolite. Anal Bioanal Chem 405: 1695-1704.

9. Delavenne X, Moracchini J, Laporte S, Mismetti P, Basset T (2012) UPLC MS/MS assay for routine quantification of dabigatran-a direct thrombin inhibitor-in human plasma. J Pharm Biomed Anal 58: 152-156.

10. Prathap B, Dey A, Rao GHS, Johnson P, Arthanariswaran P (2013) Review - Importance of RP-HPLC in Analytical Method international journal of novel trends in pharmaceutical sciences 3: 15-23

11. Shwetha P, Kumar YR, Chander MR (2014) A Validated ion-chromatography method for determination of bromoacetate in a key starting Material of Dabigantran. Int J Pharma Biomed 5: 74-77

12. EMEA European medicine agency (2008), CHMP assessment report No. EMEA\HC\82a

13. Medichem SA (2014) Dabigatran etexilate and related substance Process and composition and Use of the substance as reference Standards and Markers.

14. Khan NA, Ahir KB, Shukla NB, Mishra PJ (2014) A Validated Stability-Indicating Reversed Phase High Performance Liquid Chromatographic Method For The Determination Of Dabigatran Etexilate Mesylate . Inventi Rapid: Pharm Analysis & Quality Assurance).

15. Li J, Fang J, Zhong F, Li W, Tang Y, et al (2014) Development and Validation of A Liquid Chromatography Tandem Mass Spectrometry Assay for the Simultaneous Determination of Dabigatran Etexilate, Intermediate Metabolite And Dabigatran In 50 μl Rat Plasma and its Application to Pharmacokinetic Study. Journal of Chromatography B 973: 110.

16. Damle MC, Bagwe RA (2014) Development and Validation of Stability-Indicating RP-HPLC Method for Estimation of Dabigatran Etexilate . J Adv Sci Res 5: 39-44.

17. Synder RD and Regan, JD 1981 Mutation Research 91: 307.

18. Lee CR, Guivarch F, Van Dau CN, Tessiera D, Krstulovic AM (2003) Determination of Polar Alkylating Agents as Thiocyanate/Isothiocyanate Derivatives by Reaction Headspace Gas Chromatography. Analyst 128: 857-863.

19. Bakshi M, Singh S (2002) Development of validated stability-indicating assay methods—critical review. J Pharm Biomed Anal 28: 1011-1040.

20. Kasture AV, Vadodkar SG, Mahadik KR, More HN (2008) Pharmaceutical Analysis. Introduction to Pharmaceutical Analysis 2nd (Edn). Nirali Prakashan, New Delhi: 1-2.

21. Khopkar SM (2008) Basic concepts of Analytical chemistry 3rd (Edn) New Age Science Publisher: 7-15.

22. United States Pharmacopoeia (USP-NF XXIV) (1985) Rockville MD 20852, United States Pharmacopoeial Convention Inc: 2149-2151.

23. Text on validation of analytical Procedures Q2 (R1) (2005) ICH, Harmonised Tripartite Guideline.

24. Text on Stability Testing of new Drug Substances and Products Q1A (R2) (2003) ICH, Harmonised Tripartite Guideline.

25. Guidance for industry (1998) Stability testing of drug substances and drug products. Food and Drug Administration, Rockville, MD, USA.

25

Selective Methods for Cilostazol Assay in Presence of its Oxidative Degradation Product and Co Formulated Telmisartan Application to Tablet Formulation

Ibrahim F, Sharaf El-Din M and Heba Abd El-Aziz*

Department of Analytical Chemistry, Faculty of Pharmacy, University of Mansoura, Mansoura, 35516, Egypt

Abstract

A high performance liquid chromatographic method characterized by its rapidness and sensitivity was developed and validated for quantitation of Cilostazol (CIL) and Telmisartan (TEL) in raw material, their synthetic mixture using isocratic technique and monolithic C8 column (3 mm × 4.6 mm i.d., 2 μm pore size highly porous). The mobile phase composed of acetonitrile:0.03 M dihydrogen phosphate buffer (40:60, v/v) at pH 4.5. Quantification was achieved through UV detection at 257 nm using flow rate of 1 mL/min, and Dipyridamole (DIP) was used as internal standard. DIP, CIL, and TEL, retention times were 2.2, 3.9 and 5.1 min. respectively. Peak area ratios of each drug to Dipyridamol internal standard were plotted against concentration of each drug and linear relations were obtained in the range of 0.5-15 μg/mL for CIL 0.25-20 μg /mL for TEL. The method was successfully utilized for the assay of CIL and TEL synthetic mixture. Stability-indicating assay methods (SIAM) were mentioned for separation of CIL in presence of its degradation products. Cilostazol was subjected to acid and alkali hydrolysis, oxidation and photochemical degradation. It was stable under acidic, basic and ultraviolet degradation conditions, but undergoes oxidative degradation, therefore the drug was separated from its oxidative degradation product using our proposed high performance liquid chromatographic and derivative ultraviolet spectrophotometric methods. The first derivative method (D1) depend on measuring the amplitude values at 227 and 257 nm for Cilostazol and oxidative degradate, respectively. From calibration plots, linearity was obtained in the range of 1-35 μg/mL for Cilostazol and 2-50 μg/mL for oxidative degradate. Chromatographic separation of Cilostazol from its oxidative degradate was proceeded using the same mobile phase at pH 3.3. All methods were validated statistically as per International Conference on Harmonization (ICH) recommendations for the studied drugs and Cilostazol oxidative degradate in the concentration range of the suggested methods and applied on pure materials and synthetic mixture.

Keywords: Cilostazol; Telmisartan; High performance liquid chromatography; Stability-indicating assay method (SIAM); Oxidative degradation; First derivative spectrophotometric method

Introduction

Cilostazol (CIL) (Figure 1a) 6-[4-(1-Cyclohexyl–1H-tetrazol–5–yl) butoxyl]-3,4–dihydro–2(1H)-quinolinone [1], is a phosphodiesterase inhibitor with anticoagulant and vasodilating behavior applied for peripheral vascular disease treatment and for reduction of symptoms of intermittent claudication [2]. Telmisartan (TEL) (Figure 1b) 4′-[(1,4′-Dimethyl–2′-propyl[2,6′-bi–1H-benzimidazol]-1′-yl) methyl]-[1,1′biphenyl]-2–carboxylic acid [1], is an angiotensin receptor blocker used for management of hypertension [2]. CIL and TEL are official in USP [3], TEL is official in BP [4]. Literature survey yielded some published methods for CIL assay include: Spectrophotometry as single drug [5-8], as mixture with other drugs [9-11], HPLC methods as single drug [12-15], and with other drugs [16,17]. Several analytical methods were reported for Telmisartan assay include: Spectrophotometry as single drug [18-20], as mixture with other drugs [21-27], Spectrofluorimetric methods as a mixture with other drugs [28,29]. Also many HPLC procedures have been published for its quantitation in its single form [30-33] and as mixture with others [34-40]. Therefore, it was desirable to develop simple, accurate, cheap and rapid procedure that could be utilized for simultaneous assessment of Cilostazol and telmisartan. It is not the first time for simultaneous HPLC determination of CIL and TEL, since there was a previously reported method [17], but our proposed method was superior in that it possess the advantages of high sensitivity over the reported one, LOQ of 0.11 and 0.2 ug/ml for CIL and TEL respectively and with smaller analysis time of 5 min than the reported one (10 min). Also, our proposed HPLC method was extended for determination of both drugs in their single tablet formulation. Moreover, stability study was not recomended in the reported method, but our proposed method could be used for determination of CIL in presence of its oxidative degradation. The developed proposed method presents an economic,

simple, sensitive procedure for simultaneous assay of cilostazol and telmisartan in their synthetic mixture and also used for separation of cilostazol from its oxidative product of degradation as an application to stability testing. First derivative spectrophotometric method was used for quantitation of cilostazolin presence of its oxidative degradate. Validation of the obtained results as per ICH guidelines showed that the proposed methods are of high accuracy and good precision.

Experimental

Apparatus

Perkin Elmer TM Series 200 Chromatograph joined with a Rheodyne injector valve with a 20 μL loop and UV/VIS detector was utilized for HPLC assay. Chromatograms were recorded on a PC attached to the device. Solvent degasser was used for mobile phases degassing. Spectrophotometric analysis was accomplished using a Shimadzu (Kyoto, Japan) UV-1601 PC, UV-Visible double-beam spectrophotometer with matched 1 cm path-length quartz cells. The spectra of the studied drugs were measured in 1 cm quartz cells and recorded on a fast scan speed, setting slit width to be 1 nm against solvent blank using wavelength range (200-400) nm using Δλ=8 nm

***Corresponding author:** Heba Abd El-Aziz, Department of Analytical Chemistry, Faculty of Pharmacy, University of Mansoura, Mansoura, 35516, Egypt
E-mail: dr_heba_abdelaziz@hotmail.com

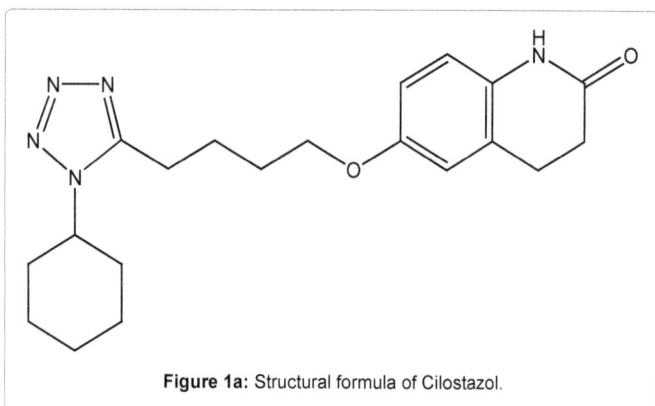

Figure 1a: Structural formula of Cilostazol.

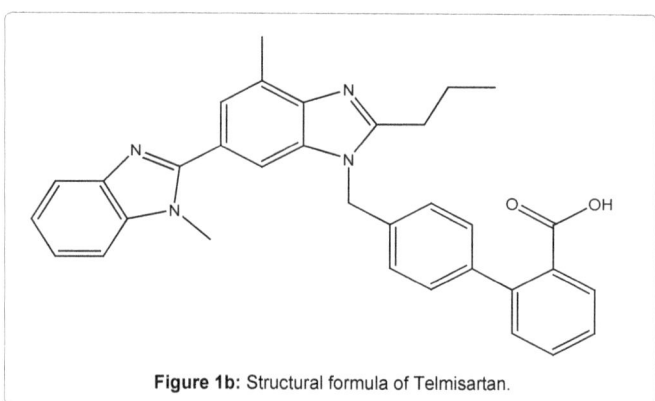

Figure 1b: Structural formula of Telmisartan.

and scaling factor=10 to obtain smoothly spectra. A Consort NV P-901 pH Meter (Belgium) was utilized for pH adjustment.

Materials and reagents

All used chemicals and reagents must be of high analytical grade and solvents used must be HPLC grade. Cilostazol, pure sample (Batch # 061500080) was kindly provided by European Egyptian Pharmaceuticals Company, Cairo, Egypt. Its purity was determined by official method (3)100.18%. Telmisartan was kindly supplied by biopharma Pharmaceutical Company, Cairo, Egypt. Its purity was determined by official method (3)100.02%. Laboratory prepared mixture consists of 40 mg CIL, 10 mg TEL, 20 mg talc powder, 15 mg maize starch, 15 mg lactose and 7 mg magnesium stearate Dipyridamol was kindly provided by Mepaco Pharmaceutical Corporation, Cairo, Egypt, with purity of 100.05%. Claudol tablet (100 mg of cilostazol) was manufactured by Sabaa International Company for Pharmaceuticals and Chemical Industries S.A.E. (M.O.H. Reg. No: 2440/2006) was obtained from commercial sources in local pharmacy. Micardis® tablet labeled to contain 40 mg of telmisartan, product of Boehringer Ingelheimpharma GmbH & Co. HG (Reg. No: 303028-01) was purchased from commercial sources in the local market. Acetonitrile (HPLC grade) was obtained from Sigma-Aldrich (Germany). -Orthophosphoric acid (85% w/v) was supplied from Riedel-deHäen (Germany). Sodium hydroxide and sodium dihydrogen phosphate were purchased from Adwic Co. (Cairo, Egypt).

Chromatographic conditions

The chromatographic assay was done using monolithic C8 column (3 mm × 4.6 mm i.d., 2 μm pore size highly porous column), A mixture of acetonitrile and 0.03 M dihydrogen phosphate buffer (40:60, v/v) was the selected mobile phase for elution, the pH was adjusted to 4.5

with 0.02 M orthophosphoric acid and filtered through a 0.45 μm membrane filter (Millipore, Ireland) and pumped at a flow rate of 1 mL/min. Such experiment was performed at room temperature and the UV detection was carried out at 257 nm. Dipyridamol was picked as the most appropriate internal standard

Stock solutions preparation

Standard stock solution of 100.0 μg/mL was used for both drugs and internal standard, using methanol as dissolving solvent for the three drugs. By suitable dilution of the stock solutions using the mobile phase, working solutions were obtained. Preserving the solutions in the refrigerator make them stable for at least a month without alteration.

Assay of cilostazol and telmisartan in raw materials

Construction of calibration plots: Cilostazol and Telmisartan working solutions were prepared from the previously mentioned stock solutions by serial dilution using the eluent to obtain final concentrations of (0.5-15) μg/mL for CIL and (0.25-20 μg/mL) for TEL into a series of 10 mL measuring flasks. 3 μg/mL of Dipyridamol stock solution was added to each flask as internal standard. Then, the final volume was adjusted using the mobile phase (pH=4.5) to the mark and mixed well. 20 μL aliquots were injected (triplicate) using the mobile phase as the eluent under the optimized chromatographic parameters. The ratios of peak area (drug/I.S.) against the final concentration of each drug in μg/ml were plotted. Alternatively, the corresponding regression equations were established.

Analysis of CIL/TEL in their synthetic mixtures: Aliquots of standard solutions of CIL and TEL keeping the curative ratio of 4:1, respectively were Transferred into a series of 10.0 mL volumetric flasks, diluted using mobile phase and mixed well then analyzed using the procedure stated under "Construction of calibration plots". From the obtained calibration plots, the percentage recoveries were calculated. Also percentage recoveries can be obtained from the corresponding regression equations.

Assay of the studied drugs in single tablet formulations: Ten Claudol tablets were exactly weighed, finely pulverized and mixed well, and then quantity of the powder equivalent to 100 mg of CIL was transferred into a 100.0 mL volumetric flask and about 70.0 mL of methanol were added. The flask contents were subjected to sonication for 30 min, completed with methanol to the final volume and filtered. Working solutions were prepared using the same solvent for dilution and analyzed using the general procedure as illustrated under "construction of calibration graph". Simultaneously, Ten Micardis tablets were accurately weighed, finely grinded and thoroughly mixed, then a quantity of the resulted powder equivalent to 100 mg of TEL was transferred into a 100.0 mL volumetric flask and continue the procedure as described in case of Claudol tablet. From the previously plotted calibration graph or by employing the corresponding regression equation, the nominal content of the tablets can be calculated.

Procedure for Cilostazol degradation products preparation: One mL aliquots of CIL standard stock solution (10 mg/mL) were transferred into a series of 25 mL volumetric flask to reach final concentration equal 400.0 μg/mL, and then 5 ml of 2M hydrochloric acid, 2M sodium hydroxide, or 15% hydrogen peroxide was added to prepare acidic, alkaline and oxidative degradation product respectively. The solutions (acidic and alkaline degradation) were refluxed at 100°C for 6 hrs, left in boiling water bath for 30 minutes at 80°C for oxidative degradation. In case of the UV degradation, the methanolic solution of CIL was subjected to Deuterium lamp at 254 nm in a wooden cabinet distance of 15 cm for 9 hours. Blank experiment was done simultaneously. After cooling the solutions were neutralized with 2M sodium hydroxide or 2M hydrochloric acid for acidic and alkaline degradation respectively, and completed with the mobile phase to the final volume. And then apply the proposed HPLC method for separation. It was found that no peaks were separated in case of acidic, alkaline and photochemical degradation and the peak area of CIL was not affected so, the drug is stable against such mentioned conditions. In case of oxidative degradation, noticeable increase in Cilostazol peak area was occurred. On changing the pH to 3.3 oxidative degradates was separated from Cilostazol. Also, derivative UV spectrophotometric method was established for quantitations of CIL with existence of its oxidative degradate.

Separation of cilostazol in presence of its oxidative degradation product using the suggested HPLC method

Cilostazol and its oxidative degradate can be separated with good resolution using our proposed HPLC method at pH 3.3. Such modification in pH is due to the great overlap between cilostazol and its oxidative degradate at pH 4.5. The retention times are 3.67 and 4.39 min. for oxidative degradate and cilostazol simultaneously. Cilostazol

can be assayed in presence of different percentages of oxidative degradate.

First derivative spectrophotometric method for quantitation of cilostazol with existence of its oxidative degradation product

Aliquots of CIL and oxidative degradate standard solutions covering the concentration range of 1.0-35.0 µg/mL and 2.0-50.0 µg/mL, simultaneously were transferred into two series of 10 mL volumetric flasks and the solutions were completed with methanol to the final volume and mixed well. The zero-order absorption spectra of CIL and its oxidative degradate were scanned against methanol showing great overlap. The first derivative spectra (^{1}D) of cilostazol and its oxidative degradate were scanned in the wavelength range (200-350) nm using $\Delta\lambda=8$ nm with zero crossing point recorded at 228 and 257 nm, simultaneously. The peak amplitude of the first derivative experiment was plotted against final concentration (µg/mL) to get the calibration graphs. Alternatively, the corresponding regression equations were derived (Figure 2).

Results and Discussion

CIL and TEL were separated using accurate and selective HPLC method and show well-resolved symmetrical peaks with good resolution in a short time nearly 5 min. using the optimized chromatographic conditions. Figure 3 shows typical chromatogram of a synthetic mixture of both drugs at their curative pharmaceutical ratio (4:1) using the proposed HPLC procedure. It is also permitted the quantitation of both drugs separately in their tablets (Figure 4A and 4B). Also, HPLC method was used for selective determination cilostazol in prescence of its oxidative degradate with satiafactory results at pH 3.3 (Figure 5). The drug was found to be completely degraded, and it was confirmed by injecting Cilostazol (10 µg/mL) and oxidative degradate (any concentration) on a separate runs then injecting a mixture of Cilostazol (10 µg/mL) and oxidative degradate (any concentration). It was regarded that peak area of CIL was the same in both cases. For spectrophotometric method of assay the UV spectrum of CIL solution in methanol showed absorption maxima at 220, 260, while that of oxidave degradate showed absorption maxima at 270 nm (Figure 6). So, great overlap between absorption spectra of CIL and degradation

product in Zero order that prevent the direct quantitation of CIL in presence of its oxidative degradation product, therefore the first derivative was carried out for the quantification of CIL in the presence of its degradation product. Figure 7a and 7b shows the first derivative spectra (D1) of CIL and its oxidative product and shows zero crossing point for each. The effect of different $\Delta\lambda$ [2,4,6,8] on the developed first derivative spectra was studied, showing that $\Delta\lambda=8$ was the selected

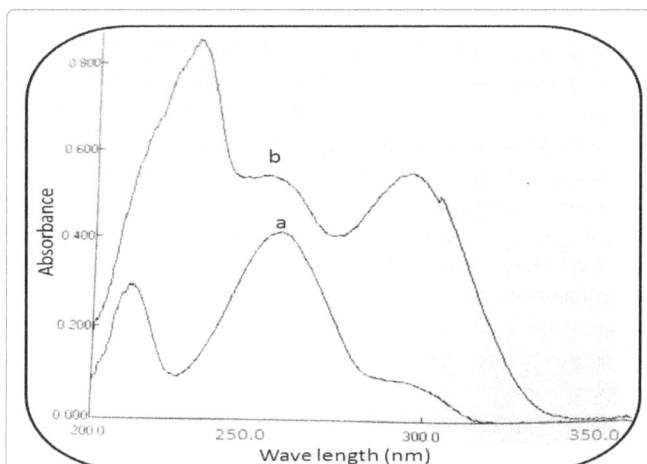

Figure 2: Zero order absorption spectra overlay of 10 µg/mL CIL (a) and 10 µg/mL TEL (b) in methanol.

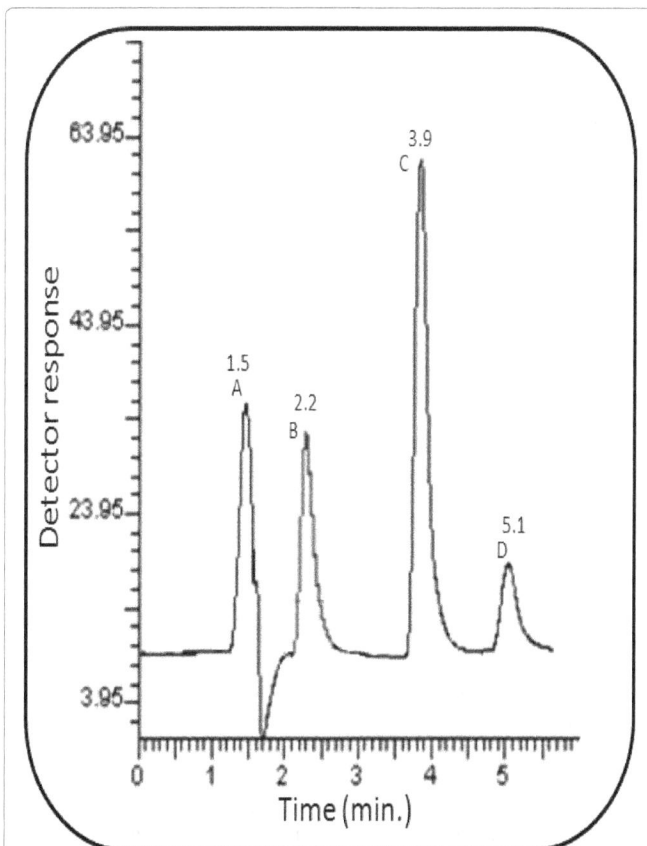

Figure 3: Typical chromatogram of synthetic mixtures under the described chromatographic conditions; acetonitril: phosphate buffer (40:60, v/v) at pH 4.5 and a flow rate of 1 mL/min. where: A: Solvent front; B: Dipyridamol (20 µg/mL); C: Cilostazol (12 µg/mL); D: Telmisartan (3 µg/mL)

one for best determination of both analytes. Using Δλ=8, CIL was quantified in presence of its oxidative degradate at zero crossing point (228) and quantitation of the degradation product in presence of CIL at zero crossing point (275). Figure 8 shows first derivative absorption spectra of CIL and its oxidative degradate in different synthetic ratios.

Chromatographic conditions study

After many experimental trials, good resolved symmetrical peaks were resulted. Summary of those trials can be summarized as follows

Column selection: Symmetry® C18 column (250 mm × 4.6 mm i.d., 5 μm particle size); shim-pack VP- ODS column (250 mm × 4.6 mm i.d., 5 μm particle size), Shimadzu, Kyoto, Japan and monolithic C8 column (3 mm × 4.6 mm i.d., 2 μm pore size highly porous column) were used to evaluate chromatographic performance and achieve good separation. Experimental studies proved that monolithic column (3 mm × 0.6 mm i.d., 2 μm pore size highly porous monolithic column) was the column of choice for this study giving well defined symmetrical peaks of both studied drugs with good resolution and short retention time.

Choice of appropriate detection wavelength: UV absorption spectra of CIL/TEL against blank methanol showed that maxima for CIL was obtained at 220 nm and 260 nm and for Telmisartan at 240, 257 and 300 nm (Figure 2). It was found that 257 nm was the wave length of choice for [3] CIL and TEL showing reasonable absorbance. Therefore, The UV detection was recorded at 257 nm allowing the assay of both drugs with high sensitivity.

Mobile phase composition: Many modifications in the eluent constituents were done to reach the conditions which lead to high chromatographic performance. Based on peak symmetry, tailing factor and retention time, the selection process was done. These modifications included; the change of mobile phase pH, the type of the organic modifier and its ratio, phosphate buffer concentration and also the flow rate. The produced results are illustrated in Table 1.

Mobile phase pH: The mobile phase pH was studied and showed that pH 4.5 was selected as the most appropriate one providing fully symmetrical peaks with high theoretical plates count, good resolution and less tailing factor value for both drugs within a short time as illustrated in Table 1. At pH above 5.5 a great overlapping between the two peaks occurs, and the two drugs eluted at the same retention time (2.4 min) at pH 6.

Ratio of organic modifier: Acetonitrile: 0.03 M phosphate buffer (40: 60, v/v) was selected as the most suitable mobile phase, it allowed the separation of CIL and TEL within a short run time with high resolution, high sensitivity for both drugs. By decreasing the acetonitrile ratio less than 40, the peak symmetry not affected but the retention time of both drugs increase and the run time became 15 min. By elevating organic modifier ratio, overlap between the peaks of both drugs occurred.

Molar concentration of phosphate buffer: The influence of molar concentration changing of phosphate buffer on the chromatographic efficiency was studied using eluents containing concentrations of 0.01 M-0.15 M of phosphate buffer. It was found that, the molar concentration of phosphate buffer is not a critical item affecting the peak separation of both studied drugs as by changing the molar strength the peak sensitivity, resolution and asymmetry not greatly affected. 0.03M was selected as the most appropriate concentration due to highly symmetrical peaks.

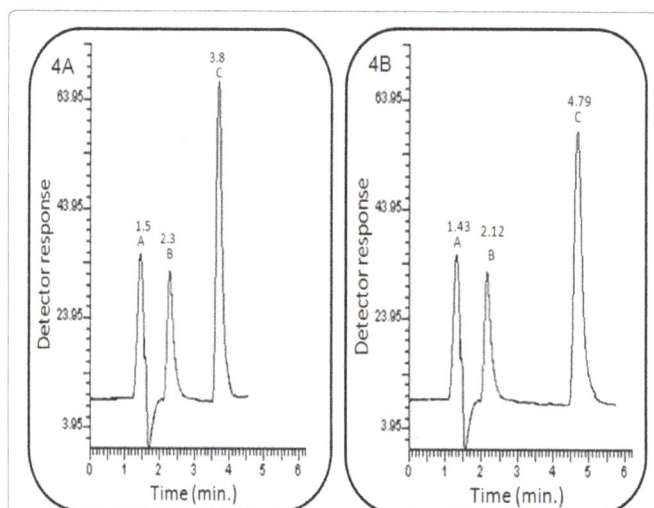

Figure 4: 4A: Typical chromatogram of the Cilostazol determination in Claudol tablet where: A: Solvent front B: Dipyridamol I.S (3.0μg/mL) C: Cilostazol (10μg/mL). **4B:** Typical chromatogram of the Telmisartan determination in Micardis tablet where: A: Solvent front B: Dipyridamol I.S (3.0μg/mL) C: Telmisartan (10μg/mL).

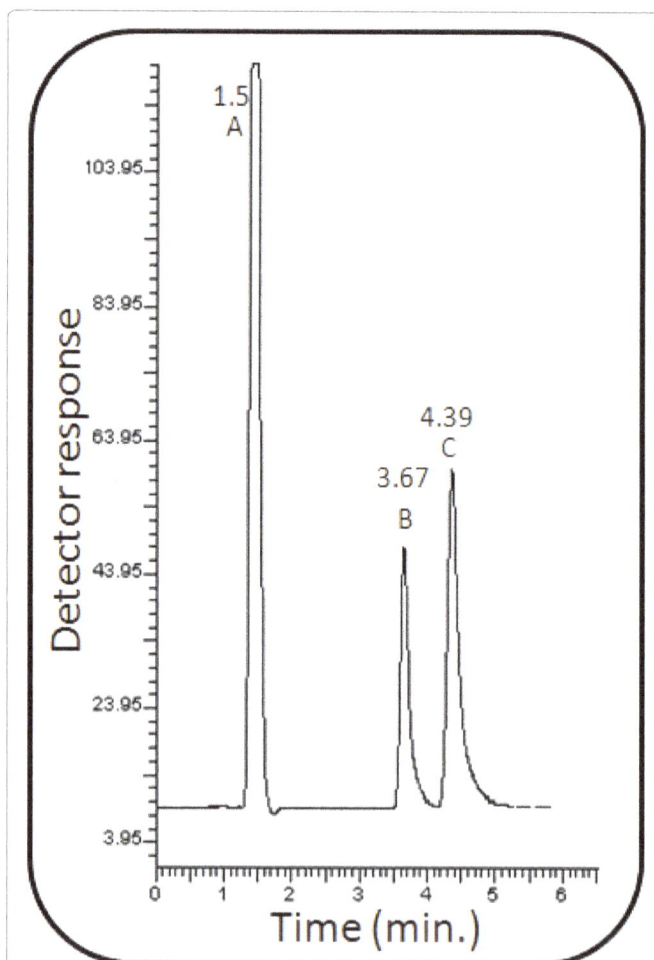

Figure 5: Typical chromatogram of cilostazol in presence of 100% of its oxidative degradation product (10 μg/mL of each) under the described chromatographic conditions; acetonitrile: phosphate buffer (40:60, v/v) at pH 3.3 and a flow rate of 1 mL/min. at 257 nm where: A: Solvent front B: Oxidative degradate (10 μg/mL) C: CIL (10 μg/mL).

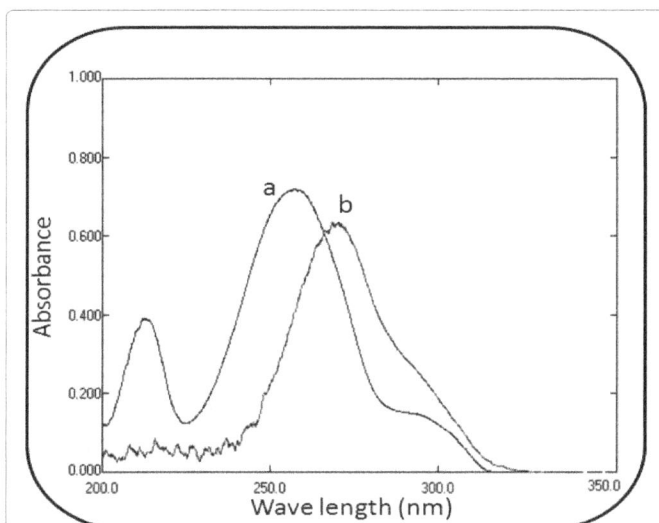

Figure 6: Zero order absorption spectra overlay of 20 µg/mL CIL (a) and 20 µg/mL Oxidative degradate (b) in methanol.

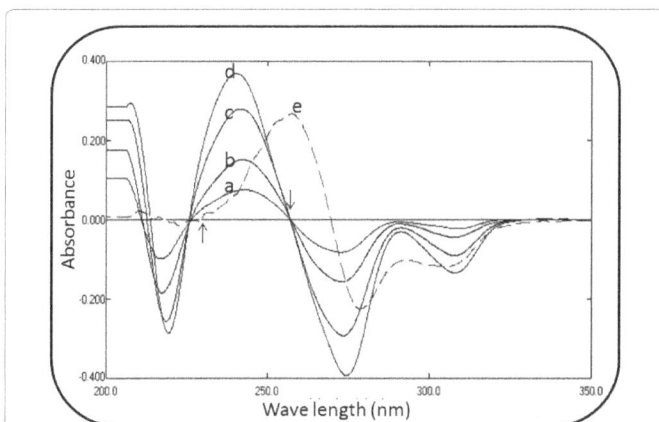

Figure 7a: First derivative absorption spectra of cilostazol and oxidative degradation product in methanol showing the zero crossing point of each, where a,b,c,d. different increasing concentrations of cilostazol and e: is 20 µg/mL oxidative degradate.

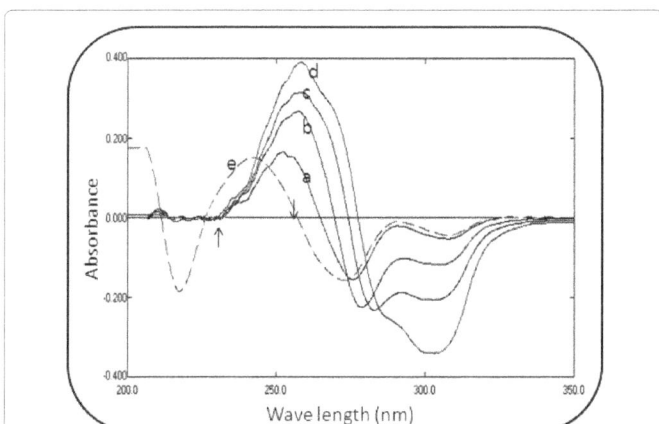

Figure 7b: First derivative absorption spectra of cilostazol and oxidative degradation product in methanol showing the zero crossing point of each, where a,b,c,d: different increasing concentrations of oxidative degradate and e: is 10µg/mL cilostazol.

Mobile phase replacement: When acetonitrile was replaced with methanol using the selected ratio methanol: phosphate buffer (40:60, v/v), no peaks were eluted till 30 min and upon replacement of phosphate buffer with water to become acetonitrile: water (40:60, v/v), the peak of TEL not appear till 30 min. So, acetonitrile: phosphate buffer (40:60, v/v) was the selected mobile phase for this study to obtain perfect chromatographic performance.

Flow rate influence: The influence of flow rate in the range of 0.8-1.2 mL/min on the separation of both drugs peaks was studied; 1.0 mL/min was the selected flow rate to obtain good separation in a small retention time.

Choice of internal standard: Several drugs were studied as internal standard; Dapoxitine, Clopidogrel, Dipyridamol, and Lamotrigene. Dipyridamol was selected as the most suitable internal standard with well resolved peak with no interference with peaks of both studied drugs and giving a well-defined peak with concentration (3 µg/mL).

Validation of the proposed methods

The proposed HPLC method was validated as per ICH guide lines using the following items: Linearity, specificity, accuracy, precision, LOD and LOQ.

Linearity and range: The calibration graphs for CIL and TEL assay by the developed HPLC method were produced from the peak area ratio [drug/I.S.] versus the concentration in µg/mL plots for both drugs. The high values of correlation coefficient showed high linearity for the concentration ranges stated in Table 2. Regression

analysis of the data yielded the corresponding equations:

P=0.084+ 0.430 C (r=0.9999) for CIL

P=0.350+0.390 C (r=0.9999) for TEL

Where: P represents the peak area ratio, C is corresponding to the concentration of the drug in µg/mL and r refers to

the correlation coefficient. Statistical analysis [41] of the results obtained from the mentioned regression equations produced high value of the correlation coefficient (r), small values of the standard deviation of residuals (Sy/x), of intercept (Sa), and of slope (Sb), and also percentage relative standard deviation and percentage error were very small value (Table 2). This analysis proved that the developed HPLC method was of high linearity.

Detection limit (LOD) and quantitation limit (LOQ): The limit of detection (LOD) was defined as the minimum level at which the drug can be reliably detected [41]. The limit of quantitation (LOQ) was determined by finding the lowest concentration that can be measured according to ICH guidlines [41,42] below which the calibration curve become nonlinear. LOD and LOQ were calculated from the following equations.

LOD=3.3 Sa /b LOQ=10 Sa /b

Where Sa=standard deviation of the intercept of the calibration curve and b=slope

of the calibration curve. The acquired data of the developed HPLC method were cited in Table 2.

Accuracy and precision: By comparing the developed HPLC method assay data with those obtained by the authorized one described by the USP [3], the accuracy of the developed method was confirmed. Statistical assessment of the obtained results using Student's t-test and

Parameter		No. of theoretical Plates (N)		Tailing factor (T$_f$)		Resolution (Rs)	Relative Retention (α)
		CIL	TELMI	CIL	TELMI		
pH of mobile phase	3	4670	6410	2.6	2.28	3.243	1.29
	4	5513	6603	2.26	2.16	2.648	1.24
	4.5	5618	6366	0.085	0.092	2.671	1.45
	5	5461	5487	2.405	2.404	3.52	1.38
	5.5	5612	5380	2.3	2.146	2.557	1.23
ACN: buffer ratio	70:30	2685	3264	0.78	3.59	1.33	2.62
	60:40	4500	4942	2.4	2.24	2.619	1.57
	50:50	5618	6366	0.085	0.092	2.671	1.28
	40:60	6686	6444	2.1	1.97	5.54	1.55
	30:70	7791	7286	2.0	1.774	6.285	1.41
Conc. of phosphate buffer (M)	0.01	7438	7098	2.056	1.778	5.66	1.48
	0.03	7110	6580	1.508	1.738	5.763	1.52
	0.05	6686	6444	2.1	1.97	5.54	1.54
	0.07	6912	6577	2.129	2.00	5.37	1.44
	0.1	6746	6385	2.24	2.055	4.578	1.45
	0.15	7023	7551	2.13	1.768	5.202	1.42
Effect of flow rate (mL/min.)	0.8	6233	6475	1.97	2.00	5.957	1.57
	1.0	7110	6580	1.508	1.738	5.763	1.52
	1.2	6028	5790	1.342	1.554	5.80	1.35

Where: Number of theoretical plates (N)= $5.54(\frac{t_R}{W_{h/2}})^2$; Resolution (R)= $\frac{2\Delta t_R}{W_1+W_2}$

Tailing factor (T$_f$), Selectivity factor (α)=tr$_2$-tm/tr$_1$-tm

Table 1: Optimization of the chromatographic conditions for separation of mixture of CIL and TELMI by the proposed HPLC method.

Parameter	CIL	TEL
Linearity and range (µg/mL)	(0.5-15)	(0.25-20)
LOD (µg/mL)	0.04	0.06
LOQ (µg/mL)	0.11	0.20
Intercept (a)	0.084	0.35
Slope (b)	0.430	0.39
Correlation coefficient(r)	0.9999	0.9999
S.D. of residuals (S$_{y/x}$)	8.5×10^{-3}	1.5×10^{-2}
S.D. of intercept (S$_a$)	4.9×10^{-3}	8.3×10^{-3}
S.D. of slope (S$_b$)	6.0×10^{-4}	8.0×10^{-4}
S.D.	0.56	0.63
%RSD	0.55	0.63
%Error	0.21	0.24

Where: (LOD) Limit of detection, (LOQ) Limit of quantitation, (% RSD) Percentage relative standard deviation, (% Error) Percentage relative error.

Table 2: Analytical performance data for Cilostazol and Telmisartan assay using the proposed HPLC methods.

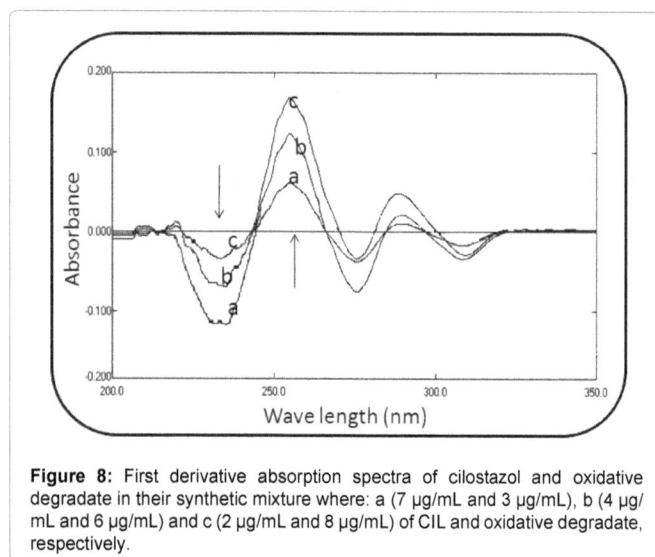

Figure 8: First derivative absorption spectra of cilostazol and oxidative degradate in their synthetic mixture where: a (7 µg/mL and 3 µg/mL), b (4 µg/mL and 6 µg/mL) and c (2 µg/mL and 8 µg/mL) of CIL and oxidative degradate, respectively.

variance ratio F-test [42] proved that there is no significant difference between the performance of the two methods regarding the accuracy and precision, respectively (Table 3). Inter-day and intra-day assay for the studied drugs were established to indicate that the developed method was precise.

Intra-day precision: Intra-day precision was performed for the developed HPLC method through replicate analysis of three concentrations of both studied drugs on three successive times within a day. Small values of % Error and % RSD were resulted from the obtained results which indicate that the developed method is of high accuracy and precision, respectively. The results are summarized in Table 4.

Inter-day precision: Inter-day precision was performed via replicate analysis of three concentrations of the studied drugs on three successive days. Table 4 represents the obtained results. The small values of % Error and % RSD indicate high accuracy and precision of the developed method, respectively.

Drug	Proposed HPLC method			Reference method [3]		
	Amount taken (µg/mL)	Amount found (µg/mL)	% Found	Amount taken (µg/mL)	Amount found (µg/mL)	% Found
CIL	0.5	0.500	100.02	0.2	0.202	101.15
	1.0	0.999	99.91	0.5	0.496	99.25
	2.0	2.024	101.24	1.0	1.001	100.14
	4.0	3.981	99.53			
	8.0	7.972	99.66			
	10.0	10.015	100.15			
	15.0	15.004	100.03			
Mean ± S.D.			100.08 0.56			100.18 0.95
t-test				0.22 (2.30)		
F-test				2.90 (5.14)		
TEL	0.25	0.248	99.44	5.0	5.011	100.24
	1.0	0.995	99.55	8.0	7.970	99.63
	2.0	1.972	98.65	10.0	10.017	100.18
	5.0	5.005	100.11			
	10.0	10.068	100.68			
	15.0	14.960	99.74			
	20.0	19.993	99.97			
Mean ± S.D.				99.73 0.63		100.02 0.34
t-test				0.72 (2.30)		
F-test				3.52 (19.32)		

N.B.*the figures between parentheses are the tabulated t and F values at P=0.05 (40)

Table 3: Assay results for the determination of CIL and TEL drugs in pure form using the proposed HPLC procedure.

	Parameters		CIL concentration (µg/mL)			TEL concentration (µg/mL)		
			2.0	8.0	10.0	1.0	5.0	10.0
Intraday	% Found		100.85	99.55	99.40	99.80	99.16	99.25
			99.90	100.03	100.50	100.55	100.50	100.40
			99.55	100.44	99.66	99.20	99.60	99.68
	x ± SD		100.10 0.67	100.01 0.45	99.85 0.58	99.85 0.68	99.75 0.68	99.78 0.58
	% RSD		0.67	0.45	0.58	0.68	0.68	0.58
	% Error		0.39	0.26	0.33	0.39	0.40	0.34
Interday	% Found		100.15	99.15	100.00	100.40	100.11	99.40
			99.57	99.66	98.59	99.55	100.29	100.80
			101.24	100.80	99.60	99.03	101.76	99.96
	x ± SD		100.32 0.85	99.87 0.85	99.40 0.73	99.66 0.69	100.72 0.91	100.05 0.71
	% RSD		0.85	0.85	0.73	0.69	0.90	0.71
	% Error		0.49	0.49	0.42	0.40	0.52	0.41

Table 4: Precision data for the determination of CIL and TEL by the proposed HPLC procedure.

Robustness of the developed HPLC method: The robustness of an analytical method is defined as the method capability to remain unaffected by deliberate minor changes in experimental parameters. The proposed HPLC method showed robustness upon minor changes of chromatographic conditions such as; mobile phase pH (4.5 ± 0.2), phosphate buffer strength (0.030 ± 0.01 M). The ratio of acetonitrile: 0.03 M phosphate buffer in the mobile phase is a critical parameter for the separation process, where small changes in organic modifier ratio from (40: 60, v/v) greatly affect both drugs resolution and run time.

Specificity: The proposed HPLC method was specific for assay of CIL and TEL in tablets. Common tablet excipients did not affect the separated peaks in the developed HPLC method.

Application of the developed HPLC method for CIL and TEL assay in their tablets

The proposed HPLC method was successfully utilized for simultaneous quantitation of CIL and TEL in their synthetic mixtures that medicinally recommended in ratio of 4:1 (w/w) (Table 5). Also, the developed method was used for the determination of the studied drugs separately each in its own tablet (Table 6). Good agreement of the resulted data with those of the authorized methods stated by USP [3] was found. Statistical analysis of the produced results using Student's t-test and variance ratio F test [42] proved that there is no significant difference between the performance of the two methods regarding the accuracy and precision, respectively.

CIL/TEL ratio	Proposed HPLC method						Reference method	
	Amount taken (µg/mL)		Amount found (µg/mL)		% Found		3	
							%Found	
	CIL	TEL	CIL	TEL	CIL	TEL	CIL	TEL
4:1	4.0	1.0	3.972	1.005	99.32	100.53	101.15	100.24
	8.0	2.0	8.053	1.989	100.66	99.46	99.25	99.63
	12.0	3.0	11.971	3.005	99.77	100.18	100.14	100.18
x					99.92	100.06	100.18	100.02
± SD					0.68	0.55	0.95	0.34
%RSD					0.68	0.54	0.94	0.33
%Error					0.39	0.31	0.54	0.19
T					0.38	0.11		
F					1.94	2.63		

N. B. Each result is the average of three separate determinations.
The values of tabulated t and F tests are 2.78 and 19.00, respectively at *p*=0.05 [29].

Table 5: Assay results for the determination of the CIL and TEL in synthetic mixtures in ratios of 4:1 (w/w) by the proposed HPLC method.

Tablet	Proposed method		Official method 3	
	Amount taken (µg/mL)	%Found	Amount taken (µg/mL)	%Found
Claudol Tablet (100.00 mg CIL)	1.0	98.86	0.2	101.65
	4.0	100.42	0.5	98.94
	10.0	99.94	1.0	100.20
Mean		99.74		100.26
± S.D.		0.80		1.36
t-test	0.57 (2.77)			
F-test	2.88 (19.0)			
			Official method 3	
	Amount taken (µg/mL)	%Found	Amount taken (µg/mL)	%Found
Micardis Tablet (40.00 mg TEL)	1.0	100.96	5.0	100.27
	5.0	99.64	8.0	99.58
	10.0	100.07	10.0	100.20
Mean		100.22		100.02
± S.D.		0.67		0.38
t-test	0.46 (2.77)			
F-test	3.14 (19.0)			

N.B.*each result is the average of three separate determinations.
*The figures between parentheses are the tabulated t and F values at P=0.05 [40].
Table 6: Assay results for the determination of both drugs in their tablets, separately.

Application of the developed HPLC method for assay of cilostazol with existence of its oxidative degradate

The proposed HPLC method was applied for CIL assay and its oxidative degradate, producing successful results. The results of assay were abridged in Table 7. Under the optimized experimental conditions, linearity of the method was achieved by plotting the peak area against the concentration in µg/mL for CIL and oxidative degradate, simultaneously. Table 8 showed assay results of 10 µg/mL CIL different times in presence of different percentages of oxidative degradation product. The regression equations were as follows:

P=9.09+38.53 C (r=0.9999) (Cilostazol)

P=-7.72+21.32 C (r=0.9999) (Oxidative degradate)

Where: P represents the peak area, C is the concentration in µg/mL and r refers to the correlation coefficient. The results obtained indicated the high linearity of the calibration curve performed for determination of cilostazol and oxidative degradation product.

First derivative spectrophotometric method for assay of cilostazol and its oxidative degradation product

The calibration plots for the determination of CIL and its oxidative degradate by the proposed derivative spectrophotometric method were obtained by plotting the amplitude of the derivative peaks against the concentration in µg/mL. Table 9 shows the linearity range of CIL and its oxidative degradate using derivative spectrophotometricmethod. The following equations illustrate the regression analysis of data:

$1D_{228}=-5 \times 10^{-4}+6.5 \times 10^{-3}C$ (r=0.9999) for (Cilostazol)

$1D_{257}=0.05+8.7 \times 10^{-3}C$ (r=0.9999) for (Oxidative degradate)

Where: (D wavelength) is the amplitude of the first derivative spectra at the stated wave

length, C is the concentration in µg/mL and r is the correlation coefficient. Also, Table 10 shows assay results of CIL and its oxidative degradate in their synthetic mixture using derivative spectrophotometric method.

| Item | Synthetic mixture of cilostazol and its oxidative degradation product | | | |
| | Amount taken (μg/mL) | | %Recovery | |
	CIL	Oxidative degradate	CIL	Oxidative degradate
	8	2	99.85	99.03
	7	3	100.06	100.58
	6	4	100.13	100.30
	4	6	100.23	99.86
	2	8	99.53	99.99
x̄			99.96	99.95
%RSD			0.28	0.59

Table 7: Assay results for determination of Synthetic mixture of Cilostazol and its oxidative degradation product using the proposed HPLC method.

| CIL | Oxidative degradate | %Recovery | |
		CIL	Oxidative degradate
(10 μg/mL)	10%	100.03	101.70
	20%	98.13	100.05
	40%	98.40	99.65
	60%	99.45	99.40
	80%	99.09	100.40
x̄		99.02	100.24

Table 8: Assay results for determination of oxidative degradation product in presence of 10 μg/mL CIL using the proposed HPLC method.

Parameter	CIL	Oxidative degradate
Linearity and range (μg/mL)	(1.0-35)	(2.0-50)
LOD (μg/mL)	0.23	0.41
LOQ (μg/mL)	0.71	1.23
Intercept (a)	-5×10^{-4}	0.05
Slope (b)	6.5×10^{-3}	8.7×10^{-3}
Correlation coefficient(r)	0.9999	0.9999
S.D. of residuals ($S_{y/x}$)	7×10^{-4}	1.4×10^{-3}
S.D. of intercept (S_a)	5×10^{-4}	1.1×10^{-3}
S.D. of slope (S_b)	0.00	0.00
S.D.	0.60	0.37
%RSD	0.60	0.37
%Error	0.25	0.15

Table 9: Analytical performance data for Cilostazol assay in presence of its oxidative degradation product using derivative spectrophotometric method.

| Item | Proposed HPLC method | | | | | |
| | Amount taken (μg/mL) | | Amount found (μg/mL) | | %Found | |
	CIL	Degradate	CIL	Degradate	CIL	Degradate
	7.0	3.0	6.986	2.987	99.81	99.59
	4.0	6.0	4.013	6.036	100.34	100.61
	2.0	8.0	1.986	7.987	99.33	99.85
x̄					99.83	100.02
± SD					0.51	0.53
%RSD					0.50	0.53
%Error					0.29	0.31

Table 10: Assay results for the determination of the CIL and its oxidative degradate synthetic mixtures using derivative spectrophotometric method.

Conclusion

Very simple, rapid, accurate and precise HPLC method was developed for simultaneous determination of CIL and TEL. The method was applied to the analysis of both drugs in their pure and tablet formulation without interference from common excipients. The results obtained were of good agreement with the authorized method. The developed HPLC method was accurate and highly precise with low percentage error and relative standard deviation so, it could be applied in quality control laboratories. HPLC method was applied for a selective assay of cilostazol in presence of its oxidative degradate with good linearity for both. Also selective first derivative spectrophotometric method was applied for assay of cilostazol in presence of its oxidative degradate showing high linearity.

References

1. Moffat AC, Osselton MD, Widdop B (2010) Clarkes Analysis of Drugs and Poisons (eds. 3). Pharmaceutical Press, London, pp: 1371-1641.

2. Sweetman SC (2009) Martindale the Complete Drug Reference (eds. 35). The Pharmaceutical Press, London, pp: 1245-1409.

3. The United States Pharmacopoeia 34 (2011) The National Formulary 29. US Pharmacopeial Convention Rockville, pp: 2336, 4358.

4. The British Pharmacopoeia London (2015) The stationary office. Electronic version 2: 979-981.

5. Basniwal PK, Kumar V, Shrivastava PK, Jain D (2010) Spectrophotometric Determination of Cilostazol inTablet Dosage Form. Trop J Pharm Res 9: 499-503.

6. Hoballah SA (2015) Spectrophotometric methods for determination of cilostazol in pure and dosage forms. IJRPC 5: 17-26.

7. Chaitanya T, Ramya B, Archana M, Himavani KV, Ramalingam P, et al. (2011) Forced degradation studies of cilostazol by validated UV-spectrophotometric method. J of Pharm Res 4: 4094.

8. Pandeeswaran M, El-Mossalamy EH, Elango KP (2011) Spectroscopic studies on the interaction of cilostazole with iodine and 2, 3-dichloro-5,6-dicyanobenzoquinone. Spectrochimica Acta Part A 78: 375-382.

9. Patel JV, Patel CN, Anand IS, Patel PU, Prajapati PH (2008) Simultaneous spectrophotometric estimation of cilostazol and aspirin in synthetic mixture. Int J Chem Sci 6: 73-79.

10. Brahmbhatt PM, Prajapati LM, Joshi AK, Kharodiya ML (2015) Q-absorbance ratio method for simultaneous estimation of cilostazol and imipramine in combined dosage form. WJPR 4: 1786-1794.

11. Damor DK, Mashru RC (2015) Method development and validation of simultaneous estimaton of cilostazol and telmisartan. Inter Bull of Drug Res 5: 12-22.

12. Chaitanya KK, Sankar DG, Samson D, Kumar CH, BalaKumar K (2013) Isocratic Reverse Phase HPLC Method Determination and Validation of Cilostazol. IJCP 1: 257-262.

13. Basniwal PK, Shrivastava PK, Jain D (2008) Hydrolytic Degradation Profile and RP-HPLC Estimation of Cilostazol in Tablet Dosage Form. Indian J Pharm Sci 70: 222-224.

14. Kurien J, Jayasekhar P (2014) Stability indicating HPLC determination of cilostazol in pharmaceutical dosage forms. Int J Pharm Bio Sci 5: 176-186.

15. Pareek D, Jain S, Basniwal PK, Jain D (2014) RP-HPLC Determination of Cilostazol in Human Plasma: Application to Pharmacokinetic Study in Male Albino Rabbit. Acta Chromatographica 26: 283-296.

16. Ambekar AM, Kuchekar BS (2014) A validated new gradient stability-indicating LC method for the simultaneous estimation of cilostazol and aspirin in bulk and tablet formulation. EJBPS 1: 149-164.

17. Patel RR, Maheshwari DG (2015) A new RP-HPLC method for simultaneous estimation of telmisartan and cilostazol in synthetic mixture. IJRSR 6: 3306-3310.

18. Sunil S, Kumar YA, Hemendra G (2012) First order derivative spectrophotometric determination of telmisartan in pharmaceutical formulation. Bulletin of Pharmaceutical Research 2: 83-86.

19. Padmavathi M, Reshma MSR, Sindhuja YV, Venkateshwararao KC, NagaRaju K (2013) Spectrophotometric methods for estimation of Telmisartan bulk drug and its dosage form. International Journal of Research in Pharmacy and Chemistry 3: 320-325.

20. Tulja RG, Prashanthi S, Srinivas N (2012) Two simple extractive spectrophotometric methods for the estimation of telmisartan in pharmaceutical formulation using bromothymol blue and orange G. Intl J of Pharm and Pharma Sci 4: 382-384.

21. Badran AI, El-Fatatry HM, Hammad SH (2015) Simultaneous estimation of telmisartan and amlodipine by second derivative spectrophotometric method and first derivative ratio-spectrophotometric method. American Journal of PharmTech Res 5: 171-189.

22. Patel R, Maheshwari DG (2015) UV spectrophtometric methods for simultaneous estimation of telmisartan and cilostazol in synthetic mixture. Pharma Science Monitor 6: 276-291.

23. Solanki BK, Dave JB, Raval PP (2014) Development and validation of UV spectrophotometric methods for simultaneous estimation of rosuvastatin and telmisartan in tablet dosage form. World Journal of Pharmacy and Pharmaceutical Sciences 3: 2030-2041.

24. Behera CC, Joshi V, Pillai S, Gopkumar P (2014) Validated spectrophotometric methods for estimation of telmisartan and hydrochlorothiazide in combined tablet dosage form. Research & Reviews: Journal of Pharmaceutical Analysis 3: 16-21.

25. Haripriya M, Antony N, Jayasekhar P (2013) Development and validation of UV spectrophotometric method for the simultaneous estimation of Cilnidipine and Telmisartan in tablet dosage form utilising simultaneous equation and absorbance ratio method. I J of Pharm and Bio Sci 3: 343-348.

26. Agarkar AR, Bafana SR, Mundhe DB, Suruse SD, Jadhav KB (2013) Spectrophotometric determination and validation of telmisartan and hydrochlorthiazide in pure and tablet dosage form. Current Pharma Research 3: 777-783.

27. Pillai S, Manore D (2012) Simultaneous spectrophotometric estimation of Telmisartan and Indapamide in capsule dosage form. International Journal of Pharmacy and Pharmaceutical Sciences 4: 163-166.

28. Belal TS, Mahrous MS, Abdel-Khalek MM, Daabees HG, Khamis MM (2014) Validated spectrofluorimetric determination of two pharmaceutical antihypertensive mixtures containing amlodipine besylate together with either candesartan cilexetil or telmisartan. Luminescence 29: 893-900.

29. Anumolu PD, Neeli S, Anuganti H, Puvvadi SBR, Subrahmanyam CVS (2013) First derivative synchronous spectrofluorimetric quantification of telmisartan/ amlodipine besylate combination in tablets. Dhaka University Journal of Pharmaceutical Sciences 12: 35-40.

30. Seshagiri Rao JVLN, Vijayasree V, Palavan Ch (2013) A validated RP-HPLC method for the estimation of telmisartan in tablet dosage forms. American Journal of PharmTech Research 3: 764-769.

31. Reddy K, Narendra K, Rao GD, Pratyusha PH (2012) Isocratic RP-HPLC method validation of telmisartan in pharmaceutical formulation with stress test stability evaluation of drug substance. Journal of Chemical and Pharmaceutical Sciences 5: 16-21.

32. Vijay KR, Arun KK, Manjunath Shetty KS, Hima NM, Sandya RD, et al. (2012) Analytical method development and validation of Telmisartan hydrochloride by RP-HPLC. Journal of Scientific Research in Pharmacy 1: 100-101.

33. Patel BA, Captain AD (2014) Development and validation of RP-HPLC method for estimation of TELMISARTAN in bulk and formulation using fluorescence detector. J of Bio and Pharm Res 3: 44-48.

34. Thapaswini B, Reddy TR, Ajitha A, Rao V, Uma Maheswara SSR (2015) A validated RP-HPLC method for estimation of telmisartan and metoprolol in its bulk and pharmaceutical dosage forms. World Journal of Pharmaceutical Research 4: 2373-2382.

35. Eswarudu MM, Venupriya K, Eswaraiah MC, Dipankar B, Raja B (2015) Validated RP-HPLC method for simultaneous estimation of rosuvastatin and telmisartan in bulk and pharmaceutical dosage form. Pharmanest 6: 2648-2653.

36. Gopi RT, Vanitha PK (2015) Stability indicating RP-HPLC method development and validation for the estimation of Telmisartan, Cilnidipine and Chlorthalidone in pharmaceutical dosage form. Int J of Pharm 5: 624-630.

37. Raj PSP, Venkateswarao P, Thangabalan B (2015) Development and validation of RP-HPLC method for the simultaneous determination of amlodipine besylate

and telmisartan in bulk and pharmaceutical dosage form. World Journal of Pharmaceutical Research 4: 1753-1761.

38. Jadhav S, Rai M, Kawde PB, Farooqui M (2015) Simultaneous determination of Telmisartan, Amlodipine Besylate and Hydrochlorothiazide in tablet dosage form by using stability-indicating HPLC method. Pharmacia Sinica 6: 13-18.

39. Gayathri P, Jayaveera KN, Goud S, Reddy NS (2015) Analytical method development and validation for the simultaneous estimation of metformin and

telmisartan in bulk and pharmaceutical dosage forms using RP-HPLC method. World Journal of Pharmacy and Pharmaceutical Sciences 4: 753-762.

40. Parihar Y, Kotkar T, Mahajan MP, Sawant SD (2014) Development and validation of RP-HPLC method for simultaneous estimation of telmisartan and cilnidipine in bulk and tablet dosage form. Pharmanest 5: 2321-2325.

41. International Conference on Harmonization ICH (2007) Guidelines.

42. Miller JN, Miller JC (2010) Statistics and Chemometrics for Analytical Chemistry. pp: 39-73.

26

HPLC-MS/MS Based Time Course Analysis of Morphine and Morphine-6-Glucoronide in ICU Patients

Eisenried A*, Peter J, Lerch M, Schüttler J and Jeleazcov C

Department of Anesthesiology, University of Erlangen-Nurnberg, Krankenhausstrabe 12, 91054 Erlangen, Germany

Abstract

Morphine is a widely-used opioid analgesic to treat post-operative pain in the intensive care unit. For the quantification of morphine and its metabolite Glucuronide (M6G) concentrations a sensitive and specific liquid chromatography–tandem mass spectrometry (LC-MS/MS) method was developed and validated according to Food and Drug Administration (FDA) guidelines. Plasma samples were extracted with solid-phase extraction and substituted with deuterated morphine and M6G as internal standards. Separation was performed by gradient elution using UPLC-like system and analyzed by MS/MS consisting of an electrospray ionization source. The lower limit of quantification was 500 "pg/ml" for morphine and 50 "pg/ml" for M6G. Intra- and interassay precision and accuracy did not exceed ± 15%. The method was applied to a clinical study during intensive care treatment of patients after coronary artery bypass grafting and can serve for further pharmacokinetic studies.

Keywords: Morphine; Morphine-6-glucuronide; High performance liquid chromatography; Tandem mass spectrometry; Pharmacokinetic; Intensive care

Introduction

Morphine is an opioid analgesic used for the treatment of moderate to severe pain. It is also commonly used to treat post-operative pain in the intensive care unit (ICU) [1]. Morphine is metabolized by conjugation to morphine-3-glucuronide (M3G) and morphine-6-glucuronide (M6G) [2]. Whereas, the main metabolite M3G, has little or no analgesic effect, M6G has been shown to be very effective and is believed likely to contribute significantly to the overall effectiveness of morphine. Impairments of the hepatic and renal function [3] may alter morphine elimination and conjugation, especially in critically ill and ICU patients [4]. To date, there are only a few reports on quantification of morphine and its metabolites in ICU patients [5]. Nevertheless, morphine has been widely investigated in terms of efficacy and side effects in other patient groups. In critical ill patients, metabolism of morphine through UDP-glucuronosyltransferase 2B7 to M6G and renal excretion of morphine and M6G may be affected with major surgery such as cardiac surgery with cardiopulmonary bypass or critical illness like septic shock or multiple organ failure [5].

The quantification of morphine and its metabolites may present several analytical problems in biological fluids. HPLC-MS/MS is the method of choice for simultaneous analysis of morphine and M6G in human serum reviewed by Bosch et al. [6] and was used for this setting. Our investigations were intended to provide a robust and validated, according to Food and Drug Administration (FDA) guidelines [7], HPLC-MS/MS protocol for determination of morphine and M6G in human plasma and give example for clinical use in intensive care patients. The method was applied to a clinical study in ICU patients after cardiac surgery with cardiopulmonary bypass. Pain treatment were performed via patient controlled analgesia (PCA) and plasma samples were drawn frequently. Deuterated morphine and M6G were used as internal standards to improve the prediction and accuracy of the method as well as the robustness of the quantification against matrix effect. Tandem mass spectrometry was used for specific and reliable detection and quantification of the analytes.

Materials and Methods

Drugs and chemicals

Morphine, M6G and the internal standards morphine–D6 and M6G-D3 (Figure 1) were purchased from LGC-Standards (Wesel, Germany). Ammoniac (30%), Methanol (LC-MS grade), ultra LC-MS water and formic acid were purchased from Roth (Karlsruhe, Germany). Drug free human plasma was obtained from Recipe Chemicals (München, Germany).

Standard solution and calibration standards

Stock solutions of morphine, morphine–D6, M6G and M6G-D3 were 1000 µg/ml, and calibration standards were prepared with methanol and stored at -20°C. Quality control (QC) samples were prepared to drug free plasma and blank study plasma samples (zero samples) by adding 50 µl standards to 450 µl drug free plasma. The calibration samples for the determination of drug concentrations (morphine 0.5–100 ng/ml, M6G 0.05–10 ng/ml) were prepared in the same way using drug free plasma.

Extraction procedure

For the assay samples were supplied together with IS (50 µl of 500 ng/ml deuterated internal standard) and 1 ml formic acid 0.1% after centrifugation at 5000 rpm for 1 min to OASIS MCX solid phase extraction cartridges (Waters, USA), which had been conditioned with 2 ml of methanol followed by 2 ml formic acid 0.1%. The cartridges were rinsed twice with 1 ml formic acid 0.1% under a slight vacuum. For descaling the residual water, the cartridges were centrifuged again for 1 minute at 5000 x g. The analytes were then eluted with 1000 µl

*Corresponding author: Eisenried A , Department of Anesthesiology, University of Erlangen-Nurnberg, Erlangen, Krankenhausstrabe 12, 91054 Erlangen, Germany
E-mail: andreas.eisenried@kfa.imed.uni-erlangen.de

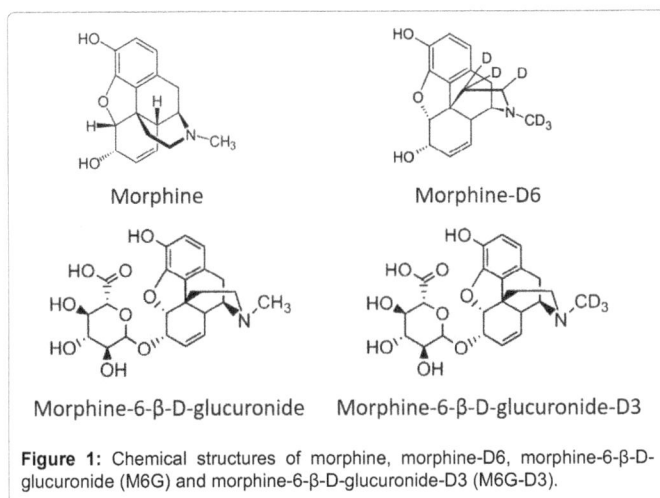

Figure 1: Chemical structures of morphine, morphine-D6, morphine-6-β-D-glucuronide (M6G) and morphine-6-β-D-glucuronide-D3 (M6G-D3).

methanol, containing 5% ammoniac. After evaporating to dryness, the extracts were solved in 150 µl formic acid 0.1%.

Equipment

The extracted samples were analyzed with HPLC and tandem mass spectrometry. Chromatographic separation was carried out with the Waters Alliance HPLC system (Waters, Eschborn, Germany). The system was upgraded and modified by Fischer Analytics (Bingen, Germany) to allow UPLC-like pressures of more than 400 bars. For chromatographic separation, a Kinetex (Biphenyl, 50 × 2.1 mm, 1.7 µm, 100 Å) analytical column (Phenomenex, Aschaffenburg, Germany) protected by a HPLC Guard Cartridge system (Biphenyl Security Guard, Phenomenex) was used. The analytes were detected with a Waters Quattro Micro tandem mass spectrometer equipped with an electrospray ionization interface (ESI). Data were collected and analyzed with MassLynx™ V4.1 software.

HPLC conditions

The chromatographic separation of the analytes was accomplished by gradient elution. The separation was started with mobile phase consisting of 5% methanol, containing 0.1% formic acid and 95% formic acid 0.1% (5:95, v/v). After 0.3 minutes the fraction of methanol was raised to 80% and held constant for 2.5 minutes (80:20, v/v). The system was re-equilibrated with starting conditions (5:95, v/v) for the next 4 minutes. Total chromatographic time was 8 minutes. Changes in mobile phase composition were achieved using step gradient. The mobile phase flow was kept at 400 µl/min and 20 µl of the extracted sample was injected for. With a column temperature of +30°C, the retention times for morphine and morphine–D6 were about 1.88 respectively 1.96 minutes for M6G and M6G-D3.

Mass spectrometry conditions

Ionization and analysis were accomplished in positive electrospray mode with multiple-reactions monitoring (MRM). The following settings for mass spectrometry were used morphine/M6G: ion source temperature was set at 120°C, the capillary and cone voltages were set to 3.8 kV and 45 V/65 V, respectively. The cone gas flow was 50 l/min. Desolvation gas temperature was 500°C and the flow was set to 900 l/min. Nitrogen was used as the desolvation and cone gas, and argon was used as the collision gas. The product ions were 286 -> 165 for morphine, 292 -> 165 for morphine-D6, 462 -> 286 for M6G and 465 -> 289 for M6G-D3. Dwell time was 200 ms for all ion transitions monitored.

Method validation

The method was validated according to the Food and Drug Administration (FDA) guidelines on recovery, linearity, precision, accuracy, selectivity, and specificity [7].

Loss of morphine and M6G during extraction recovery: The efficiency of the extraction procedure of morphine, M6G and internal standards from human plasma was analyzed in triplicates at concentrations of 0.5, 10 and 100 ng/ml for morphine and 0.05, 1, 10 ng/ml for M6G in human plasma. Recovery was calculated by comparison of the results obtained for samples extracted according to the analytical procedure of total plasma concentration with those obtained for the standard solutions added to the blank plasma extracts, corresponding to 100% recovery.

Linearity: Assay linearity was evaluated on 3 consecutive days by constructing freshly prepared plasma calibration samples over the concentration range of 0.5 to 100 ng/ml (morphine) and 0.05 to 10 ng/ml (M6G). The calibration curves were generated with Target Lynx software (v. 4.1; Waters) using linear regression.

The lower limit of quantification (LLOQ) for morphine and M6G was defined as the lowest standard on the calibration curves with a peak response at least five times that of the blank response (S/N ratio) and with precision of 20% and accuracy of 80-120%.

Precision and accuracy: Intraday accuracy and precision were evaluated by analysis of 3 calibration standards per range and 5 replicates of each QC sample concentration in one day. For the assessment of the interday accuracy and precision of the method, sets consisting of calibration standards and 3 different concentrations of QC samples (0.5, 10, 100 ng/ml for morphine and 0.05, 1, 10 ng/ml for M6G) were run on 4 to 16 days. Each set contained 6 calibration standards and 2 to 5 replicates were analyzed. The precision was determined as the intra-assay and inter-assay relative standard deviations (%RSD) of the determined concentrations: $\%RSD = 100 \cdot SD/M$, where M is the mean and SD is the standard deviation. Accuracy was expressed as the relative error from nominal concentration: %RE=100 (experimentally obtained concentration–nominal concentration)/(nominal concentration). The acceptance criteria for accuracy were ± 15% deviation from nominal values and for precision %RSD less than 15%.

Selectivity and specificity: Test for selectivity was carried out in six lots of blank patient plasma. In the first set, the blank human plasma was directly injected after extraction (without analyte and IS), while the other set was spiked with IS and LLOQ concentration of morphine or M6G. The acceptance criterion requires that at least 90% of selectivity samples should be free from any interference at the retention time of analyte and IS. The specificity of the assay was evaluated by comparing chromatograms of a blank plasma sample, a blank plasma sample spiked with 0.5 ng morphine or 0.05 ng M6G and IS; and a patient's plasma sample containing morphine and M6G.

Stability

Short-term stability, post-processing stability, stability after three freeze–thaw cycles, and stock solution stability were evaluated according to the analytical procedure of plasma concentration. For this purpose, 3QC samples per drug range were prepared: 0.5, 10, 100 ng/ml for morphine and 0.05, 1, 10 ng/ml for M6G. The samples were maintained at room temperature (+20°C) for 24 h for the analysis of short-term stability. Post-processing stability was evaluated by maintaining the samples in the autoinjector at 25 °C for six days, followed by injection into the chromatographic system. For the evaluation of stability after three freeze–thaw cycles, the QC samples were frozen at −75°C for 24 h. After this period, the samples were again thawed and frozen for 24 h and this process was repeated until the third thawing cycle when the samples were extracted and analyzed. The stability of stock solutions was evaluated after maintaining them at room temperature (+20°C) for 12 h. Long term stability was tested after storing the QC-samples at -75°C for 12 months, after which the samples were analyzed. The results were compared to those obtained for freshly prepared samples and are expressed as relative error (%RE). The solutions were considered stable if the deviation from nominal value was within ± 15.0%.

Application to clinical study

The method was applied to investigate the concentration time courses of morphine and its metabolite M6G during intensive care treatment of 25 patients after cardiac surgery with cardiopulmonary bypass. These patients participated in a clinical study registered with the ClinicalTrials.gov and EudraCT databases (identifiers NCT02483221 and 2014-004088-19, respectively). Anesthesia was induced and maintained with target controlled infusions (TCI) of propofol (Disoprivan® 2%, AstraZeneca, Wedel, Germany) using the pharmacokinetic model of Marsh et al. [8] and targeting plasma concentrations between 2.5 and 4 μg/ml. After the end of the surgery, the patients were transferred to the ICU where the propofol infusion was continued for further 2-3 h until weaning from mechanical ventilation with an infusion rate of 2-3 mg/kg/h. Post-operative pain therapy was performed with morphine PCA.17 blood samples for the pharmacokinetic measurements were drawn at baseline and 1, 3, 5, 7, 10, 30, 60, 120 minutes and two hourly after starting the PCA. To determine the arterial morphine and M6G concentrations, 4 ml blood per sample were drawn from an artery line (S-Monovette® Kalium EDTA, Sarstedt, Nürnbrecht, Germany), after 1 ml blood had been drawn and discarded. After each sample collection, the intra-arterial catheter was flushed with 2 ml of heparinized NaCl-solution. The samples were kept on ice and plasma was separated within 15 minutes and stored at -75°C until analysis. Samples with concentrations above the linear calibration range (>100 ng/ml for morphine and >10 ng/ml for M6G) were diluted to fit within the range and were reanalyzed.

Results and Discussion

The aim of our investigation was to develop and validate an analytical method to measure concentrations of morphine and its metabolite M6G from human plasma samples, and the application of this method to a clinical study where morphine was used for pain therapy after cardiac surgery with cardiopulmonary bypass. To our knowledge there are no validated investigations in intensive care patients, which describe the analysis of morphine and M6G plasma concentrations in humans.

The previously published methods were modified also by using HPLC-system upgraded to be used in UPLC-like pressures to further improve the analytical performance of the system. Furthermore, deuterated morphine and M6G was utilized as internal standards to increase the robustness against possible matrix effects and to increase the accuracy and precision of the method. The method was validated according to the Food and Drug Administration (FDA) guidelines [7].

Recovery of morphine and M6G

There was no observed inconsistency in the recovery of morphine and M6G. The relative extraction recoveries of morphine and M6G were determined at 3 QC samples per drug: 0.5, 10, 100 ng/ml with 98.9, 103.0 and 100.4% for morphine and 0.05, 1, 10 ng/ml with 96.7, 99.4 and 99.1% for M6G.

Chromatography and mass spectrometry

The best chromatographic separation was obtained with a mobile phase containing a low acid concentration and high percentage of organic modifier. To get an ESI compatible mobile phase the analytes were injected with 0.1% formic acid. The fraction of acetonitrile was raised to 80% during the separation process. Ionic strength and pH were kept stable by adding formic acid to all eluents. A flow rate of 0.4 ml/min were used during the whole run without any loss in quality of chromatographic separation.

Analytical conditions for mass spectrometry were tested to obtain a high intensity of [M+H]+ ions of the analytes. The positive product ion mass spectra showed high abundances of fragment ions. The product ion for morphine was 286, for M6G 462. The precursor -> product ion transformations of m/z 286 -> 165 (Figure 2A) and m/z 462 -> 286 (Figure 2B) were used for MRMs for morphine and M6G, respectively. No interference from endogenous substances in the blank plasma lots or from other metabolites was observed in the retention time of morphine and M6G.

Linearity and limits of detection and quantification

Linear calibration curves were identified for morphine and M6G concentrations within the range from 0.5 to 100 ng/ml (y=61.9x; r²=0.995) and from 0.05 to 10 ng/ml (y=2.03x+0.018; r²=0.999). The Limit of Detection (LOD) for morphine was 10 pg with a S/N ratio of 21.91 and 5 pg with a S/N ratio of 32.13 for M6G. The lower limit of quantification was set at 0.5 ng/ml for morphine and 0.05 ng/ml for M6G. Representative chromatograms of plasma spiked with the LOD concentrations of morphine and M6G is presented in Figure 3.

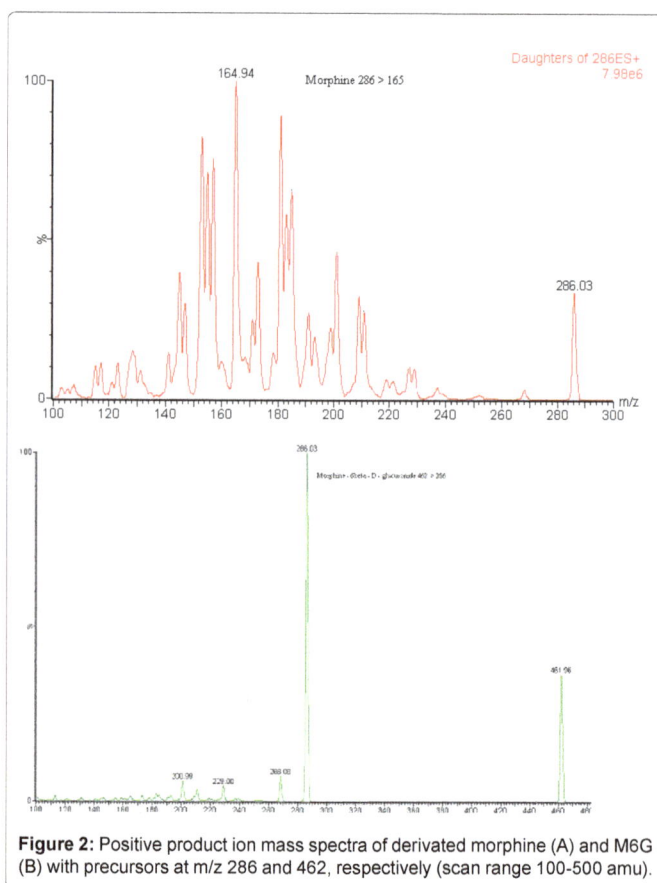

Figure 2: Positive product ion mass spectra of derivated morphine (A) and M6G (B) with precursors at m/z 286 and 462, respectively (scan range 100-500 amu).

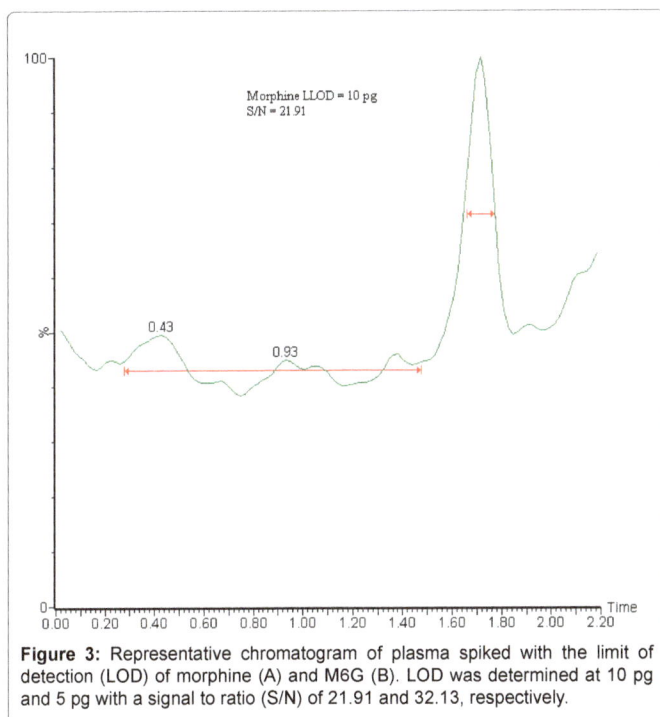

Figure 3: Representative chromatogram of plasma spiked with the limit of detection (LOD) of morphine (A) and M6G (B). LOD was determined at 10 pg and 5 pg with a signal to ratio (S/N) of 21.91 and 32.13, respectively.

Precision and accuracy

The validation data regarding intraday and interday precision and accuracy are summarized in Table 1. The measurement of morphine and M6G by our method showed good intraday and interday precision and accuracy. Compared to previous studies, the utilization of deuterated morphine and deuterated M6G improved the intra- and interday accuracy and precision.

Selectivity and specificity

Specificity was evaluated by comparing chromatograms of blank patient plasma with corresponding drug-free plasma sample spiked with morphine or M6G and IS and with a blank patient samples from the clinical study. None of the six blank plasma lots caused a significant interference at the retention times of the analytes or internal standards. The signal response ratios (blank interference/LLOQ standard) were within the acceptable criteria for all studied compounds (Figures 4 and 5). The matrix effect at QC concentration level of 0.5 ng/ml (n=5) was 0.51 ± 0.04 (%RSD 8.30) for morphine and QC concentration level of 0.05 ng/ml (n=5) was 0.06 ± 0.005 (%RSD 8.49) for M6G indicating that no co-eluting substance influenced the ionization of the analytes.

Stability

No significant degradation of the analytes were observed during the short-term (24-hour) storage at room temperature, three freeze-thaw cycles, or long-term (>2-month) storage at –75°C, suggesting that no stability related problems are to be expected during routine analysis (Table 2). The degradation was less than 15% for all compounds. Also, the stock solutions showed no significant degradation after short term (12 hour) storage at room temperature.

Pharmacokinetic application

This method was utilized in a clinical trial to investigate the time course of morphine and its active metabolite morphine-6-glucuronide (M6G) in patients after cardiac surgery with cardiopulmonary bypass. Following a total intravenous anesthesia with propofol and

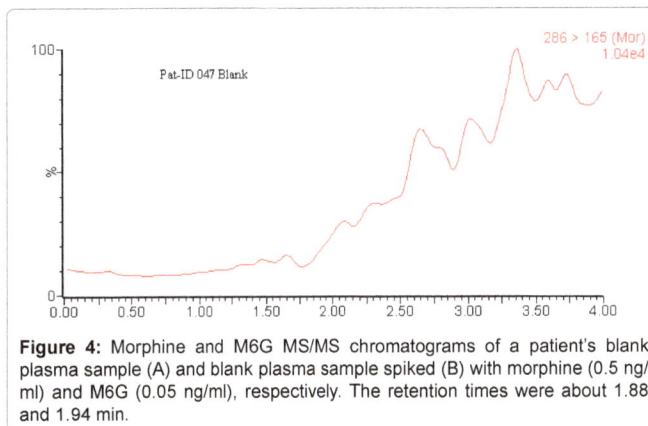

Figure 4: Morphine and M6G MS/MS chromatograms of a patient's blank plasma sample (A) and blank plasma sample spiked (B) with morphine (0.5 ng/ml) and M6G (0.05 ng/ml), respectively. The retention times were about 1.88 and 1.94 min.

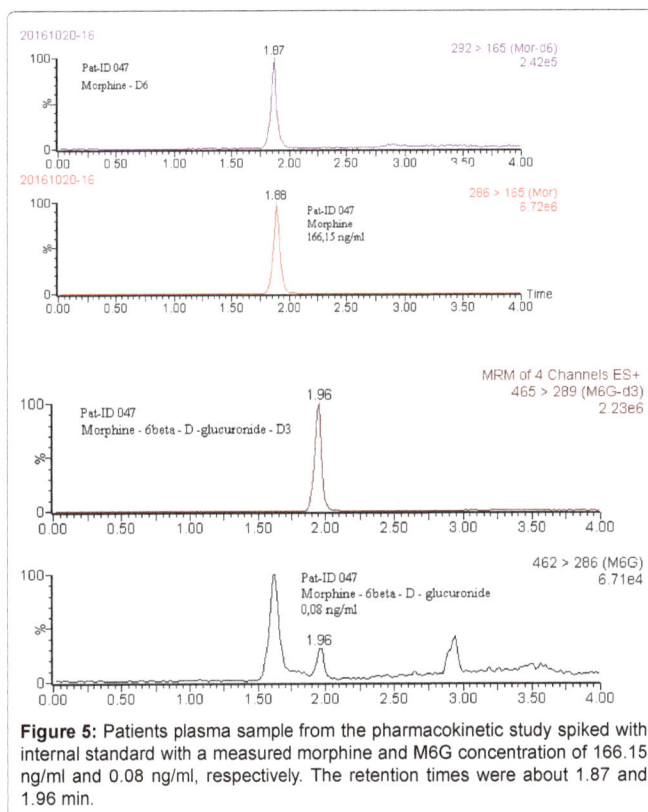

Figure 5: Patients plasma sample from the pharmacokinetic study spiked with internal standard with a measured morphine and M6G concentration of 166.15 ng/ml and 0.08 ng/ml, respectively. The retention times were about 1.87 and 1.96 min.

intraoperative analgesia with sufentanil a morphine based patient controlled analgesia (PCA) was applied for post-operative pain therapy. During the post-operative phase at the ICU the morphine and M6G plasma concentrations were measured in 17 blood samples of 6 patients over a time period of up to 18 h between arrival at the ICU and the following post-operative day. The morphine concentrations varied between 0.77 and 324.4 ng/ml, the M6G concentrations varied between 0.18 and 20.86 ng/ml. All samples were above the LLOQs and shows a time course of morphine and M6G concentration in a representative patient, respectively.

Conclusion

Our investigation demonstrates a sensitive and specific analytical method for determining the concentration of morphine and M6G. The method was used successfully to measure morphine and M6G plasma

	Nominal concentration (ng/ml)	Intraday				Interday			
		Mean ± SD	%RSD	%RE	n	Mean ± SD	%RSD	%RE	n
Morphine	0.5	0.51 ± 0.04	8.3	2.2	5	0.52 ± 0.04	7.74	4.08	20
	10	10.42 ± 0.52	4.96	4.2	5	10.2 ± 0.66	6.52	1.64	11
	100	99.30 ± 1.51	1.52	-0.7	5	97.8 ± 1.85	1.9	-2.15	11
Morphine-6-glucoronide	0.05	0.06 ± 0.005	8.49	11.2	5	0.05 ± 0.006	10.64	5.64	11
	1	1.03 ± 0.054	5.21	2.8	5	0.99 ± 0.067	6.84	-1.43	11
	10	9.51 ± 0.27	2.83	-4.9	5	9.7 ± 0.24	2.51	-3.47	11

%RSD: Relative standard deviation; %RE: Relative error from nominal concentration

Table 1: Accuracy and precision of the analytical method for the concentration measurement of morphine and morphine-6-glucoronide in human plasma.

	Mean ± SD (% RE)					
	Morphine			Morphine-6-glucoronide		
	0.5 ng/ml	10 ng/ml	100 ng/ml	0.05 ng/ml	1 ng/ml	1 ng/ml
Freeze and thaw stability (n=5)	0.5 ± 0.04 (-3.88)	9.8 ± 0.64 (-2.36)	96.8 ± 1.12 (-3.22)	0.049 ± 0.01 (-1.2)	0.96 ± 0.06 (-3.58)	9.8 ± 0.12 (-2.12)
Short-term stability[a] (n=4)	0.50 ± 0.04 (-0.05)	9.9 ± 0.64 (-0.52)	98.5 ± 2.51 (-1.53)	0.050 ± 0.01 (0.50)	0.95 ± 0.05 (-5.08)	9.68 ± 0.15 (-3.17)
Long-term stability[b] (n=4)	0.52 ± 0.02 (4.5)	NA	100.3 ± 5.09 (0.3)	0.058 ± 0.003 (15.5)	NA	10.0 ± 0.34 (0.3)

[a]Short-term stability, 24h in room temperature (20°C); [b]Long-term stability, 12 months in -75°C; %RE: Relative error from nominal concentration.

Table 2: Stability analysis of the analytical method for the measurement of morphine and morphine-6-glucoronide concentrations in human plasma.

concentration from patients during intensive care treatment and receiving several other drugs concomitantly.

Acknowledgements

The authors thank Rainer Knoll, Dipl. Bioingenieur (Department of Anesthesiology, University of Erlangen-Nürnberg, Erlangen, Germany) for his kind help in conducting the drug analysis.

References

1. Pandharipande PP, Patel MB, Barr J (2014) Management of pain, agitation, and delirium in critically ill patients. Pol Arch Med Wewn 3: 114-123.

2. Lotsch J (2005) Opioid Metabolites. J Pain Symptom Manage 5: 10-24.

3. Andersen G, Christrup L, Sjogren P (2003) Relationships among morphine metabolism, pain and side effects during long-term treatment: an update. J Pain Symptom Manage 1: 74-91.

4. Sessler CN, Grap MJ, Ramsay MA (2008) Evaluating and monitoring analgesia and sedation in the intensive care unit. Crit Care Suppl 3: S2.

5. Ahlers SJGM, Välitalo PAJ, Peeters MYM, van Gulik L, van Dongen EPA, et al. (2015) Morphine Glucuronidation and Elimination in Intensive Care Patients. Anesth Analg 5: 1261-1273.

6. Bosch ME, Sánchez AR, Rojas FS, Ojeda CB (2007) Morphine and its metabolites: Analytical methodologies for its determination. J Pharm Biomed Anal 3: 799-815.

7. Food and Drug Administration (2001) Guidance for Industry: Bioanalytical Method Validation. US Dep Heal Hum Serv May, pp: 4-10.

8. Marsh B, White M, Morton N, Kenny GN (1991) Pharmacokinetic model driven infusion of propofol in children. Br J Anaesth 1: 41-48.

27

Development of On-Line Solid-Phase Extraction-Liquid Chromatography Coupled with Tandem Mass Spectrometry Method to Quantify Pharmaceutical, Glucuronide Conjugates and Metabolites in Water

Lea B[1], Nicolas C[1]*, Dominique W[1] and Guy R[2]

[1]Ecole National de Chimie de Rennes, CNRS, UMR 6226, 11 Allee de Beaulieu, CS 50837, 35708 Rennes Cedex 7, France
[2]Veolia Eau, Direction Technique Region Ouest, 8 allee Rodolphe Bopierre, 35020 Rennes Cedex 9, France

Abstract

The present work describes the development of an analytical method, based on automated on-line solid phase extraction followed by ultra-high-performance liquid chromatography coupled with tandem mass spectrometry (SPE-LC-MS/MS) for the quantification of 37 pharmaceutical residues, covering various therapeutic classes, and some of their main metabolites, in surface and drinking water. A special attention was given to some glucuronide conjugates and metabolites of active subtances. Multiple Reaction Monitoring (MRM) was chosen and two transitions per compound are monitored (quantification and confirmation transitions). Quantification is performed by standard addition approach to correct matrix effect. The method provides limit of quantification inferior to 20 ng.L^{-1} for all compounds. The methodology was successfully applied to the analysis of surface water and drinking water of 8 drinking water treatment plant in west of France. The highest drug concentrations in surface water and drinking water were reported for ketoprofen, hydroxyibuprofen, acetaminophen, caffeine and danofloxacin.

Keywords: Pharmaceuticals; Automated on-line solid phase extraction; Liquid chromatography; Tandem mass spectrometry; Water analysis

Introduction

Pharmaceuticals are an important group of emerging contaminants in the environment [1]. In recent years many reports have been made on the occurrence of the large, differentiated group of pharmaceuticals in wastewater, surface water, ground water and drinking water in many countries [2-9]. After administration, most pharmaceuticals are not completely metabolized. The unmetabolized parent pharmaceutical and some metabolites are subsequently excreted from the body via urine and faeces [10]. Reports have shown that many pharmaceuticals do not totally degrade during conventional wastewater treatment [11,12]. The concentrations of individual compounds in wastewater, surface water, ground water and drinking water are typically in the range of ng/L to μg/L. The effect on long-term pharmaceutical residues in aquatic environments remains largely unknown. In addition, the risks to the environment are evaluated for a particular drug, while we find a mixture of all these compounds in aquatic environments. Studies have shown that combinations of drugs may be more powerful than the simple addition of two drugs individually toxic effects [13,14].

Wastewater effluent is a major source for the input of pharmaceuticals to the environment [11,12], which can then migrate through water systems and into source water intended for drinking water supplies. Advanced wastewater treatment processes have been shown to significantly reduce the concentrations of emerging contaminants. However, some compounds are not completely removed even if treatment techniques are used [15]. Moreover, most of the WWTP do not include these specifically designed treatment units.

In this context, sensitive analytical methods allowing the quantification of many pollutants at trace concentration is essential. Solid Phase Extraction (SPE) is the most commonly used technique to prepare sample before analysis. SPE allows the concomitance of analyte concentration and interferences removal [16,17]. To date, most of the published multi-residue methods for the determination of ultra traces of pharmaceuticals compounds in surface and drinking water use off-line SPE followed by gas chromatography mass spectrometry (GC–MS) or by liquid chromatography-tandem mass spectrometry (LC–MS/MS) [2-5,7,9,12]. However, On-line Solid Phase Extraction is an emerging method for analysis of the trace compounds of organic micropollutants (reactive drugs, pesticides). This technique has many advantages: saving time, automated method, reproducibility, very low solvent consumption, small sample handling, SPE cartridges reuse [17]. The cartridges used to concentrate pharmaceuticals residues are usually OasisTM HLB or hydrophobic resins. [18,19]. This technique is generally coupled to liquid chromatography with UV, MS or MS/MS detector with reversed phase column [20-24].

The objectives of this work has been to develop a fully automated method to analyze a number of target compounds belonging to different therapeutical classes and some by product using on-line SPE directly coupled to liquid chromatography tandem mass spectrometry (LC–MS/MS). This analytical technique limits matrix effect impact. However remaining, interfering species can affect the analytical train, especially natural organic matter may coeluate with targeted compounds which leads to a signal disturbance causing over/underestimation or false positive results, or some compounds may react with targeted molecules during sampling and storage [25].

This method was evaluated in different water matrices: UltraPure Water (UPW) to develop the analytical method, surface water and drinking water for validation.

*****Corresponding author:** Nicolas Cimetiere, Ecole National de Chimie de Rennes, CNRS, UMR 6226, 11 Allee de Beaulieu, CS 50837, 35708 Rennes Cedex 7, France, E-mail: Nicolas.cimetiere@ensc-rennes.fr

Materials and Methods

Compound selection

32 pharmaceuticals and 3 metabolites and 2 glucuronide conjugates were selected for this study (Table 1 and S1). These molecules were chosen based on the following criteria: (i) selected compounds should exhibit a variety of physical properties, such as functional groups and polarity, (ii) they should represent of a diversity of pharmaceutical classes, (iii) high frequencies of environmental occurrence, (iv) low removal efficiencies by drinking water and wastewater treatment techniques in France or others countries [2-9]. Table 1 lists the 37 molecules selected for our study and their optimized parameters for quantification, chemical structure is provided in the Figure S1 in Supporting Information. Thereafter, the molecules will be called by the short identifiers which are given in the Table 1. The pharmaceutical classes represented are cardiovascular drugs, anticancer agents, human or veterinary antibiotics, neuroleptics, non-steroidal anti-inflammatory drugs and hormones.

Pharmaceutical standards and reagents

All pharmaceutical compounds have minimum 90% purity, used as received in solid form and were obtained from Sigma Aldrich (France). Ultra-pure water (UPW) was delivered by a Elga Pure Lab System (resistivity 18.2 MΩ.cm, COT <50 µg C/L). Chromatographic and SPE solvents, acetonitrile (ACN) with or without 0.1% formic acid (FA) and methanol (MeOH) were purchased from JT Baker (LC-MS grade) and were used in association with UPW in also or not with 0.1% formic acid.

All concentrated stock solution of individual pharmaceuticals were prepared in methanol with a concentration of 500 mg.L^{-1} and stored at -20°C. The mixed spiking solutions were prepared in methanol at 500 µg.L^{-1} and stored at 4°C during 15 days maximum. This mixed spiking solution is daily diluted in water to obtained 500 ng.L^{-1} before use for standard addition. Concentrations prepared for analytical development and to quantify the target compounds in the different matrices are: 5, 10, 20, 50, 100, 250 and 500 ng.L^{-1}.

On-line solid phase extraction and liquid chromatography

The analytical system consists of an automated SPE sampler coupled with an LC-MS/MS. The online extraction was carried out using a 2777 auto sampler equipped with two parallel Oasis™ HLB cartridge (Direct Connect HP 20 µm, 2.1 mm × 30 mm) working sequentially. The switching from the loading flow pattern, to elution, then conditioning and back to loading is performed using two six positions Everflow™ valves. Loading eluent (UPW) and conditioning eluent (methanol) were provided by a quaternary pump (Acquity™ QSM). Elution of the analytes from the SPE cartridge to LC system was achieved by connected the cartridge to the inlet of the separation column and using the initial chromatographic elution solution.

Separation was carried out using a reversed phase column (Acquity™ BEH C18, 100 mm × 2.1 mm ID, 17 µm) placed in an oven (45°C). The elution gradient was produced by a binary pump (Acquity™ BSM) and was optimized and will be described later in the manuscript.

Mass spectrometry

The mass spectrometer (Quattro Premier, Micromass™) operates with the following conditions: cone gas (N$_2$,50 L.h^{-1},120°C), desolvation gas (N$_2$,750 L.h^{-1}, 350°C), collision gas (Ar,0.1 mL.min^{-1}), capillary voltage (3000 V). The ionization source of the mass spectrometer is an electrospray (ESI) used either in the positive or the negative mode according to pharmaceutical compounds structure (Table 1). All the analysis, are made in "multiple reaction monitoring" (MRM) mode, the parent ion from the ESI source is selected in the first quadrupole (pseudo molecular ion in most cases) and fragmented in the collision cell. One or more fragments (quantification ion and, when available, confirmation ions) are then selected by the third quadrupole before being detected by a photomultiplier. This mode allows high sensitivity and selectivity.

Results and Discussion

Mass spectrometry optimization

The selection of optimum detection parameters (collision energy, cone voltage, ionization mode) for each targeted compound was carried out by introducing a standard diluted single solute solution at 5 mg.L^{-1} directly in the mass spectrometer (without separation). The pseudo-molecular ion [M+H]$^+$ or [M-H]$^-$ was selected as the parent ion. Acetaminophen-glucuronide was ionized as sodium adducts [M+Na]$^+$ and the daughter ion correspond to the sodium adduct of paracetamol obtained by the loss of glucuronic acid. Similar fragmentation pattern with loss of carbohydrate group was observed with Glu-OZP [M+H]$^+$ → [M-Glu+H]$^+$. In some cases, the standard molecules were purchased as sodium or chloride salt so molecular weight of the commercial product indicated in the Table 1 does not correspond to the formula of active compounds. So the molecular weights indicated in the Table 1 do not correspond to the mass of the pseudo molecular ion (AML, LOS, NAF, PRA, TRI, DOX, ERY, LINCO and TYL). Positive mode was selected for most of the molecules and 8 analytes were ionized under negative mode because of their tendency to lose a proton. Two transitions are chosen for quantification and confirmation. If possible transition corresponding to the loss of simple's fragments (i.e., –H$_2$O or –CO$_2$) has been preferred for quantification or confirmation transition. Only one transition could be found to 4 molecules: Ibuprofen, Gemfibrozil, Tamoxifen and Hydroxy-Tamoxifen. The results are presented in Table 1.

On-line SPE method development

The efficiency of the SPE step was studied using two different types of SPE cartridge phases: Oasis HLB (Direct Connect HP 20 µm, 2.1 mm × 30 mm) and X Bridge C18 (Direct Connect HP 10 µm, 2.1 mm × 30 mm). The low energy interactions are predominant with the C18 phases, unlike for HLB phases where the dipole-dipole interactions are brought into play. Table 2 presents characteristics (log(Kow), pka, coefficient of dissociation, dipolar moment) of molecules. The extraction yield was then calculated according to the following equation:

$$\text{Extraction yield (\%)} = 100 \times \text{Area}_{\text{SPE mode}} / \text{Area}_{\text{conventional mode}}$$

For each compounds, the area obtained with the injection of 5 mL of solution at 100 ng.L^{-1} in SPE mode was compared to the area obtain in conventional mode (Vinj=5 µL; C=100 µg.L^{-1}).

The results are presented in Figure 1. In a global overview the extraction yields are better with the Oasis HLB phase in comparison to the C18 phase. 11 molecules have slightly better extraction yields with the XBridge C18 media. Given these results, Oasis HLB phase was chosen for the SPE cartridges. The extraction yields are between 24% and 96%. Six molecules, among them three hormones (ATE, TRI, DOX, EE, βE and EO) have extraction yields inferior or equal to 50% but the signal is sufficient for our analysis given the reproducibility of the extraction step. The loading time and flow rate influence the analyte

Pharmaceutical class	Molecule (short identifier)	N°CAS	MW (g/mol)	Formula of the active substance	ESI	Parents ion	Daughter ion(Q)	Cones (V)	Collisions (V)	Confirmation ion	Collisions (V)	Dwell time (ms)	Tr (min)
Cardiovascular drugs	Amlodipin (AML)	111470-99-6	567.05	$C_{20}H_{25}ClN_2O_5$	+	409.6	238.1	18	11	409.6	13	50	4.03
	Atenolol (ATE)	29122-68-7	266.34	$C_{14}H_{22}N_2O_3$	+	267	145	34	26	74	23	50	1.18
	Losartan (LOS)	124750-99-8	461	$C_{22}H_{23}ClN_6O$	+	423.6	405.2	30	12	207	22	50	4.25
	Naftidrofuryl (NAF)	03200-6-4	473.56	$C_{24}H_{33}NO_3$	+	384.6	99.7	40	21	84.7	25	50	4.29
	Pravastatin (PRA)	81131-70-6	446.51	$C_{23}H_{36}O_7$	-	423.2	100.6	34	23	321.1	16	50	2.63
	Propanolol (PRO)	525-66-6	259.4	$C_{16}H_{21}NO_2$	+	260.2	116	34	18	183	18	50	3.33
	Gemfibrozil (GEM)	25812-30-0	250.33	$C_{15}H_{22}O_3$	-	249	121	34	23			50	4.95
	Trimetazidin (TRI)	13171-25-0	339.26	$C_{14}H_{24}Cl_2N_2O_3$	+	267.4	180.9	21	16	165.8	26	50	1.18
Anticancer agent	Tamoxifen (TAM)	10540-29-1	371.5	$C_{26}H_{29}NO$	+	372.5	72	45	14			50	5.42
	Hydroxytamoxifen (OH-TAM)	68047-06-3	387.2	$C_{26}H_{29}NO_2$	+	388.2	72	45	14			50	4.58
	Ifosfamide (IFO)	3778-73-2	261	$C_7H_{15}Cl_2N_2O_2P$	+	261.02	153.95	25	22	92.04	25	75	3
Human Antibiotic	Doxycycline (DOX)	24390-14-5	512.94	$C_{22}H_{24}N_2O_8$	+	445.5	428.2	30	18	153.8	28	50	2.95
	Erythromyicin (ERY)	114-07-8	769.96	$C_{37}H_{67}NO_{13}$	+	734.2	158	28	30	576.2	19	50	3.68
	Ofloxacin (OFX)	82419-36-1	361.37	$C_{18}H_{20}FN_3O_4$	+	362	318	34	19	261	28	80	1.35
	Sulfaméthoxazole (SUL)	723-46-6	253.278	$C_{10}H_{11}N_3O_3S$	+	254	92	26	28	156	16	50	2.74
	Trimetoprime (TRP)	738-70-5	290.3	$C_{14}H_{18}N_4O_3$	+	291.2	230	24	24	261.1	26	50	1.18
Veterinarian Antibiotic	Danofloxacin (DANO)	112398-08-0	357.38	$C_{19}H_{20}FN_3O_3$	+	358.5	314	35	19	283	25	50	1.53
	Lincomycin (LINCO)	859-18-7	461.37	$C_{18}H_{34}N_2O_6S$	+	407.6	125.9	40	26	359.3	18	50	1.23
	Sulfadimerazine (SFZ)	57-68-1	278.33	$C_{11}H_{12}N_4O_2S$	+	279.4	185.9	29	16	91.7	26	50	1.91
	Tylosin (TYL)	74610-55-2	1066.19	$C_{46}H_{77}NO_{17}$	+	917	174	60	37	773	29	50	3.84
Neuroleptic	Carbamazepine (CBZ)	298-46-4	236.27	$C_{15}H_{12}N_2O$	+	237.1	194	28	19	179	39	50	3.85
	Epoxycarbamazepine (Ep-CBZ)	36507-30-9	252.27	$C_{15}H_{12}N_2O_2$	+	253.3	179.9	28	28	236	12	50	3.2
	Oxazepam (OZP)	604-75-1	286.71	$C_{15}H_{11}ClN_2O_2$	+	287.4	241	34	20	269.1	14	50	4.08
	Oxazepam (Glu-OZP)	6801-81-6	462.84	$C_{21}H_{19}ClN_2O_8$	+	463.2	287.1	26	15	269	26	15	3.34
Non-steroidal anti-inflammatory drugs (NSAID)	Diclofenac (DICLO)	15307-79-6	294.14	$C_{14}H_{11}Cl_2NO_2$	+	296.1	250	22	10	214.1	25	100	5.5
	Ibuprofen (IBU)	15687-27-1	206.28	$C_{13}H_{18}O_2$	-	205	161	17	7			50	4.06
	Hydroxyibuprofen (OH-IBU)	51146-55-5	222.28	$C_{13}H_{18}O_3$	-	221.2	177	19	9	158.7	13	50	1.2
	Ketoprofen (KETO)	22071-15-4	254.28	$C_{16}H_{14}O_3$	+	255	209	29	12	105	22	100	4.14
	Salicylic acid (SCA)	69-72-7	138.12	$C_7H_6O_3$	-	137	92.6	30	14	64.7	28	70	1.16
Miscellaneous	Acetaminophen (PARA)	103-90-2	151.16	$C_8H_9NO_2$	+	152	110	25	15	90	10	50	1.24
	Acetaminophen Glucuronide (Glu-PARA)	16110-10-4	327.29	$C_{14}H_{17}NO_8$	+	350	173.8	33	15				1.64
	Caffeine (CAF)	58-08-2	194.19	$C_8H_{10}N_4O_2$	+	195.1	137.7	37	18	109.7	22	50	1.35
	Hydrochlorothiazide (HCTZ)	58-93-5	297.74	$C_7H_8ClN_3O_4S_2$	-	296.2	77.6	42	28	204.8	22	50	1.5
Hormone	Ethyinylestradiol (EE)	57-63-6	296.4	$C_{20}H_{24}O_2$	-	295.2	144.9	54	40	183	35	50	4.07
	17β-Estradiol (βE)	50-28-2	272.38	$C_{18}H_{24}O_2$	-	271.1	145	50	38	183	41	70	3.89
	Estrone (EO)	53-16-7	270.37	$C_{18}H_{22}O_2$	-	269.1	145	53	35	183	36	70	4.14
	Progesterone (PGT)	57-83-0	314.46	$C_{21}H_{30}O_2$	+	315.2	97	32	24	109	26	50	5.77

Table 1: List of the 35 pharmaceuticals with pharmaceutical class Molecule (short identifier), N°CAS, MW (g/mol), formula, mass parameter and retention time.

retention onto the pre concentration cartridge. If the loading time is too short, a part of the molecules of interest will not be collected in the cartridge. MeOH is used for the cartridge conditioning during 3 minutes and UPW for the loading sample during 5.5 minutes at 2 mL/min. 5 mL of sample are injected onto the cartridge. Elution of our compounds is made using the initial chromatographic conditions. The pre concentration method takes 8.5 minutes. The pH of samples and eluents was also optimized to try to improve the extraction yields. The Figure 2 shows the effect of pH (3,7 and 9) on molecule's recovery yields. Most of the targeted compounds were efficiently extracted at neutral pH values. The recovery yields of thirteen molecules (LOS, GEM, TAM, OH-TAM, IFO, TYL, DICLO, PARA, CAF, CBZ, OZP, PGT and ERY) do not show significant pH dependence. ATE, NAF and LINCO were comparatively more recovered under neutral condition due to the amine/ammonium repartition for the low pH values. DANO and OFX are amphoteric molecules and exhibit higher recovery yields under acid extraction than under neutral conditions.

AML and OFX have extraction yields superior to 100%, the differences may be included within the experimental errors. Three hormones have a better extraction yields at basic pH while below 23% for an acid pH. The SPE appears globally controlled by the carboxylic functions. The best compromise to our analytical method is the neutral pH.

Chromatographic conditions

Three chromatographic columns packed with different stationary phases were studied, two using the reversed phase mode: Acquity BEH C18 (100 mm × 2.1 mm ID, 1.7 μm) and Acquity HSST3 (100 mm × 2.1 mm ID, 1.7 μm). These two columns have the same stationary phase but Acquity HSST3 should allow for better separation of polar molecules due to the greater proportion of residual silanol groups. The third column has a polar stationary phase: BEH amide (100 mm × 2.1 mm ID, 1.7 μm) in order to separate the analyte using hydrophilic interaction liquid chromatography (HILIC). Comparing the chromatograms obtained for the C18 and HSST3 column, the

Molecule	Log (Kow)	pka	Coefficient of dissociation	Dipolar moment
AML	3	8.6	$5.00\ 10^{-5}$	
ATE	0.16	9.6	$1.50\ 10^{-5}$	5.71
LOS	1.19	5,5	$8.80\ 10^{-3}$	
NAF	4.56	8.7	$4.70\ 10^{-5}$	2.83
PRA	1.35	4.5	$5.60\ 10^{-3}$	
PRO	3.48	9.5	$1.70\ 10^{-5}$	
GEM	4.77	4.7	$4.40\ 10^{-3}$	
TRI	1.04	4.3/8.9	$7.00\ 10^{-3}$	
TAM	3.24	8.76	$4.20\ 10^{-5}$	
OH-TAM	4.74	3.2/6.4	$6.30\ 10^{-4}$	
IFO	0.86	13.2	$2.50\ 10^{-7}$	
DOX	2,37	3.5/7.7	$1.40\ 10^{-4}$	
ERY	3,02	8.8	$3.90\ 10^{-5}$	
OFX	0.65	6.1	$9.40\ 10^{-4}$	7.2
SUL	0.79	5.7	$1.40\ 10^{-3}$	
TRP	0.91	7.1	$2.80\ 10^{-4}$	
DANO	0,44	6.0	$9.90\ 10^{-4}$	
LINCO	0,56	7.6	$1.60\ 10^{-4}$	
SFZ	0.19	7	$3.20\ 10^{-4}$	7.34
TYL	1.63	7.7	$1.40\ 10^{-4}$	
CBZ	2,77	7	$1.00\ 10^{-7}$	3.66
Ep-CBZ	1.58	15.9	$1.00\ 10^{-8}$	
OZP	2,24	1.7/11.6	$1.30\ 10^{-1}$	
DICLO	4,51	4	$8.00\ 10^{-3}$	4.55
IBU	3,79	4.5	$5.30\ 10^{-3}$	4.95
OH-IBU	3,97	4.8	$3.90\ 10^{-3}$	
KETO	3.12	4.45	$6.00\ 10^{-3}$	
SCA	1,19	3	$3.10\ 10^{-2}$	
PARA	0,49	9.5	$1.80\ 10^{-5}$	4.55
CAF	-0.091	14	$2.10\ 10^{-1}$	3.71
HCTZ	-0,07	7.9	$1.00\ 10^{-4}$	
EE	3,67	10.3	$7.00\ 10^{-6}$	
βE	3.57	10.71	$4.40\ 10^{-6}$	1.56
EO	3.69	10.4	$6.00\ 10^{-6}$	3.45
PGT	4	18.9	$3.50\ 10^{-10}$	

Table 2: log (Kow), pka, coefficient of dissociation and dipolar moment of molecules.

results are quite similar. Seven minutes are required to obtain sufficient separation. It should be underlined that the resolution between two consecutive peaks was quite low. However, because the quantification was done using different MRM channels this poor resolution does not affects the analytical performances.

Figure 3 summarizes the results by plotting the polarity (log Kow) as function of the capacity factor of the molecule, molecules with k'<1 form the un retained groups with no log(kow) dependences. For the others, correlation between k' and log(kow) shows two adverse behaviors in relation with the different stationary phase, BEH and HSST3 on the one part and HILIC on the second part. Reversed phase HPLC columns (BEH C18 and HSST3) provide a satisfactory separation with k' ranging from 0.93 to 9.91 according to the polarity of the considered compounds. However numerous analytes exhibit a high polarity and were poorly retained using reversed-phase HPLC. Normal phase HPLC column (BEH Amide) provides separation with k' ranging from 0.1 to 9.6. Molecules retained by the reversed phase HPLC column are not retained in normal phase HPLC with k'<1. Moreover, peak tailing are observed for some molecules with HSST3 (SUL, GEM, DOX) and with HILIC column (PARA, DANO, HCTZ, TRI). The best compromise for our analyses is to use the BEH C18 column.

The mobile phase flow rate was $0.4\ mL.min^{-1}$, corresponding to the optimum zone of the Van Deemter curve with this column [26]. The elution conditions were optimized. Two chromatographic separation methods were needed to quantify all the target analytes. Indeed, analytes with ESI+ detection have better sensitivity with acidified eluents (with 0.1% of formic acid) unlike molecules with ESI- detection which have better sensitivity with neutral eluents. Moreover, the combination of both positive and negative ionization mode during the same run does lead to a decrease of the sensibility.

The elution conditions start with 20% ACN/80% UPW during 1 minute followed by a gradient 90% ACN within 6 minutes and remain constant for 1 min before returning to initial conditions, details of the method are presented in Supporting information (Section B – Figures S1-S3)

Examples of chromatograms obtained with a solution of 50 $ng.L^{-1}$ in UPW and the eluent program are presented in Figure 4. 12

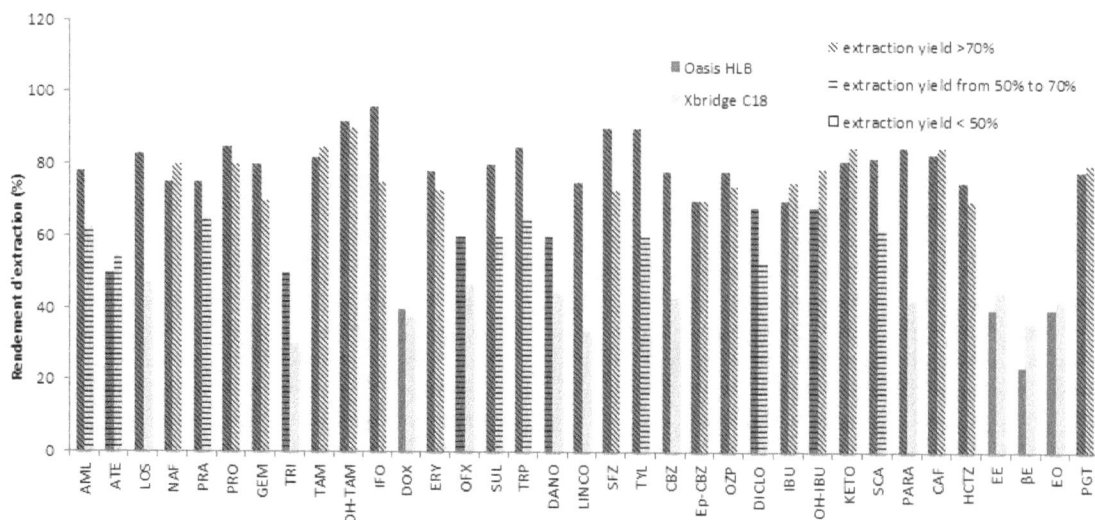

Figure 1: Extraction yields calculated for the two cartridges (Oasis HLB and Xbridge C18) tested for all molecules in neutral pH.

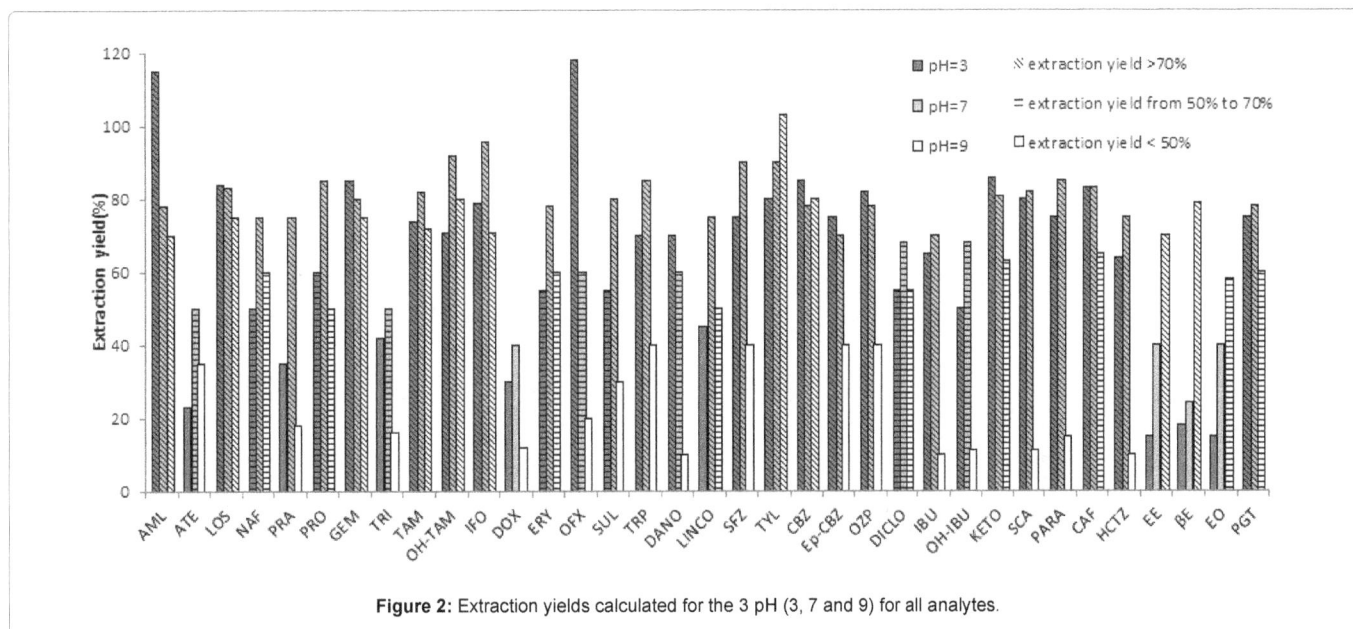

Figure 2: Extraction yields calculated for the 3 pH (3, 7 and 9) for all analytes.

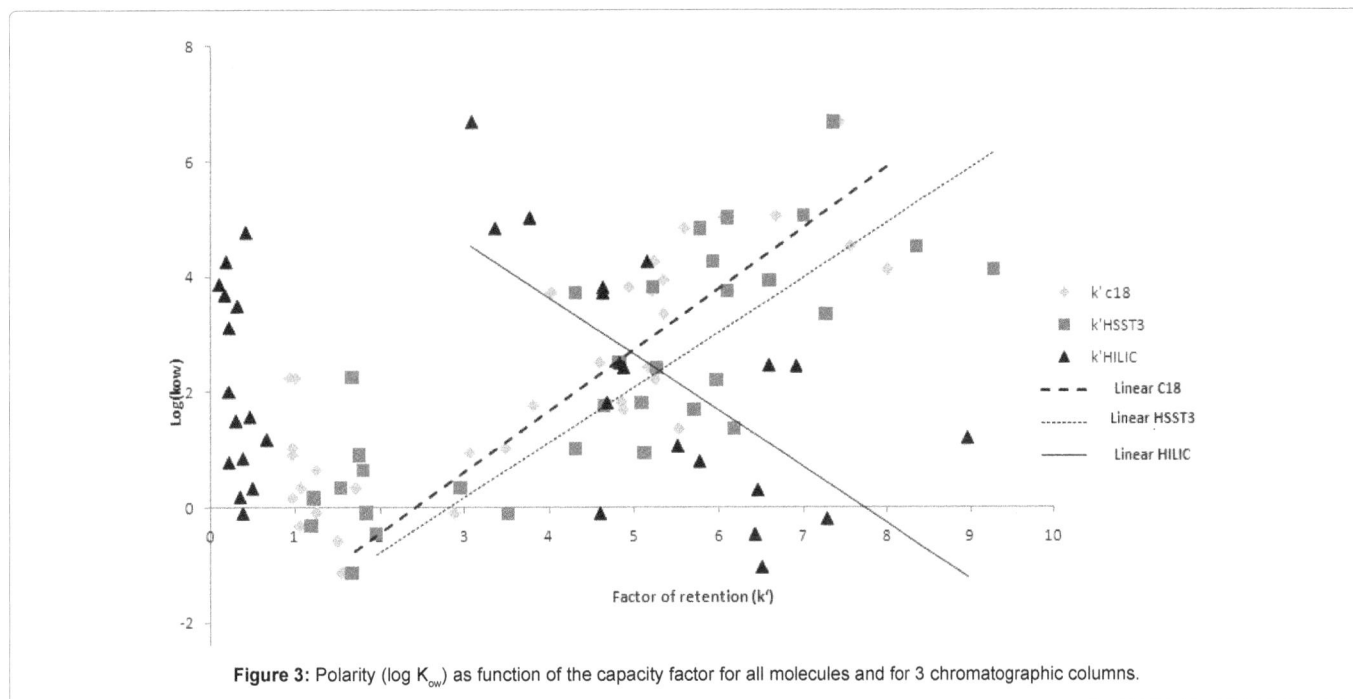

Figure 3: Polarity (log K_{ow}) as function of the capacity factor for all molecules and for 3 chromatographic columns.

molecules elute within two minutes for the ESI+/acid eluent method. As mentioned above, the detection mode (MRM) allows an accurate quantification even if the resolution is low.

Quantification limit and matrix effect

Standard addition method was selected for calibration method in order to minimize or eliminated matrix effects. Figures 5 present examples of calibration curve for CBZ in UPW, Groundwater (GW), Drinking water (DW) and Surface water (SW). Limit of quantification (LOQ) were determined for all targeted compounds in UPW and GW with the equation given in Figure 5a, in accordance with the AFNOR NF-T-90-210 norm for all analytes. GW could be considered free of pharmaceuticals residues because GW is drawn from a well recovering

the waters on a small watershed without collective or on-site sanitation water release, and UPW can be considered as a matrix blank. Negatively ionized molecules (EO, BE, EE, HCTZ, SCA, IBU, OH-IBU, GEM, PRA) have higher limits of quantification because the background noise is more important than for ESI+. The values of the quantification limit of targeted compounds are presented in Figure 6a. LOQ values obtained range from (5 to 17) ng/L. These limits of quantification are sufficient for our purpose.

Measurement errors were incorporated by defining the 90% confidence intervals (Figure 5b). Figures 5c and 5d show standard addition calibration lines of CBZ in GW and DW. Comparisons of the slopes obtained with real waters to the slope obtain in the blank (aGW/aUPW and aDW/aUPW) allow a comprehensive approach of

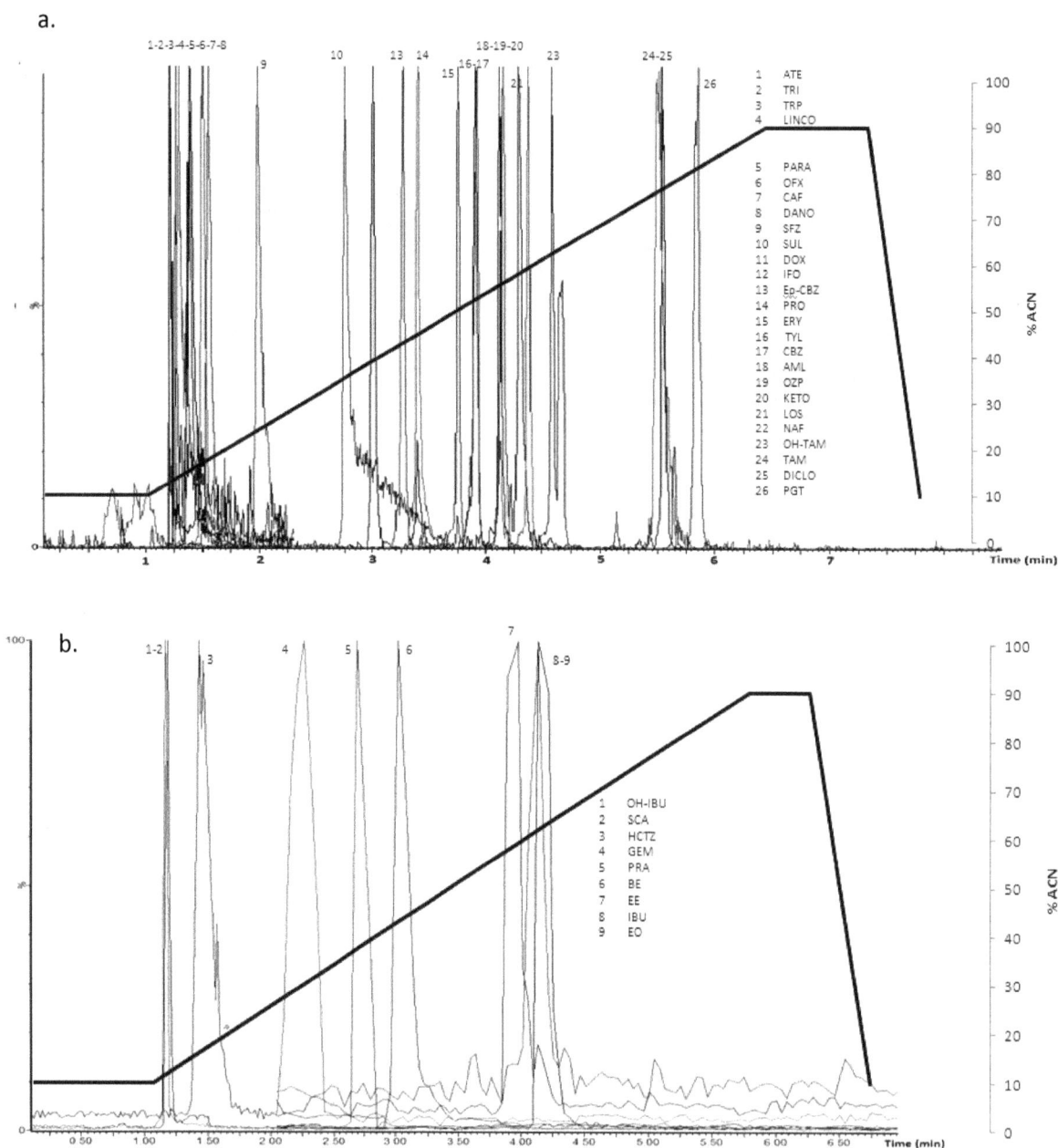

Figure 4: Chromatogram obtained at 50 ng/L in UPW, **a.** first method with ESI+; **b.** Second method with ESI-.

the matrix effects. These slope ratios are presented in Figure 6b for all analytes. The matrix effect is a classical phenomenon which can be very important in liquid chromatography coupled with mass spectrometry because of the ionization process may be drastically influenced by the presence of interfering species. Many studies have already described this phenomenon especially with wastewaters. The presence of organic or inorganic substance can cause inhibition (<1) or enhancement (>1) of a compound's signal [27-29]. In our case, natural organic matter may disturb the SPE step or mass ionization so the rationalization of the slopes provides a global overview of matrix effect but do not allow to identify the critical step.

In Figure 6b, matrix effects are not significant when the ratio is close to 1. In drinking water this ratio was close to 1 for most of the analytes, only AML has a ratio superior to 5.

Analysis of surface water and drinking water

The developed method was used to determine the concentration of 37 pharmaceuticals substances in inflow and outflow waters of 8 drinking water treatment plants (DWTP) in west of France. The samples were collected once a month between October 2013 and April 2015, resulting in an average of 100 inflow and 100 outflow concentration values for each molecule. Nine pharmaceuticals have not been detected or with concentrations below the LOQ (AML, TAM, OH-TAM, IFO,

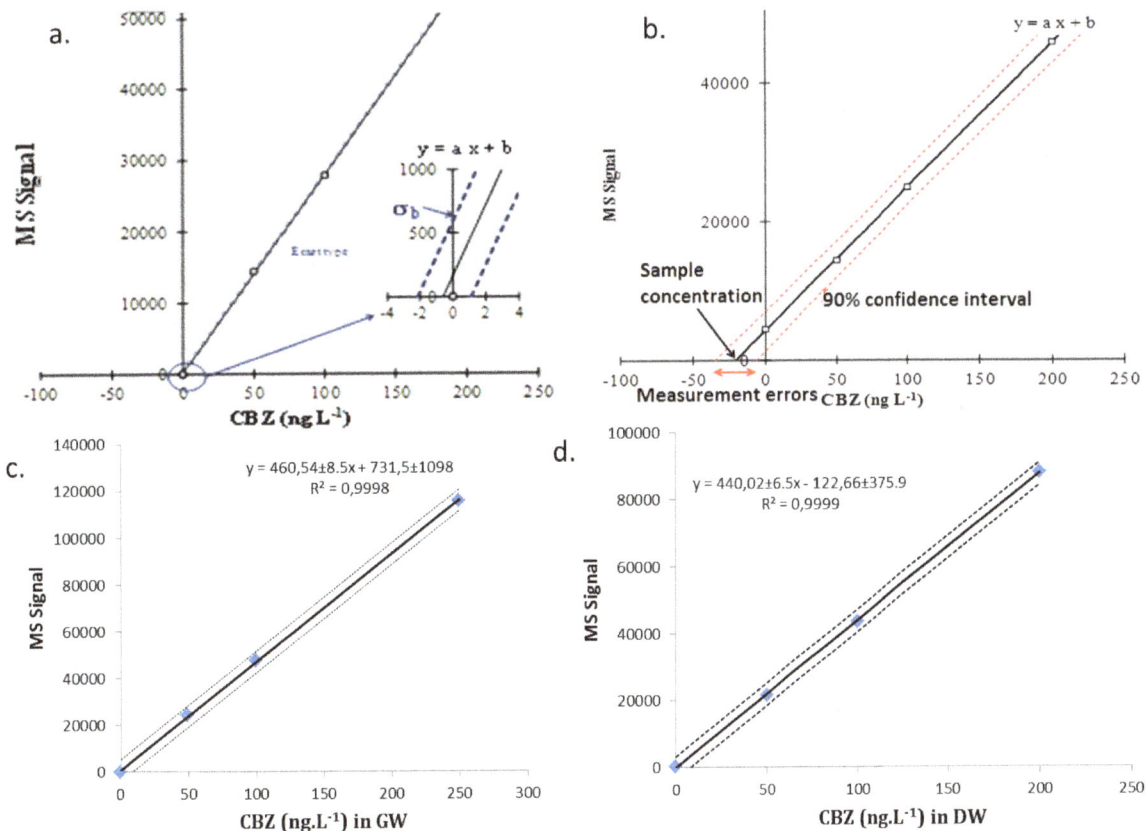

Figure 5: a. Equation of LOQ determanation, **b.** Exemple of standard addition for CBZ with 90% confidence interval, **c and d.** Example of standard addition in GW and DW for CBZ.

ERY, LINCO, EE, βE and PGT). Figure 7 shows the concentrations of 27 pharmaceuticals or metabolites in surface water as a box plot; this statistical representation summarizes the data, for each compound, by the mean values, median value, first and third quartiles and observed extrema. 7 molecules (PARA-GLU, KETO, OH-IBU, DANO, PARA, SCA, CAF) have a mean concentration greater than 50 ng.L^{-1}. 10 molecules were quantified with mean concentrations higher than 10 ng.L^{-1} (GEM, CBZ, DICLO, OZP, OFX, IBU, HCTZ, ATE, PRO and DOX). The last detected 10 molecules exhibit mean concentration lower than 10 ng. L^{-1} (SFZ, SUL, TRI, PRA, Ep-CBZ, TRP, EO, NAF, TYL, LOS). For some molecules, large differences between the extrema are observed (PARA-Glu, KETO, OH-IBU, SCA). These differences depend on the sampling date essentially. It should be underlined that median values are close to mean values indicating that extrema values do not play an important role. The maximum observed concentration in surface water was 650 ng.L^{-1} for KETO. Detection frequencies depend on compounds and range from 100% occurrence for CAF and PARA and 9% for TYL. 13 molecules (PARA-Glu, KETO, OH-IBU, DANO, PARA, CAF, SCA, DICLO, GEM, CBZ, OZP, OFX and ATE) were quantified in more than 50% of surface water samples. In drinking water (Figure 8), six molecules (KETO, PARA-Glu, OH-IBU, DANO, PARA and CAF) were quantified in 90% or more of the drinking water samples. These 6 molecules were also the most quantified molecules in surface water. The overall mean concentration values are between 4 (OZP) and 327 ng/L. The maximum concentration found was 650 ng/L for KETO. For drinking water, the same remark than for surface water may be made concerning the gap between minimum and maximum

concentrations: the eight drinking water treatment plants operate different treatment chains with different type of water resources.

Conclusion

A multi residue analysis was developed using on-line solid phase extraction connected to liquid chromatography coupled with tandem mass spectrometry in order to quantify residue trace levels 35 pharmaceuticals compounds in surface and drinking water. The short implementation time needed to achieve the pre concentration and the analysis, 17 minutes for the positive mode method and 15 minutes for the negative mode method is among the most significant advantages of this method compared to off-line solid phase extraction. The developed method with a pre concentration factor of one thousand showed detection limits compatible with the study of environmental matrices with very low analyte concentrations. The limits of detection and quantification are between 1.5 and 4 ng/L and 4 and 17 ng /L, respectively. Standard addition was chosen for the quantification of molecules in water samples to overcome the matrix effects and provide an accurate determination of targeted compounds. Among all studied substances, doxicycline appeared to be the most affected by a matrix effect. The developed methods were applied to eight surfaces and drinking water. In surface water, 12 molecules could be quantified in almost all analyzed samples with a maximum concentration value of 650 ng/L for Ketoprofen. In drinking water, 5 molecules could be regularly detected, with overall mean concentration values between 20 a 120 ng/L.

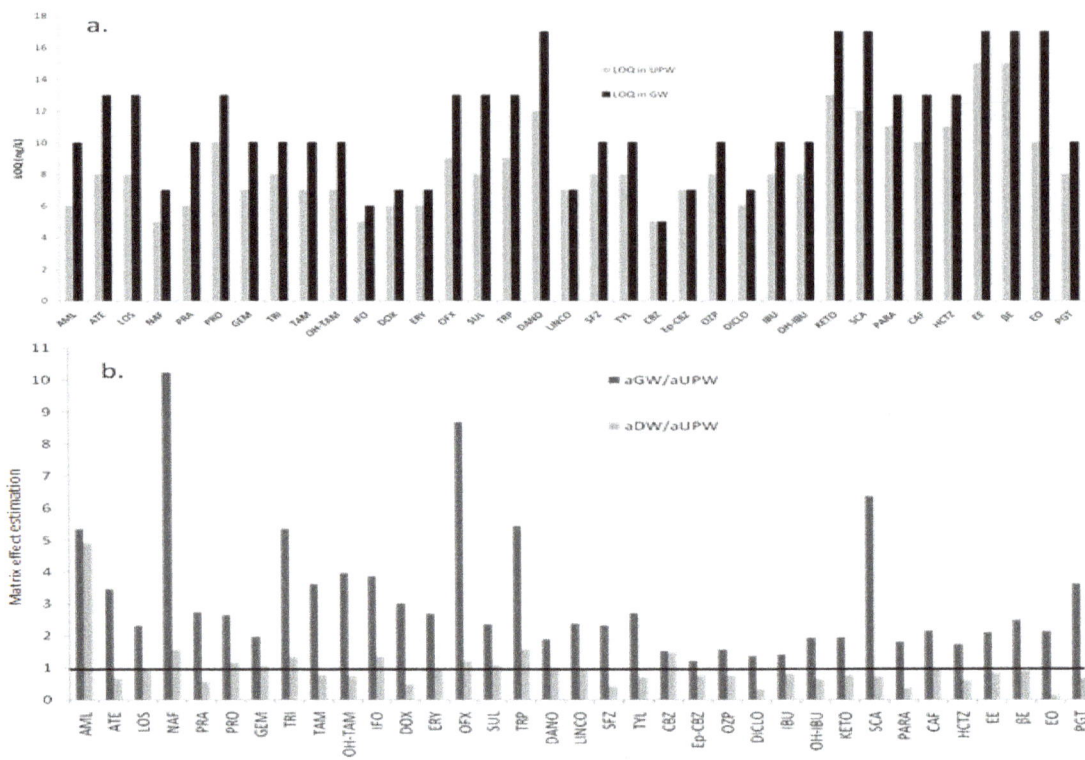

Figure 6: a. LOQ in UPW and GW for all molecules, **b.** Matrix effects of all analytes.

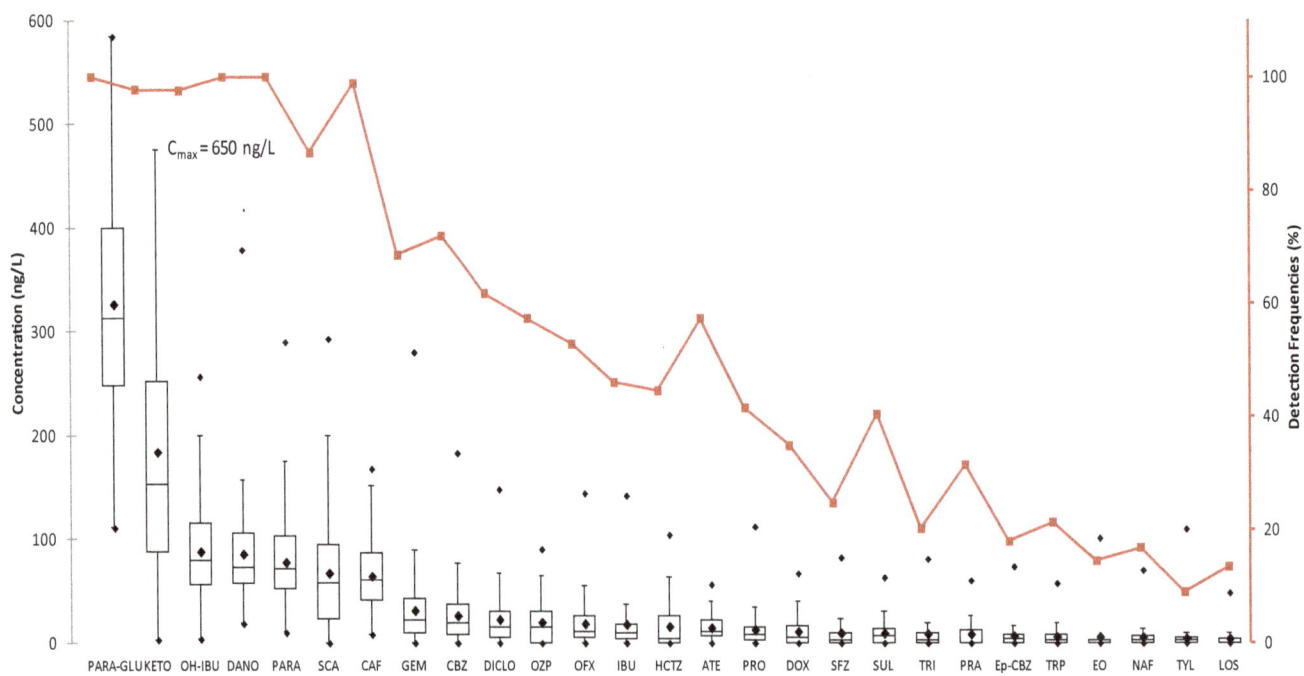

Figure 7: Overall mean concentrations (♦), median value, first and third quartiles and extrema of 27 molecules detected on average above LOQ in surface waters and detection frequencies (%, broken line).

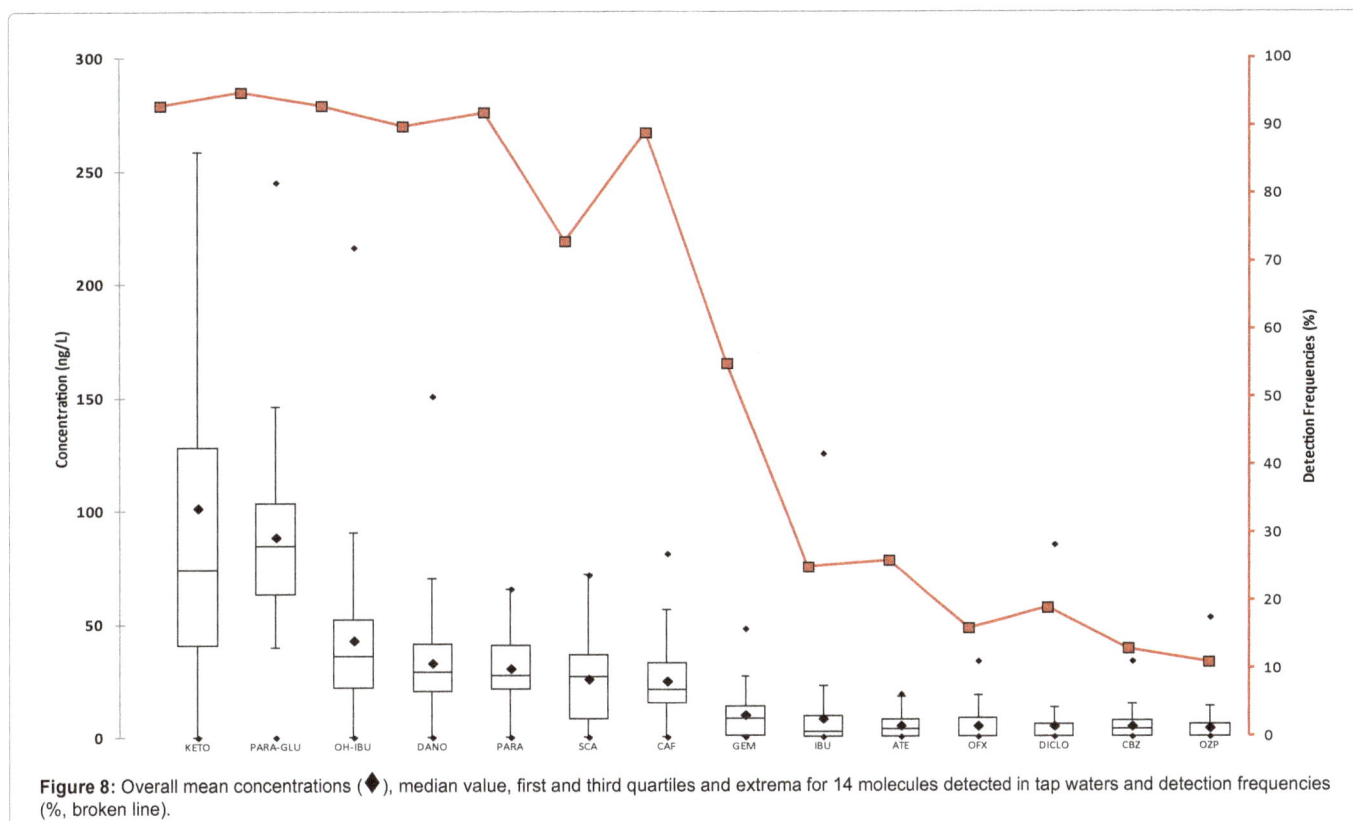

Figure 8: Overall mean concentrations (◆), median value, first and third quartiles and extrema for 14 molecules detected in tap waters and detection frequencies (%, broken line).

References

1. Jorgensen SE, Halling-Sorensen B (2000) Drugs in the Environment. Chemosphere 40: 691.

2. Gros M, Rodriguez MS, Barceló D (2012) Fast and comprehensive multi-residue analysis of a broad range of human and veterinary pharmaceuticals and some of their metabolites in surface and treated waters by ultra-high-performance liquid chromatography coupled to quadrupole-linear ion trap tandem mass spectrometry. Journal of Chromatography A 1248: 104-121.

3. Togola A, Budzinski H (2008) Multi-residue analysis of pharmaceutical compounds in aqueous samples. Journal of Chromatography A 1177: 150-158.

4. Viglio L, Aboulfald K, Deneshvar A, Prevost M, Sauve S (2008) On-line solid phase extraction and liquid chromatography/tandem mass spectrometry to quantify pharmaceuticals, pesticides and some metabolites in wastewaters, drinking, and surface waters. J Environ Monit 10: 482-489.

5. Kim SD, Cho J, Kim IS, Vanderford BJ, Snyder SA (2007) Occurrence and removal of pharmaceuticals and endocrine disruptors in South Korean surface, drinking, and waste waters. Water Research 41: 1013-1021.

6. Miege C, Choubert JM, Ribeiro L, Eusebe M, Coquery M (2009) Fate of pharmaceuticals and personal care products in wastewater treatment plants - Conception of a database and first results. Environmental Pollution 157: 1721-1726.

7. Moldovan Z (2006) Occurrences of pharmaceutical and personal care products as micropollutants in rivers from Romania. Chemosphere 64: 1808-1817.

8. Mompelat S, Le Bot B, Thomas O (2009) Occurrence and fate of pharmaceutical products and by-products, from resource to drinking water. Environment International 5: 803-814.

9. Stackelberg PE, Gibs J, Furlong ET, Mayer MT, Zaugg SD, et al. (2007) Efficiency of conventional drinking-water-treatment processes in removal of pharmaceuticals and other organic compounds. Science of the Total Environment 377: 255-272.

10. De Graaff MS, Vieno NM, Kujawa-Roeleveld K, Zeeman G, Temmink H, et al. (2011) Fate of hormones and pharmaceuticals during combined anaerobic treatment and nitrogen removal by partial nitritation-anammox in vacuum collected black water. Water Research 45: 375.

11. Daughton CG, Ternes TA (1999) Pharmaceuticals and personal care products in the environment: agents of subtle change. Environ Health Perspect 107: 907-938.

12. Vieno NM, Tuhkanen T, Kronberg L (2006) Analysis of neutral and basic pharmaceuticals in sewage treatment plants and in recipient rivers using solid phase extraction and liquid chromatography–tandem mass spectrometry detection. Journal of Chromatography A 1134: 101-111.

13. Viglio L, Aboulfald K, Deneshvar A, Prevost M, Sauve S (2008) On-line solid phase extraction and liquid chromatography/tandem mass spectrometry to quantify pharmaceuticals, pesticides and some metabolites in wastewaters, drinking, and surface waters. J Environ Monit 10: 482-489.

14. Carlsson C, Johansson AK, Alvan G, Bergman K, Kuhler T (2006) Are pharmaceuticals potent environmental pollutants? Part I: Environmental risk assessments of selected active pharmaceutical ingredients. Science of the Total Environment 364: 67-87.

15. Cleuvers M (2003) Aquatic ecotoxicity of pharmaceuticals including the assessment of combination effects. Toxicology Letters 142: 185-194.

16. Gerrity G, Gamage S, Holady SJ, Mawhinney DB, Quinones O, et al. (2011) Pilot-scale evaluation of ozone and biological activated carbon for trace organic contaminant mitigation and disinfection. Water research 45: 2155-2165.

17. Rodriguez-Mozaz S, Lopez de Alda MJ, Barcelo D (2007) Advantages and limitations of on-line solid phase extraction coupled to liquid chromatography-mass spectrometry technologies versus biosensors for monitoring of emerging contaminants in water. Journal of chromatography A 1152: 97-115.

18. Trenholm R, Vanderford B, Snyder S (2009) On-line solid phase extraction LC-MS/MS analysis of pharmaceutical indicators in water: A green alternative to conventional methods. Talenta 79: 1425-1432.

19. Miège C, Favier M, Brosse C, Canler JP, Coquery M (2006) Occurrence of betablockers in effluents of wastewater treatment plants from the Lyon area (France) and risk assessment for the downstream rivers. Talenta 70: 739-744.

20. Postigo C, Alda MJLD, Barcelo D (2008) Fully Automated determination in the low nanogram per liter level of defferent classes of drugs and abuse in sewage water by On-line solid-phase extraction-liquid-chromatography-electrospray-tandem mass spectrometry. Anal Chem 80: 3123-3134.

21. Viglio L, Aboulfadl K, Mahvelat AD, Prevost M, Sauve S (2008) On-line solid phase extraction and liquid chromatography/tandem mass spectrometry to quantify pharmaceuticals, pesticides and some metabolites in wastewaters, drinking, and surface waters. J Environ Monit 10: 482-489.

22. Alnouti Y, Srinivasan K, Wadell D, Bi H, Kavetshaia O, et al. (2005) Development and application of a new on-line SPE system combined with LC-MS/MS detection for high throuput direct analysis of pharmaceutical compounds in plasma. Journal of Chromatography A 1080: 99-106.

23. Wang S, Huang W, Fang G, He J, Zhang Y (2008) On-line coupling of solid-phase extraction ti high-performance liquid chromatography for determination of eostrogens in environment. Anal Chim Acta 606: 194-201.

24. Feitosa FJ, Temime B, Chiron S (2007) Evaluating on-line solid-phase extraction coupled to liquid chromatography ion trap mass spectrometry for reliable quantification and confirmation of several classes of antibiotics in urban wastewaters. Journal of Chromatography A 1164: 95-104.

25. Van De Steene J, Lambert W (2008) Comparison of matrix effects in HPLC-MS/MS and UPLC-MS/MS analysis of nine basic pharmaceuticals in surface waters. Journal of The American Society for Mass Spectrometry 19: 713-718.

26. Swartz M, Murphy B (2007) The Ultra Performance LC: the future of liquid chromatography. Laboratory Technology, pp: 20-22.

27. Cimetiere N, Soutrel I, Lemasle M, Laplanche A, Crocq A (2013) Standard addition method for the determination of pharmaceutical residues in drinking water by SPE–LC–MS/MS. Environmental Technology 34: 3031-3041.

28. Kasprzyk-Hordern B, Dinsdale RM, Guwy AJ (2008) The effect of signal suppression and mobile phase composition on the simultaneous analysis of multiple classes of acidic/neutral pharmaceuticals and personal care products in surface water by solid-phase extraction and ultra-performance liquid chromatography–negative electrospray tandem mass spectrometry. Talanta 74: 1299-1312.

29. Balakrishnan VK, Terry KA, Toito J (2006) Determination of sulfonamide antibiotics in wastewater: A comparison of solid phase microextraction and solid phase extraction methods. Journal of Chromatography A 113: 1-10.

28

Stability-Indicating Chromatographic Methods for the Determination of a Skeletal Muscle Relaxant and an Analgesic in their Combined Dosage Form

El-Saharty YS, Riad SM, Yehia AM* and Sami I

Department of Analytical Chemistry, Faculty of Pharmacy, Cairo University, Kasr El-Aini Street, ET 11562, Cairo, Egypt

Abstract

Two stability indicating chromatographic methods have been proposed for the determination of Dantrolene sodium (DNT) and Paracetamol (PAR). TLC was applied for the separation of the proposed drugs and their degradation products. The method employed silica gel as stationary phase and chloroform- methanol- glacial acetic acid (9:1:0.35 by volume) as the mobile phase, and the chromatograms were scanned at 230 nm. Determination of DNT and PAR was successfully achieved over the concentration ranges of 0.3–8 µg/band and 5– 36 µg/band with mean percentage recoveries of 100.04 ± 0.970 and 100.23 ± 0.922 for DNT and PAR, respectively. In addition, an isocratic RP-HPLC method was developed for the separation of the studied components on a C18 column using phosphate buffer pH 3.0-methanol (35:65 by volume) at ambient temperature. The chromatograms were detected at 230 nm for optimum sensitivity of both drugs with running time less than 10 min. DNT and PAR were determined by HPLC in concentration ranges of 0.5-50 and 1-150 µg mL^{-1} with mean percentage recoveries of 100.33 ± 0.843 and 100.14 ± 1.084, respectively. Both TLC and HPLC were applied successfully for the assay of pharmaceutical dosage form. The proposed chromatographic methods were validated as per ICH guidelines and statistically compared with a reported gradient RP-HPLC method.

Keywords: Dantrolene sodium; Paracetamol; Stability indicating method; TLC; HPLC

Introduction

Musculoskeletal disorders are better treated with combination of skeletal muscle relaxant (SMR) and non-steroidal anti-inflammatory drugs (NSAID) rather than single agent [1]. Dantrolene Sodium (DNT), the hemiheptahydrate of the sodium salt of 1-[5-(4-nitrophenyl) furfurylideneamino] imidazolidine-2, 4-dione [2], (Figure 1a), is a muscle relaxant with a direct action on skeletal muscle. It is also given for the treatment of malignant hyperthermia [3]. Paracetamol (PAR), 4′-Hydroxyacetanilide; N-(4-Hydroxyphenyl) acetamide [2], (Figure 1b), has analgesic, antipyretic properties and weak anti-inflammatory activity. The mechanism of analgesic action remains to be fully elucidated, but may be due to inhibition of prostaglandin synthesis both centrally and peripherally [3].

Literature review revealed that several electrochemical methods [4-6] and HPLC methods [7–14] have been reported for the determination of DNT either in pharmaceutical or biological samples. Degradation behavior of DNT was reported in different buffers and temperature [15]. The study showed that base catalysis is more predominant where degradation of the neutral species is catalyzed by hydroxyl ion. Hydrolysis of hydantoin ring took place to form an opened ring structure (DNT-DEG); 2-(1-carbamoyl-2-((5-(4-nitrophenyl) furan-2-yl) methylene) hydrazinyl) acetic acid [15], (Figure 1c). Several methods for the determination of PAR in combination with other drugs; such as spectrophotometric [16-24], electrochemical [25,26], HPLC [27-30] and TLC [31-33] methods have been reported. Para-Aminophenol (PAP), (Figure 1d) was reported as a primary hydrolytic degradation product of PAR or it may be originated from the synthesis [17].

Few analytical methods have been described for the simultaneous determination of DNT and PAR including spectrophotometric [34-36] and HPLC [36,37] methods. Concerning stability indicating methods for determination of DNT and PAR binary mixture, there was only a reported gradient RP-HPLC method [37]. The reported method separated DNT and PAR from their degradation products using HS C18 analytical column with gradient elution using a mixture of 50 mM sodium dihydrogen phosphate, 5 mM heptane sulfonic acid sodium salt, pH 4.2 and acetonitrile as mobile phase. The flow rate was 1.5 mL/min and detection at 214 nm with 30 min run time. Since no reported TLC, the aim of our study was to develop inexpensive stability indicating TLC method. In addition, the work proposed an isocratic HPLC method as stability indicating assay with better sensitivity and shorter run time for the determination of DNT and PAR.

Experimental

Materials and reagents

Pure authentic sample: Dantrolene sodium authentic sample was kindly supplied by Chemipharm Pharmaceutical Industries, 6th October, Giza, Egypt. Its purity was found to be 99.67 ± 0.913 according to a reported RP-HPLC method [37].

- Paracetamol authentic sample was obtained from El-Nile pharmaceutical Company, Cairo, Egypt. Its purity was checked to be 100.04 ± 0.641 according to a reported RP-HPLC method [37].

- Para-Aminophenol Pure sample, labeled with 99% purity, was purchased from Sigma Aldrich, Germany.

*Corresponding author: Ali M. Yehia, Department of Analytical Chemistry, Faculty of Pharmacy, Cairo University, Kasr El-Aini Street, ET 11562, Cairo, Egypt E-mail: aliyehia00@hotmail.com/ali. yehia@pharma.cu.edu.eg

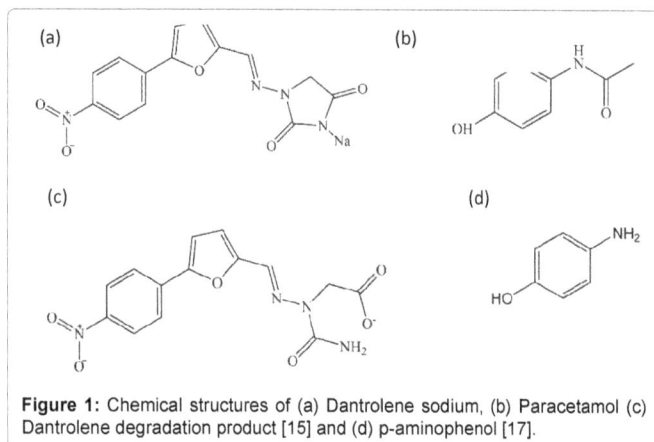

Figure 1: Chemical structures of (a) Dantrolene sodium, (b) Paracetamol (c) Dantrolene degradation product [15] and (d) p-aminophenol [17].

Degradation product: Degradation product of dantrolene was laboratory prepared by dissolving 50 mg of pure DNT in 50 mL of 2 M aqueous NaOH then reflux in a 500-mL flask for 2 hr and neutralized with 2 M aqueous HCl. The formed precipitate was filtered, washed with distilled water and dried. The dried residue was used as the DNT-DEG.

Pharmaceutical dosage form: Dantrelax Compound˙ capsules manufactured by Chemipharm Pharmaceutical Industries, 6th October, Giza, Egypt, batch No. 130859A and 131186A, and labeled to contain 25 mg Dantrolene sodium salt and 300 mg Paracetamol per capsule.

Chemicals and reagents: All chemicals used were of pure spectroscopic analytical grade or HPLC grade including methanol and chloroform (Sigma Aldrich, Germany), Glacial acetic acid (Adwic, Egypt), hydrochloric acid and sodium hydroxide (Merck, Germany), sodium dihydrogen phosphate (El-Nasr Pharmaceutical Chemicals Co., Egypt), and double distilled deionized water.

Instrument and software

Thin Layer Chromatography (TLC) system consisted of a Camag Linomat auto sampler (Muttenzl, Switzerland), a Camag micro syringe (100 mL) and a Camag 35/N/30319 TLC scanner with winCATS software; an ultraviolet (UV) lamp with a short wavelength at 254 nm (Desaga, Wiesloch, Germany); and TLC plates pre-coated with silica gel G.F254 10 × 20 cm, 0.25 mm thickness (Merck, Darmstadt, Germany).

The HPLC system consisted of Agilent pump (model 1100 series; Agilent, Germany) equipped with a variable wavelength detector and a 20 μL injection loop. Agilent C18 column (250 mm × 4.6 mm I.D., particle size 5 μm) was used as stationary phase. The samples were injected with a 50 μL Hamilton analytical syringe. The detector was operating at 230 nm and isocratic elution with 1 ml/min flow rate.

Solutions

Standard solutions: Stock standard solution (1.0 mg mL⁻¹) of DNT was prepared by accurately weighing 25 mg of its bulk powder into 25-mL volumetric flask. About 10 mL methanol was added, sonicated for few minutes, and diluted to the volume with methanol.

Stock standard solution (10.0 mg mL⁻¹) of PAR was prepared by accurately weighing 250 mg of its bulk powder into 25-mL volumetric flask. About 10 mL methanol was added, sonicated for few minutes, and diluted to the volume with methanol.

Working standard solutions (100.0 μg mL⁻¹) of DNT and PAR were prepared by suitable dilutions of their corresponding stock standard solutions.

Pharmaceutical formulation solution: Twenty Dantrelax Compound˙ capsules were weighed, carefully evacuated and reweighed again. The contents of capsules were mixed then accurate amount of the powder equivalent to one capsule was transferred into 100-mL beaker; 25 mL methanol was added, sonicated for 30 minutes then filtered into 50-mL volumetric flask. The residue was washed three times each using 5 mL methanol and the solutions were completed to the mark with the same solvent to have a dosage form solution containing 0.5 mg mL⁻¹ of DNT and 6.0 mg mL⁻¹ of PAR.

Procedures

TLC-spectrodensitometric method: Aliquots equivalent to 0.3–8.0 mg of DNT and 5.0–36.0 mg of PAR were transferred from their respective stock standard solutions into separate series of 10-mL volumetric flasks; the volumes were completed to the mark with methanol. A 10-μL aliquot of each solution was applied to a TLC plate using 50 μL Hamilton syringe. Bands of 6 mm width were spaced 1 cm from the bottom edge of the plate. The plate was developed in a chromatographic tank, previously saturated for at least 1 hr with the developing mobile phase; to a distance of about 8 cm by the ascending technique using chloroform: methanol: glacial acetic acid (9:1:0.35 by volume) as the mobile phase. The plate was removed, air-dried, and the bands were visualized under a UV lamp at 254 nm. The chromatogram was scanned at 230 nm and calibration curves representing the relationship between the recorded area under the peak and the corresponding concentrations of DNT and PAR in micrograms per band were plotted and the regression equations were computed.

HPLC: Suitable aliquots were accurately transferred from DNT and PAR standard solutions into 10-mL volumetric flasks and the volumes were completed to the mark with mobile phase to construct calibrations for DNT and PAR in the range of 0.5-50.0 μg mL⁻¹ and 1.0 -150.0 μg mL⁻¹, respectively. Twenty microliters of these solutions were injected in triplet into the HPLC system. Agilent C18 column was used as stationary phase with isocratic elution using a mobile phase consisting of phosphate buffer pH 3.0 (50 mM of sodium dihydrogen phosphate adjusted to pH 3 using ortho-phosphoric acid) - methanol (35:65 by volume) at flow rate 1 ml/min. The chromatograms were recorded at 230 nm and calibration curves were plotted between peak area and the corresponding concentrations of each drug.

Assay of synthetic mixtures: Synthetic mixtures containing different ratios of DNT and PAR were prepared from their respective standard solutions. The previously described chromatographic conditions were followed for the analysis of these mixtures by TLC or HPLC. The concentrations of DNT and PAR were calculated from the corresponding regression equations.

Assay of pharmaceutical formulation: For TLC assay, 2.0 μL of the pharmaceutical solution was directly applied, whereas 0.2 mL aliquot was accurately transferred into 10-mL volumetric flask and diluted to the volume with the mobile phase for HPLC assay. The descried chromatographic conditions under the calibration procedure of each method were followed for the determination of DNT and PAR.

Results and Discussion

Many pharmaceutical compounds undergo degradation during storage or even during the different processes of their manufacture. Several chemical or physical factors can lead to the degradation of drugs [38]. Without exception, the studied drugs are liable to hydrolysis, where DNT is liable to alkaline degradation [15] and PAP is reported as a primary hydrolytic degradation product of PAR or it may be originated from the synthesis [17]. So in this work stability indicating methods were developed. The applied chromatographic systems provided simple and accurate methods for quantitative determination of the studied drugs in presence of their degradation products.

TLC

Densitometry offers a simple way of quantifying directly on a TLC plate by measuring the optical density of the separated bands. The amounts of compounds are determined by comparing them to a standard curve from reference materials chromatographed simultaneously under the same condition [39]. In this study, TLC was applied for the separation of DNT and PAR from their degradation products depending on the difference in R_f values.

Several developing systems were applied for the separation of the proposed components. First attempt using toluene-ethyl acetate mixture in different ratios provided poor separation particularly for DNT and PAP bands. Sufficient separation was achieved upon using chloroform-methanol as a developing system; however PAR and PAP showed overlapped bands. Additions of ethyl acetate to the latter system worsen the separation and PAP showed a tailed band. Finally, optimum separation was achieved when chloroform-methanol- glacial acetic acid (9:1:0.35 by volume) was used as a developing system. Four well resolved bands of DNT, PAR, DNT-DEG and PAP were obtained having R_f values of 0.70, 0.40, 0.23 and 0.14, respectively, (Figure 2).

A polynomial relationship was found to exist between the integrated area under the peak of the separated bands at the selected wavelength (230 nm) and the corresponding concentration of DNT (0.3–8.0 µg/band) and PAR (5.0– 36.0 µg /band) (Figure 3).

RP-HPLC method

Looking for simple isocratic RP-HPLC method with optimum separation parameters necessitated studying of several critical factors that may affect the separation of DNT and PAR in the presence of their degradation products. Several trials were done using buffer pH 3.0 with several organic modifiers (methanol and acetonitrile) as mobile phase on Agilent C18 column. The elution order in all cases was PAP, PAR, DNT-DEG then DNT. Among these trials, the last eluted peak required significantly large retention volume with acetonitrile in the mobile phase, since it is the least polar organic modifier in our trials. Further trials revealed that buffer and methanol mixture was appropriate for

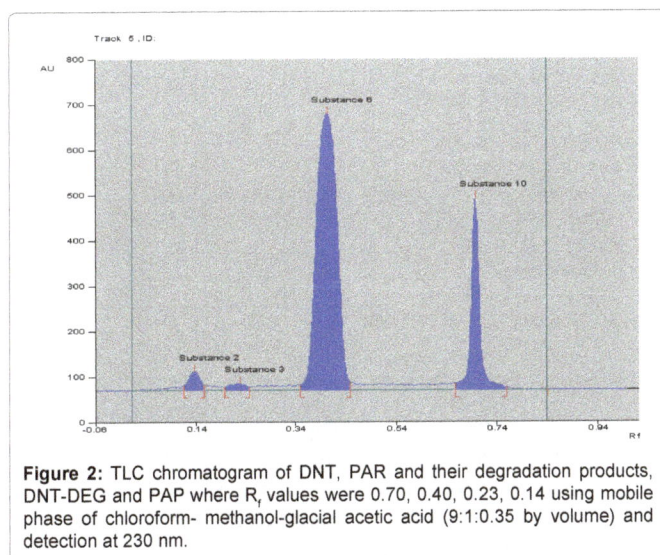

Figure 2: TLC chromatogram of DNT, PAR and their degradation products, DNT-DEG and PAP where R_f values were 0.70, 0.40, 0.23, 0.14 using mobile phase of chloroform- methanol-glacial acetic acid (9:1:0.35 by volume) and detection at 230 nm.

such separation. However buffer ratio revealed to play a fundamental role in the separation. At high buffer ratio, resolution between PAR and its degradation product increased but high retention for DNT was observed. For instance, using phosphate buffer pH 3.0-methanol (50:50 by volume) as mobile phase increased both resolution of PAR and capacity factor of DNT. In order to have optimum resolution of PAR along with lower DNT capacity factor, stepwise decrease in buffer ratio was done. Trials revealed that using phosphate buffer pH 3.0-methanol (35:65 by volume) as mobile phase achieve sufficient PAR resolution (1.57) and least applicable DNT capacity factor (3.85) with minimum run time. Since the studied drugs contain ionizable function groups with pka 7.5 and 9.5 for DNT and PAR, respectively, buffer pH should be considered in optimization. The optimum buffer ratio for separation was then tried using different buffer pH values (3.0, 4.0 and 5.0). Changes in separation performance were observed, where symmetry of PAP peak and capacity factor of DNT were enhanced using buffer pH 3.0. Also the effect of heptane sulfonic acid sodium salt was studied, negligible improvement in peak symmetry was observed. Therefore it was excluded in the proposed system.

Finally, optimization provided that using phosphate buffer pH 3.0-methanol (35:65 by volume) as mobile phase with 1 mL/min flow rate, UV detection at 230 nm, on Agilent C18 column was optimum for the separation of studied components, (Figure 4). The suggested chromatographic system allows complete baseline separation in less than 10 min run time. It is worth noting that the previously mentioned optimum chromatographic conditions was applied on different commercially available C18 columns (Kinetex® C18 and XBridge® C18) having the same dimensions of the used Agilent C18. Same separation profiles were obtained without significant changes in chromatographic parameters.

Robustness is a validation parameter that should be tested for HPLC method. Minor changes in buffer ratio, buffer pH and flow rate was studied. Levels of buffer ratio in mobile phase (35± 3 mL,), buffer pH (3.0 ± 0.2) and the flow rate (1.0 ± 0.05 units) were varied while keeping the other chromatographic conditions at optimal specified levels. Robustness was expressed by RSD% values of PAR and DNT peak areas as quantitative responses. All values were less than 5%. These results agree with the method robustness for the quantitative analysis of studied drugs. System suitability parameters [40] was checked and method performance parameters including capacity factor (k), selectivity (α), resolution (R_s), tailing factor (T) and column efficiency (N) are listed in Table 1. All calculated parameters are satisfactory to indicate good specificity of the method for the stability assessment of DNT and PAR.

The results obtained during analysis of laboratory prepared mixtures, (Table 2), show that the proposed chromatographic methods were valid for the determination DNT and PAR in different laboratory prepared mixtures. Validation of the proposed methods was performed according to ICH guidelines [41]. Selectivity of the proposed methods was assisted by synthetic mixture analyses. Good mean recoveries were obtained, (Table 3). Linear regression parameters, concentration ranges and satisfactory results of the accuracy of the proposed methods along with LOD and LOQ were shown in Table 3. The proposed methods repeatability and intermediate precision were evaluated and satisfactory RSD% values were obtained in the same table. The proposed chromatographic methods were applied successfully for the determination of DNT and PAR in Dantrelax Compound® capsules, (Table 4). Moreover, standard addition technique has been applied and the obtained results proved the validity of the method and absence of interference of excipients, (Table 4). A statistical comparison of the results obtained by the proposed and a reported [37] methods for

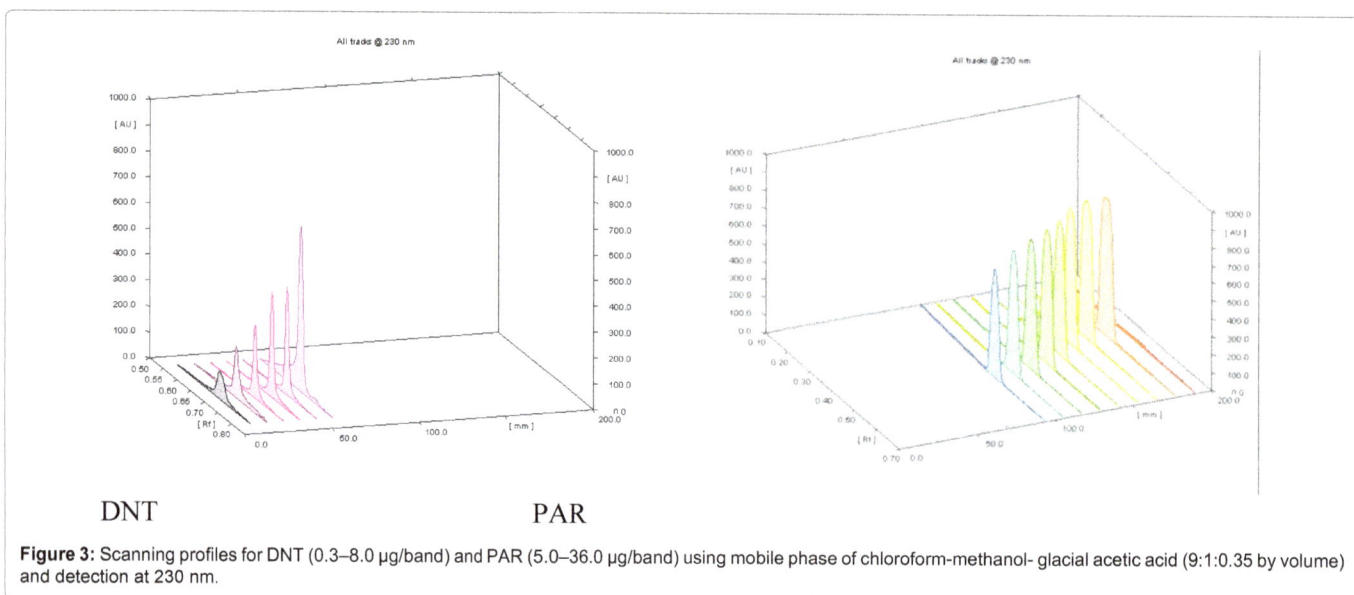

DNT

PAR

Figure 3: Scanning profiles for DNT (0.3–8.0 µg/band) and PAR (5.0–36.0 µg/band) using mobile phase of chloroform-methanol- glacial acetic acid (9:1:0.35 by volume) and detection at 230 nm.

Parameter	TLC				HPLC				Reference value[a] [40]
	PAP	DEG	PAR	DNT	PAP	PAR	DEG	DNT	
Retention factor (mm) / Retention time (min)	0.14	0.23	0.40	0.70	2.89	3.18	5.26	7.28	-----
Capacity factor (k')	6.14	3.35	1.50	0.43	0.93	1.12	2.51	3.85	1–10 acceptable
Selectivity (α)	-----	1.84	2.23	3.50	-----	1.2	2.2	1.5	> 1
Resolution (R$_s$)	-----	2.14	2.51	3.69	-----	1.57	7.46	5.95	> 1.5
Tailing factor (T)	1.00	1.00	1.00	1.00	1.00	1.00	1.00	1.20	T=1 for a typical symmetric peak
Efficiency (N)	169	324	272	2079	2791	8049	2558	12453	Increases with efficiency of the separation
Height equivalent to theoretical plates (mm)	0.5917	0.3086	0.3673	0.0480	0.0896	0.0031	0.0098	0.0020	The smaller the value. The higher the column efficiency

[a]Reference values for HPLC system suitability parameters.

Table 1: The system suitability test results of the developed RP-HPLC method.

TLC				HPLC			
Concentration (µg mL⁻¹)		Recovery%		Concentration (µg mL⁻¹)		Recovery%	
DNT	PAR	DNT	PAR	DNT	PAR	DNT	PAR
1.0	12.0	100.23	99.75	40.0	20.0	99.63	100.13
2.0	24.0	101.53	98.74	1.0	10.0	101.91	99.34
6.0	12.0	98.72	101.05	10.0	10.0	102.01	99.34
1.0	4.0	101.76	98.28	10.0	20.0	102.01	99.15
1.0	24.0	100.23	99.49	20.0	10	100.47	100.49
6.0	3.0	100.47	100.57	10.0	40.0	102.01	99.10
5.0	5.0	99.85	99.49				
Mean ± RSD%		100.39 ± 1.021	99.62 ± 0.966	Mean ± RSD%		101.68 ± 0.677	99.59 ± 0.580

Table 2: Determination of DNT and PAR in laboratory-prepared mixtures by the proposed methods.

DNT and PAR is shown in Table 5. The calculated t and F values were less than the corresponding tabulated ones, which revealed that there was no significant difference with respect to accuracy and precision between the proposed methods and the reported one.

Conclusion

The developed methods were simple, accurate and precise for the determination DNT and PAR in presence of their degradation products. Good resolution between studied components and their degradation justified the importance of the proposed liquid

Parameters	TLC		HPLC	
	DNT	PAR	DNT	PAR
Selectivity (Mean ± RSD%)	100.39 ± 1.021	99.62 ± 0.966	101.68 ± 0.677	99.59 ± 0.580
Linearity Range (µg mL⁻¹)	0.3- 8.0	5.0-36.0	0.5-50.0	1.0-150.0
Intercept	2419.3	14212	- 1.911	2.1276
SE of intercept	0.0094	0.0079	1.4508	3.2087
Slope (X coefficient)	4082.1	1478.6	7.9703	11
SE of slope (X coefficient)	0.0054	0.0009	0.0517	0.0402
Slope (X² coefficient)	-186.33	-21.5	0	0
SE of slope(X² coefficient)	0.0006	0.00002	----	----
Correlation coefficient (r)	0.9997	0.9997	0.9997	0.9999
Accuracy (Mean ± RSD%)	100.66 ± 0.608	99.43 ± 0.621	100.33 ± 0.843	100.14 ± 1.084
Precision (RSD %) Repeatability	0.593	0.299	0.308	0.753
Intermediate precision	1.267	0.874	0.814	0.988
LOD (µg mL⁻¹)	0.01	0.27	0.19	0.32
LOQ (µg mL⁻¹)	0.02	0.83	0.25	0.94

Table 3: Validation parameters for the proposed chromatographic methods.

chromatographic methods as stability indication one. The proposed TLC offered an inexpensive simple liquid chromatographic technique for the separation. Additionally, the proposed HPLC with isocratic elution showed higher sensitivity and shorter retention time compared to the reported method.

TLC						HPLC					
DNT			PAR			DNT			PAR		
	Standard addition			Standard addition			Standard addition			Standard addition	
Assay Mean ± RSD%	Added concentration[a] (µg/band)	Recovery%	Assay Mean ± RSD%	Added concentration[a] (µg/band)	Recovery%	Assay Mean ± RSD%	Added concentration[b] (µg mL⁻¹)	Recovery%	Assay Mean ± RSD%	Added concentration[b] (µg mL⁻¹)	Recovery%
101.12 ± 0.701	0.5	101.11	100.24 ± 1.678	6.0	101.14	100.78 ± 0.090	5.0	100.81	99.38 ± 0.581	60.0[c]	102.10
	1.0	98.61		12.0	102.33		10.0	101.41		120.0[c]	100.45
	2.0	98.33		24.0	98.90		15.0	100.00		180.0[c]	99.84
	Mean	99.35		Mean	100.79		Mean	100.74		Mean	100.80
	± RSD%	1.540		± RSD%	1.728		± RSD%	0.704		± RSD%	1.160

[a]Added concentration to pharmaceutical formulation containing 1.0 and 12.0 µg/band of DNT and PAR, respectively.
[b]Added concentration to pharmaceutical formulation containing 10.0 and 120.0 µg mL⁻¹ of DNT and PAR, respectively.
[c]Two fold dilutions were applied to the analysed solutions.

Table 4: Assay results and application of standard addition technique for the analysis of DNT and PAR in their pharmaceutical dosage form by the proposed chromatographic methods.

Value	TLC		HPLC		Reported method [37][a]	
	DNT	PAR	DNT	PAR	DNT	PAR
Mean	100.66	99.43	100.33	100.14	99.67	100.04
SD	0.612	0.617	0.846	1.086	0.910	0.641
RSD%	0.608	0.621	0.843	1.084	0.913	0.641
n	5	5	6	7	5	5
V(Variance)	0.374	0.381	0.716	1.180	0.828	0.411
Student's t Test	2.019 (2.306)[b]	1.533 (2.306)[b]	1.236 (2.262)[b]	2.00 (2.228)[b]	--------	-------
F-value	2.21 (5.05)[c]	1.07 (5.05)[c]	1.15 (6.26)[c]	2.87 (6.16)[c]	--------	------

[a]Hadad et al.; RP-HPLC using HS C18 column (250 mm × 4.6 mm i.d., 5 µm particle size), with gradient mobile phase; consisting of (A) 50 mmol L⁻¹ sodium dihydrogen phosphate, 5 mmol L⁻¹ heptane sulfonic acid sodium salt, pH 4.2 and (B) acetonitrile. The flow rate was 1.5 mL/min, UV detection at 214 nm and run time for 30 min.
[b]The corresponding theoretical values of t at (P=0.05)
[c]The corresponding theoretical values of F at (P=0.05)

Table 5: Statistical comparison between the proposed and reported HPLC method for the determination of DNT and PAR in their bulk powder.

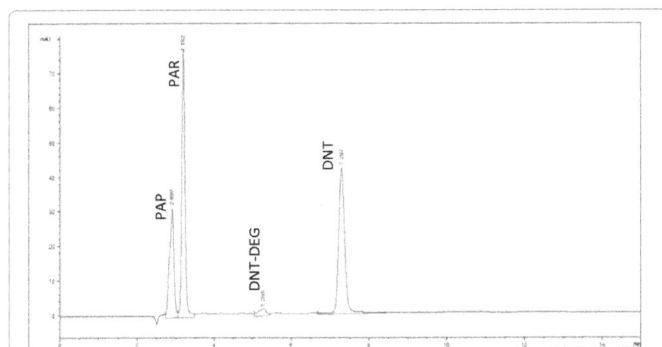

Figure 4: HPLC chromatogram of DNT, PAR (10.0 µg mL⁻¹) and their degradation products, DNT-DEG and PAP (20.0 µg mL⁻¹) using Agilent C18 column, mobile phase of phosphate buffer pH 3.0-methanol (35:65 by volume), flow rate 1 ml/min and detection at 230 nm.

In conclusion, the proposed methods were successfully applied and validated for simultaneous determination of DNT and PAR in pure powder form and in pharmaceutical formulation and therefore can be used as stability-indicating procedures in quality control laboratories where economy and time are essential.

References

1. Beebe FA, Barkin RL, Barkin S (2005) A clinical and pharmacologic review of skeletal muscle relaxants for musculoskeletal conditions.Am J Ther 12: 151-171.

2. British Pharmacopoeia (2011) Her Majesty Stationery Office, London.

3. Martindale (2007) The Complete Drug Reference. 35th edition, Pharmaceutical Press, London,UK.

4. Cox PL, Heotis JP, Polin D, Rose GM (1969) Quantitative determination of dantrolene sodium and its metabolites by differential pulse polarography.J Pharm Sci 58: 987-989.

5. Reddy JCS, Reddy SJ, Suresh Reddy C (1992) Differential pulse polarographic determination of dantrolene. Indian J Pharm Sci 54: 41–44.

6. Ghoneim EM (2007) Electroreduction of the muscle relaxant drug dantrolene sodium at the mercury electrode and its determination in bulk form and pharmaceutical formulation.Chem Pharm Bull (Tokyo) 55: 1483-1488.

7. Hollifield RD, Conklin JD (1973) Determination of dantrolene in biological specimens containing drug-related metabolites.J Pharm Sci 62: 271-274.

8. Saxena SJ, Honigberg IL, Stewart JT, Keene GR, Vallner JJ (1977) Liquid chromatography in pharmaceutical analysis VI: determination of dantrolene sodium in a dosage form.J Pharm Sci 66: 286-288.

9. Saxena SJ, Honigberg IL, Stewart JT, Vallner JJ (1977) Liquid chromatography in pharmaceutical analysis VII: determination of dantrolene sodium in biological fluids.J Pharm Sci 66: 751-753.

10. Hackett LP, Dusci LJ (1979) Determination of dantrolene sodium in human plasma using high-performance liquid chromatography.J Chromatogr 179: 222-224.

11. Katogi Y, Tamaki N, Adachi M, Terao J, Mitomi M (1982) Simultaneous determination of dantrolene and its metabolite, 5-hydroxydantrolene, in human plasma by high-performance liquid chromatography.J Chromatogr 228: 404-408.

12. Wuis EW, Grutters AC, Vree TB, Kleyn Evd (1982) Simultaneous determination of dantrolene and its metabolites, 5-hydroxydantrolene and nitro-reduced acetylated dantrolene (F 490), in plasma and urine of man and dog by high-performance liquid chromatography. J Chromatogr B 231: 401–409.

13. Lalande M, Mills P, Peterson RG (1988) Determination of dantrolene and its reduced and oxidized metabolites in plasma by high-performance liquid chromatography.J Chromatogr 430: 187-191.

14. Wuis EW, Janssen MG, Vree TB, van der Kleijn E (1990) Determination of a dantrolene metabolite, 5-(p-nitrophenyl)-2-furoic acid, in plasma and urine by high-performance liquid chromatography.J Chromatogr 526: 575-580.

15. Khan SR, Tawakkul M, Sayeed VA, Faustino P, Khan MA (2012) Stability Characterization, Kinetics and Mechanism of Degradation of Dantrolene in Aqueous Solution. Sci Res 3: 281–290.

16. Yesilada A, Erdogan H, Ertan M (1991) Second Derivative Spectrophotometric Determination of p- Aminophenol in the Presence of Paracetamol. Anal Lett 24: 129–138.

17. Bouhsain Z , Garrigues S , Morales-Rubio A , Guardia M (1996) Flow injection spectrophotometric determination of paracetamol in pharmaceuticals by means of on-line microwave-assisted hydrolysis and reaction with 8-hydroxyquinoline (8-quinolinol). Anal Chim Acta 330: 59–69.

18. Rodenas V, García MS, Sánchez-Pedreño C, Albero MI (2000) Simultaneous determination of propacetamol and paracetamol by derivative spectrophotometry.Talanta 52: 517-523.

19. Bloomfield MS (2002) A sensitive and rapid assay for 4-aminophenol in paracetamol drug and tablet formulation, by flow injection analysis with spectrophotometric detection.Talanta 58: 1301-1310.

20. Filik H, Hayvali M, Kilic E (2005) Sequential spectrophotometric determination of paracetamol and p-aminophenol with 2, 2'-(1, 4-phenylenedivinylene) bis-8-hydroxyquinoline as a novel coupling reagent after microwave assisted hydrolysis. Anal Chim Acta 535: 177–182.

21. Abdellatef HE, Ayad MM, Soliman SM, Youssef NF (2007) Spectrophotometric and spectrodensitometric determination of paracetamol and drotaverine HCl in combination.Spectrochim Acta A Mol Biomol Spectrosc 66: 1147-1151.

22. El-Yazbi FA, Hammud HH, Assi SA (2007) Derivative-ratio spectrophotometric method for the determination of ternary mixture of aspirin, paracetamol and salicylic acid.Spectrochim Acta A Mol Biomol Spectrosc 68: 275-278.

23. Khoshayand MR, Abdollahi H, Shariatpanahi M, Saadatfard A, Mohammadi A(2008) Simultaneous spectrophotometric determination of paracetamol, ibuprofen and caffeine in pharmaceuticals by chemometric methods. Spectrochim Acta A 70: 491–499.

24. Yehia AM, Abd El-Rahman MK (2015) Application of normalized spectra in resolving a challenging Orphenadrine and Paracetamol binary mixture. Spectrochim Acta A Mol Biomol Spectrosc 138: 21-30.

25. Lourenção BC, Medeiros RA, Rocha-Filho RC, Mazo LH, Fatibello-Filho O (2009) Simultaneous voltammetric determination of paracetamol and caffeine in pharmaceutical formulations using a boron-doped diamond electrode. Talanta 78: 748-752.

26. Atta NF, Galal A, Azab SM (2011) Electrochemical Determination of Paracetamol Using Gold Nanoparticles – Application in Tablets and Human Fluids. Int J Electrochem Sci: 5082–5096.

27. Celma C, Allué JA, Pruñonosa J, Peraire C, Obach R (2000) Simultaneous determination of paracetamol and chlorpheniramine in human plasma by liquid chromatography-tandem mass spectrometry.J Chromatogr A 870: 77-86.

28. Németh T, Jankovics P, Németh-Palotás J, Koszegi-Szalai H (2008) Determination of paracetamol and its main impurity 4-aminophenol in analgesic preparations by micellar electrokinetic chromatography.J Pharm Biomed Anal 47: 746-749.

29. Arayne MS, Ahmed F (2009) Simultaneous Determination of Paracetamol and Orphenadrine Citrate in Dosage Formulations and in Human Serum by RP-HPLC. J Chin Chem Soc 56: 169–174.

30. Vignaduzzo SE, Kaufman TS (2013) Development and validation of a HPLC method for the simultaneous determination of bromhexine, chlorphenramine, paracetamol and pseudoephedrine in their combined cold medicine formulations. Journal of Liquid Chromatography & Related Technology 36: 2829-2843.

31. Abdellatef HE, Ayad MM, Soliman SM, Youssef NF (2007) Spectrophotometric and spectrodensitometric determination of paracetamol and drotaverine HCl in combination.Spectrochim Acta A Mol Biomol Spectrosc 66: 1147-1151.

32. Mostafa NM (2010) Stability indicating method for the determination of paracetamol in its pharmaceutical preparations by TLC densitometric method. J Saudi Chem Soc 14: 341–344.

33. Solomon WDS, Anand PRV, Shukla R, Sivakumar R, Venkatnarayanan R(2010) Application of TLC- Densitometry Method for Simultaneous Estimation of Tramadol Hcl and Paracetamol in Pharmaceutical Dosage Forms. Int J ChemTech Res 2: 1188–11932.

34. El-Bagary RI, Elkady EF, Hegazi MA, Amin NE (2014) Spectrophotometric methods for the simultaneous determination of paracetamol and dantrolene sodium in pharmaceutical dosage form. Eur J Chem 5: 96–100.

35. Salem H, Mohamed D (2015) A comparative study of smart spectrophotometric methods for simultaneous determination of a skeletal muscle relaxant and an analgesic in combined dosage form. Spectrochim Acta A 140: 166–173.

36. Rashed NS, Abdallah OM, Farag RS, Awad SS (2014) Validated Bivariate Calibration Spectrophotometric and High Performance Liquid Chromatographic Methods for Simultaneous Determination of Dantrolene Sodium and Paracetamol in Pharmaceutical Dosage Form. Adv Anal Chem 4: 1–8.

37. Hadad GM, Emara S, Mahmoud WM (2009) Development and validation of a stability-indicating RP-HPLC method for the determination of paracetamol with dantrolene or/and cetirizine and pseudoephedrine in two pharmaceutical dosage forms.Talanta 79: 1360-1367.

38. Henry MB, Charles OL, Wait RL (1963) Physical and Technical Pharmacy. Mc Graw-Hill Book Co., Inc., New York, Toronto, London, pp 621–644.

39. Grinberg N (1990) Modern Thin-Layer Chromatography. Marcel Dekker Inc, New York.

40. United States Pharmacopoeia Commission (2004) United States pharmacopeia – national formulary. United States Pharmacopeial Inc: 2280–2282.

41. International Conference on Harmonization (1996) Validation of Analytical Procedures: Methodology (Q2B), Retrieved October 26, 2014.

29

Influence of Ionization and Sample Processing Techniques on Matrix Effect of a Pulmonary Artery Antihypertensive Drug

Al Asmari AK[1]*, Ullah Z[1], Al Rawi AS[2], Ahmad I[1], Al Eid AS[3] and Al Nowaiser NA[4]

[1]Research Center, Prince Sultan Military Medical City, Riyadh, Saudi Arabia
[2]Department of Medicine, Prince Sultan Medical Military City, Riyadh, Saudi Arabia
[3]Department of Pharmacy, Prince Sultan Medical Military City, Riyadh, Saudi Arabia
[4]Department of Family and Community Medicine, Prince Sultan Medical Military City, Riyadh, Saudi Arabia

Abstract

The issue of selectivity due to ion suppression or enhancement caused by the sample matrix has become a challenge in mass spectrometric (MS) analysis. It has been pragmatic that an efficient sample preparation method can reduce the matrix effect significantly. The study was planned to assess the effect of ionization type and sample preparation techniques on the presence or absence of matrix effect in quantitative liquid chromatography-tandem mass spectrometric (LC-MS/MS) analysis of bosentan (BSN) by post extraction addition method. Different sample processing techniques, i.e., protein precipitation (PPT), liquid–liquid extraction (LLE), and solid phase extraction (SPE) were evaluated. The sample were analyzed by both LC-electrospray ionization (ESI)-MS/MS and LC-atmospheric pressure chemical ionization (APCI)-MS/MS. Chromatographic partition achieved on an Aquasil C$_{18}$ column (100 × 2.1 mm, five μm). The mobile phase consisted of ammonium formate (pH 4.5) and methanol (10:90 v/v) in non-gradient elution mode. Our results demonstrated that both ionization sources showed matrix effect, but ESI was more prone to matrix effect than APCI. Sample processing techniques could decrease or increase matrix effect. PPT was found to be least efficient sample preparation technique, due to the presence of many remaining matrix constituents and often resulting in significant matrix effects. LLE also provided clean final extracts. However, the efficiency of LLE was lower than SPE. SPE methods resulted in cleaner extracts and considerably reduced the levels of residual matrix components from plasma samples that ultimately leads to significant reduction in matrix effects. LC-APCI-MS/MS was found to be the ionization of choice for quantitative analysis of BSN and other drugs with similar physicochemical properties. The combination of SPE and the ionization source offer a major advantage in reducing matrix effects resulting from plasma components and in improving the selectivity and sensitivity of drug analysis.

Keywords: Bosentan; Matrix effect; Ionization source; Plasma extraction; Mass spectrometry

Introduction

LC-MS-MS is a powerful analytical tool for quantitative drug analysis due to its high sensitivity, selectivity, and specificity. It is allowing for the analysis of trace amounts of intended analytes in complex mixtures [1-3]. By these characteristics, one can expect simplification of sample preparation and less time-consuming method and sample analysis time during routine analysis. However, the limitation associated with LC-MS analysis is its susceptibility to a matrix effect [4]. Matrix effect is the impact of eluting remaining matrix components on the ionization of the compound of interest [5,6]. Typically, diminished precision and accuracy of subsequent measurements are the results of suppression or enhancement of analyte response [5]. Matrix effect caused by several factors, including endogenous phospholipids, dosing media, formulation agents, and mobile phase modifiers [7]. Matrix effects cause either in ion suppression, or, in some cases, ion enhancement. It can be highly unpredictable and can be difficult to control or foresee [8].

The magnitude and nature of suppression or enhancement are the functions of the concentration of the co-eluting matrix constituents. Moreover, matrix effects are analyte specific, and several ways have been tried to minimize/eliminate the matrix effect [9]. These include changes in sample preparation techniques, chromatographic conditions, selection of ionization source, changes in polarity, and stable isotope-labeled internal standard and decreasing the amount of sample for extraction [10]. The use of inefficient extraction process, improper ion polarity, and ionization source are the major contributors to the matrix

effect [11]. However, the data on this important and synergistic effect of sample preparation and ionization source on matrix effect is limited [12]. Dams et al. determined the matrix effect for morphine by post-column infusion method where they continuously injected analyte after infusing an extract from a sample into the system. However, the drawback of this technique is that it does not provide a quantitative perceptive of the matrix effect observed by exact analytes [4].

Here in this study, we tried to provide a complete evaluation of both sample preparation methods and ion source optimization for reducing or eliminating matrix effects by post extraction addition method in rat plasma. In this study, we evaluated the three most widely used sample preparation techniques, i.e., SPE, PPT, and LLE. The observations obtained from this study will be used to pair the optimal sample preparation technique and ionization type for the quantitation of BSN (Figure 1) and other drugs with similar physicochemical properties.

*Corresponding author: Abdulrahman K Al-Asmari, Research Center, Prince Sultan Military Medical City, PO Box 7897 (S-775), Riyadh 11159, Saudi Arabia
E-mail: abdulrahman.alasmari@gmail.com

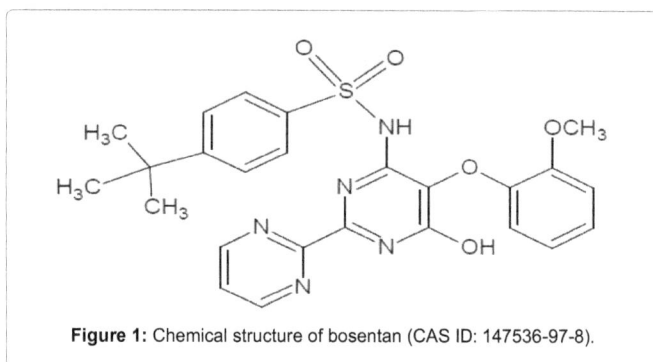

Figure 1: Chemical structure of bosentan (CAS ID: 147536-97-8).

Materials and Methods

Materials

BSN and internal standard (IS) azithromycin obtained from Clearsynth Ltd. Mumbai India. HPLC-MS grade acetonitrile and methanol (purity 99.9%) purchased from Sigma-Aldrich, Germany. Ethyl acetate was procured from Merck Specialties Pvt. Ltd. MS grade ammonium acetate and ammonium formate obtained from Fluka Analytical, Sigma-Aldrich, The Netherlands. Formic acid (purity >98%) procured from Fluka Analytical, Sigma-Aldrich, Germany. Water used in the analysis was prepared in-house with Milli-Q water purification system procured from Millipore (Millipore Corporation, USA). Other chemicals and materials used in the study were of ACS grade from commercial sources.

Liquid chromatography conditions

Chromatographic partition was achieved on an Aquasil C_{18} column (100 × 2.1 mm, 5 μm particle size), attached to UHPLC of Thermo Scientific-Dionex Ultimate 3000 (Serial # 7248679, Part # 5035.9200). The UHPLC was equipped with a quaternary solvent system, an autosampler, solvent manager (Serial # 8074857, Part # 5082.0010). The mobile phase consisted of ammonium formate (pH 4.5) and methanol (10:90 v/v) in non-gradient elution mode, which was degassed. The flow rate of mobile phase was kept at 0.3 ml/min. A fixed amount of 10 μL of sample solution injected into the MS detector in each run. The total chromatographic run time was found to be 5 min. The column maintained at 35 ± 5°C, and the pressure of the system was 70 bar.

Mass spectrometry conditions

Mass spectrometric measurements were performed on Thermo Scientific, LCQ Fleet, Ion Trap mass spectrometer (Serial# LCF 10356, San Jose, CA, USA). The accurate mass and composition for the parent ions and the daughter ions were calculated using the Xcalibur software. Ionization of analytes of interest was done using electrospray ionization (ESI) and atmospheric pressure chemical ionization (APCI). All MS/MS data for BSN and IS were collected in positive ion polarity mode by selected reaction monitoring (SRM). The optimized parameter for ESI were: spray voltage 5.0 kV, sheath gas 40, auxiliary gas 5, sweep gas 0 (highly pure nitrogen). The APCI parameter settings were: vaporizer temperature 300°C, corona discharge needle voltage 5 kV, sheath gas (high-purity nitrogen) 50, and no use of auxiliary gas. Transfer capillary temperature was set at 270°C for both ESI and APCI respectively.

Sample processing

Stock solutions of BSN and IS prepared in methanol (1 mg/ml). Working solutions of BSN was prepared at eight different concentrations (10.04-2002.42 ng/mL) in methanol–water (50:50;

v/v). These samples were spiked in plasma (2% v/v) to yield calibration standards. QC samples prepared at four concentrations (10.04, 25.37, 907.10, and 1645.64 ng/mL) levels. We investigated the following three different sample processing techniques.

Liquid–Liquid extraction

For sample preparation, 200 μL of plasma was mixed with 50 μL of IS working solution (1000.0 ng/mL). To this solution, 200 μL of 5% orthophosphoric acid and 200 μL of milli-Q water was added and vortexed. This mixture was extracted with 4 mL of t-butyl methyl ether, by placing the tubes on reciprocating shaker for 20 min at 100 rpm. These samples were subjected to flash freezing (liquid nitrogen), and 3 mL of clear supernatant was separated and then dried at 45°C under a stream of nitrogen at 25 psi.

Protein precipitation

For sample preparation, 200 μL of plasma was mixed with 50 μL of IS working solution (1000.0 ng/mL). To this solution, 200 μL of 5% orthophosphoric acid and 200 μL of milli-Q water was added and vortexed. Acetonitrile (1 mL) was transferred to this mixture for protein precipitation and then vortexed and centrifuged for 5 min at 4000 rpm. The clear supernatant was separated and then dried at 45°C under a stream of nitrogen at 25 psi.

Solid phase extraction

For sample preparation, 200 μL of plasma was mixed with 50 μL of IS working solution (1000.0 ng/mL). To this solution, 200 μL of 5% orthophosphoric acid and 200 μL of milli-Q water was added and vortexed. The diluted samples were transferred to Agilent Bond Elute Plexa˙ cartridges (30 mg/cc) which were preconditioned with 1 ml methanol and followed by 1 ml of water. The cartridges containing the samples were centrifuged for 2 minutes at 1500 rpm. Subsequently, the cartridges were washed twice with 1 mL of water and eluted twice with 1 mL of methanol. The eluate was dried at 45°C under a stream of nitrogen at 25 psi.

Finally, the every dried residue obtained by employing above stated three methodologies was reconstituted with 500 μL of the mobile phase. Ten μL of the reconstituted sample was injected onto the column for analysis.

Matrix effect

Post-extraction addition technique was used to evaluate matrix effect. Plasma samples were prepared by reconstituting the extracted blank plasma samples with reference solution containing analyte and IS. Since extraction protocol involved terminal drying step, biological at concentrations representing the quality control (QC) concentration at low (LQC), medium (MQC), and high (HQC) levels. The neat sample was reference solution prepared at an appropriate concentration in reconstitution solution (at LQC, MQC, and HQC level) [8,12,13].

Matrix effect was calculated as per the following equation

$$\% \text{ Ion supression or enhancement} = 1 - \frac{\text{Mean peak area response of the analyte after post extracted samples}}{\text{Mean peak area response of the analyte in aqueous samples}} \times 100$$

The negative value stands for the % of ion enhancement, and the positive value stand for the % of ion suppression.

Method efficiency (ME)

ME was determined by measuring the mean peak area response of six replicates of extracted QC samples (at LQC, MQC, and HQC level) against the average peak area response of aqueous solutions. ME of BSN was estimated by using the following equation:

$$\% \ ME = \frac{\text{Mean peak area response of the analyte after extraction}}{\text{Mean peak area response of the analyte in aqueous samples}} \times 100$$

Results and Discussion

Variables control

All the variables in the processing of samples were strictly controlled to avoid any discrepancies in the comparison of outcomes. The volume of the plasma used for the each measurement and composition of the reconstituted solvents was kept constant for all sample preparation techniques. Other variables taken into consideration were pretreatment of the samples before extraction, the volume of the reconstituted solvent, injection volume, and analytical column. Control of dilution and reconstituting solvents and their volume is necessary to keep the lipid concentrations consistent across all the sample processing techniques [1,6,14]. Pretreatment of the samples was kept uniform across processing methods to avoid any impact on final composition. The single analytical column was used throughout the analysis to achieve a uniform chromatographic separation for all the techniques. The volume of injection was constant throughout the process to ensure the same amount of sample loaded onto the column. The composition of rinsing solution and mobile phase were also kept identical throughout the analysis.

Optimizing mass spectrometric parameters

Negative matrix effect, signal suppression, and elevated background in MS analysis are due to the less purity of biological samples. To overcome these conditions the only possibility is to increase S/N and simultaneously maintaining high throughput performance by SRM mode. In the study the matrix effect was observed with ESI than APCI, i.e., with both ionization types, but was more ubiquitous with the ESI [15-17]. The ESI type of ionization was found to be highly susceptible to matrix effect due to its wide polarity range. In the presence of hydrophobic interference, the APCI was found to be less sensitive to the matrix effect. In APCI source, the BSN and IS formed positive ions under acidic conditions, induced by the mobile phase due to the addition of a proton [4,12]. Several daughter ions were observed for both, i.e., analyte as well of internal standard in the daughter ion spectra. The data for BSN and IS were collected in positive ion polarity mode at a transition of m/z 552.3-202.0 and 749.6-591.6 respectively, as these ions were copious, selective and produced unwavering responses.

Optimizing liquid chromatographic parameters

Since the study was about to compare the matrix with different conditions, it was necessary to set LC conditions to get sharp peak shape and adequate response. The selection of mobile phase buffer pH, flow rate, column type and the injection volume was utmost important. Different ratios of ACN-water and methanol-water combination tried as the mobile phase. The buffer tried were ammonium acetate, ammonium trifluoroacetate, and ammonium formate in varying strength. The final mobile phase consisted of ammonium formate (pH 4.5) and methanol (10:90 v/v) in non-gradient elution mode was found to be most appropriate for faster elution, better efficiency and peak shape (Figure 2). Aquasil C$_{18}$ column (100 × 2.1 mm, 5 µm particle size) was useful in the separation and elution of both compounds in a short time (5 min.).

Optimizing sample preparation techniques

The growing focus on sample analysis has shown the way to the

regular practice of preparing samples by the simplest, fastest way possible. All the three previous stated extraction methodologies were optimized to avoid matrix effect and to get high ME across all the QC levels. During optimization, plasma was processed with different aliquot volumes to attain desired S/N ratio at LOQ level. Finally, 200 µL aliquot volumes were selected for processing through various extraction techniques.

Liquid–Liquid extraction

LLE was included in extraction techniques as it is an efficient means of sample preparation and is a cleaner option. We evaluated the reconstituted LLE extracts for matrix effects and analyte recovery [8,18,19]. As per the observations shown in Table 1, an extracting solvent for LLE process was selected based on higher % ME attained for an extracting solvent compared to the other solvents and/or solvent mixture. Thus, as per the observations the tertiary butyl methyl ether was selected as an extracting solvent. The data for optimization of pretreatment solution and volume are presented in Table 2. The pretreatment solution, i.e., 200 µL of 5% orthophosphoric acid and 200 µL of Milli-Q water was found to be suitable to attain higher % PE. However, the LLE is less realistic for many reasons. Sometimes the final extraction solvent is not compatible with initial reversed-phase LC mobile phase ratios and requires gradual removal of the supernatant [20,21].

Protein precipitation

The PPT is the common and fastest method possible for sample extraction from a biological fluid. During PPT extraction, there was a massive reduction in %ME and enhancement of ion suppression with acetonitrile. Acidified acetonitrile with 1% formic acid resulted in substantial enhancement in the ion suppression. However, pretreatment of plasma with 200 µL of 5% orthophosphoric acid and 200 µL of milli-Q water led to decrease in matrix effect and improved ME Table 3. Although the PPT procedure is quick and easy, it does not produce an immaculate final extract. This procedure fails to remove enough of the plasma components, specifically phospholipids, which are known to cause inconsistency in analyte signal intensity in MS. The

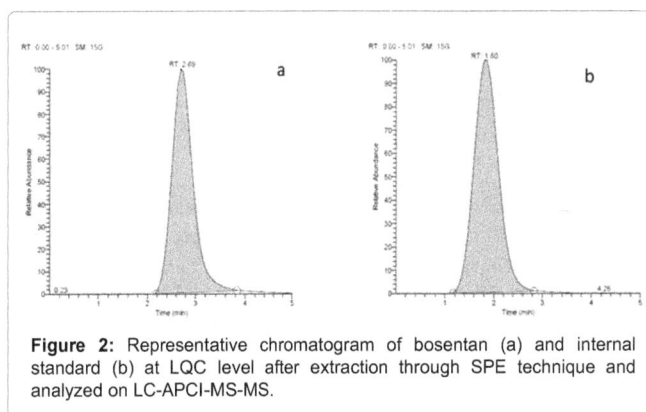

Figure 2: Representative chromatogram of bosentan (a) and internal standard (b) at LQC level after extraction through SPE technique and analyzed on LC-APCI-MS-MS.

Extraction Solvent	(%) Recovery
Tertiary butyl methyl ether	55
Ethyl Acetate	43
Diethyl ether	30
Dichloromethane/n-hexane (50:50,v/v)	33
Ethyl acetate/N-hexane (50:50,v/v)	39

Table 1: Trial for selection of solvent during LLE.

Extraction condition	ESI		APCI	
	% RE	% Ion suppression	% RE	% Ion suppression
100 µL of 2% OPA and 100 µL of Milli-Q water	43	39	53	27
100 µL of 5% OPA and 100 µL of Milli-Q water	69	21	74	19
200 µL of 5% OPA and 200 µL of Milli-Q water	73	17.5	76	11.23

Table 2: Effect of extraction condition of LLE on ion suppression and method efficiency.

Extraction condition	ESI		APCI	
	% RE	% Ion suppression	% RE	% Ion suppression
Acetonitrile	46	41	52	36
1% formic acid in acetonitrile	42	52	48	47
Acetonitrile/200 µL 5% OPA/200 Milli Q Water	67	28	78	19

Table 3: Effect of extraction condition of PPT on ion suppression and method efficiency.

particular organic solvent used in PPT procedures has a remarkable effect on the overall cleanliness of the final extract [5,22].

Solid phase extraction

For SPE, we selected an Agilent Bond Elute Plexa˚ (30 mg/cc) polymeric sorbent. Conventional procedure for conditioning of sorbent with methanol followed by water was carried out. The optimization of washing and elution steps was performed, which are presented in Table 4. As it is earlier reported that SPE offers cleaner extracts than PPT and LLE. Our results also demonstrated that matrix cleanup was found to be good with SPE, and pre-concentration step involved in the process did not affect the ME [23]. However, omitting a pre-concentration step results in loss of sensitivity [11,24,25].

Matrix effect and method efficiency

Post-extraction addition method was used to assess the ionization suppression or enhancement of the analyte at three QC concentration levels. The results in detail are presented in Table 5. Among all three techniques, minimum matrix effect was observed in SPE technique when analyzed through APCI or ESI ion source. The SPE technique combines with APCI source was found to be the best combination wherein to avoid maximum possible ion suppression. In each technique and on each ion source the matrix effect was consistent across all QC levels.

The hemolysed and lipemic plasma did not cause many variations at LOQQC and HQC levels. The data for the different type of plasma are presented in Figures 3 and 4. It can be observed that SPE and LLE samples having minor variation in precision and accuracy of QC samples in comparison to PPT samples. Ionization sources the APCI was found to be having least matrix effect on the target analyte.

In all the processing techniques, maximum ion suppression was noticed in PPT, which was not affected by the type of ionization technique (APCI or ESI). In PPT, the ion suppression was mainly due to co-eluting substances in the samples during extraction that ultimately results in decreased ME in comparison to other processing techniques. The detailed observation for ME is given in Table 6. The LLE and SPE techniques were found to be having least ion suppression and higher ME, however as per the ME the best technique was found top be SPE.

For the SPE process, Bond Elute Plexa cartridges were used, and

various ratios of aqueous and organic and solvents were tried during the selection of washing and elution solvents. The detailed observation is given in Table 4. 100% water was selected as the washing solvent as the incorporation of the little amount of organic solvent (methanol) resulted in decreased ME. Pure methanol was used as the eluting solvent on the basis of highest recovery achieved. Washing and elution steps were repeated twice.

SPE technique was found to be the best technique in comparison to all other processes by ion suppression and ME. Our results are by earlier reported results that SPE is the technique of choice to increase ME and to eliminate matrix effects. The factor that could have led the higher ME in the SPE technique may be that we used water for washing step to get good recovery and sensitivity. Sample loss was avoided at this step as no organic solvent was included [4,26]. In LLE technique, the factors responsible for the slight decrease in ME could be attributed

Washing optimization	Elution optimization	ESI		APCI	
		% RE	% Ion suppression	% RE	% Ion suppression
with 1 mL water	with 1 mL methanol	83	8.9	85	4.2
with 1 mL water	with 1 mL acetonitrile	72	18	75	15.2
with 1 mL water twice	with 1 mL methanol twice	86	3.4	89	2.6
with 1 mL 2% methanol in water twice	with 1 mL methanol twice	69	16	73	13
with 1 mL 2% methanol in water twice	with 1 mL 2% ammoniated methanol twice	63	17.6	57	19

Table 4: Effect of extraction condition of SPE on ion suppression and method efficiency.

Matrix effect/Ion suppression (%)				
Ion Source	Extraction method	QC samples		
		LQC	MQC	HQC
ESI	Solid Phase Extraction	7.8	6.7	9.4
	Liquid Liquid Extraction	17.1	18.6	24.4
	Protein Precipitation	22.5	19.2	28.0
APCI	Solid Phase Extraction	4.6	5.2	5.9
	Liquid Liquid Extraction	14.1	16.8	20.2
	Protein Precipitation	20.8	17.4	22.0

Table 5: Ion suppression of bosentan across all QC levels after extraction with different techniques and analyzed on different ion source in MS.

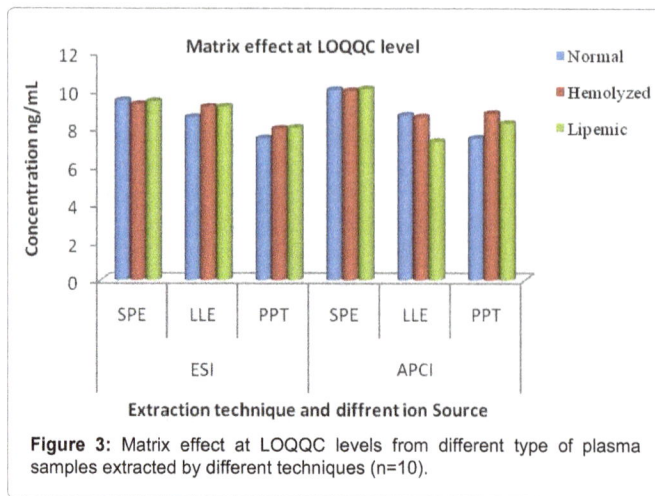

Figure 3: Matrix effect at LOQQC levels from different type of plasma samples extracted by different techniques (n=10).

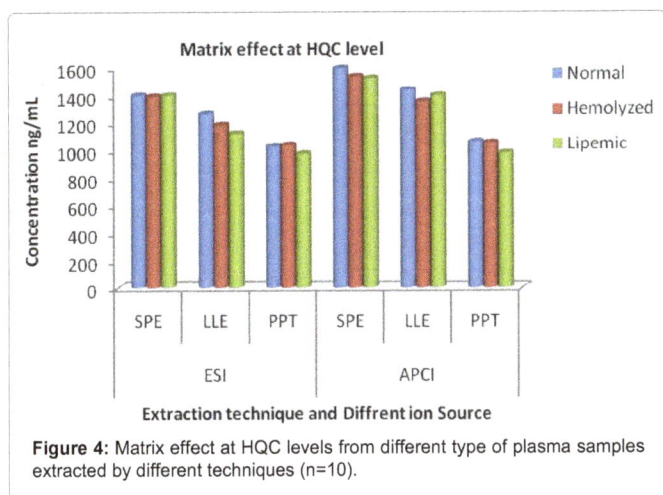

Figure 4: Matrix effect at HQC levels from different type of plasma samples extracted by different techniques (n=10).

Method efficiency (%)				
Ion Source	Extraction method	QC samples		
		LQC	MQC	HQC
ESI	Solid Phase Extraction	80.9	82.2	79.4
	Liquid Liquid Extraction	66.8	56.8	63.0
	Protein Precipitation	15.2	10.2	28.0
APCI	Solid Phase Extraction	84.4	88.7	83.2
	Liquid Liquid Extraction	65.9	54.1	62.7
	Protein Precipitation	14.4	10.6	13.8

Table 6: Method efficiency of bosentan across all QC levels after extraction with different techniques and analyzed on different ion source in MS (n=6).

Solid Phase Extraction method		QC Sample Nominal Concentration (ng/mL)			
		LOQQC (10.04)	LQC (25.37)	MQC (907.10)	HQC (1645.64)
% Accuracy	Intra day*	102.39	103.88	101.46	101.34
	Inter day#	104.66	104.40	100.70	100.92
% Precision	Intra day*	4.61	3.64	1.34	1.94
	Inter day#	5.38	3.74	1.19	1.62

*n=12, Accuracy expressed as (mean calculated concentration/nominal concentration) × 100; #n=18, Precision or coefficient of variation (CV), expressed as the ratio of the standard deviation to the mean.

Table 7: Precision and Accuracy of QC Samples Analyzed on Ion Trap MS (APCI Source).

to the solubility of the internal matter of phospholipids and protein of the plasma in t-butyl methyl ether used in the extraction process.

The best results regarding matrix effect for all the three processes were observed on APCI ion source. A possible reason for observing the best results on APCI, as compared to ESI, could be the little lower sensitivity of APCI, and because of this, the internal components of the plasma and other external components responsible for matrix effect could not be ionized in the ion source. Other reason for differences in matrix effect could be likely due to the different ionization process, with APCI, as it is based on gas phase reactions and ESI mainly based on liquid-phase reactions. However, samples of PPT analyzed on APCI also showed the vast extent of matrix effect; that might be due to the use of a supernatant liquid that contained a higher concentration [4] of endogenous matrix components [12]. In PPT, the cleanup of biofluid is minimal due to the non-selective nature of the technique [3].

Inter and intra-day precision and accuracy of the analyte were performed after completion of the experiment to suitably use the

developed method for the analysis purpose. The combination used was APCI-MS-MS with SPE extraction technique. The calibration curve in plasma was found to be linear from 10.04 ng /mL to 2002.42 ng/mL for BSN. Calibration curve was drawn using peak area ratio of analyte to the IS and by using linear, weighted least squares regression analysis with a weighting factor of $1/(x)^2$. The value of correlation coefficient (r) was found to be greater than 0.99 during analysis of precision and accuracy batches (Table 7). The Intra-day and inter-day accuracy ranged from 100.70-104.66%. The Intra-day and inter-day precision (% CV) ranged from 1.19-5.38.

Conclusion

Our results demonstrated that matrix effect was dependent on ionization source type and sample preparation techniques. The matrix effect observed for both types of ionization sources, but the ESI was found to be more vulnerable in comparison to APCI. Sample preparation techniques were having a direct influence on matrix effect. Matrix suppression data indicated that SPE was the most efficient technique to avoid ion suppression followed by LLE and PPT respectively. LC-APCI-MS/MS was found to be the ionization of choice for quantitative analysis of BSN and other drugs with similar physicochemical properties. However, in order to develop an analytical method for a molecule with minimal matrix effect, selection of sample processing technique and selection of ionization source of particular sensitivity should be considered as key points.

References

1. Xu RN, Fan L, Rieser MJ, El-Shourbagy TA (2007) Recent advances in high-throughput quantitative bioanalysis by LC-MS/MS. J Pharm Biomed Anal 44: 342-355.

2. Vogeser M, Seger C (2010) Pitfalls associated with the use of liquid chromatography-tandem mass spectrometry in the clinical laboratory. Clin Chem 56: 1234-1244.

3. Patel D (2011) Matrix effect in a view of LC-MS/MS: an overview. Int J Pharm Bio Sci 2: 559-564.

4. Dams R, Huestis MA, Lambert WE, Murphy CM (2003) Matrix effect in bio-analysis of illicit drugs with LC-MS/MS: influence of ionization type, sample preparation, and biofluid. J Am Soc Mass Spectrom 14: 1290-1294.

5. Van Eeckhaut A, Lanckmans K, Sarre S, Smolders I, Michotte Y (2009) Validation of bioanalytical LC-MS/MS assays: evaluation of matrix effects. J Chromatogr B Analyt Technol Biomed Life Sci 877: 2198-2207.

6. Antignac JP, de Wasch K, Monteau F, Hubert De B, François Andre, et al. (2005) The ion suppression phenomenon in liquid chromatography-mass spectrometry and its consequences in the field of residue analysis. Analytica Chimica Acta 529: 129-136.

7. Ghosh C, Shinde CP, Chakraborty BS (2012) Influence of ionization source design on matrix effects during LC-ESI-MS/MS analysis. J Chromatogr B Analyt Technol Biomed Life Sci 893-894: 193-200.

8. Chambers E, Wagrowski-Diehl DM, Lu Z, Mazzeo JR (2007) Systematic and comprehensive strategy for reducing matrix effects in LC/MS/MS analyses. J Chromatogr B Analyt Technol Biomed Life Sci 852: 22-34.

9. Chiu ML, Lawi W, Snyder ST, et al. (2010) Matrix effects-a challenge toward automation of molecular analysis. JALA 15: 233-242.

10. Wang S, Cyronak M, Yang E (2007) Does a stable isotopically labeled internal standard always correct analyte response?: A matrix effect study on a LC/MS/MS method for the determination of carvedilol enantiomers in human plasma. Journal of pharmaceutical and biomedical analysis 43: 701-707.

11. Trufelli H, Palma P, Famiglini G, Cappiello A (2011) An overview of matrix effects in liquid chromatography-mass spectrometry. Mass Spectrom Rev 30: 491-509.

12. Vats P, Verma SM, Khuroo A, Monif T (2015) Effects of Sample Processing Techniques and Instrument Model on Matrix Effect in LC-MS/MS for a Model Drug. Journal of Liquid Chromatography & Related Technologies 38: 1315-1323.

13. Taylor PJ (2005) Matrix effects: the Achilles heel of quantitative high-performance liquid chromatography–electrospray–tandem mass spectrometry. Clinical biochemistry 38: 328-334.

14. Mallet CR, Lu Z, Mazzeo JR (2004) A study of ion suppression effects in electrospray ionization from mobile phase additives and solid-phase extracts. Rapid Commun Mass Spectrom 18: 49-58.

15. Rogatsky E, Stein D (2005) Evaluation of matrix effect and chromatography efficiency: new parameters for validation of method development. J Am Soc Mass Spectrom 16: 1757-1759.

16. Pascoe R, Foley JP, Gusev AI (2001) Reduction in matrix-related signal suppression effects in electrospray ionization mass spectrometry using on-line two-dimensional liquid chromatography. Analytical chemistry 73: 6014-6023.

17. Jessome LL, Volmer DA (2006) Ion suppression: a major concern in mass spectrometry.

18. Romanik G, Gilgenast E, Przyjazny A, Kamiński M (2007) Techniques of preparing plant material for chromatographic separation and analysis. J Biochem Biophys Methods 70: 253-261.

19. Nováková L, Vlcková H (2009) A review of current trends and advances in modern bio-analytical methods: chromatography and sample preparation. Anal Chim Acta 656: 8-35.

20. van den Ouweland JM, Kema IP (2012) The role of liquid chromatography-tandem mass spectrometry in the clinical laboratory. J Chromatogr B Analyt Technol Biomed Life Sci 883-884: 18-32.

21. Khan M, Weng N (2008) Method Development for Preclinical Bioanalytical Support. Pharmaceutical Sciences Encyclopedia.

22. Pucci V, Di Palma S, Alfieri A, Bonelli F, Monteagudo E (2009) A novel strategy for reducing phospholipids-based matrix effect in LC-ESI-MS bioanalysis by means of HybridSPE. J Pharm Biomed Anal 50: 867-871.

23. Chou CC, Lee MR, Cheng FC, Yang DY (2005) Solid-phase extraction coupled with liquid chromatography–tandem mass spectrometry for determination of trace rosiglitazone in urine. Journal of Chromatography A 1097: 74-83.

24. Gosetti F, Mazzucco E, Zampieri D, Gennaro MC (2010) Signal suppression/enhancement in high-performance liquid chromatography tandem mass spectrometry. J Chromatogr A 1217: 3929-3937.

25. Kang X, Pan C, Xu Q, Yao Y, Wang Y, et al. (2007) The investigation of electrospun polymer nanofibers as a solid-phase extraction sorbent for the determination of trazodone in human plasma. Anal Chim Acta 587: 75-81.

26. Lanckmans K, Sarre S, Smolders I, Michotte Y (2007) Use of a structural analogue versus a stable isotope labeled internal standard for the quantification of angiotensin IV in rat brain dialysates using nano-liquid chromatography/tandem mass spectrometry. Rapid Commun Mass Spectrom 21: 1187-1195.

30

Gel Filtration Chromatography Technique as Tool of Simple Study Seminal Plasma Proteins in Domestic Animals

Vasconcelos André Belico de*, Oliveira Jamil Silvano de, Lagares Monique de Albuquerque

*University of Uberaba, Av. Nenê Sabino 1800, Uberaba, Minas Gerais, Brazil

Abstract

Seminal fluid is the liquid component of sperm, providing a safe surrounding for spermatozoa. The seminal plasma has the feature common to many other body fluid, characterized by a high dynamic range of proteins, which visualizes physiological and biochemical processes of semen. As in other body fluids, it is convenient to distinguish in the seminal plasma between proteins and non-proteins. The molecular exclusion chromatographic technique presents important points to constitutional proteins preservation, in addition to the conventional phenomenon; witch exclusion hydrophobic interactions can provide a higher resolution in the chromatogram and do not provide non-specific interactions between protein structures. The aim was the chromatography profile to simple study of proteins seminal plasma domestic animals (Stallions, Canine, and Goat), by the technique of gel filtration chromatography. The samples (seminal plasma proteins) were chromatographed in a Superose 12 HR 10/30, equilibrated with 25 mMTris-HCl (Sigma) with 0.15 M NaCl (Sigma), pH 7.4 at room temperature in a fast performance liquid chromatography (FPLC-system), using a flow rate of 0.5 mL.min-1. There was a molecular separation of significantly different molecular weights, which makes possible the logarithmic relationship between molecular weight and the elution volume. Calibration of the gel filtration column resulted in an equation $y = ax + b$, where the values of a and b were 5.43 and -2.175, respectively, and the corresponding errors were 0.115 and 0.256, respectively. The experimental error was less than 5% for most of the protein molecular masses. The work showed that the gel filtration chromatography technique provided an excellent analytical repeatability, and could therefore, be a valuable tool to the study preliminary of seminal plasma.

Keywords: Stallion; Canine; Goat; Protein

Introduction

Serious chemical analysis of the seminal plasma was taken up during the present century, leading to the discovery and identification of several substances as the proteins [1]. There is increasing evidence for the multifunctional nature of proteins. It is possible that they may have different functions in the reproductive process [2,3], as biochemical markers for identification of biological properties of animal semen [4,5]. The number of distinct types of proteins normally present in the seminal plasma also varies, but most of the seminal proteins is distributed among four to ten fractions, the proportions of which vary individually and in relation to the time interval intervening between ejaculation analysis [1]. In this respect the high performance the technique of gel filtration chromatography methods is a distinct advantage as study the constituent proteins seminal plasma separating by the molecular size [6-9]. The aim was the chromatography profile to simple study of proteins seminal plasma domestic animals (Stallions, Canine, and Goat), by the technique of gel filtration chromatography, using a Superose 12 HR 10/30, in a fast performance liquid chromatography (FPLC-system).

Materials and Methods

Animals

Ejaculates of six stallions (5-20 years old) were collected using an artificial vagina (model "Hannover"), the ejaculates of six canines (2-4 year old) were collected by masturbation with a glove or bare hand, and the ejaculates of two goats (2- years old) were by electroejaculator. Semen samples were evaluated for progressive motility with a bright field microscopy (100×). Ejaculates containing a minimum of 50% of spermatozoa with progressive motility were used in the study. All samples were cooled to +5°C with a cooling rate of 1°C/min; such a rate does not induce a cold-shock effect [9-11].

Seminal plasma proteins purifications

To obtain isolated seminal plasma proteins, semen was centrifuged for 30 min at 4°C (600 g). Then, the supernatant was brought to 36% (wt/vol) saturation with ammonium sulfate adjusted to pH 2.0 with 6 mol L^{-1} HCl, stirred for 30 min, and allowed to stand at 0°C for 30 min [10]. The samples were centrifuged again (600×g, 4°C, 30 min), the supernatant and the ammonium sulfate pellet were dialyzed for 24 h up to a 8.000 dilution factor with 0.5% (by volume) aqueous acetic acid solution using exclusion membrane of 1,000 Da cut-off, and finally the sample were lyophilized, freeze-dried and stored -20°C, according to Vasconcelos et al. [11].

Protein concentration assay was proceeded in test tube, mixing bradford reagent with protein sample (1: 0.1 mL portions, respectively) and strained with coomassie brilliant blue BG-250 [12]. Then, absorbance was measured at 595 nm by spectrophotometry (Shimadzu-160A).

Chromatography analysis

The seminal samples purifications (0.5 mL; 0.5 mL and 0.2 mL) of stallion, canine and goat, respectively were applied to a Superose 12

---

*Corresponding author: Vasconcelos André Belico de, University of Uberaba, Av. Nenê Sabino 1800, Uberaba, Minas Gerais, Brazil, CEP 38061-500
E-mail: devasconcelos.a.b@gmail.com, andre.vasconcelos@uniube.br

HR 10/30 column, molecular exclusion chromatography, equilibrated with 25 mM Tris-HCl (Sigma) with 0.15 M NaCl (Sigma), pH 7.4 at room temperature in a fast performance liquid chromatography (FPLC-system), using a flow rate of 0.5 mL.min⁻¹ and collected fraction volume of 1.5 mL. The mobile phase was the same solution used for equilibration. The Column was calibrated according to the methods of Andrews [13]. Molecular weight markers (β-amilase 200 kDa, Alcohol desidrogenase 150 kDa, Bovine Albumin 66 kDa, Anidrase carbonic 29 kDa, Cytochrome C 12,4 kDa, Aprotinin 6,5 kDa). Molecular weight markers were like a chromatography marker. Total protein was estimated in each seminal plasma sample by spectrophotometry (Shimadzu-160A); absorbencies were read at 215 nm [14].

Statistical analysis

The linear regression between seminal plasma protein concentration, relative mass and retention time were calculated using the software (Origin version 5').

Results and Discussion

Chromatographic calibration

Calibration results of the gel filtration column: $y = ax + b$; parameters (value, error): $a = 5.43, 0.115$; $b = -2.175, 0.256$; $R2 = 0.973$; ($p < 0.001$), and exclusion volume (Vo) 7.64 min. Our results were according to Andrews methods [15] the experiment error was smaller than 5% for more in the molecular mass the protein.

Chromatography proteins

The proteins chromatography of the stallion seminal plasma were represent by 12 fractions with four different retention time peaks (t_M), according to the chromatogram (Figure 1) in six animals studied. Only four picks graphic areas were selected by showing the relative mass (M_R) increase.

The results emphasized the molecular exclusion technique, characterized by the proteins separation of high molecular weight, proceeding to low molecular weight. The proteins relative mass (M_R), the retention picks value among proteins samples of the seminal plasma were presented (Table 1). This stallion seminal plasma proteins were statistically similar in all animals ($p < 0.05$), as related by Vasconcelos et al. [11].

The chromatogram of canine seminal plasma proteins (Figure 2) described several retention peaks. The relative mass were calculated according to retention time (23 min; 30.6 min; 37 min), in view of the greater proteins concentration in this range, by spectrophotometer analysis.

The retention time and molecular weight averages calculation, of each sample, related to each canine ejaculate suggests the repeatability of the technique represented by the equivalence between the results measurement (Table 2). Where molecular weights can be uncertain, the most likely values indicated often correspond to values suggested by the gel-filtration results. The technique repeatability represented among the measurements results, because there were no statistical difference between animals ($p < 0.05$). This repeatability theory described by Vial and Jardy [15] defined that the equipment calibration were characterized by the use of standard proteins, according to this work, and the effective selective separation protocol of seminal plasma proteins.

Chromatographic samples analyze of goat seminal plasma were presented in five peaks (Figure 3). The results showed that proteins

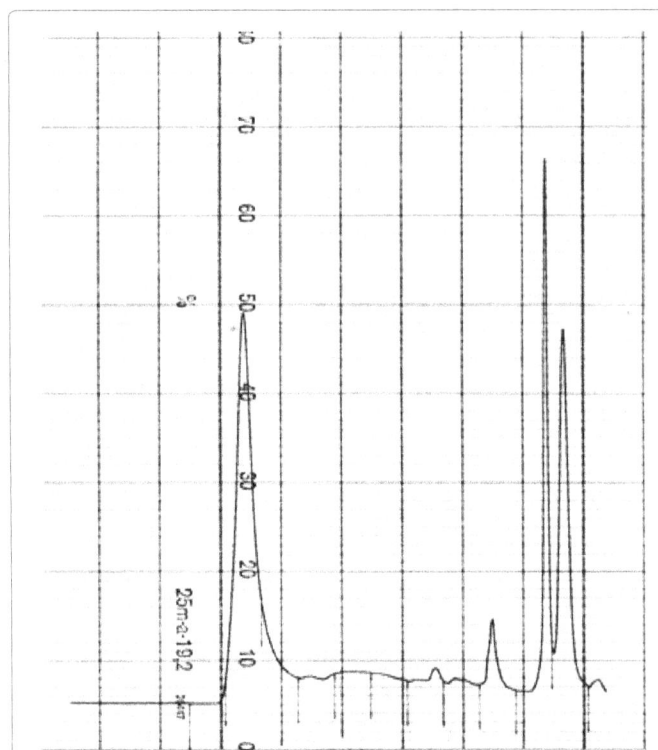

Figure 1: Stallion seminal plasma proteins by chromatography analysis. Superose 12 Column HR 10/30 (FPLC system; 0.025M Tris-HCl buffer pH 7.4; flow rate 0.5 ml/min, sensibilitiy 0.5).

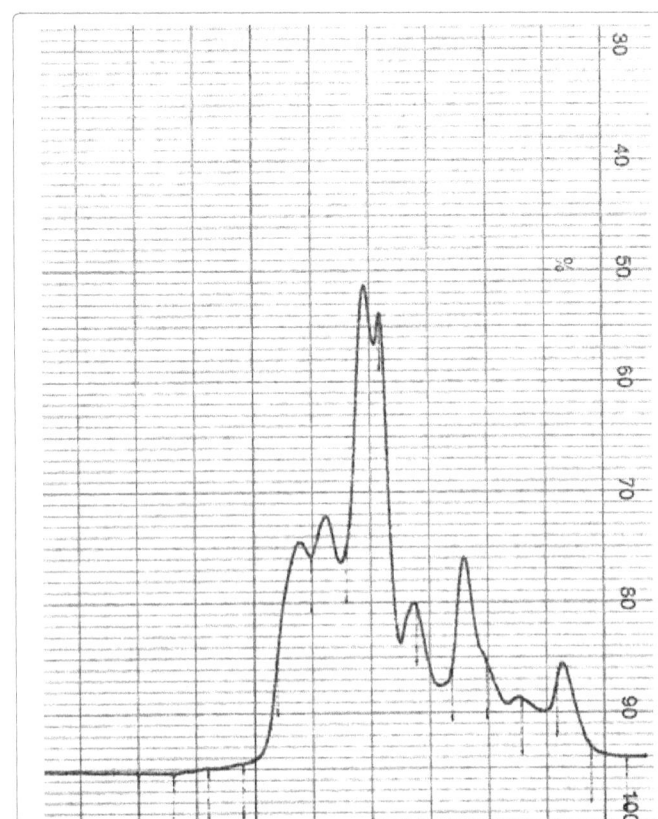

Figure 2: Canine seminal plasma proteins by Chromatography analysis. Superose 12 Column HR 10/30 (FPLC system; 0.025M Tris-HCl buffer pH7.6; flow rate 0.5 ml/min, sensibilitiy 0.5).

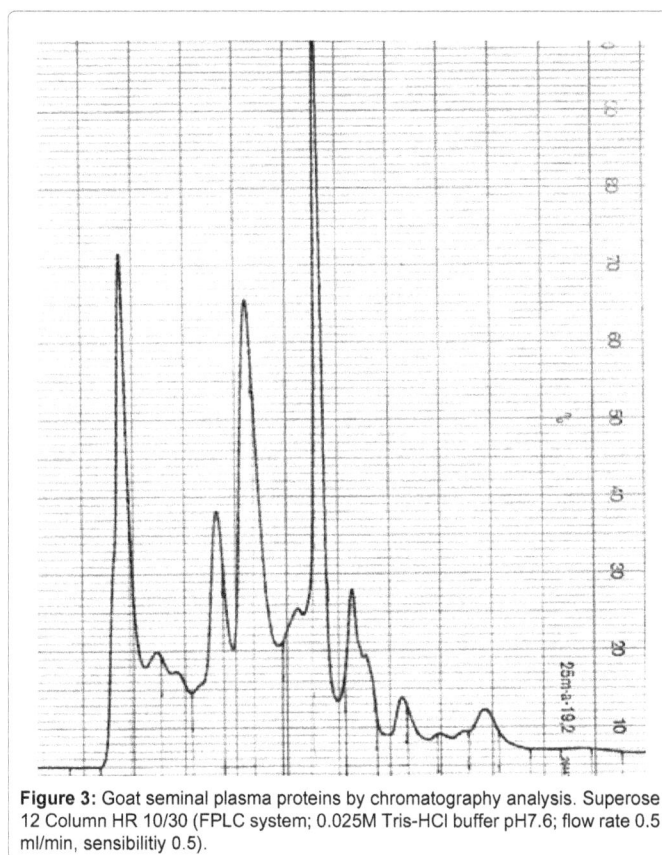

Figure 3: Goat seminal plasma proteins by chromatography analysis. Superose 12 Column HR 10/30 (FPLC system; 0.025M Tris-HCl buffer pH7.6; flow rate 0.5 ml/min, sensibilitiy 0.5).

variability were seen between the two goats evaluated and the five picks were maintained. This variability can occur since the proteins may have different structural characteristics as a free carbon [15]. However, this technique described a good correlation between relative mass (M_R) and gel-filtration behavior.

The similarity among the chromatographic profiles obtained from different animals was evaluated through the comparison of the retention time of the signs chromatographically observed and quantified in the Table 3. The related in work with gel exclusion chromatography on G-200 Shephadex performed by Mann [1] showed that two of the basic proteins were purified to apparent homogeneity.

During the analysis of a new sample, it is possible that not all frequently detected components will be. The failure of their detection indicates that there is a variation in the identifications process concerning mainly technical aspects, like spot excision, digestion, peptide recovery, mass spectra acquisition, and identity assignment [16]. This technique of gel filtration chromatography can be a tool for studying the initial fraction seminal plasma proteins.

A proteomic analysis involves proteins separation and proteins identification as well as characterization of the post-translational modifications. The limitations in a proteomic analysis are of two kinds: (i) those related to the composition of the proteome to be analyzed, mainly concerning protein expression levels and (ii) limitations of the analytical methods [14]. The chromatographic methods can reduce the complexity of the proteins mixtures because of different binding's principles, and every approach adds a unique resolving power. The standardization using known proteins enable one reason for study area in chromatogram and define a relative protein samples mass present in

Stallion	w_b 1		w_b 2		w_b 3		w_b 4	
	(t_M) (min)	M_R (Da)	(t_M) (min)	M_R (Da)	(t_M) (min)	M_R (Da)	(t_M) (min)	M_R (Da)
1	14.7	295120.9	34.8	13803.8	38.7	7585.8	40.7	5495.4
2	15.1	275422.9	34.5	14454.4	38.9	7244.4	40.9	5370.3
3	15.2	275422.9	35.0	13182.6	39.0	7244.4	40.0	6165.9
4	14.9	288403.2	35.7	12022.6	39.7	6456.5	40.7	5495.4
5	14.6	301995.2	35.1	13182.6	39.6	6606.9	40.6	5623.4
6	14.9	288403.2	34.9	13489.8	39.1	7079.5	40.1	6025.9

Data the same column were not statistically different ($p<0.05$).
Table 1: Values of retention time (t_M) and relative mass (M_R) in the stallion seminal plasma proteins after chromatography of molecular exclusion.

Canine	w_b 1		w_b 2		w_b 3	
	(t_M) (min)	M_R (Da)	(t_M) (min)	M_R (Da)	(t_M) (min)	M_R (Da)
1	23.2	81283.1	30.6	26302.7	37.6	8912.5
2	23.2	81283.1	30.6	26302.7	37.7	8709.6
3	23.1	81283.1	30.8	25118.9	37.5	8912.5
4	23.1	81283.1	30.7	25703.9	37.8	8709.6
5	23.2	81283.1	30.9	24547.1	37.3	9332.5
6	23.2	81283.1	30.6	26302.7	37.7	8709.6

Data the same column were not statistically different ($p<0.05$).
Table 2: Values of the estimate of total protein and of retention times (t_M) of the samples of canine seminal plasma.

Goat	w_b 1		w_b 2		w_b 3		w_b 4		w_b 5	
	(t_M) (min)	M_R (Da)	(t_M) (min)	M_R (Da)	(t_M) (min)	M_R (Da)	(t_M) (min)	M_R (Da)	(t_M) (min)	M_R (Da)
1	15.5	251188.9	24.5	63095.7	27.9	39810.7	34.6	12589.3	43.9	3162.3
2	15.9	251188.6	24.7	63095.7	28.2	39810.7	35.1	12589.3	43.7	3162.3

Data the same column were not statistically different ($p<0.05$).
Table 3: Results of time retention (t_M) and values of the estimate of relative mass (M_R) of goat seminal plasma samples.

biological fluids. This technique allows that these molecules interaction with the column matrix do not occur [17,18], concluded that the physico-chemical properties of the technique does not interfere with the protein structure sample, allowing to monitor the proteins variability presented in the plasma seminal of individuals and species studied.

Acknowledgements

The authors wish to acknowledge the financial support given to us by the FAPEMIG and the Department of Biochemistry and Immunology of UFMG. *In memoriam* Marcelo Matos Santoro[2].

References

1. Mann T, Lutwak-Mann C (1981) Male Reproductive. Springer-Verlag Berlin Heidelberg, New York.

2. Moura AA, Koc H, Chapman DA, Killian GJ (2006) Identification of proteins in the accessory sex gland fluid associated with fertility indexes of dairy bulls: a proteomic approach. J Androl 27: 201-211.

3. Martínez-Heredia J, Estanyol JM, Ballescà JL, Oliva R (2006) Proteomic identification of human sperm proteins. Proteomics 6: 4356-4369.

4. Perreault SD, Zirkin BR, Rogers BJ (1982) Effect of trypsin inhibitors on acrosome reaction of guinea pig spermatozoa. Biol Reprod 26: 343-351.

5. Fraser LR, Beyret E, Milligan SR, Adeoya-Osiguwa SA (2006) Effects of estrogenic xenobiotics on human and mouse spermatozoa. Hum Reprod 21: 1184-1193.

6. Laurent TC, Killander J (1964) A theory of gel filtration and its experimental verification. Journal of chromatography 14: 317-330.

7. Determann H, Michel W (1966) The correlation between molecular weight and elution behavior in the gel chromatography of proteins. Journal of chromatography 25: 303-313.

8. Porath J, Flodin P (1959) Gel filtration: a method for desalting and group separation. Nature 183: 1657-1659.

9. Silva-Junior JG (2004) Cromatografia de proteínas: guia teórico e prático. (1stedn). Interciência, Rio de Janeiro, pp: 44-46.

10. Kutzler MA (2005) Semen collection in the dog. Theriogenology 64: 747-754.

11. Vasconcelos AB, Santos AM, Oliveira JS, Lagares Mde A, Santoro MM (2009) Purification and partial characterization of proteinase inhibitors of equine seminal plasma. Reprod Biol 9: 151-160.

12. Bradford M (1976) Rapid and sensitive method for the quantization of microgram quantities of proteins utilizing the principle of protein-dye binding. Analytical Biochemistry 72: 248-254.

13. Andrews P (1965) The gel-filtration behaviour of proteins related to their molecular weights over a wide range. Biochem J 96: 595-606.

14. Harris DA, Bashford CL (1987) In: Spectrophotometry e spectrofluorimetry. (1nd ed) England: Practical approach series, pp: 49-59.

15. Vial J, Jardy A (2001) Interlaboratory studies: the best way to estimate dispersion of an HPLC method and a powerful tool for analytical transfers. Chromatographia 53: 141-148.

16. Garbis S, Lubec G, Fountoulakis M (2005) Limitations of current proteomics technologies. J Chromatogr A 1077: 1-18.

17. Krull I, Swartz M (1998) Quantitation in method validation. LC-GC. 16: 1084-1090.

18. Cuadros-Rodríguez L, Gámiz-Gracia L, Almansa-López EM, Bosque-Sendra JM (2001) Calibration in chemical measurement processes. II. A methodological approach. TrAC Trends in Analytical Chemistry 20: 620-636.

31

Optimization of an Enrichment and LC-MS/MS Method for the Analysis of Glyphosate and Aminomethylphosphonic Acid (AMPA) in Saline Natural Water Samples without Derivatization

Dana Pupke[1]*, Lea Daniel[1] and Daniel Proefrock[2]

[1]*Institute of Coastal Research/Molecular Recognition and Separation, Helmholtz-Zentrum Geesthacht, Max-Planck-Straße 1, D-21502 Geesthacht, Germany*
[2]*Institute of Coastal Research/Marine Bioanalytical Chemistry, Helmholtz-Zentrum Geesthacht, Max-Planck-Straße 1, D-21502 Geesthacht, Germany*

Abstract

Glyphosate represents the main ingredient of commercially available total-herbicides such as Roundup®, which is frequently applied in agricultural areas. Leaching of the herbicide to rivers also results in its transport to the estuarine and coastal environment. The analysis of low concentrations of glyphosate (µg/L range) in sea water samples is a challenge due to the high salinity as well as the overall complex matrix composition of such samples. To overcome these difficulties an optimized two-step method for the enrichment of glyphosate and its main degradation product aminomethylphosphonic acid (AMPA) from saline natural water samples using an ion exchange based procedure is presented in this study. The determination of both analytes was performed using an optimized LC-MS/MS MRM method, which does not require any derivatization of the analytes. The optimized method was further characterized in terms of achievable LODs and LOQs, robustness and repeatability. The achievable enrichment factor dependent on the sample volume was investigated. Finally, the developed method was applied for the analysis of samples from the tidal influenced Elbe region between Hamburg and the Elbe estuary near Cuxhaven. While the glyphosate concentrations determined in the different samples were below the LOQ, AMPA concentrations at almost all sites were above the method´s limit of quantification (6 µg/L) indicating a decreasing gradient from the Hamburg area to the estuary region.

Keywords: Glyphosate; Aminomethylphosphonic acid; Herbicide; Sea water; River Elbe

Abbreviations: AMPA: Aminomethylphosphonic acid; GLY: Glyphosate; LOQ: Limit of quantification; LOD: Limit of detection.

Introduction

The herbicide glyphosate [N-(Phosphonomethyl)glycine] represents the main ingredient of commercially available products such as Roundup® or Vorox®, which are widely applied for weed control in agriculture and on railway systems as well as in private households [1] on a large scale all over the world. The use of glyphosate containing herbicides increased during recent years in particular due to the widespread application of genetically modified soybean and wheat. The global glyphosate application was over 825 kilo tons in the year 2014. The glyphosate use in the United States of America is reported with 125 kilo tons in 2014 [2].

Also in Germany glyphosate is used in agriculture, forestry and by private persons. The amount of glyphosate applied in Germany was estimated by the German Federal Government to be 5 to 6 kilo tons in agriculture and private gardens in the year 2014 [3]. This represents about 25% of all used herbicides [4].

Glyphosate is degraded in various ways in the environment due to biotic / microbial [5,6] and abiotic/chemical [7] processes. Also the photodegradation / photolysis of glyphosate is described in the literature. In particular its binding to solid particle matter and sediments, evacuation by infiltrating water and uptake / sorption by macrophytes (plants) lead to a reduction of the glyphosate concentrations in the water phase [5].

The main metabolite of glyphosate in all degradation processes is aminomethylphosphonic acid (AMPA). Under appropriate conditions glyphosate is metabolized to AMPA in particular by microorganisms, which are present in the soil [8]. In soil glyphosate is degraded faster

than in water due to the large number of microbes. In soils half-lives of 56 days [9] to 217 days [10] are reported depending on the temperature and the content of organic matter. The biological degradation can proceed under either aerobic but also anaerobic conditions [8].

As a consequence of its widespread application glyphosate can be detected in natural waters during the whole year, with maximum concentration levels in April and October, correlating with agricultural application events [11]. Even in drinking water glyphosate concentrations in the low µg L^{-1} range have been detected [12]. The degradation of glyphosate and AMPA in natural ground and river waters is described by Mallat et al. with 60 to 100 h [13].

A study of the Bavarian State Institute of Agriculture showed that glyphosate can be detected in surface runoff waters of treated agricultural areas due to the leaching caused by rain events. Concentrations of up to 40 µg/L have been found in the analyzed runoff water samples after the application of 3 L glyphosate per hectare (amount of active ingredient corresponds to 360 g/L). Even after a period of seven months an average glyphosate concentration of 1.84 µg/L has been detected in the runoff

---

*Corresponding author: Dana Pupke, Institute of Coastal Research/Molecular Recognition and Separation, Helmholtz-Zentrum Geesthacht, Max-Planck-Straße1, D-21502 Geesthacht, Germany, E-mail: dana.pupke@hzg.de

water. For AMPA the mean measured concentration was 0.33 µg/L after the same time period. The correlation with rain events during the investigated time frame showed an un-delayed and fast transport of the contaminant and its metabolite from the soil into the runoff [14] and finally from the treated field into the draining river systems. Run-off modeling [15] for different pesticides e.g., at the Great Barrier Reef already showed the great contribution of such events in terms of matter and pollutant transport from agricultural areas to rivers and the sea. In comparison to soil environments simulation tests in natural sea water of the Australian Great Barrier Reef showed the high persistence of glyphosate in saline water, depending on light conditions and water temperature [16]. At low light conditions and 25°C the degradation of glyphosate proceeded most rapidly and the herbicide could not be detected after 180 days anymore. The slowest degradation was observed at complete darkness and 31°C. In that case glyphosate was still detectable after 330 at a level of 52%. All these reported values are above the reported half lives in soils [17].

In several publications negative effects of glyphosate and AMPA to water organisms are described. It has been figured out that glyphosate and its formulation Roundup® respectively have negative effects onto microbial communities in coastal waters at concentrations of 1 µg/L, reported from microcosm experiments [18]. Another study describes the influence of glyphosate application for weed control onto a nearby river ecosystem. In the case of the water flea *Daphnia magna* a decreased food intake as well as oxidative stress were observed after glyphosate application on adjacent fields. Environmental concentration levels of 20-60 µg/L for glyphosate have been reported in this study [19]. The application of Roundup® to zebrafish e.g., reduced the egg production of female fish and increased the mortality even at low concentration levels of 10 µg/L. The same low glyphosate concentrations inhibit the cholinesterase activity of mussels and fish [20]. Among others the effects [21] due to chronic exposure of glyphosate on three-spined stickleback (*Gasterosteus aculeatus*) have been investigated [22].

In Germany the application and discharge of glyphosate into the environment is still not regulated. Even though environmental authorities in e.g., Hamburg or Schleswig-Holstein detect glyphosate and AMPA in rivers routinely, however to our knowledge the monitoring programs did not include a systematic control of the occurrence of glyphosate and AMPA in coastal waters until now. The lack of information regarding the occurrence of glyphosate and AMPA in the coastal environment might be caused by the challenging detection of both compounds in such complex, matrix rich natural water samples.

Several analytical methods are described for the determination of glyphosate and AMPA. Predominantly HPLC-MS based methods with post-column derivatization with o-phthaldialdehyde (OPA) [23] or pre-column derivatization with 9-fluorenylmethyl chloroformate (FMOC-Cl) [24] have been described in the literature. Also direct LC-MS/MS methods in aqueous samples [25] (instrument detection limits (IDLS) of 1.2 for glyphosate and 0.9 µg/L for AMPA) have been reported. A method for equal complex samples, in this case rat plasma, is applying FMOC-derivatization and ESI-MS [26], resulting in LOQs of 5 (GLY) and 10 µg/L (AMPA) with only fluorescence detection (FLD) and 0.4 and 2 µg/L respectively with ESI-MS detection. Recently the application of an automated detection system based on sequential injection analysis for glyphosate and AMPA with OPA derivatization has been described [27].

The high and sometimes varying salt content of natural water samples represents the main analytical challenge in glyphosate and AMPA determination. Saline water samples may not be directly injected into the LC-MS/MS system due to destruction of the orifice of the mass spectrometer and corrosion of metal parts in the analytical devices. The salinity is not only problematic for the detection in sea water, but also in estuary samples which show the typical tidal influenced salt gradients. To overcome these issues, an optimized two-step enrichment process will be presented in this study. The enrichment process is coupled to a sensitive direct LC-MS/MS method, which requires no derivatization steps, resulting in a less time-consuming procedure compared to other methods.

Therefore, different anion and cation exchange materials have been used during the sample preparation process allowing the efficient matrix separation of the targeted analytes from the saline water matrix. In order to show the applicability of the optimized method, samples from different locations along the Elbe river and within its estuary, representing a possible transport pathway for glyphosate from agriculturally treated urban areas into the coastal areas, have been analyzed with the optimized method.

Materials and Methods

Chemicals and reagents

All chemicals applied in this study are of analytical grade or higher purity. Glyphosate (N-(Phosphonomethyl)glycine) and AMPA (Aminomethylphosphonic acid) have been obtained as crystalline substances (Sigma-Aldrich, Seelze, Germany). The isotopically labelled standards of Glyphosate (2-13C, 99%; 15N, 98%+, 100 µg/mL) and AMPA (13C, 99%; 15N, 98%, methylene-D2, 100 µg/mL) have been purchased as solution (Cambridge Isotope Laboratories Inc., MA USA).

The mobile phase consists of acetonitrile and water. 500 mL of each solution are supplemented with 2 Vol% (10 mL) formic acid (98-100% suprapur, Merck KGaA, Darmstadt, Germany).

An amount of 10 mg of glyphosate or AMPA respectively is dissolved in 10 mL purified water to prepare a stock solution of 1 g L^{-1}. These solutions were stored at 4°C and have been used for a maximum period of four weeks. Appropriate dilutions are prepared weekly, dissolving the appropriate amount of the stock solution in ultrapure water or artificial/natural sea water respectively. Stability experiments showed that working solutions and calibration solutions are stable for a minimum of 4 weeks when stored at 4°C in the dark.

For the method validation in ultrapure and artificial sea water 10 equidistant concentration levels of glyphosate and AMPA are prepared for both concentration ranges (1, 2, 3, 4, 5, 6, 7, 8, 9, 10 µg/L and 10, 20, 30, 40, 50, 60, 70, 80, 90, 100 µg/L respectively)

Hydrochloric acid solutions are prepared by the dilution of 37% HCl solution (Merck KGaA, Darmstadt, Germany) with ultrapure water resulting in 6 M and 0.1 M solutions respectively.

Artificial sea water is prepared after the recipe of Zobell [28]. Therefor the different required salts were dissolved in two thirds of the end volume (1 L) of ultrapure water and then filled up to the end volume. The prepared sea water is stored at 4°C in the dark until usage.

The pH value of this solution is 6. For the examination of the influence of natural pH values, solutions with a pH value of 8 have been also prepared. The pH value is adjusted to 8 using a 0.1 M NaOH solution. All solutions are stored in Nalgene HDPE bottles (Thermo Fisher Scientific, MA, USA) at 4°C in the dark.

Preparation of the sample enrichment columns

An amount of 400 ± 5 mg of Chelex 100 resin is transferred into an empty PolyPrep column (Bio-Rad Laboratories, Inc., CA, USA) and filled up with a volume of 5 mL of pure water. The resin is carefully stirred with a plastic stirrer before the removal of the water with a pipette. Afterwards further 2 mL of pure water are added to the column followed by careful agitation of the column to remove any remaining air bubbles within the resin bed. Then the resin is left to settle and again the supernatant is removed with a pipette without running the resin dry. A slice of clean filter cotton (Fluval, HAGEN Deutschland GmbH and Co. KG, Holm DE) is placed on the top of the resin using a clean tweezer. Such prepared columns are stored in the fridge at 4°C after addition of 10 mL pure water. The water is then drained away directly before use.

In the same way the AG 1-X8 column (resin amount 900 ± 5 mg) is prepared.

Two-step preparation of water samples

Before the application of the matrix separation procedure, the saline water samples are filtrated using hydrophilic syringe filters. Three independent samples with a volume of 10 mL are prepared for each sampling site and each calibration point. The whole sample preparation process is schematically described in Figure 1.

A sample volume of 10 mL is applied to the Chelex column. Then the column is rinsed with 10 mL of ultrapure water. Subsequently 2 mL of 0.1 M hydrochloric acid are applied to the column. Then the sample is eluted with 1.8 mL of 6 M hydrochloric acid. The eluate is collected in a 2 mL Eppendorf tube.

The AG 1-X8 column is preconditioned with 2 mL of 6 M hydrochloric acid. Then the sample is applied and collected in a 2 mL Eppendorf tube again.

The sample preparation with the two-step process is conducted on an EluVac vacuum manifold SPE system (LC-Tech, Dorfen, Germany),

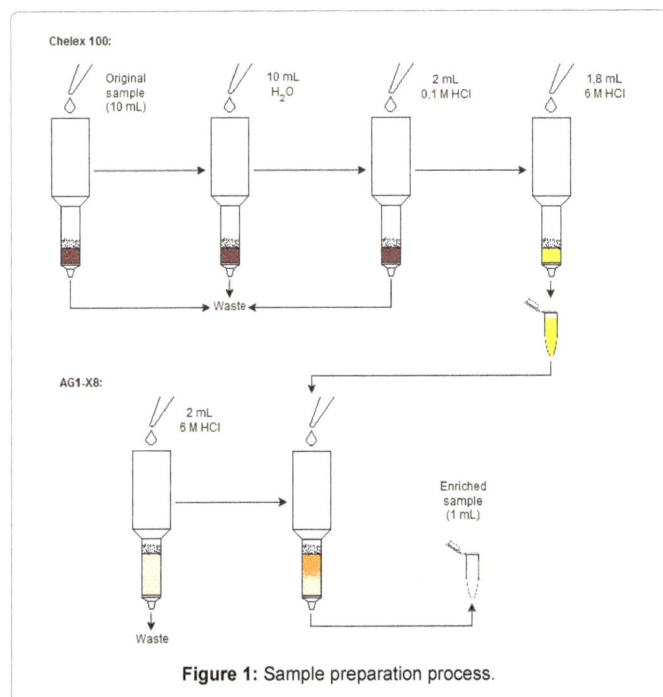

Figure 1: Sample preparation process.

allowing the simultaneous treatment of up to twenty samples.

After preparing the samples with the two-step enrichment process, they are freeze-dried in a SpeedVac vacuum concentrator (45°C, run time 8.5 h, heat time 7.5 h, cooling trap at -100°C) in order to evaporate the hydrochloric acid. Then the residue is filled up to a volume of 1 mL with ultrapure water and stored at 4°C until measurement with LC-MS/MS. 198 μL of this preprocessed sample is applied to a LC vial. A volume of 1 μL of the isotopically labelled standard solutions of glyphosate and AMPA each and 4 μL formic acid are added to the sample directly before measurement, according to RL 2005/6/EG for organic compounds. Then the samples are homogenized for one minute using a lab shaker.

Glyphosate and AMPA analysis by LC/MS-MS method

The detection of glyphosate and AMPA is conducted using a tandem mass spectrometer (API 4000, AB Sciex) coupled to an HPLC system (Agilent type 1100/1200). The chromatographic separation is performed on a Thermo Hypercarb column (100 mm × 3 mm with a particle size of 3 μm). In order to protect the column from pollution due to particles or residues of the sample preparation procedure, a guard cartridge is used (Hypercarb 10 mm × 3 mm). Hypercarb columns consist of 100% porous graphitic carbon and are pH stable from 0 to 14, so they are eminently suitable for acidic samples.

An injection volume of 10 μL of the sample is used during all experiments at a solvent flow rate of 0.2 mL/min. The column temperature is constantly set to 25°C. The separation is conducted under isocratic conditions. The mobile phase consists of acetonitrile (A) and water (B), both supplemented with 2% formic acid. For the removal of analyte residues a gradient is applied at the end of each measurement cycle before re-equilibration of the column (Table 1). All HPLC conditions are summarized in Table 2. The parameters for the MS method are listed in Table 3.

Ionization takes place in the positive mode (MRM). One measurement cycle has the duration of 19.5 min, delay time is set to 0.2 s and the cycle time is set to 1.64 s.

The following MRM settings and transitions are used for the detection of glyphosate and AMPA. In order to verify the presence of the both compounds two mass transitions are utilized each (compare Table 4).

For quantification the mass fragments m/z=88 for glyphosate and m/z=30 for AMPA have been used respectively.

All MRM parameters have been optimized for optimum sensitivity.

Sampling and preparation of the tested natural water samples

The analyzed samples have been collected along a transect from the Elbe estuary upstream to the city of Hamburg during a ship campaign in August 2015. A volume of 100 mL water sample was taken on each of the following sampling sites (compare Figure 2, Table 5). Two samples with a volume of 10 mL were prepared for each sampling site.

Regarding the aim of detecting low concentrations of glyphosate and AMPA, also the influence of the initial sample volume on the enrichment factor of the sample preparation method was investigated. Therefore, sample volumes of 10 mL and 20 mL sampled at the Elbe station Seemannshöft were spiked with 10 μg/L glyphosate and AMPA. Each sample was prepared as triplicate using the two step enrichment method.

Another sample from the same location was prepared with the

Time t (min)	Phase B (%)	Flow (mL/min)
0.00	100	0.2
3.00	100	0.2
10.00	0	0.2
10.10	100	0.4
15.00	100	0.4
15.10	100	0.2

Table 1: Cleaning gradient scheme.

Separation column	Hypercarb 100 mm × 3 mm, 3 μm
Guard cartridge	Hypercarb 10 mm × 3 mm, 3 μm
Injection volume	10 μL
Flow rate	0.2 mL/min
Temperature	25°C

Table 2: HPLC conditions.

Source/Gas Parameters	
Curtain Gas	40 L/min
Collision Gas	6 L/min
Ion Spray Voltage	4500 V
Temperature	450°C
Ion Source Gas 1	40 L/min
Ion Source Gas 2	35 L/min
Compound Specific Parameters	
Declustering Potential	25 V
Entrance Potential	10 V
Collision Energy	15 V
Collision Cell Exit Potential	12 V

Table 3: Parameters of MS analysis.

	Glyphosate	Glyphosate isotope standard	AMPA	AMPA isotope standard
Quantifier	170.1/88.1	172.1/90.1	112.1/30.1	116.1/34.1
Qualifier	170.1/60.1	172.1/62.1	112.1/83.2	116.1/83.2

Table 4: Mass transitions glyphosate and AMPA (m/z).

No	Place	Elbe (km)
1	Lt. Vogelsand	746.3
2	Cuxhaven, Kugelbake	727.0
3	Neufeld	721.6
4	Glameyer	716.1
5	Otterndorf	710.0
6	Brunsbüttel Elbhafen	693.0
7	Glückstadt	675.5
8	Pinnau	660.3
9	Blankenese	636.1
10	Seemannshöft	628.8
11	Köhlbrandbrücke	622.6
12	Zollenspieker	598.7

Table 5: Sampling sites along the river Elbe (map created with ArcGis).

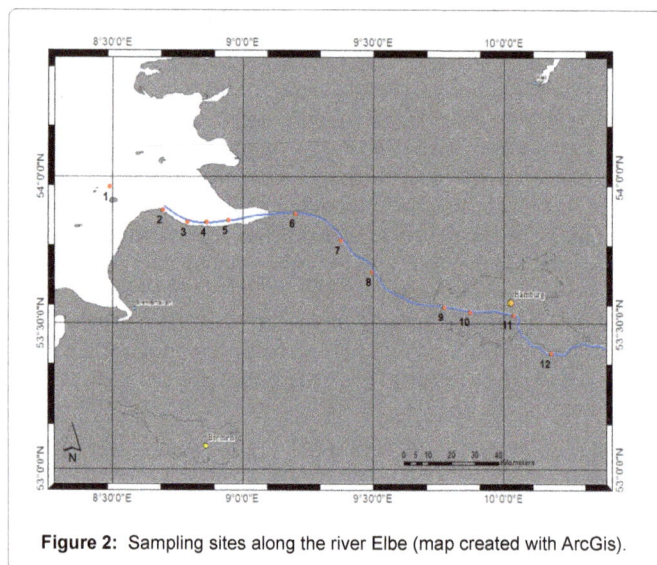

Figure 2: Sampling sites along the river Elbe (map created with ArcGis).

described method to compare the effects when processing either filtered or unfiltered samples. In this case the samples have been spiked with 5 and 10 μg/L glyphosate and AMPA respectively.

Results and Discussion

Method validation

Linearity: In order to investigate the linearity of the method, the calibration curves of the method including the enrichment process of glyphosate and AMPA in ultrapure water and artificial sea water samples were calculated. These curves were determined for ten different concentration levels in the range of 10 to 100 μg/L, as shown in Figure 3. The ratio of the analyte (glyphosate/AMPA) to its isotope standard is plotted across the adjusted concentration of the analyte. The ratio of the analyte to the analyte standard is directly proportional to the concentration, so the method is linear in the examined range.

Sensitivity: As to learn from the calibration curves, the sensitivity of our optimized method is better for glyphosate than for AMPA and expectably better in ultrapure water than in artificial sea water for both analytes, in both examined concentration ranges. The regression lines show good coefficients of determination for glyphosate and for AMPA. The increases of the regression lines for glyphosate are 2 to 4 times higher than for AMPA, ensuing the presented method being more sensitive for glyphosate than for AMPA. It has to be taken into consideration that there are still some matrix effects. The worse sensitivity for AMPA may be contributed to the lower m/z ratio.

Robustness: In order to validate the optimized method for natural water samples, additional samples from the Elbe have been collected at the measuring station "Seemannshoft" in June 2015. Filtered and unfiltered samples were compared with spiked samples of the same sampling site in the Elbe. In both cases a sample volume of 10 mL was tested.

As shown in Figure 4 we interestingly observed AMPA findings at this sampling site. In the filtered samples an AMPA concentration of 18 ± 2 μg/L was found, while in the unfiltered samples 18 ± 6 μg/L AMPA has been detected respectiveley. The slight difference between filtered and unfiltered samples is within the standard deviation of the method. Due to this fact, all further samples have been filtered before the final enrichment process. In both samples the glyphosate levels were below the LOD.

In the spiked samples (5 µg/L) glyphosate is detected at a concentration level of 36 ± 14 µg/L and AMPA at 48 ± 13 µg/L after the enrichment process. For the spike level of 10 µg/L the mean glyphosate concentration after enrichment was 202 ± 81 µg/L while an AMPA concentration of 225 ± 35 µg/L has been quantified. The slightly higher relative standard deviations in the higher spiked samples may point out a non-linear adsorption behavior of the Chelex 100 resin for glyphosate and AMPA. This corresponds to the different sensitivities for different concentration ranges and our batch experiments concerning multilayer adsorption in another study (not published yet).

Accuracy and repeatability: Three independently preprocessed samples for each sample volume were analyzed and compared to the nominal values in order to examine the accuracy and repeatability of the described method.

The good accuracy and repeatability of the method can be seen from Figure 5. The mean values for all three samples are within the standard deviation of the expected concentrations (dashed and dotted lines in the Figure 5). For glyphosate the accuracy and repeatability are better than for AMPA due to a lower standard deviation for glyphosate.

The accuracy is moreover defined as a combination of the selectivity and the linearity of a method, and the calibration curve must meet the origin point of that curve. All of these three requirements are fulfilled, showing the method presented here being accurate.

Detection and quantification limits: The limits of detection and quantification for the sample preparation process coupled to the optimized LC-MS/MS method were determined using the software tool DIN Test, which calculates both parameters according to DIN 32645 [29-31]. Three samples for each of ten equidistant concentration levels in the range of 1 ... 10 µg/L glyphosate and AMPA in a common mixture were prepared with the presented two step enrichment process and analyzed with LC-MS/MS. The detection limits of the entire preparation and detection process are 4 µg/L for glyphosate and 0.8 µg/L for AMPA, the limits of quantification are 18 µg/L for glyphosate and 6 µg/L for AMPA. Equal LODs for glyphosate and AMPA are reported in literature, but in that study the water samples were prepared with a much more time-consuming procedure, followed by a derivatization step using trifluoroacetic anhydride and trifluoroethanol before the detection by GC-MS. MDLs of 6 µg/L for glyphosate [19] in reagent water and 9 µg/L for ground water are reported in the EPA method for the determination of glyphosate with post-column derivatization and LC-MS/MS. Other authors describe MDLs of 0.8 µg/L for AMPA and 0.5 µg/L for glyphosate with IC-ESI-MS. But in that method the separation time with about 30 minutes was longer than that of our presented method [32]. The earlier reported limit of detection for glyphosate attained with an equal LC-MS method on a Hypercarb column is obviously worse providing a LOD of 40 µg/L in drinking water [33]. In fishpond water a LOD of 2 µg/L was achieved with an LC-MS method on a HILIC column. To our knowledge for saline water matrices no comparable [34] studies do exist. After the much more time consuming FMOC derivatization LODs in the ng/L range were achieved in saline water samples [35]. Further method optimization would be necessary to detect concentrations in the ng/L range with this method as well. Higher sample volumes would either result in a longer sample preparation time or would require other dimensions of the sample preparation columns.

Additionally, the hydrochloric acid leads to massive corrosion of metal parts, probably resulting in metal complex forming with glyphosate. Glyphosate that has formed a metal complex can no longer be detected with LC/MS-MS. This may explain our higher

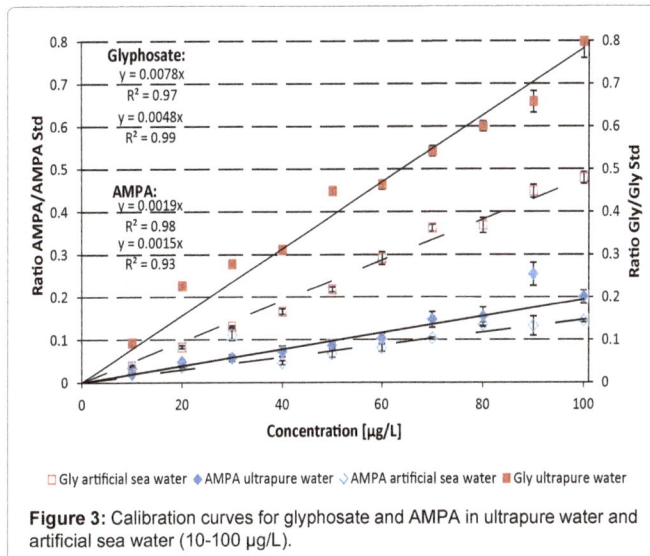

Figure 3: Calibration curves for glyphosate and AMPA in ultrapure water and artificial sea water (10-100 µg/L).

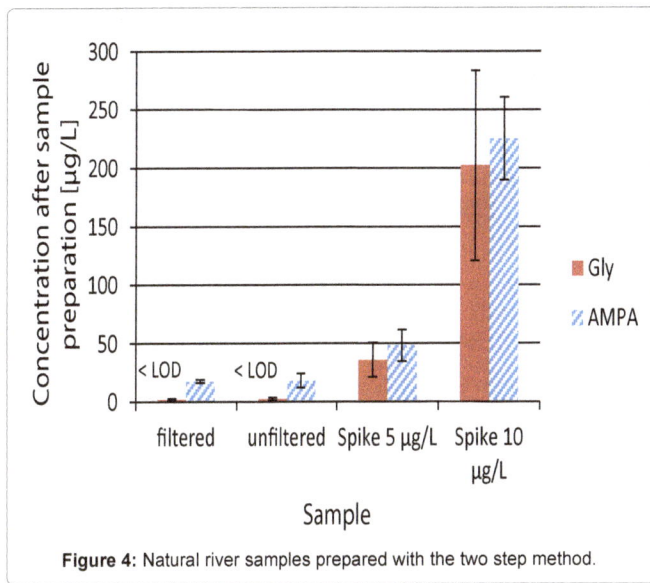

Figure 4: Natural river samples prepared with the two step method.

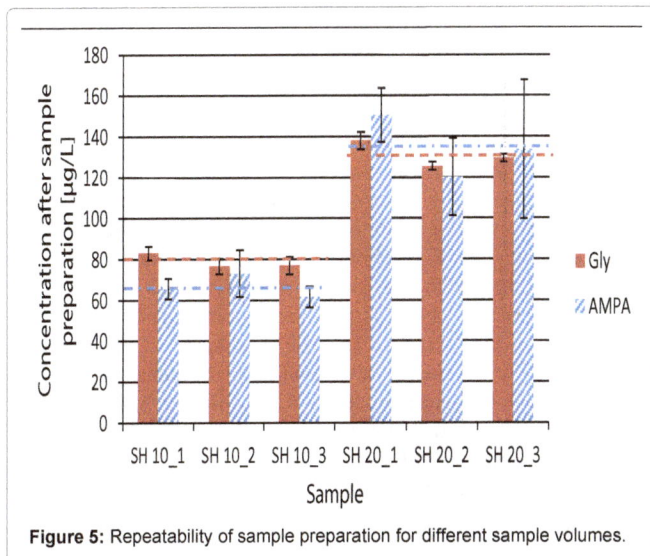

Figure 5: Repeatability of sample preparation for different sample volumes.

limits of detection and quantification compared to reported levels. Nevertheless, the obtained limits of detection and quantification enable the measurement of environmentally relevant concentration levels.

Enrichment factors: Due to the low expected concentration levels of both analytes in natural water samples, a pre concentration of glyphosate and AMPA is essential. Figure 6 shows the influence of the initial sample volume on the enrichment factor. The enrichment factor is defined as the ratio of the concentration after the two-step sample preparation process towards the adjusted concentration before preparation. It was determined in three different samples for 10 mL and 20 mL each. For glyphosate enrichment factors of approximately 8 with sample volumes of 10 mL and 13 with sample volumes of 20 mL were attained, while for AMPA factors of about 7 and 14 have been achieved depending on the sample volume.

So with the sample preparation process not only matrix reduction in complex natural water samples is possible, but also a satisfying detection improvement for both analytes becomes possible. This enables the detection of environmentally relevant concentration levels.

With larger sample volumes even higher enrichment factors and lower concentration levels could probably be attained in the future. However, for the application of larger sample volumes the column dimensions may need to be adjusted to ensure a sufficient binding capacity of the ion exchange resins, as well as to maintain a good passage of the higher sample volume through the columns.

Application of the optimized method to natural water samples from the River Elbe

Having proven the applicability of the method for natural water samples, a sample transect along the river Elbe between the Elbe estuary in Cuxhaven and the city of Hamburg has been taken. As one can see from Figure 7, AMPA could be detected at all sampling sites.

A trend of decreasing concentrations between the sampling sites 7 (Gluckstadt, industry site) with a value of 23 ± 6 µg/L and sampling site 1 (Cuxhaven, North Sea) with a concentration of 1 ± 1 µg/L has been observed. This is mainly a result of the increasing dilution effects along the river Elbe in the direction of the North Sea.

At the Hamburg sampling sites 8 to 12 AMPA concentrations between 7 ± 5 and 14 ± 3 µg/L have been found. At Seemannshöft an AMPA concentration of 12 ± 3 µg/L has been found. This value is comparable to data obtained during an earlier sampling campaign in June 2015.

The high AMPA concentrations that have been found may not only be contributed to the degradation of glyphosate. As described by other authors, AMPA is also formed from other nitrogen containing organic phosphonates. Certain methylenephosphonic acids as ATMP, EDTMP, DTPMP, HDTMP or 1-Hydroxy-(ethandiphosphonic acid) are used in detergents, as corrosion inhibitors and in textile and paper industry in large amounts [36-38]. Therefore, an identification of the particular source for our AMPA findings is not obvious. The high findings at the sampling sites 6 and 7 perhaps may result from the paper and color industry located in Glückstadt [3]. The concentrations of glyphosate are under the LOD of the whole method on all sampling sites except 6 and 7. The mean concentration levels at these sites are even above the limit of quantification with mean values of 73 and 30 µg/L. Due to the high standard deviation within those sample sites we have to consider these values as outliers (results not shown).

To our knowledge no transect along the river Elbe between

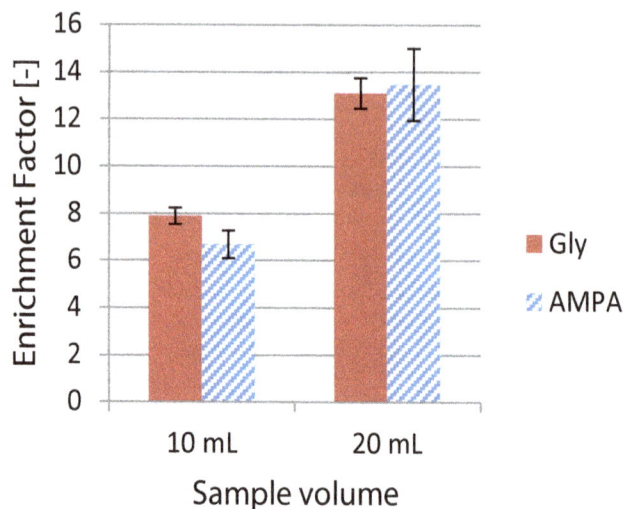

Figure 6: Influence of the sample volume on to the enrichment factor.

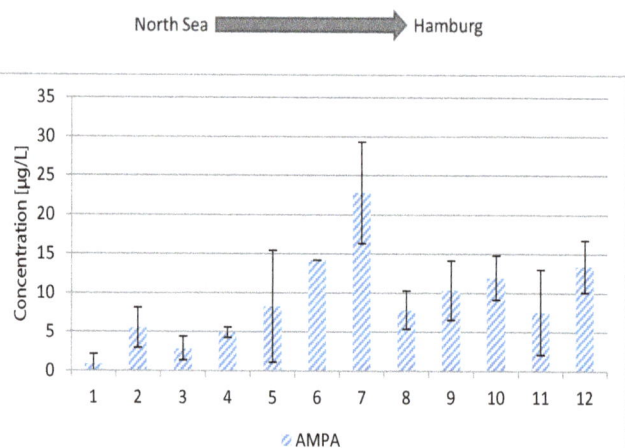

Figure 7: AMPA concentrations in the Elbe estuary.

Hamburg and Cuxhaven has been investigated regarding glyphosate and AMPA concentration levels before.

However, the detected AMPA concentrations in the estuary near Cuxhaven in this study correspond good with reported findings in the estuaries of the Baltic sea. The highest concentration levels of 2 µg/L for glyphosate and 4 µg/L for AMPA have been found in the Muehlenfliess estuary near Rostock. In that paper no transects along rivers were investigated [35].

Another study shows high glyphosate and AMPA concentration levels in extracted sediment samples along the same Elbe transect in May 2014. In the Elbe estuary near Cuxhaven the AMPA concentration level in the sediment was below the LOD, increasing to a level of about 35 µg/kg in the direction of Hamburg city. Glyphosate concentrations between approximately 5 and 12 µg/kg have been reported for all sampling sites. Water samples have been investigated only for the city of Hamburg during the time period of November 2012 to November 2013 in that study. The highest concentration level of glyphosate has been about 3 µg/L in June 2013. The AMPA concentration levels in any of the water samples in the city of Hamburg have been below the LOD [39].

Conclusion

A two-step enrichment process without time-consuming derivatization steps was developed and validated in this study. The determination of glyphosate and AMPA with the described LC/MS-MS method in these preprocessed samples is possible in different complex water matrices. Overall the developed two step enrichment process with subsequent LC-MS/MS detection shows a good accuracy and repeatability with a satisfying sensitivity depending on the water matrix.

Enrichment factors between 7 and 14 for sample volumes of 10 and 20 mL respectively have been achieved for both analytes in natural water samples from the river Elbe.

The applicability of the method for the detection of glyphosate and AMPA in natural water samples was shown. Therefore, during a ship campaign in August 2015 the Elbe estuary between Hamburg and Cuxhaven has been sampled. The detected glyphosate concentrations were below the calculated limit of detection for almost all sampling sites. However, at all stations AMPA could be detected at decreasing concentration levels with increasing distance to Hamburg. This may be contributed to dilution effects of the water body. Our findings may be contributed to a high discharge of glyphosate and AMPA in the industry region Hamburg.

As described above, even concentration levels of glyphosate in the low and middle μg/L range may have negative effects to water organisms. The concentration levels detected in this study along a transect of the river Elbe may consequently affect the local aquatic and estuarine fauna as well. Further sampling campaigns and the regular monitoring of glyphosate and AMPA concentrations in the Elbe estuary by the responsible authorities should be taken into consideration.

Acknowledgements

We would like to thank the Prandtl ship´s crew for the possibility to sample the river Elbe and any involved colleagues in sampling and preparation work.

References

1. Brauchli J (2004) BUWAL Bundesamt fur Umwelt, WuL, Bestimmung von Glyphosat und AMPA auf Bahnanlagen.

2. Benbrook CM (2016) Trends in glyphosate herbicide use in the United States and globally. Environmental Sciences Europe 28: 1-15.

3. Bundestag D (2011) Risikobewertung und Zulassung des Herbizid-Wirkstoffs Glyphosat.

4. Bundestag D (2015) Drucksache 18/6490 - Antwort der Bundesregierung auf die Kleine Anfrage der Abgeordneten Harald Ebner, Steffi Lemke, Nicole Maisch, weiterer Abgeordneter und der Fraktion Bündnis 90 / DIE GRÜNEN ,Folgen aus der Gefährdung von Bestäubern und der Umwelt durch Neonikotinoide und andere Pestizidwirkstoffe.

5. Newton M, Horner LM, Cowell JE, White DE, Cole EC (1994) Dissipation of glyphosate and aminomethylphosphonic acid in North-American forests. J Agr Food Chem 42: 1795-1802.

6. Hallas LE, Adams WJ, Heitkamp MA (1992) Glyphosate degradation by immobilized bacteria: field studies with industrial wastewater effluent. Applied and environmental microbiology 58: 1215-1219.

7. Barrett KA, McBride MB (2005) Oxidative degradation of glyphosate and aminomethylphosphonate by manganese oxide. Environ Sci Technol 39: 9223-9228.

8. Schuette J (1998) Environmental fate of glyphosate. Environmental Monitoring & Pest Management, Department of Pesticide Regulation, Sacramento, CA, USA.

9. Aubin AJ, Smith AE (1992) Extraction of [C-14] Glyphosate from Saskatchewan soils. J Agr Food Chem 40: 1163-1165.

10. Komossa D, Gennity I, Sandermann H (1992) Plant-metabolism of herbicides with C-P bonds - glyphosate. Pestic Biochem Phys 43: 85-94.

11. Mandiki SNM, Gillardin V, Martens K, Ercken D, De Roeck E, et al. (2014) Effect of land use on pollution status and risk of fish endocrine disruption in small farmland ponds. Hydrobiologia 723: 103-120.

12. Delmonico EL, Bertozzi J, de Souza NE, Oliveira CC (2014) Determination of glyphosate and aminomethylphosphonic acid for assessing the quality tap water using SPE and HPLC. Acta Sci-Technol 36: 513-519.

13. Mallat E, Barcelo D (1998) Analysis and degradation study of glyphosate and of aminomethylphosphonic acid in natural waters by means of polymeric and ion-exchange solid-phase extraction columns followed by ion chromatography post-column derivatization with fluorescence detection. J Chromatogr A 823: 129-136.

14. Henkelmann G (2001) Das Verhalten von Glyphosat in der Umwelt - Forschungsergebnisse zum Austrag und zur Verlagerung. In: Institut für Agrarokologie, OLuB, an der Bayerischen Landesanstalt für Landwirtschaft, Arbeitsbereich IAB (eds.).

15. Coupe RH, Kalkhoff SJ, Capel PD, Gregoire C (2012) Fate and transport of glyphosate and aminomethylphosphonic acid in surface waters of agricultural basins. Pest Manag Sci 68: 16-30.

16. Shaw M, Silburn DM, Thornton C, Robinson B, McClymont D (2011) Modelling pesticide runoff from paddocks in the Great Barrier Reef using HowLeaky. 19th International Congress on Modelling and Simulation, Australia pp: 2057-2063.

17. Mercurio P, Flores F, Mueller JF, Carter S, Negri AP (2014) Glyphosate persistence in seawater. Marine pollution bulletin 85: 385-390.

18. Stachowski-Haberkorn S, Becker B, Marie D, Haberkorn H, Coroller L, et al. (2008) Impact of Roundup on the marine microbial community, as shown by an in situ microcosm experiment. Aquat Toxicol 89: 232-241.

19. Puertolas L, Damasio J, Barata C, Soares AMVM, Prat N (2010) Evaluation of side-effects of glyphosate mediated control of giant reed (Arundo donax) on the structure and function of a nearby mediterranean river ecosystem. Environ Res 110: 556-564.

20. Webster TMU, Laing LV, Florance H, Santos EM (2014) Effects of glyphosate and its formulation, Roundup, on reproduction in zebrafish (Danio rerio). Environ Sci Technol 48: 1271-1279.

21. Sandrini JZ, Rola RC, Lopes FM, Buffon HF, Freitas MM, et al. (2013) Effects of glyphosate on cholinesterase activity of the mussel Perna perna and the fish Danio rerio and Jenynsia multidentata: In vitro studies. Aquat Toxicol 130: 171-173.

22. Le Mer C, Roy RL, Pellerin J, Couillard CM, Maltais D (2013) Effects of chronic exposures to the herbicides atrazine and glyphosate to larvae of the threespine stickleback (Gasterosteus aculeatus). Ecotox Environ Safe 89: 174-181.

23. Oppenhuizen ME, Cowell JE (1991) Liquid chromatographic determination of glyphosate and aminomethylphosphonic acid (AMPA) in environmental water: collaborative study. Association of Official Analytical Chemists 74: 317-323.

24. Roseboom H, Berkhoff CJ (1982) Determination of the herbicide glyphosate and its major metabolite aminomethylphosphonic acid by high-performance liquid-chromatography after fluorescence labeling. Anal Chim Acta 135: 373-377.

25. Hao CY, Morse D, Morra F, Zhao XM, Yang P (2011) Direct aqueous determination of glyphosate and related compounds by liquid chromatography/tandem mass spectrometry using reversed-phase and weak anion-exchange mixed-mode column. J Chromatogr A 1218: 5638-5643.

26. Bernal J, Bernal JL, Martin MT, Nozal MJ, Anadon A, et al. (2010) Development and validation of a liquid chromatography-fluorescence-mass spectrometry method to measure glyphosate and aminomethylphosphonic acid in rat plasma. J Chromatogr B 878: 3290-3296.

27. Jahnke B, Frank C, Fernández JF, Niemeyer B (2014) A sequential injection analysis method for the determination of glyphosate and aminomethylphosphonic acid in water samples. American Chemical Science Journal 5: 163-173.

28. ZoBell CE (1946) Marine Microbiology - A monograph on hydrobacteriology.

29. Karl M, Telgheder (2008) Berechnung der Verfahrensstandardabweichung und Nachweis-Erfassungs-und Bestimmungsgrenze aus einer Kalibrierung gemäß DIN 32645.

30. Molt K, Telgheder U (2010) Berechnung der Verfahrensstandardabweichung und Nachweis-, Erfassungs- und Bestimmungsgrenze aus einer Kalibrierung gemäß DIN 32645 in Universität Duisburg-Essen, F.I.A.C.

31. Kolb M, Bahr A, Hippich S, Schulz W (1993) Ermittlung der Nachweis-, Erfassungs- und Bestimmungsgrenze nach DIN 32645 mit Hilfe eines Programms Calculation of Detection Limit, Identification Limit and Determination Limit according to DIN 32645 with the Aid of a Computer Program. Acta Hydrochimica Et Hydrobiologica 21: 308-311.

32. Winfield TW (1990) EPA Method 547 Determination of glyphosate in drinking water by direct aqeous injection, HPLC, post-column derivatization and fluorescence detection.

33. Maurer R, Henday S, Wang L, Schnute B (2009) Analysis of glyphosate and AMPA in environmental samples by ion chromatography mass spectrometry (IC-ESI-MS). Dionex Corporation, Sunnyvale, CA, USA and Olten, Switzerland.

34. Ding J, Guo H, Liu W, Zhang W, Wang J (2015) Current progress on the detection of glyphosate in environmental samples. J of Sc and App Bio Medicine 3: 88-95.

35. Skeff W, Neumann C, Schulz-Bull DE (2015) Glyphosate and AMPA in the estuaries of the Baltic Sea method optimization and field study. Marine pollution bulletin 100: 577-585.

36. Reupert R, Fuchs S (1997) HPLC-Analytik von Glyphosat und AMPA durch Nachsäulenderivatisierung. GIT Laborfachzeitschrift 5: 468-747.

37. William EG, Tom CJF (1992) Environmental properties and safety assessment of organic phosphonates used for detergent and water treatment applications. Anthropogenic Compounds pp: 261-285

38. Steber J, Wierich P (1987) Properties of aminotris (methylenephosphonate) affecting its environmental fate: Degradability, sludge adsorption, mobility in soils, and bioconcentration. Chemosphere 16: 1323–1337.

39. Sanders T, Lassen S (2015) The herbicide glyphosate affects nitrification in the Elbe estuary, Germany. European Geoscience Union General assembly, Konferenz, Wien.

32

High Performance Liquid Chromatographic Determination of Naproxen in Prepared Pharmaceutical Dosage Form and Human Plasma and its Application to Pharmacokinetic Study

Saiqa Muneer[1,2,3*], Iyad Naeem Muhammad[1], Muhammad Asad Abrar[4], Iqra Munir[5], Iram Kaukab[5], Abdullah Sagheer[6], Habiba Zafar[7], Kishwar Sultana[2]

[1]University of Karachi
[2]The University of Lahore
[3]The university of Queensland
[4]Islamia University Bahawalpur
[5]Bahauddin Zakariya University Multan
[6]University of Veterinary and Animal Sciences Lahore
[7]Comsats Institute of Technology, Abottabad

Abstract

A simple, fast, economical and precise reverse phase liquid chromatographic method was developed, optimized and validated for quantification of naproxen sodium according to the standard guidelines. The separation of the analyte and internal standard was achieved over C-18 column using acetonitrile: water: glacial acetic acid as mobile phase in a ratio of 50:49:1 (v/v) in isocratic mode at a flow rate of 1.2 mL/min at a wavelength of 254 nm at ambient temperature. The retention time was found to be less than 10 minutes. The limit of detection (LOD) was 10 ng/mL and limit of quantification (LOQ) was 15 ng/mL. Naproxen sodium was extracted from biological samples by using acetonitrile as extraction solvent. The linearity was found to be 0.5 to 80 ppm. All the parameters were validated for accuracy, precision, linearity, sensitivity, reproductivity and stability. The method was successfully applied for the quantification of the naproxen sodium in animal and human plasma. Therefore, the method was found to be accurate, reproducible, sensitive, cost effective, less time consuming and can be successfully applied on routine analysis of naproxen sodium in pharmaceutical formulations and pharmacokinetic studies in human and animal models.

Keywords: Naproxen sodium; RP-HPLC; Biological samples; Pharmacokinetics

Introduction

Naproxen sodium is chemically sodium (2S)-2-(6-methoxynaphthalen-2-YL) propionate. (National Center for Biotechnology Information 2016). Naproxen is a propionic acid derivative that belongs to the aryl acetic group of non-steroidal anti-inflammatory agents [1,2]. They act by inhibiting the prostaglandin synthesis thereby producing several side effects such as ulceration and gastrointestinal bleeding [3-5]. Although many high-performance liquid chromatographic methods have been developed for the estimation of naproxen sodium alone [2,5] or in combination for the simultaneous determination of naproxen with other drugs in different matrices. These methods involved derivatization and post-column photochemical reactions, mostly used buffer solution as mobile phase due to which the mobile phase becomes complex. But there is always a need of a meaningful cost-effective method for estimation of naproxen sodium because the present methods do not provide the sensitivity [1,2,5]. The aim of the present study was to develop a simple and effective RP-HPLC method for quantification of naproxen in pharmaceutical preparations [5-9] as well as in spiked human plasma and animal plasma (rabbits) [7,10,11] in a very short time and to validate that method according to ICH and FDA guidelines (Food and drug administration 2012; United states Pharmacopoeia 2007; International Conference on Harmonization (ICH) 2005;). The method was successful due to its suitability and reproducibility. This method is more precise, accurate, highly sensitive, rapid and cost effective when observed under different experimental conditions [12-15]. The dominance of the new assay for its intended use in the human pharmacokinetic study was upheld for its validation and applications over the other methods of naproxen estimation found in literature (Figure 1).

Experimental

Chemicals and reagents

Naproxen sodium was gifted by Life Pharma, Multan, Pakistan. Acetonitrile and water (HPLC grade) and glacial acetic acid was purchased from Merck (Germany). Other chemicals used were of analytical grade and Naproxen sodium tablet formulations used in this experiment were prepared during research work and a commercial brand was purchased from local market [16].

Instrumentation

High performance liquid chromatographic system equipped with an auto sampler and UV visible detector was used for analysis. The data was recorded using Liquid chromatography software.

Chromatographic conditions

The liquid chromatograph is equipped with a 254-nm detector

*Corresponding author: Saiqa Muneer, PhD Student, School of Pharmacy, University of Queensland, 20 Cornwall Street, Woollongabba, 4109, Brisbane, Queensland 4102, Australia
E-mail: saiqamunir77@gmail.com (or) s.muneer@uq.edu.au

and a 4.6-mm × 15-cm column C-18 that contains 5-μm packing L1. The flow rate is about 1.2 mL per minute. Chromatograph of the standard preparation, and record of the peak responses as directed under procedure: the column efficiency, determined from the analyte peak, is not less than 4000 theoretical plates [17,18].

Mobile phase: A suitable mixture of acetonitrile, water, and glacial acetic acid (50:49:1) was prepared (Make adjustments if necessary. Increased resolution may be achieved by increasing the proportion of water in the mobile phase). The mobile phase was membrane filtered (pore size 0.45 μm), sonicated and degassed before use.

Solvent mixture: A suitable mixture of acetonitrile and water with a ratio (90:10) was prepared (Figure 2).

Preparation of assay

For estimation of naproxen sodium, twenty tablets of 275 mg naproxen sodium tablets were taken randomly, crushed and powdered and the average weight was calculated. The powder equivalent to 2.75 mg of naproxen sodium was taken and transferred to 100 mL volumetric flask and initially dissolved in 10 mL water, and sonicated for 10 minutes until the material was completely dispersed. Then added about 80 mL of acetonitrile, and sonicated for an additional 5 minutes. The flask was allowed to reach room temperature, diluted with acetonitrile to volume, and mix. Any insoluble matter was allowed to settle, then 1.0 mL of the clear supernatant was transferred to a 100 mL volumetric flask; diluted with mobile phase to volume, and mix [19,20].

Preparation of standard solution

An accurately weighed quantity of Naproxen WS was dissolved in solvent mixture to obtain a solution having a known concentration of about 2.5 mg per mL. Then transferred 1.0 mL of the resulting solution to a 100 mL volumetric flask, diluted with mobile phase to volume, and mix. This solution contains about 25 μg of Naproxen WS per mL.

Procedure

Separately injected equal volumes (about 20 μL) of the Standard preparation and the Assay preparation into the chromatograph, recorded the chromatograms, and measured the responses for the major peaks. The relative retention times are around 0.6 for naproxen. The quantity, in mg, of Naproxen ($C_{14}H_{14}O_3$) was calculated in the portion of tablets taken by the formula:

$$10C\ (R_U/R_S)$$

where, C is the concentration (μg per mL) of Naproxen WS in the Standard preparation, and R_U and R_S are the ratios of the response of the naproxen peak to the response of peak obtained from the Assay preparation and the Standard preparation, respectively.

The peak area was determined and plotted against concentration μg/mL in plasma. 20 μL was injected into chromatograph and each chromatogram is recorded. The results of calibration curve and system suitability are shown in Table 2 and Figure 3.

Optimization of chromatographic conditions

Various experimental conditions such as different mobile phase, flow rate, detection wavelength and column over temperature were optimized for the analysis of RP-HPLC in isocratic mode. Good separation and shorter run time was obtained by the mobile phase used in varying concentrations of acetonitrile and water to adjust the flow rate to 1.2 mL/min where it gives the improvement in peak shapes of

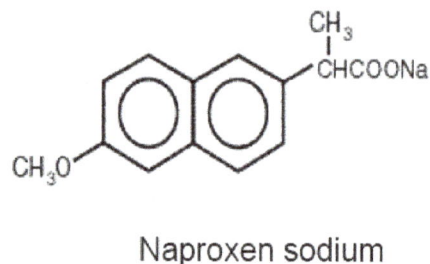

Figure 1: Chemical structure of Naproxen sodium.

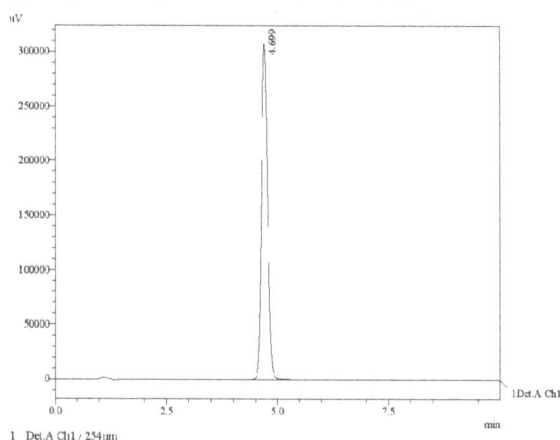

Figure 2: A chromatogram of naproxen sodium.

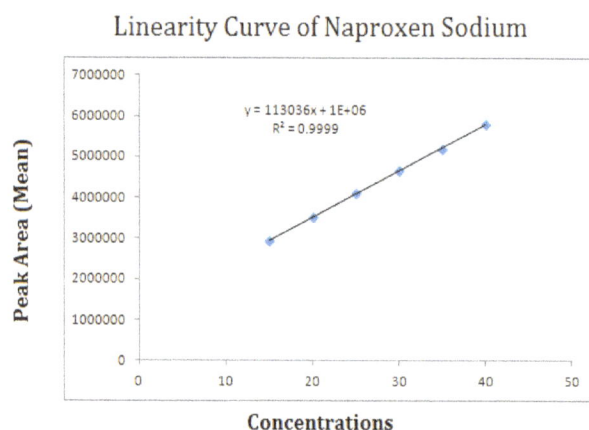

Figure 3: Calibration Curve for Naproxen Sodium.

naproxen. Detection wavelength was also varied in the experiment to obtain different chromatograms to achieve better peaks and 254 nm was selected as optimized. Column C-18 well the naproxen by reverse phase chromatography.

Preparation of human plasma samples

Blood samples were collected in heparinized glass tubes from healthy volunteers in University of Lahore, Pakistan after obtaining the informed consent. They were then centrifuged at 3500 rpm for 10 minutes and then filtered through 0.45 μm membrane filter and the plasma is separated and stored at -20°C till analysis. At the time of analysis, the serum was spiked with the naproxen sodium drug solution in a ratio of 1:1 and diluted ten folds to obtain 100 µg/mL. 20 µL of reconstituted samples were injected into the HPLC-UV system [20,21].

Extraction: Naproxen sodium (internal standard) was extracted using acetonitrile ACN. The human plasma was then spiked with the internal standard of naproxen sodium (0.5 µg/mL). The samples were then vortexed for 2 mins and then centrifuged at 3500 rpm for 10 mins. The supernatant was collected and volume make up to 1 mL and 20 µL was injected directly into the system.

Preparation of rabbit plasma samples

Same procedure was adapted for rabbits and samples were collected at a predefined time from the rabbits in heparinized test tubes.

Method validation

The developed method was validated for specificity, accuracy, repeatability, precision, LOD, LOQ, robustness as per ICH guidelines.

Specificity: The difference between API and other excipients present in the sample is called specificity. The specificity was determined by injecting the excipients and standard solution of naproxen sodium.

Accuracy: Accuracy is defined as the nearest of the nominal and the actual results. The accuracy is expressed as percentage recovery of the analyte recovered by the assay. Accuracy was determined by the applying the proposed method to different concentrations of drug product mixtures. The injections were taken in triplicate and mean peak area is taken to calculate the concentration.

The accuracy of the proposed method is calculated by the following equation

$$\text{Accuracy}(\%) = \frac{\text{actual Concentration}}{\text{nominal concentration}} \times 100$$

The average accuracy was 98.5% Mean recoveries for naproxen sodium for the specific formulation are shown in table. The results show that are the parameters were within the expected range (Table 1).

Repeatability: The repeatability was done on 3 replicates of each concentration of the standard solution.

Linearity: The linearity plots were obtained over a concentration range of 15 µL to 40 µL to encompass the expected concentration in measured samples. The average peak areas were plotted against respective concentration. The linearity of the proposed method was studied by using the calibration curves (Figure 1) for the determination of the coefficient of correlation and intercept values (Tables 2 and 3).

Limit of Detection and Limit of Quantification: The limit of detection (LOD) was and limit of quantification (LOQ) were determined as per ICH guidelines. The limit of detection (LOD) was 10 ng/mL and limit of quantification (LOQ) was 15 ng/mL (Table 4).

Precision: Precision was determined through both reproducibility and repeatability. It is defined as the degree of agreement among individual test results where the method is applied repeatedly to multiple readings. Precision is determined by interday and intraday precision. Precision is done by applying the data of three different concentrations on consecutive three days throughout the linearity range thrice in a day. The relative standard deviation for the response factor was calculated. The %RSD value measured during assessment of precision was <2% for naproxen sodium indicating that the method is precise as per ICH guidelines (Tables 5 and 6).

Sample analysis: To quantify the amount of drug in the finished tablet dosage form, the sample solution of naproxen sodium was analyzed by the proposed method. The contents were determined by the use of calibration curves and the formula is given by,

Y=mx+c

Where, Y=peak area of the analyzed sample; m=slope of the analyzed sample; x=concentration of the analyzed sample and c=intercept of calibration curve.

System suitability: System suitability was conducted in order to confirm the effectiveness of the chromatographic system. Fr this purpose, different suitability parameters such as retention time, tailoring factor, capacity factor (k), selectivity factor (α), resolution, number of theoretical plates were checked by repeatedly injecting the solution of naproxen sodium. The values are given in Table 7.

Robustness: The robustness of the method was determined by minor deliberate changes in experimental conditions such as column

Injected conc. (µg/mL)	Recovered conc. (µg/mL)	% Recovery
107	105.82511	98.9
120	119.1235	99.26
230	224.4162	97.56

Table 1: Accuracy (% recovery) of Naproxen.

Conc. of Standard	Standard ID		Peak Area	Mean
15 µg/mL	A	A1	2942222	2941321.67
		A2	2936694	
		A3	2945049	
20 µg/mL	B	B1	3513865	3528324.00
		B2	3524865	
		B3	3546242	
25 µg/mL	C	C1	4098615	4091748.67
		C2	4087587	
		C3	4089044	
30 µg/ml	D	D1	4630074	4650964.00
		D2	4659300	
		D3	4663518	
35 µg/mL	E	E1	5104615	5197315.00
		E2	5244524	
		E3	5242806	
40 µg/mL	F	F1	5801346	5784360.00
		F2	5753372	
		F3	5798362	

Table 2: Linearity Details of Naproxen sodium.

Standard ID	Concentrations	Peak area (Mean)
A	15 µg/mL	2941321.67
B	20 µg/mL	3528324.00
C	25 µg/mL	4091748.67
D	30 µg/mL	4650964.00
E	35 µg/mL	5197315.00
F	40 µg/mL	5784360.00

Table 3: Linear regression data for naproxen sodium calibration curve.

Spiked conc.	Conc. observed	Accuracy	(%RSD)
20	20.381	101.90	1.48
10	10.202	102.02	2.72
5	4.93124	98.62	2.06
1	1.014	101.4	1.35
0.5	0.49878	99.75	0.21

Table 4: Determination of naproxen in tablet dosage form by suggested HPLC method.

Spiked conc.	Conc. observed	Accuracy	(%RSD)
20	20.212	101.06	1.4
10	10.43	104.3	2.3
5	4.80	96	1.89
1	1.23	123	1.24
0.5	0.54	108.2	2.15

Table 5: Determination of naproxen in human plasma by suggested HPLC method.

Spiked conc.	Conc. observed	Accuracy	(%RSD)
20	20.421	102.1	2.1
10	9.464	94.6	0.87
5	5.023	100.46	1.02
1	0.985	98.5	1.13
0.5	0.49	98	0.912

Table 6: Determination of naproxen in rabbit plasma by suggested HPLC method.

Validation parameter	Naproxen
Recovery (%)	98.56
Repeatability (%RSD)	1.564
Precision range	0.87-2.87
Interday (n=3)	1.79
Intraday (n=3)	1.95
Limit of detection (ng/mL)	10
Limit of quantification (ng/mL)	15

Table 7: System suitability parameters.

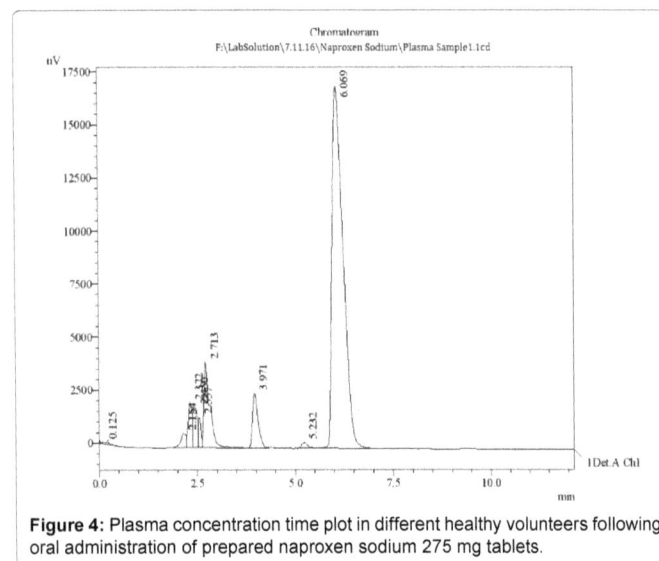

Figure 4: Plasma concentration time plot in different healthy volunteers following oral administration of prepared naproxen sodium 275 mg tablets.

oven temperature (± 2°C), flow rate of mobile phase (± 0.2 mL/min), composition of mobile phase (± 1). The data obtained after the minor changes in the experimental conditions did not affect the recoveries, peak area and the retention time which showed that the method is robust.

System stability: Stability was tested in an auto sampler for 0, 6, 12 and 24 hr and triplicates of low and high control samples were analyzed and deviation was calculated. Long term stability was tested for two weeks by taking low and high-quality control samples in triplicate. The mean concentration of sample was taken and compared with the samples taken at 0 hr [22] (Figure 4).

Application of the Methods

This method was a part of an extensive research, "Establishment of *in-vitro in-vivo* correlation (IVIVC) studies for developed naproxen sodium tablet formulations". The method was successfully applied to the *in-vitro in-vivo* evaluation of the pharmaceutical formulations of naproxen in animal models as well as in human plasma. In pharmaceutical development, it is important to employ *in-vitro* release to select formulations that can provide the adequate therapy. Drug release kinetics provide valuable data on the structural behavior of the drug and its release pattern. The developed method was successfully employed for *in-vitro* evaluation and pharmacokinetic data of naproxen sodium following IV administration of this drug. The drug content of each sample was determined and the pharmacokinetic parameters were determined by the use of the software.

Conclusion

A reverse phase HPLC method of Naproxen sodium has been developed in tablet dosage form and validated as per ICH guidelines, USP and FDA by using mobile phase of acetonitrile and water in a ratio of 50:49:1 at pH 7.8 using 4.6 mm × 15 cm column that contains 5 μm packing L1 at ambient temperature. The flow rate was 1 ml/min and the eluent was determined at a wavelength of 254 nm. The retention time of naproxen was 4.699 min (Figure 2). Precision was reflected by mean %RSD of 1.564 which showed the sensitivity of the developed method. A simple extraction procedure was implemented resulting in good recovery of both drug and the internal standard. All the analyte was separated and well resolved from less than 7 minutes. Various experimental conditions such as column over temperature, pH of the mobile phase, detector wavelength, mobile phase ratio and flow rate were optimized for quantitative analysis of naproxen sodium in spiked human plasma and pharmaceuticals and biological samples. The current method is simple, economic, accurate, precise, specific, and rapid and will be conveniently applied for routine analysis in pharmaceutical dosage form, spiked human plasma and biological samples with slight modification in extraction procedures.

References

1. Botting RM (2006) Inhibitors of cyclooxygenases: mechanisms, selectivity and uses. Journal of Physiology and Pharmacology 57: 113-124.

2. Tanjin S, Islam F, Sultan MZ, Rahman A, Chowdhury SR, et al. (2013) Development and Validation of a Simple RP-HPLC Method for Determination of Naproxen in Pharmaceutical Dosage Forms. Bangladesh Pharmaceutical Journal 16: 137-141.

3. US Food and Drug Administration (2001) FDA guidance for industry: bioanalytical method validation. US Department of Health and Human Services, Food and Drug Administration Center for Drug Evaluation and Research.

4. United States Pharmacopoeia (2007).

5. Gajraj NM (2003) Cyclooxygenase-2 inhibitors. Anesthesia & Analgesia 96: 1720-1738.

6. Deshpande P, Gandhi S, Mantri RA, Luniya KP, Dubey SU, et al. (2012) Validated Method Development for Estimation of Naproxen sodium as Bulk Drug and in Tablet Dosage Form by HPTLC. Eurasian Journal of Analytical Chemistry 7.

7. Elsinghorst PW, Kinzig M, Rodamer M, Holzgrabe U, Sörgel F (2011) An LC–MS/MS procedure for the quantification of naproxen in human plasma:

Development, validation, comparison with other methods, and application to a pharmacokinetic study. Journal of Chromatography B 879: 1686-1696.

8. Ahmadi M, Madrakian T, Afkhami A (2015) Enantioselective solid phase extraction prior to spectrofluorometric determination: a procedure for the determination of naproxen enantiomers in the presence of each other. RSC Advances 5: 5450-5457.

9. Mehta P, Sharma CS, Nikam D, Ranawat MS (2012) Development and validation of related substances method by HPLC for analysis of naproxen in naproxen tablet formulations. Int J Pharm Sci Drug Res 4: 63-69.

10. Khan MF, Zuthi SS, Kayser MS, Islam MS, Asad S, et al. (2016) A Simple RP-HPLC Method Development and Validation for the Simultaneous Estimation of Naproxen and Rabeprazole. Journal of Applied Pharmaceutical Science 6: 147-152.

11. Upton RA, Buskin JN, Guentert TW, Williams RL, Riegelman S (1980) Convenient and sensitive high-performance liquid chromatography assay for ketoprofen, naproxen and other allied drugs in plasma or urine. Journal of Chromatography A 190: 119-128.

12. Solanki SD, Patel PU, Suhagiya BN (2011) Development and validation of spectrophotometric method for simultaneous estimation of sumatriptan succinate and naproxen sodium in pharmaceutical dosage form. JPSBR 1: 50-53.

13. Sloka SN, Gurupadayya BM, Kumar CA (2011) Simultaneous spectrophotometric determination of Naproxen and Pantoprazole in pharmaceutical dosage form. J Appl Chem Res 17: 65-74.

14. Lian H, Hu Y, Li G (2013) Novel metal ion-mediated complex imprinted membrane for selective recognition and direct determination of naproxen in pharmaceuticals by solid surface fluorescence. Talanta 116: 460-467.

15. Khan A, Iqbal Z, Khadra I, Ahmad L, Khan A, et al. (2016) Simultaneous determination of domperidone and Itopride in pharmaceuticals and human plasma using RP-HPLC/UV detection: Method development, validation and application of the method in in-vivo evaluation of fast dispersible tablets. Journal of Pharmaceutical and Biomedical Analysis 121: 6-12.

16. Croitoru O, Spiridon AM, Belu I, Turcu-Ştiolică A, Neamţu J (2015) Development and validation of an HPLC method for simultaneous quantification of clopidogrel bisulfate, its carboxylic acid metabolite, and atorvastatin in human plasma: application to a pharmacokinetic study. J Anal Methods Chem.

17. Haque A, Shahriar M, Parvin MN, Islam SM (2011) Validated RP-HPLC method for estimation of ranitidine hydrochloride, domperidone and naproxen in solid dosage form. Asian Journal of Pharmaceutical Analysis 1: 59-63.

18. Tanam AA, Khan MF, Rashid RB, Sultan MZ, Rashid MA (2015) Validation and Optimization of a Simple RP-HPLC Method for the Determination of Paracetamol in Human Serum and its Application in a Pharmacokinetic Study with Healthy Bangladeshi Male Volunteers. Dhaka University Journal of Pharmaceutical Sciences 13: 125-131.

19. Shah Y, Iqbal Z, Ahmad L, Khan A, Khan MI, et al. (2011) Simultaneous determination of rosuvastatin and atorvastatin in human serum using RP-HPLC/UV detection: method development, validation and optimization of various experimental parameters. Journal of Chromatography B 879: 557-563.

20. Sultana N, Arayne MS, Iftikhar B (2008) Simultaneous Determination of Atenolol, Rosuvastatin, Spironolactone, Glibenclamide and Naproxen Sodium in Pharmaceutical Formulations and Human Plasma by RP-HPLC. Journal of the Chinese Chemical Society 55: 1022-1029.

21. Sultan M, Stecher G, Stoggl WM, Bakry R, Zaborski P, et al. (2005) Sample pretreatment and determination of non-steroidal anti-inflammatory drugs (NSAIDs) in pharmaceutical formulations and biological samples (blood, plasma, erythrocytes) by HPLC-UV-MS and µ-HPLC. Current Medicinal Chemistry 12: 573-588.

22. Yilmaz B, Asci A, Erdem AF (2014) HPLC method for naproxen determination in human plasma and its application to a pharmacokinetic study in Turkey. Journal of Chromatographic Science 52: 584-589.

Permissions

The contributors of this book come from diverse backgrounds, making this book a truly international effort. This book will bring forth new frontiers with its revolutionizing research information and detailed analysis of the nascent developments around the world.

We would like to thank all the contributing authors for lending their expertise to make the book truly unique. They have played a crucial role in the development of this book. Without their invaluable contributions this book wouldn't have been possible. They have made vital efforts to compile up to date information on the varied aspects of this subject to make this book a valuable addition to the collection of many professionals and students.

This book was conceptualized with the vision of imparting up-to-date information and advanced data in this field. To ensure the same, a matchless editorial board was set up. Every individual on the board went through rigorous rounds of assessment to prove their worth. After which they invested a large part of their time researching and compiling the most relevant data for our readers.

The editorial board has been involved in producing this book since its inception. They have spent rigorous hours researching and exploring the diverse topics which have resulted in the successful publishing of this book. They have passed on their knowledge of decades through this book. To expedite this challenging task, the publisher supported the team at every step. A small team of assistant editors was also appointed to further simplify the editing procedure and attain best results for the readers.

Apart from the editorial board, the designing team has also invested a significant amount of their time in understanding the subject and creating the most relevant covers. They scrutinized every image to scout for the most suitable representation of the subject and create an appropriate cover for the book.

The publishing team has been an ardent support to the editorial, designing and production team. Their endless efforts to recruit the best for this project, has resulted in the accomplishment of this book. They are a veteran in the field of academics and their pool of knowledge is as vast as their experience in printing. Their expertise and guidance has proved useful at every step. Their uncompromising quality standards have made this book an exceptional effort. Their encouragement from time to time has been an inspiration for everyone.

The publisher and the editorial board hope that this book will prove to be a valuable piece of knowledge for researchers, students, practitioners and scholars across the globe.

List of Contributors

Pinto J, Mendes VM, Coelho M and Manadas B
CNC - Center for Neuroscience and Cell Biology, University of Coimbra, Portugal

Baltazar G
CICS - UBI - Health Sciences Research Centre, University of Beira Interior, Portugal

Pereira J LGC
Departamento de Química, Universidade de Coimbra, Portugal

Dunn MJ
Proteome Research Centre, UCD Conway Institute of Biomolecular and Biomedical Research, University College Dublin, Portugal

Cotter DR
Department of Psychiatry, Royal College of Surgeons in Ireland

Venkata Subba Rao D and Raghuram P
ScieGen Pharmaceuticals INC, 89 Arkay Drive, Hauppauge, NY, USA

Harikrishna KA
Biological E Limited, ICICI Knowledge Park, Shamirpet, Hyderabad-500078, India

Wujian J, Kuan-wei P, Sihyung Y and Wang MZ
Department of Pharmaceutical Chemistry, The University of Kansas, Lawrence, KS, USA

Huijing S and Mario S
Eshelman School of Pharmacy, The University of North Carolina at Chapel Hill, Chapel Hill, NC, USA

Ferro P
Bioanalysis Research Group, Neuroscience Research Program, IMIM (Hospital del Mar Medical Research Institute), Barcelona, Spain

Segura J
Bioanalysis Research Group, Neuroscience Research Program, IMIM (Hospital del Mar Medical Research Institute), Barcelona, Spain

Department of Experimental and Health Sciences, Pompeu Fabra University (UPF), Barcelona Biomedical Research Park, Barcelona, Spain

Krotov G and Zvereva I
Anti-Doping Centre, Moscow, Russia

Mateus JA
Integrative Pharmacology and Systems Neuroscience Research Group, IMIM (Hospital del Mar Medical Research Institute), Barcelona, Spain

Pérez-Mana
Integrative Pharmacology and Systems Neuroscience Research Group, IMIM (Hospital del Mar Medical Research Institute), Barcelona, Spain

Department of Pharmacology, Therapeutics and Toxicology, Autonomous University of Barcelona, Cerdanyola, Spain

Reza Akramipour, Mitra Hemati and Simin Gheini
School of Medical, Kermanshah University of Medical Sciences, Kermanshah, Iran

Nazir Fattahi
Research Center for Environmental Determinants of Health (RCEDH), Kermanshah University of Medical Sciences, Kermanshah, Iran

Hamid Reza Ghaffari
Social Determinants in Health Promotion Research Center, Hormozgan University of Medical Sciences, Bandar Abbas, Iran

Department of Environmental Health Engineering, School of Public Health, Tehran University of Medical Sciences, Tehran, Iran

Ambekar Archana M and Kuchekar Bhanudas S
MAEER's Maharashtra Institute of Pharmacy, Department of Pharmaceutical Chemistry, Pune, Maharashtra, India

Cheng-Tung Chen, Cheng-Wei Cheng and Yu-Fang Hu
Pharmaceutical Development Center, TTY Biopharm Company Limited, Taipei, Taiwan

Godoy ALPC
Department of Clinical and Toxicological Analyzes, School of Pharmacy, Federal University of Bahia (UFBA) , Rua Barão de Jeremoabo, 147, Salvador, BA, Brazil
Laboratory of Veterinary Infectology, Hospital-School of Veterinary Medicine, Federal University of Bahia, Av. Adhemar de Barros, 500, Salvador, BA, Brazil

De Jesus C
Laboratory of Veterinary Infectology, Hospital-School of Veterinary Medicine, Federal University of Bahia, Av. Adhemar de Barros, 500, Salvador, BA, Brazil

Larangeira DF and Barrouin-Melo SM
Laboratory of Veterinary Infectology, Hospital-School of Veterinary Medicine, Federal University of Bahia, Av. Adhemar de Barros, 500, Salvador, BA, Brazil Department of Anatomy, Pathology and Clinics, School of Veterinary Medicine and Animal Science, UFBA, Salvador, BA, Brazil

De Oliveira ML, Rocha A, Pereira MPM and Lanchote VL
Department of Clinical, Toxicological and Bromatological Analyzes, Faculty of Pharmaceutical Sciences of Ribeirão Preto, University of São Paulo, Avenida do Café, S / N, Ribeirão Preto, SP, Brazil

Mulubwa M, Rheeders M and Viljoen M
Centre of Excellence for Pharmaceutical Sciences (Pharmacen), Division of Pharmacology, North-West University, Potchefstroom 2520, South Africa

Du Plessis L and Grobler A
DST/NWU Preclinical Drug Development Platform (PCDDP), North-West University, Potchefstroom 2520, South Africa

Ardente AJ and Walsh M
College of Veterinary Medicine, University of Florida, Gainesville, FL, USA

Wells RS
College of Veterinary Medicine, University of Florida, Gainesville, FL, USA Chicago Zoological Society's Sarasota Dolphin Research Program, C/O Mote Marine Laboratory, Sarasota, FL, USA

Garrett TJ
Southeast Center for Integrated Metabolomics, Mass Spectrometry Core Laboratory, University of Florida, Gainesville, FL, USA

Smith CR
National Marine Mammal Foundation, San Diego, CA, USA

Colee J
Department of Statistics, Institute of Food and Agricultural Sciences, University of Florida, Gainesville, FL, USA

Hill RC
Small Animal Clinical Sciences, College of Veterinary Medicine, University of Florida, 2015 SW 16th Ave., Gainesville, FL 32608, USA

Gyemant G
Department of Inorganic and Analytical Chemistry, University of Debrecen, 4032 Debrecen, Egyetem tér 1, Hungary

Andrasi M and Gaspar A
Department of Inorganic and Analytical Chemistry, University of Debrecen, 4032 Debrecen, Egyetem tér 1, Hungary Cetox Ltd., 4032 Debrecen, Egyetem tér 1, Hungary

Mishra PR and Meshram DB
Department of Quality Assurance, Pioneer Pharmacy Degree College, Sayajipura, Vadodara, Gujarat, India

Satone D
Department of Pharmaceutical Sciences, RTM Nagpur University, Nagpur, Maharashtra, India

Charagondla K
Analytical Research & Development, InvaGen Pharmaceuticals Inc., 7 Oser ave, Hauppauge, New York, 11788, USA

Meenakshi K Chauhan and Nidhi Bhatt
NDDS Research Laboratory, Department of Pharmaceutics, Delhi Institute of Pharmaceutical Sciences and Research, New Delhi, India

Gosavi SM and Tayade MA
Department of Pharmaceutical Chemistry, M. G. V.'s Pharmacy College, Panchavati, Mumbai, Agra Road, Nashik- 422003, Maharashtra, India

Karas K, Kuczynska J, Bienkowski P and Mierzejewski P
Institute of Psychiatry and Neurology, Department of Pharmacology, Sobieskiego 9, 02-957, Warsaw, Poland

Sienkiewicz-Jarosz H
Institute of Psychiatry and Neurology, Department of Neurology, Sobieskiego 9, 02-957, Warsaw, Poland

Santini DA and Sutherland CA
Center for Anti-Infective Research and Development, Hartford Hospital, 80 Seymour Street, Hartford, CT 06102, USA

Nicolau DP
Center for Anti-Infective Research and Development, Hartford Hospital, 80 Seymour Street, Hartford, CT 06102, USA

Division of Infectious Diseases, Hartford Hospital, Hartford, CT, Hartford Hospital, 80 Seymour Street, Hartford, CT 06102, USA

Nuvula Ashok Kumar, Korla Lakshmana Rao, Zinia Chakraborthy , Archana Giri and Komath Uma Devi
Centre for Biotechnology, JNTUH, Hyderabad, India

Sassi A, Hassairi A, Jaidane M and Saguem S
Metabolic laboratory of Biophysics and Applied Occupational and Environmental Toxicology, Faculty of Medicine, university of Sousse, Tunisia

Kallel M
Chemical, pharmacological and pharmaceutical drug development, Faculty of Pharmacy, Monastir, Tunisia

Yu-Sheng Lin, Wenting Liu and Chunfei Hu
Division of Nanobionic Research, Suzhou Institute of Nano-Tech and Nano-Bionics, Chinese Academy of Sciences, Suzhou, Jiangsu, China

Ashutosh KS
Department of Pharmaceutical Analysis and Quality Assurance, AKRG College of Pharmacy, Nallajerla, West Godavari, Andhra Pradesh, India

Manidipa D
Department of Pharmaceutics, AKRG College of Pharmacy, Nallajerla, West Godavari, Andhra Pradesh, India

Seshagiri RJVLN
Srinivasarao College of Pharmacy, Pothinamallayyapalem, Madhurawada, Visakhapatnam, Andhra Pradesh, India

Gowri SD
College of Pharmaceutical Sciences, Andhra University, Visakhapatnam, Andhra Pradesh, India

Alan Wong and Jan A Glinski
Planta Analytica LLC, New Milford, Connecticut, USA

Brijesh D Patel, Usmangani K Chhalotiya and Dhruv B Patel
Department of Pharmaceutical Chemistry and Analysis, Indukaka Ipcowala College of Pharmacy, New Vallabh Vidyanagar, Anand, Gujarat, India

Dare M, Jain R and Pandey A
Department of chemistry, Dr. Hari Singh Gour University Sagar India, Department of Analytical development API Micro Labs Limited-API Banglore 560001, India

Ibrahim F, Sharaf El-Din M and Heba Abd El-Aziz
Department of Analytical Chemistry, Faculty of Pharmacy, University of Mansoura, Mansoura, 35516, Egypt

Eisenried A, Peter J, Lerch M, Schüttler J and Jeleazcov C
Department of Anesthesiology, University of Erlangen-Nurnberg, Krankenhausstrabe 12, 91054 Erlangen, Germany

Lea B, Nicolas C and Dominique W
Ecole National de Chimie de Rennes, CNRS, UMR 6226, 11 Allee de Beaulieu, CS 50837, 35708 Rennes Cedex 7, France

Guy R
Veolia Eau, Direction Technique Region Ouest, 8 allee Rodolphe Bopierre, 35020 Rennes Cedex 9, France

El-Saharty YS, Riad SM, Yehia AM and Sami I
Department of Analytical Chemistry, Faculty of Pharmacy, Cairo University, Kasr El-Aini Street, ET 11562, Cairo, Egypt

Al Asmari AK, Ullah Z and Ahmad I
Research Center, Prince Sultan Military Medical City, Riyadh, Saudi Arabia

Al Rawi AS
Department of Medicine, Prince Sultan Medical Military City, Riyadh, Saudi Arabia

Al Eid AS
Department of Pharmacy, Prince Sultan Medical Military City, Riyadh, Saudi Arabia

Al Nowaiser NA
Department of Family and Community Medicine, Prince Sultan Medical Military City, Riyadh, Saudi Arabia

Vasconcelos André Belico de, Oliveira Jamil Silvano de and Lagares Monique de Albuquerque
University of Uberaba, Av. Nenê Sabino 1800, Uberaba, Minas Gerais, Brazil

Dana Pupke and Lea Daniel
Institute of Coastal Research/Molecular Recognition and Separation, Helmholtz-Zentrum Geesthacht, Max-Planck-Straße 1, D-21502 Geesthacht, Germany

Daniel Proefrock
Institute of Coastal Research/Marine Bioanalytical Chemistry, Helmholtz-Zentrum Geesthacht, Max Planck-Straße 1, D-21502 Geesthacht, Germany

Iyad Naeem Muhammad
University of Karachi

Saiqa Muneer
University of Karachi
The University of Lahore
The university of Queensland

Kishwar Sultana
The University of Lahore

Muhammad Asad Abrar
Islamia University Bahawalpur

Iqra Munir and Iram Kaukab
Bahauddin Zakariya University Multan

Abdullah Sagheer
University of Veterinary and Animal Sciences Lahore

Habiba Zafar
Comsats Institute of Technology, Abottabad

Index

www.ingramcontent.com/pod-product-compliance
Lightning Source LLC
Chambersburg PA
CBHW080626200326
41458CB00013B/4526